MW01038114

Biophysical Chemistry

Biophysical Chemistry

James P. Allen

A John Wiley & Sons, Ltd., Publication

Blackwell Publishing was acquired by John Wiley & Sons in February 2007. Blackwell's publishing program has been merged with Wiley's global Scientific, Technical and Medical business to form Wiley-Blackwell.

Registered office: John Wiley & Sons Ltd, The Atrium, Southern Gate, Chichester, West Sussex, PO19 8SQ, UK

Editorial offices: 9600 Garsington Road, Oxford, OX4 2DQ, UK
The Atrium, Southern Gate, Chichester, West Sussex, PO19 8SQ, UK
111 River Street, Hoboken, NJ 07030-5774, USA

For details of our global editorial offices, for customer services and for information about how to apply for permission to reuse the copyright material in this book please see our website at www.wiley.com/wiley-blackwell

Library of Congress Cataloging-in-Publication Data

Allen, James P.
Biophysical chemistry / James P. Allen.
p. ; cm.
Includes bibliographical references and index.
ISBN 978-1-4051-2436-2 (hardcover : alk. paper) 1. Physical biochemistry I. Title.
[DNLM: 1. Biophysics. 2. Chemistry, Physical. 3. Biochemistry. QT 34 A427b 2008]

QD476.2A44 2008
572'.43—dc22

2007038528

ISBN: 978-1-4051-2436-2

A catalogue record for this book is available from the British Library.

Set in 10/12.5pt Meridien by Graphicraft Limited, Hong Kong
Printed in Singapore by Markono Print Media Pte Ltd

1 2008

Short contents

Contents

Preface

Astronauts in orbit above the Earth have a unique and special perspective. The problems and issues concerning the world from this broad perspective may seem to be much different than those concerning the average person, especially a student studying physical chemistry. The laws of thermodynamics that were developed over 100 years ago may seem to have a limited significance compared to the issues that can alter the Earth on a large scale, such as the hydrogen economy and global climate change. The goal of this book is to provide an understanding of physical chemistry that is needed for a firm scientific understanding of such problems. It is my hope that the extensive reference to current issues will give students the opportunity to discuss relevant issues from a scientific standpoint. These sections, which are identified as Research directions, present not only the background on specific issues but also ask what the unanswered questions are and how they are being addressed by scientists.

Chapters 2–8 of the book present thermodynamics and kinetics, with biological applications ranging from global climate change and nitrogen fixation to drug design and proton transfer. Chapters 9–16 focus on quantum mechanics and spectroscopy. In this section, issues of biology are presented with an emphasis on understanding the function of proteins at a molecular level.

The last part of the book (Chapters 17–20) is written with the hope that the ideas of thermodynamics, kinetics, quantum mechanics, and spectroscopy can be integrated to understand biology on a broad scale, with the specific examples being signal transduction, ion channels, molecular imaging, and photosynthesis. These chapters are independent of each other and can be presented in any combination. The intention of these chapters is to provide the instructor with the opportunity to teach biology from a physical-chemistry viewpoint and show how the concepts of the course can be used in an integrative fashion rather than simple parts.

One of the balances in organizing this text is to present a rigorous treatment of the material without expecting an unrealistic understanding of mathematical concepts. The text has two mechanisms to maintain a proper balance. First, students have often been taught a high level of mathematics but have not used such concepts in their recent courses. Throughout the text are short math concept boxes that will remind the students of how to complete a specific step (for example, the derivative of an exponential). Second, formal derivations of expressions are included but highlighted, for example Schrödinger's equation for the hydrogen atom is solved explicitly. By providing the derivation, students can gain an appreciation of the mathematical concepts behind the expression. However, the text is written such that the derivation can be skipped without disruption. Thus, the instructor can decide on which derivations to present in class, while students can always work though the derivations as they wish.

This book was developed from a course taught by the author that is targeted primarily towards undergraduate biochemistry students but also intended for students in physics, biology, and engineering. I wish to thank those students for their comments, which helped shape this textbook. I would also like to thank my colleagues who have commented on the chapters, especially Wei-Jen Lee, who read the chapters very carefully. The reviewers and editors have all been very helpful, with special acknowledgment to Elizabeth Frank, Nancy Whilton, and Haze Humbert. The notes of Neal Woodbury served as the initial basis for several chapters, and many figures represent artwork designed by Aileen Taguchi; both of these proved to be invaluable in writing this book. Finally, I wish to thank my family, JoAnn, Hannah, and Celeste, for their love and support.

James P. Allen

I

Basic thermodynamic and biochemical concepts

Thermodynamics is the characterization of the states of matter, namely gases, liquids, and solids, in terms of energetic quantities. The states that will be considered primarily in these chapters are all macroscopic, although the same ideas apply at a microscopic level, as will be discussed in Part III. In thermodynamics, there is a set of rules that objects must obey. The beauty of these rules is that they are very general and apply to all types of objects, ranging from gas molecules to cell membranes to the world. It does not matter how complex the system is, as once the properties are established, then the rules can be applied. As an example, consider how these ideas would apply to you. It would be theoretically possible to analyze every biochemical process in your body and establish the net energy change. However, at best, this would be a very difficult task and is different for every person, with athletes such as Lance Armstrong using energy differently than the average person. Thermodynamics requires that energy is not lost and so the energy taken into your body must be all converted into biological processes (provided there is no weight change). Thus, the complex problem of analyzing all processes can be simplified by looking at the overall difference, by considering how much energy is available when you metabolize the food that you eat.

There are three basic thermodynamic ideas that are now identified as laws, as they have been found to always be applicable. The "zeroth" law provides a definition of the key energetic parameter of an object: temperature. The first law defines the conservation of energy that we have just used. The second law defines how another key parameter, entropy, which defines the order of an object, can be used to understand how energy can be generated by allowing systems to become more disordered. The next few chapters discuss in detail how these rules can be used using certain mathematical relationships.

In this chapter, the fundamental properties of matter are defined. Before considering complex biological systems, a very simple system, called an ideal or perfect gas, is described. This system is chosen as its properties can be understood without any prior knowledge of its chemical or physical nature. After the basic concepts are established, then we will go back and discuss how certain physical properties, such as the size or charge of an object, can be incorporated. The objective of this text is to focus thermodynamic concepts on biological problems. Therefore, included at the end of the chapter is a short review of the basic properties of biological systems, including the structural properties of proteins and nucleic acids.

FUNDAMENTAL THERMODYNAMIC CONCEPTS

States of matter

Matter can be considered to be in one of three states, a gas, liquid, or solid. A gas is considered to be a fluid that always fills the container that it occupies. The particles that form the gas are widely separated and move in a disordered motion. A liquid is a fluid that, in the presence of a gravitational field, occupies only the lower portion of a container and has well-defined surfaces. The particles interact with each other weakly at short distances, and the movement of any given particle is restricted by collisions with other particles. A solid has a shape that is independent of a container. The position of each particle is fixed although the particles can vibrate about their position. From a thermodynamic viewpoint, the basic difference between these three states is the difference in motion of the particles, so transitions can be made between these states if the degree of motion changes. These states are characterized by a few fundamental properties, as described below.

Pressure

Pressure conceptually is a measure of the force that an object exerts on the surface of another object. Formally, the pressure, P, is the force, F, divided by the area, A:

$$P = \frac{\text{Force}}{\text{Area}} = \frac{F}{A} \tag{1.1}$$

A balloon expands due to the increase in pressure as air is forced inside the balloon. The increase in the amount of air inside pushes on the walls of the balloon making the pressure inside the balloon greater than the pressure outside. The balloon expands until an equilibrium is reached, with the pressures balanced, as will be discussed later.

Pressure can be calculated using several different units. The force on an object is given by the product of the mass, m, and the acceleration, a, according to Newton's law:

$$F = ma \tag{1.2}$$

Remember that the rate of change in position gives the velocity, v, and acceleration is the rate of change of velocity:

$$v = \frac{dx}{dt}$$
$$a = \frac{dv}{dt} = \frac{d}{dt}\left(\frac{dx}{dt}\right) \tag{1.3}$$

Acceleration has units of distance time^{-2} or m s^{-2}, so force has units of kg(m s^{-2}). Dividing the force by area gives the standard unit for pressure called the Pascal, Pa:

$$\text{Pa} = \left[\text{kg}\frac{\text{m}}{\text{s}^2}\right]\left[\frac{1}{\text{m}^2}\right] = \frac{\text{kg}}{\text{m s}^{-2}} \tag{1.4}$$

Since thermodynamics is intimately related to energy, it is convenient to consider in terms of energy. Energy, E, is given by the product of the force exerted over a distance:

$$E = F \times d \tag{1.5}$$

The unit of energy is Joules, J, that can be written as:

$$\text{J} = \left(\text{kg}\frac{\text{m}}{\text{s}^2}\right)(\text{m}) = \frac{\text{kg m}^2}{\text{s}^2} \tag{1.6}$$

Comparing the units for pressure (eqn 1.4) and energy (eqn 1.5) allows the units for pressure to be rewritten as:

$$\text{Pa} = \frac{\text{J}}{\text{m}^3} \tag{1.7}$$

Pressure can be expressed in terms of an energy per volume. A variety of units are used to describe pressure. One convenient unit is to express the pressure in terms of the pressure that our atmosphere exerts at sea level, or 1 atm. The units for Pascals and atmospheric pressure can be converted using:

$$1 \text{ atm} = 101{,}325 \text{ Pa} \tag{1.8}$$

Table 1.1
Standard pressure in various units.

1.000 atmosphere (atm)
33.899 feet of water
14.696 pounds/square inch absolute (psia)
29.92 inches of mercury (inHg)
760.0 millimeters of mercury (mmHg)
1.013×10^5 Pascals (Pa)
1.013×10^5 Newton/square meters ($N\ m^{-2}$)
1.01325 bar

Other common units of pressure are the bar and Torr. The bar is equal to 100,000 Pa and so is very close to 1 atm:

$$1\ \text{atm} = 101.325\ \text{kPa} = 1.01325\ \text{bar} \tag{1.9}$$

The Torr represents the pressure required to push a column of mercury up by 1 mm and can be converted to Pascals using:

$$1\ \text{Torr} = 133.325\ \text{Pa} \tag{1.10}$$

Due to the range of units available, standard pressure can be expressed as any of several equivalent values and units (Table 1.1).

Scientists make use of a certain terminology when expressing measures of pressure. Atmospheric pressure is the pressure of the air surrounding us. Barometric pressure is the same as atmospheric pressure but this term is used in reference to the use of a barometer to measure the atmospheric pressure. The value of pressure can be expressed using different references. The gauge pressure is the pressure measured relative to the atmospheric pressure. Alternatively, the absolute pressure is relative to a complete vacuum, or zero pressure. Hence, the absolute pressure is equal to the sum of the gauge pressure and the barometric pressure.

For biological systems, the only relevant pressure is normally 1 atm, so why would you be concerned about pressure? Whereas most living systems are found on the surface of the planet, life is also found at the bottom of the oceans centered around deep-sea vents. The initial discovery was made in 1977 by scientists who were examining the hydrothermal fluids of seafloor vents using the submersible ALVIN. Deep-sea vents are regions where there is volcanic activity. Water can seep into cracks and reach temperatures as high as 400°C. The hydrothermal vent carries dissolved minerals and is cooled rapidly upon emergence from the vent.

Ocean vents are found at a wide range of depths; for example, vents at 30 m are found off the coast of New Zealand, and the deep-sea vents are found at 3600 m. To the great surprise of the scientists, the vents were found to contain life. The vents are rich sources of not only minerals but also heat and are surrounded by a rich variety of organisms. There is extreme pressure at the deep-sea vents; for example, a tube worm at a vent 2500 m below sea level experiences a pressure of approximately 250 atm, which is enough to collapse most objects, but the physiology of the organisms at the vents allows them to thrive.

Temperature

Temperature is usually thought of in terms of how hot or cold an object is, but in thermodynamics temperature is a measure of molecular motion. Consider how temperature is changed due to the presence of a source of heat. An electric heater gets hot when a current – that is, electrons – flows through a heating coil. The electrons lose the energy that had been provided by the voltage source, for example a battery, due to the resistance of the heating coil. Since no work is being done (the coil is not moving) the energy of the electrons is lost by converting the electrical energy into heat. Light generated by a fire or the sun is absorbed by an object causing electrons to change their relative position (orbital) around the nucleus in atoms. The electrons can return to their initial state only by releasing the energy as a vibration or heat.

Temperature is usually measured relative to certain physical processes. The most common temperature scale is the Celsius scale (°C), which is derived by assigning a value of 0°C to the temperature at which water freezes and 100°C to the temperature at which water boils, when the pressure is at atmospheric pressure. In some countries, including the USA, the Fahrenheit scale is still used. For this scale, a temperature of 100°F was originally assigned to body temperature but this was later corrected to 98.6°F.

Materials can have either negative or positive temperatures on the Celsius scale. For thermodynamics, the preferred scale is the Kelvin scale (K), in which the temperature is always positive. The temperature in Celsius can be converted into the temperature in Kelvin by adding the value of 273.15:

$$\text{Temperature (K)} = \text{temperature } (^\circ\text{C}) + 273.15 \qquad (1.11)$$

A critical concept of temperature is sometimes referred to as the "zeroth law" of thermodynamics. If two objects are separated and at different temperatures, the temperatures will become equal if they come into contact with each other and a sufficient time has passed. The equilibration of temperature arises because heat will pass from the hotter object to the cooler object until there is no net flow when the two temperatures are the same.

The preferred temperature of living organisms varies tremendously. Warm-blooded mammals prefer a fixed temperature, whereas cold-blooded animals readily live at different temperatures. Some bacteria, termed thermophiles, prefer very high temperatures such as those found at hot springs at locations such as Yellowstone National Park in the USA. By gaining the ability to live at relatively high temperatures, these organisms have gained an ecological advantage over other organisms that otherwise might compete for the same nutrients. The coloration of these organisms arises due to their incorporation of pigmentation that is used in the capture of light through a process called photosynthesis (Chapter 20).

Volume, mass, and number

The quantity of a physical system can be characterized by its volume or by the amount. The volume is a measure of the space occupied and is measured in liters. The volume is not a fixed quantity of an object. For example, puffer fish expand greatly in volume as a defense mechanism when they sense danger. These fish possess a special sac inside their bodies that can be filled with water rapidly, and the combination of a large body, their tough outer body, and the presence of toxins makes them difficult prey. The mass is a measure of the quantity of matter an object contains. The unit of mass is the kilogram, although gram or milligram are also commonly used in biochemistry. Remember that the weight is a useful measure only at a fixed gravitational-field strength. Astronauts in space experience microgravity and have a very small weight while their mass remains the same. In addition to the mass, it is useful to know the number of atoms (or molecules) present in the object. The common unit is the mole, which equals unity for 6.022136×10^{23} molecules, with the conversion constant between number and moles being termed Avogadro's constant. The molar mass is the mass per mole of a substance, with the unit Dalton, Da, being equal to 1 g/mol. The molar mass is commonly used for biological objects such as proteins. For example, myoglobin, the protein that serves as the oxygen store and carrier in muscle, has a molar mass of about 15 kDa.

PROPERTIES OF GASES

The ideal gas laws

Scientists, starting in the seventeenth century, discovered that for gases the properties of pressure, P, temperature, T, volume, V, and amount, n, are all quantitatively related to each other though the ideal gas law. When a gas in a chamber is compressed by a piston the volume decreases but the pressure is found to increase. If you heat a gas in a closed container, the pressure in the container increases. Also, as you increase the amount

Table 1.2
The gas constant *R* expressed in different units.

R	**8.314472 J K^{-1} mol^{-1}**
	8.314472 × 10^3 Pa L K^{-1} mol^{-1}
	62.364 L Torr K^{-1} mol^{-1}
	1.98722 cal K^{-1} mol^{-1}

of gas in a chamber the pressure will increase. These observations lead to the ideal gas law:

$$PV = nRT \tag{1.12}$$

In this equation, *R* is the gas constant. The value of this constant depends upon the units for the various variables, with some values given in Table 1.2.

To gain an understanding of the ideal gas law, we consider how a hot-air balloon works first at a macroscopic level (Figure 1.1). Consider the four different parameters of the gas law, *T*, *P*, *n*, and *V*. The volume is simply defined by the size of the balloon. What about the temperature? The temperature inside the balloon is much greater than the temperature outside the balloon (hence the name hot-air balloon). The pressure inside and outside the balloon must be the same because there is a large opening at the bottom of the balloon. Hot-air balloons do not work by expansion, rather the balloon material simply captures the hot air. An important aspect is that the relative number of air molecules inside the balloon (per unit volume) is less than the number outside. For the balloon to rise into the air the density of the air inside the balloon must be less than that of the air outside the balloon. In fact, the major reason for heating the air inside the balloon is to trap air that has less density, providing an upward hydrostatic force that pushes the balloon up when this force is greater than the gravitational force due to the mass of the balloon.

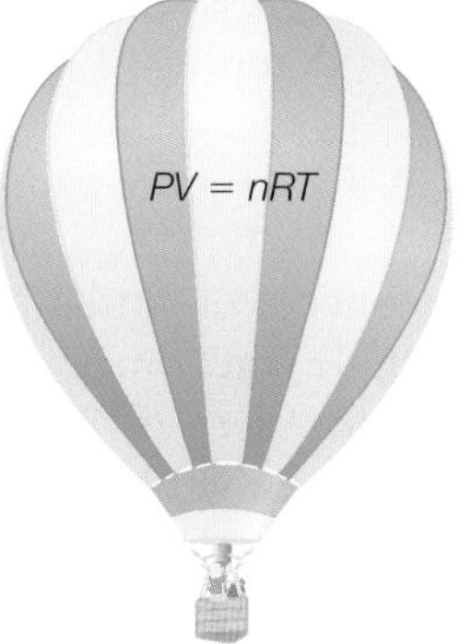

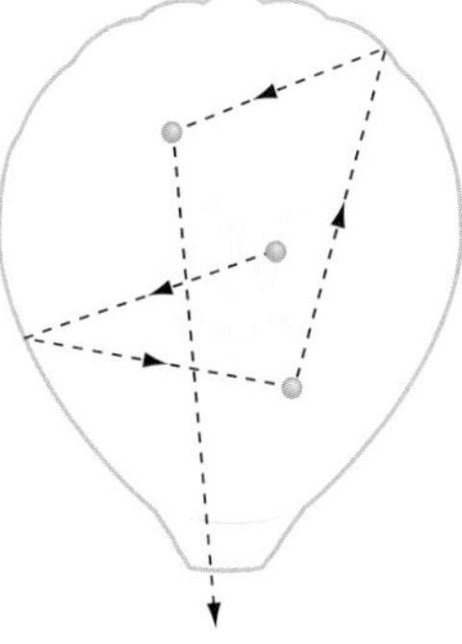

Figure 1.1 The ability of hot-air balloons to travel through the air can be understood in terms of the interactions of the air molecules against the balloon wall.

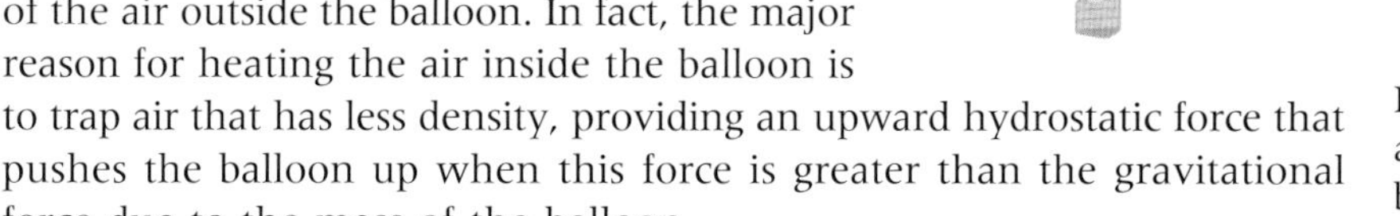

The motion upward of the hot-air balloon can also be viewed from an equivalent molecular viewpoint. Consider a single air molecule inside the balloon. This molecule will travel through the air until it hits the side of the balloon, at which point it bounces away. Eventually, the air molecule will travel down through the opening and leave the balloon. As the temperature increases, the average motion of each air molecule increases. The increase in speed of each molecule results in a decrease in the average time that it takes for a gas molecule to escape the balloon. Thus, the temperature increase leads to a decrease in the average number of gas molecules at a microscopic level.

Gas mixtures

In the ideal gas model, the precise nature of the gas molecules does not influence the properties of the gas. Thus, the properties of gases that are mixtures of two different types of gas molecules can be predicted, as was first realized based upon a series of experiments by John Dalton in the early nineteenth century. The basic idea is that since the properties of the individual gas molecules do not matter and the gas molecules are considered not to interact with each other except through collisions, the properties of the mixture are determined by the additive contribution of each gas molecule. Any given type of gas molecule, which we can identify as the ith type, is considered to have a certain partial pressure P_i that corresponds to the pressure the gas would create if it were alone in the container. For a container with a volume V, the ith type of gas has a certain number of gas molecules, n_i, and a partial pressure, P_i, that is given by:

$$P_i = \frac{n_i RT}{V} \tag{1.13}$$

The total pressure of a mixture of gases composed of i different gas molecules is determined by the sum of the individual partial pressures:

$$P = P_1 + P_2 + \ldots + P_i \tag{1.14}$$

The concept of partial pressures in a gas mixture brings us to the concept of a mole fraction. The mole fraction of a certain gas A is the number of moles of A, n_A, divided by the total number of moles of all gases in the vessel. For a gas mixture with i types of gas molecules, the mole fraction for gas A, x_A, is given by:

$$x_A = \frac{n_A}{n_A + n_B + \ldots + n_i} \tag{1.15}$$

Since the partial pressure of the gas A, P_A, is proportional to the amount of A in moles, then the partial pressure P_A will also be proportional to the mole fraction of A:

$$P_A = x_A \frac{nRT}{V} \tag{1.16}$$

KINETIC ENERGY OF GASES

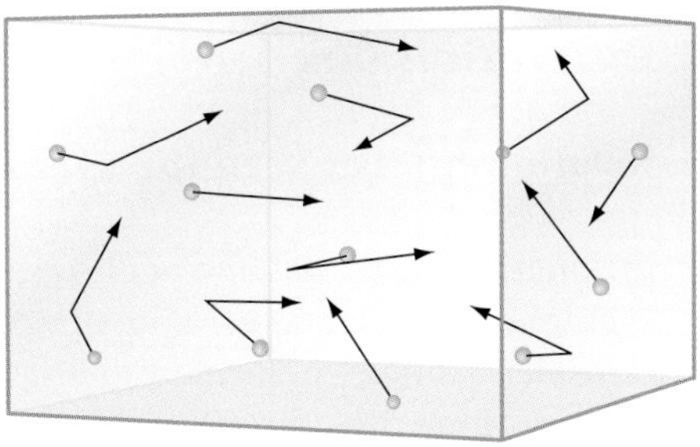

In the ideal gas model, the gas molecules are considered to move randomly with a wide range of speeds within a certain volume (Figure 1.2). The size of the molecules is neglected and the molecules are assumed not to interact except during the collision process. Pressure arises from the collisions of the gas molecules with the walls of the container. Since billions of collisions arise every second the pressure is constant with time. With this kinetic model the temperature of a particle can related to its motion. For an ideal gas with n moles and a molecular weight of M, the average velocity, $\langle v \rangle$, can be related to the product of the pressure and volume:

Figure 1.2 The molecules of an ideal gas move randomly within the enclosure.

$$PV = nRT = \frac{1}{3}nM\langle v^2 \rangle \tag{1.17}$$

The reader is invited to follow the derivation of this relationship in Derivation box 1.1. The form of this relationship can be understood if we consider the temperature, T, to reflect the energy of the system, realizing that the kinetic energy, KE, is proportional to the square of the velocity:

$$KE = \frac{1}{2}mv^2 \tag{1.18}$$

Since the temperature T is proportional to energy and hence the velocity squared, it should not be a surprise that the product PV can also be written in terms of the velocity squared. The factor of 1/3 in eqn 1.17 arises because not all of the molecules are moving in the same direction. The root mean square velocity can be determined by solving eqn 1.17 for the velocity:

$$v_{\text{rms}} = \sqrt{\frac{3RT}{M}} \tag{1.19}$$

REAL GASES

Ideal gases are assumed to be particles with no size and interactions other than the ability to collide with each other. Although many gases behave in a nearly ideal way, none are perfectly ideal and their properties deviate from those predicted based upon the ideal gas law. These deviations provide insight into the interactions between molecules and the reactions that they can undergo in chemical reactions. The simplest approach is to assume that these effects are small and that the ideal gas law can be

Derivation box 1.1

Relationship between the average velocity and pressure

In kinetic theory, gas molecules are considered to randomly collide with the wall and on average exert a certain pressure due to the sum of the collisions averaged over time (Figure 1.3). Consider a single molecule moving in three dimensions with a velocity v, with v_x being the velocity along the x direction. After collision, the molecule is assumed to have the same kinetic energy but to be moving in the opposite direction with the velocity along the x axis being $-v_x$. The linear momentum, p, of the particle is given by the product of its mass and velocity:

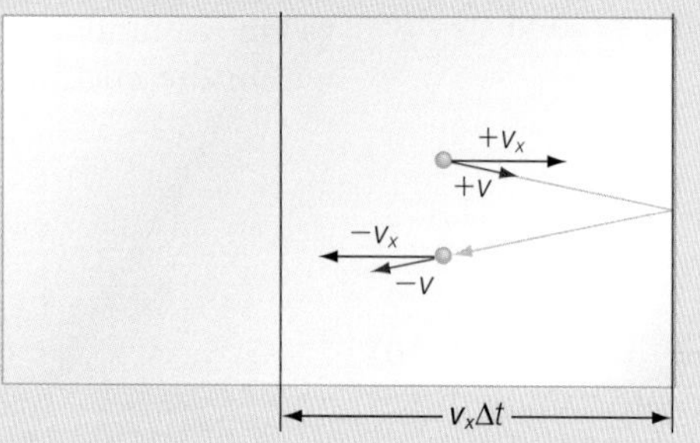

Figure 1.3 Pressure arises from the random collisions of gas molecules with the walls. A molecule traveling with the momentum mv_x will travel a distance $v_x\Delta t$ during the time Δt.

$$p = mv \qquad \text{(db1.1)}$$

so the change in momentum along the x direction, Δp_x, is given by:

$$\Delta p_x = 2mv_x \qquad \text{(db1.2)}$$

During a time interval Δt, a particle with a velocity v_x can travel a distance given by the product of the velocity and time:

$$\text{Distance} = \text{velocity} \times \text{time} = (v_x)(\Delta t) \qquad \text{(db1.3)}$$

On average, half of the particles that are within the distance $v_x\Delta t$ are moving towards the wall and will collide within the time Δt. If the area of the wall is A and the number of molecules per unit area is N, then the number that will hit the wall is:

$$\text{Number colliding with wall in } \Delta t = \left(\frac{1}{2}Nv_x\Delta t\right)(A) \qquad \text{(db1.4)}$$

and the total momentum change is given by the product of the number (eqn db1.4) and the momentum change of each molecule (eqn db1.2):

$$\text{Total momentum change} = \left(\frac{1}{2}Nv_x\Delta t\right)(A)(2mv_x) = NAmv_x^2\Delta t \qquad \text{(db1.5)}$$

and the rate of momentum change is given by the total momentum change divided by the time interval, yielding:

$$\text{Rate of total momentum change} = NAmv_x^2 \qquad \text{(db1.6)}$$

The force exerted on the walls is equal to the mass times the acceleration, or equivalently, the rate of momentum change. Since the pressure is equal to the force divided by the area, the pressure, P, can be written as the rate of the total momentum change divided by the area:

$$P = Nmv_x^2 \tag{db1.7}$$

The pressure is then proportional to the average of the square of v_x. Since the particles are traveling in three dimensions, the average total velocity, $\langle v_{av} \rangle$, is given by the sum of the squares of the three individual components:

$$\langle v^2 \rangle = \langle v_x^2 \rangle + \langle v_y^2 \rangle + \langle v_z^2 \rangle \tag{db1.8}$$

On average, each of these three components is equal, so the sum is just three times the value for one component, and the total velocity can be written as:

$$\langle v^2 \rangle = 3\langle v_x^2 \rangle \tag{db1.9}$$

and the pressure (eqn db1.7) can now be written as:

$$P = Nmv_x^2 = Nm\frac{\langle v^2 \rangle}{3} \tag{db1.10}$$

The number density N is the product of the number of moles n and Avogadro's number N_A divided by the volume, so the pressure can be rewritten as:

$$P = Nm\frac{\langle v^2 \rangle}{3} = \frac{nN_A}{V}m\frac{\langle v^2 \rangle}{3} \tag{db1.11}$$

Finally, dividing by the volume yields:

$$PV = nN_A m\frac{\langle v^2 \rangle}{3} \tag{db1.12}$$

modified with two empirical constants, a and b, to accommodate both the size of each gas molecule and interactions between the molecules, using the van der Waals' equation:

$$P = \frac{nRT}{V - nb} - a\left(\frac{n}{V}\right)^2 \tag{1.20}$$

What are these new parameters and how are they related to the size and interactions of the gas molecules? The first term of the equation appears

Table 1.3
van der Waals parameters for gases.

Gas	a (L^2 atm mol^{-1})	b (L mol^{-1})
Argon	1.35	0.032
Carbon dioxide	3.60	0.043
Ethane	5.49	0.064
Helium	0.034	0.024
Hydrogen	0.244	0.027
Nitrogen	1.39	0.039
Oxygen	1.36	0.032

to be the ideal gas equation except that the volume V is reduced by the quantity nb. The inclusion of a term that reduces the volume reflects the fact that real gases occupy a certain volume, and so the volume of space they have to occupy is less that the total volume. The reduction of volume depends both on how many molecules are present (n) and the volume of each gas molecule (b). The second term of eqn 1.20 incorporates the interactions between molecules. The fraction n/V is essentially the concentration of the gas. The effect of the interaction between gases is approximately dependent upon the product of the interaction strength, a, and the concentration of the gas molecules, n/V, squared. This dependence upon the concentration arises from the argument that the probability of two molecules being close enough together to interact depends on the probability of molecule A and molecule B both being in a small volume element. Each of these probabilities depends on the concentration, so the probability of both molecules being in the small volume element is proportional to the concentration squared. The parameter is a measure of the ion. These modifications of the ideal gas equation are not perfect but they provide a useful means of considering the effects that size and interactions should have on ideal gas behavior. Each of these parameters, a and b, has been determined for each type of gas molecule; for example, oxygen has values of 1.36 L^2 atm mol^{-1} and 0.032 L mol^{-1} for a and b respectively whereas helium, which is smaller and interacts much more weakly, has values of 0.034 L^2 atm mol^{-1} and 0.024 L mol^{-1} (Table 1.3).

Liquifying gases for low-temperature spectroscopy

The existence of interactions between gases is most evident by the fact that gases can be condensed into liquids by reducing the temperature to below a certain critical temperature. For example, we know that water boils when it is brought to a temperature of 100°C at atmospheric pressures. Gases can be liquefied by suddenly changing the pressure (Figure 1.4).

The sudden decrease in pressure causes a rapid expansion of the volume of the gas. Since the expansion is competing against the attraction of the gas molecules for each other, energy is lost by the expansion. This energy loss effectively decreases the temperature of the gas molecules. For most situations, the temperatures at which gases of other molecules become liquid are much lower than room temperature, hence repeated cycles of expansion must be used. Once a liquid gas is achieved, it serves as a valuable low-temperature bath that can be used to cool samples; for example, the temperatures of 77 and 4.2 K are obtained with the use of liquid nitrogen and helium, respectively, at 1 atm. The availability of an experimental means to poise samples at low temperatures was key to the development of many of the spectroscopic techniques used to characterize biological samples.

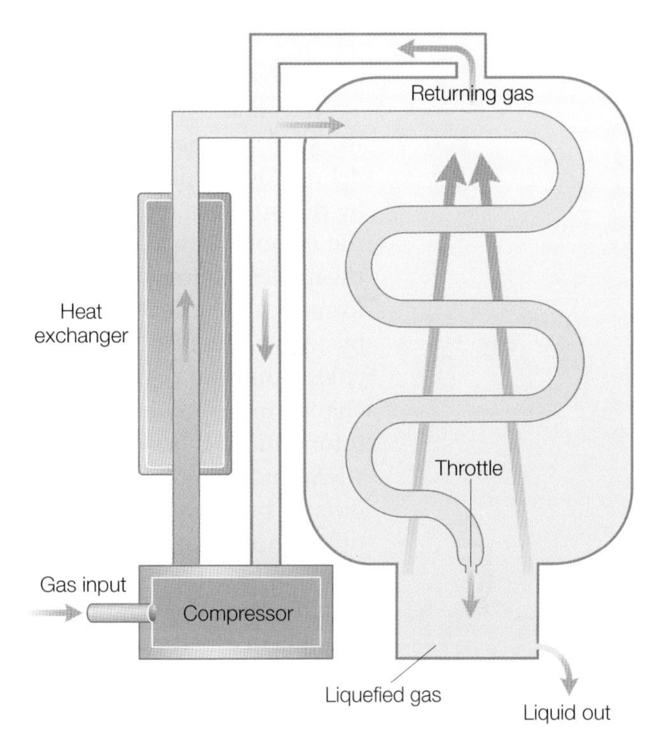

Figure 1.4 Gases can be liquefied by use of pressure changes and heat exchangers.

MOLECULAR BASIS FOR LIFE

Living organisms are composed of many different kinds of molecules. These molecules are used to extract energy from the environment and the energy is used to build intricate structures and perform various types of work. Unlike inanimate objects, living organisms can perform self-replication and self-assembly. These functions are performed by components that contribute to the entire ensemble as part of a larger, coordinated program for reproduction and perpetuation. The complexity of the components is usually reflective of the domain to which an organism belongs, with prokaryotic cells not possessing the complex architecture found in eukaryotic cells.

For all cells, energy is a key aspect, as the ability to grow and reproduce depends upon a constant supply of energy. Thermodynamics provides the framework for understanding how the energy in the form of sunlight or nutrients can be utilized by the cell to perform mechanical work, synthesize compounds, or be converted into heat. In some cases, a process may be energetically favorable but too slow to be useful, as can be understood from a consideration of the factors that control the kinetics. Since many reactions involve bond rearrangements or the transfer of electrons, the atomic properties influence the reactions as explained by quantum mechanics. In general, cells from prokaryotes or bacteria tend to be simpler than those from eukaryotes.

Cells are the structural and functional units of organisms and are designed to carry out the required chemical reactions. The periphery of a cell is defined by the plasma membrane whose properties are determined largely by the presence of molecules called lipids. The interior of the cell, termed the cytoplasm, is composed of an aqueous solution as well as specialized organelles. Among the components are proteins that are designed to perform specific functions. Examples are enzymes that can expedite reactions by many orders of magnitude, transporters that move ions and molecules from one region to another, and receptors that initiate a chemical change in response to the presence of a certain molecule. The biological information is stored and expressed using deoxyribonucleic acid, or DNA, and ribonucleic acid, or RNA. Since the biochemical reactions are dictated by the performance of these cellular components, an understanding of the molecular basis requires a detailed knowledge of the structure and function of these molecules. Whereas such an understanding is discussed in biochemistry texts, the fundamental aspects needed for this text are summarized below.

Cell membranes

Membranes form the boundaries of cells and regulate the trafficking of molecules into and out of the cell. To permit the rearrangements that occur in processes such as membrane fusion, membranes are flexible. The components of the membrane are not static but mobile, forming a fluid mosaic (Figure 1.5). Phospholipids and steroids establish the bilayer in which polar

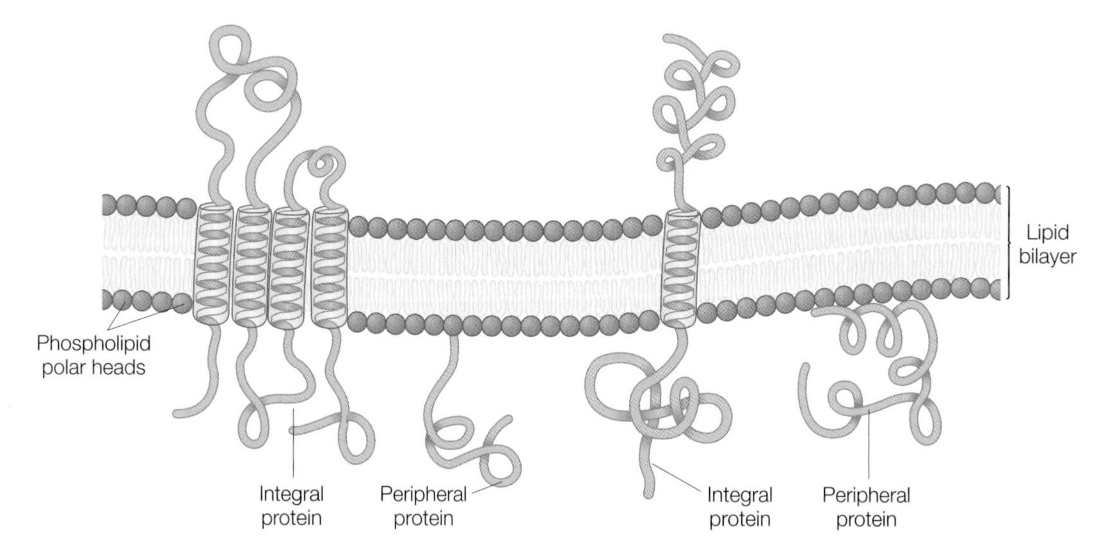

Figure 1.5 The fluid-mosaic model of cell membranes showing the lipids, and integral and peripheral proteins. The lipids and proteins can move within the bilayer.

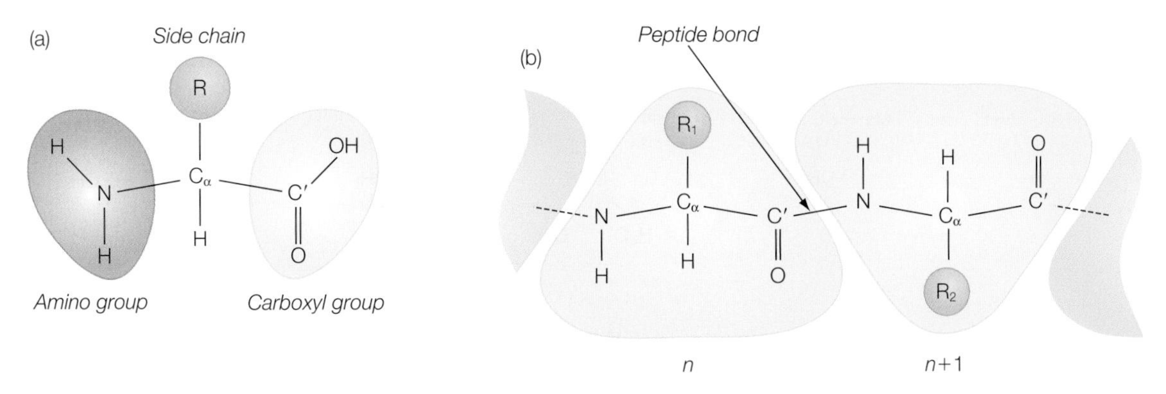

Figure 1.6 (a) The basic structure of an amino acid with the side chain represented by R. (b) Amino acids can be covalently joined through a peptide bond to form a peptide.

regions of the lipids are exposed to aqueous environments and the hydrophobic regions are buried. Proteins are embedded within the bilayer, usually forming long helices connected by loops that lie outside of the bilayer. In eukaryotic cells, carbohydrates may be attached to the exposed portions of the lipids and proteins. The membrane mosaic is fluid as the lipids and proteins are able to move within the membrane.

Amino acids

Proteins are built from amino acids. Each amino acid has the fundamental unit of a carbonyl group (COOH) and an amino group (NH_2) bonded to a central carbon atom that is designated as C_α (Figure 1.6). The amino acids differ from each other by the presence of the side chain. There are 20 different side chains found in proteins and these differ in the structure, size, and charge. Each atom of an amino acid is uniquely identified, with the atoms of the side chain being designated by the Greek letters β, γ, δ, and ε proceeding from the α carbon (the oxygen, nitrogen, and other carbon do not have letters assigned). There are two possible stereoisomers of each amino acid that are mirror images of each other, but only the L-amino acids are found in proteins (Figure 1.7). The presence of only one isomer allows cells to synthesize proteins that are asymmetric, causing reactions to be stereospecific.

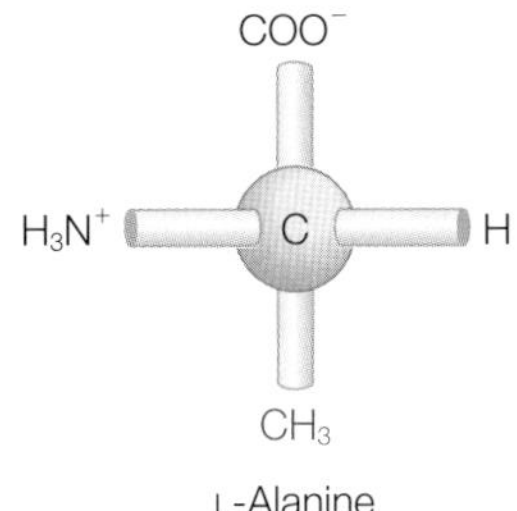

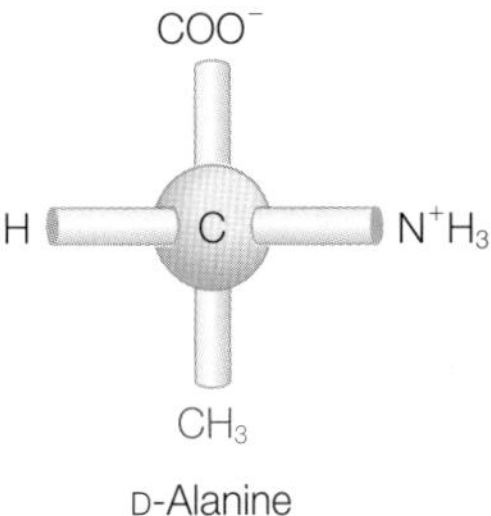

Figure 1.7 Of the two possible stereoisomers, L-amino acids are found in proteins.

Classification of amino acids by their side chains

An understanding of the properties of proteins requires knowledge of the chemical properties of the amino acids. The amino acid residues can be classified into groups according to the properties of the side chains.

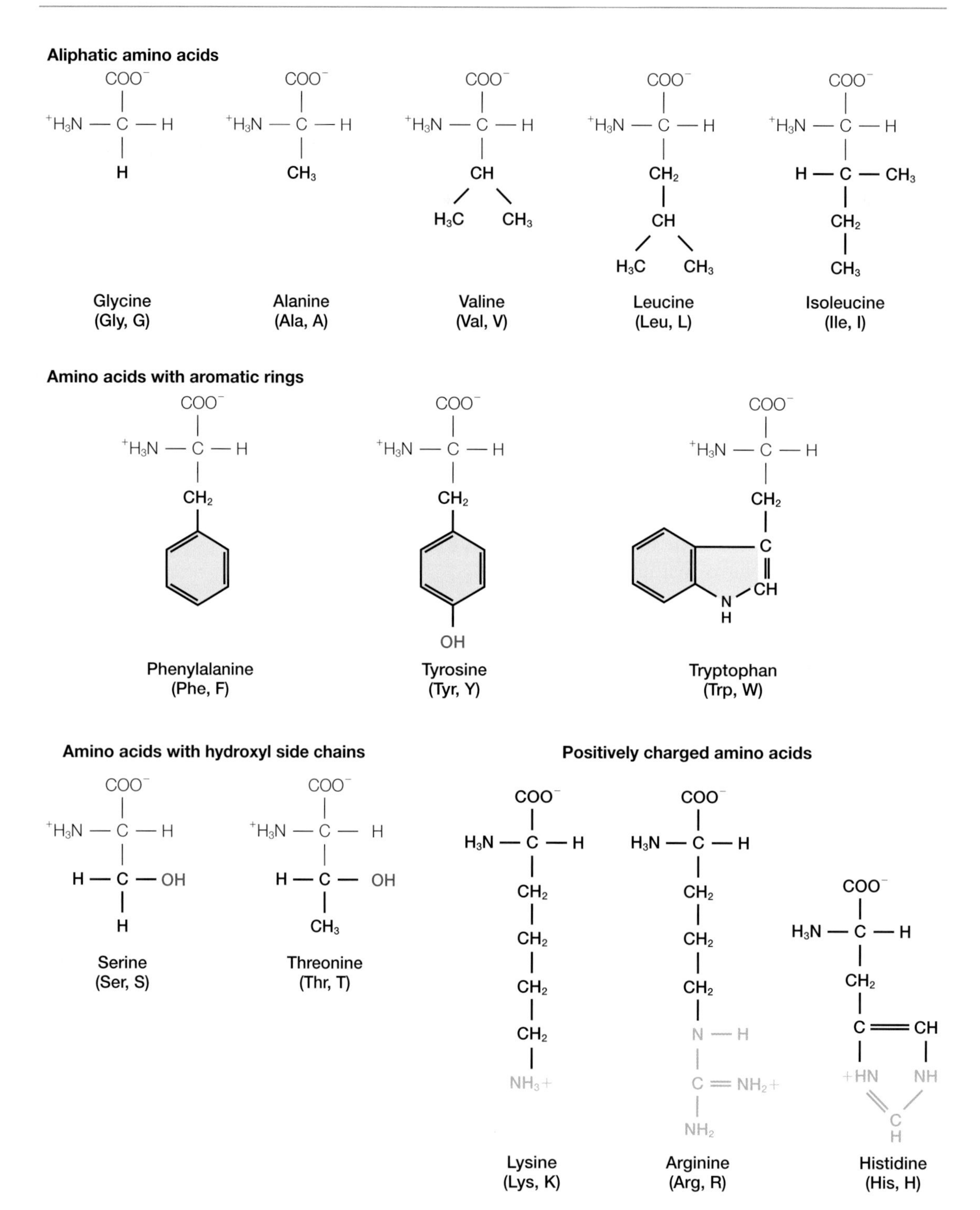

Figure 1.8 The 20 standard amino acids found in proteins.

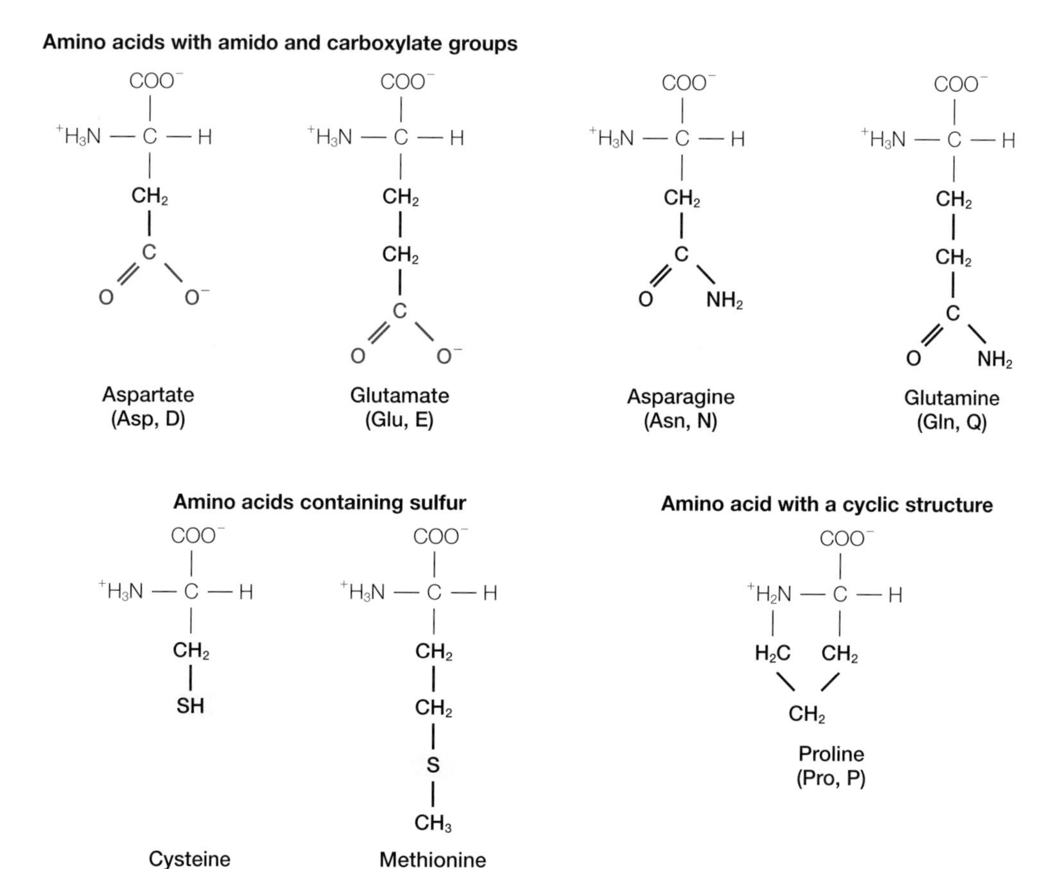

Figure 1.8 *(Cont'd)*

These groups have distinctive properties such as charge and polarity. The structure of the 20 standard amino acids are shown in Figure 1.8.

There are five amino acids with aliphatic side chains – glycine, alanine, valine, leucine, and isoleucine – as shown in Figure 1.8. These side chains are generally hydrophobic and will be usually found clustered in the interior of the protein. Glycine has the simplest structure, with a single hydrogen serving as the side chain. Alanine is next with a methyl group, with larger hydrocarbon side chains found on valine, leucine, and isoleucine. As the side chains become more extended the amino acids become more hydrophobic.

Three amino acids have aromatic side chains. Phenylalanine contains a phenyl ring attached to a methylene (-CH_2-). The aromatic ring of tyrosine contains a hydroxyl group. This group makes tyrosine less hydrophobic than phenylalanine and also more reactive as the hydroxyl can form a hydrogen bond. Tryptophan has an indole ring joined by a methylene group. Phenylalanine and tryptophan are highly hydrophobic and usually buried within proteins. Tyrosine is less hydrophobic due to the hydroxyl group, which can form hydrogen bonds. Because of the delocalized π electron distribution of these aromatic rings, these amino acid residues have an optical absorption spectrum in the ultraviolet region that can be used to characterize proteins containing these residues.

Four amino acids have hydroxyl or sulfur-containing side chains. Two amino acids, serine and threonine, have aliphatic hydroxyl side chains. The hydroxyl groups on serine and threonine make them more hydrophilic and reactive than alanine and valine. Two amino acids contain sulfur atoms: cysteine has a terminal sulfhydryl group (or thiol group) and methonine has a thioether linkage. Because of the sulfur-containing side chains, these residues are hydrophobic. The sulfhydryl group of cysteine is highly reactive and will often be found forming a disulfide bond with another nearby cysteine.

There are two groups of residues that can be charged, rendering them highly hydrophilic. Lysine and arginine are positively charged at neutral pH. The side chains of lysine and arginine are the longest ones of the 20 amino acid residues. Histidine can be uncharged or positive, depending upon the local environment, and is often found at the active site of proteins due to the reactivity of the imidazole ring. Aspartate and glutamate are amino acids that are almost always negatively charged. Asparagine and glutamine have very similar structures to aspartate and glutamate respectively except for a terminal amide group in place of the carboxylate group.

Finally, one amino acid, proline, has an aliphatic side chain but differs from the other 19 amino acids because it has a cyclic structure. The side chain shares many of the properties of the aliphatic amino acids but has the rigidity of the ring compared to the flexibility of the other amino acids. Proline is often found in bends and turns of proteins because of the uniqueness of its character.

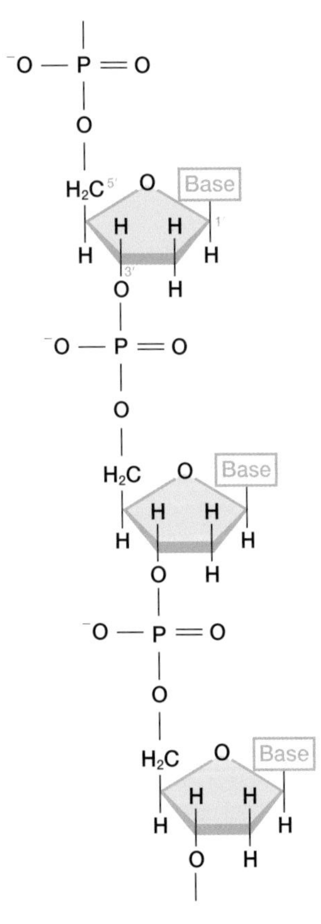

Figure 1.9 The chemical structure of a DNA chain.

DNA and RNA

Deoxyribonucleotides make up the basic units of deoxyribonucleic acid, DNA, whereas ribonucleotides form the basis for ribonucleic acid, RNA. A nucleotide consists of a nitrogenous base, a sugar, and one or more phosphate groups (Figure 1.9). In DNA the sugar is deoxyribose and the bases in DNA are either purines, adenine (A) or guanine (G), or pyrimidines, thymine (T) or cytosine (C). In DNA, the nucleotides are linked together in a defined chain. The 3′-hydroxyl of the sugar of one deoxyribonucleotide is joined to the 5′-hydroxyl of the adjacent sugar by a phosphodiester bridge. With this arrangement a backbone is created consisting of the deoxyriboses linked by the phosphate groups.

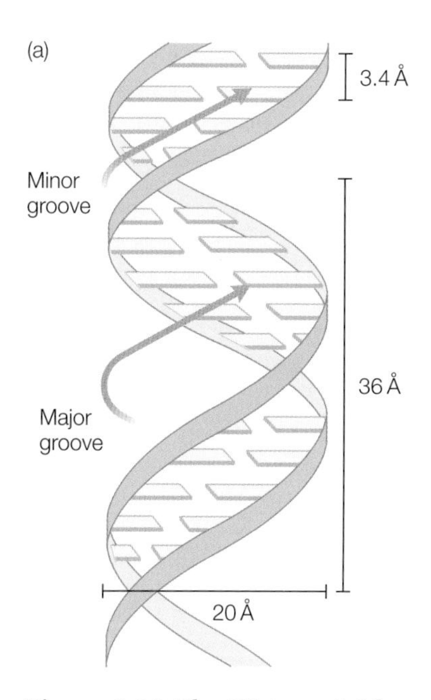

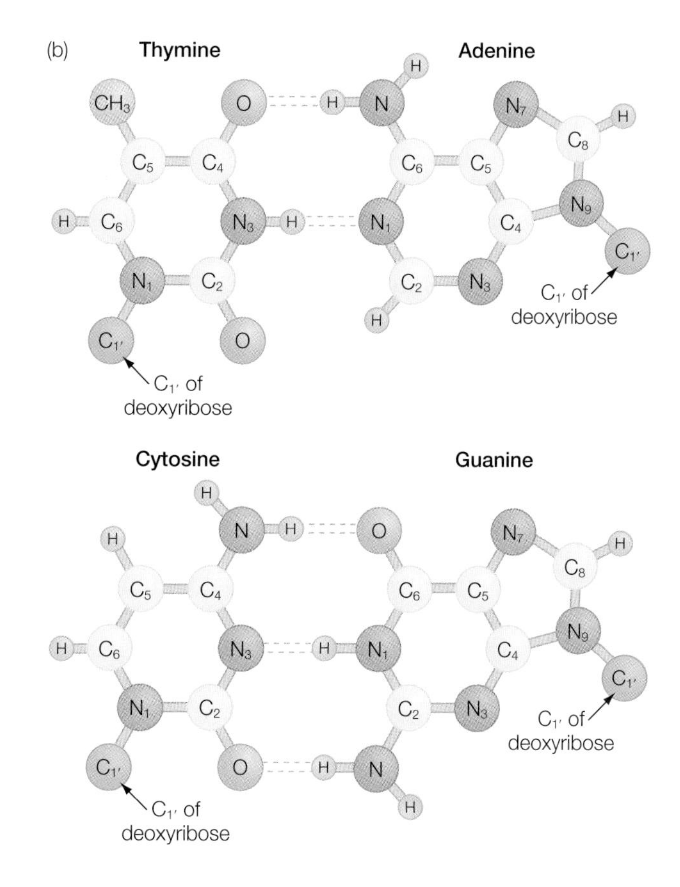

Figure 1.10 The Watson–Crick model for the structure of DNA. (a) Schematic representation of the double helix. (b) Base-pairing in the double helix is always between thymine and adenine or cytosine and guanine.

The variable part of the chain is the identity of the base at each position, or the sequence.

In 1953, James Watson and Francis Crick deduced that the three-dimensional structure of DNA was a double helix (Figure 1.10), which led to their Nobel Prize in Medicine in 1962. DNA has two chains running in opposite directions down a central axis, forming a double helix with a diameter of 20 Å. The helix repeats after 10 bases, yielding a repeat distance of 36 Å and resulting in the presence of a major and minor groove. The purine and pyrimidine bases are located in the interior of the helix perpendicular to the helix axis. The two chains are stabilized by hydrogen bonds between pairs of bases. The base pairing is specific due to consideration of the hydrogen bonding and steric factors. Thymine (T) is always paired with adenine (A) as both of these bases can form two hydrogen bonds. Cytosine (C) is always paired with guanine (G), forming three hydrogen bonds. The pairing of a purine with a pyrimidine allows the helix to maintain a fixed geometry for each pairing of bases and consequently a repeating helix. In addition to the traditional double helix, DNA can form other helical structures, known as the A and Z forms.

PROBLEMS

1.1 A cube is placed on the ground. The density of the material of the cube is 1 kg/l. What is the pressure exerted on the ground by the cube if the length of an edge is (a) 0.1 m, and (b) 0.01 m? The acceleration of gravity is 9.81 m s^{-2}.

1.2 If we were to isothermally compress an ideal gas from 4 L to 2 L, resulting in a final pressure of 6 atm, what was the original pressure?

1.3 Express a pressure of 202.6 Pa in terms of the following units: (a) atm, (b) bar.

1.4 Express the following temperatures on the Kelvin scale: (a) 0°C, (b) −270°C, (c) 100°C.

1.5 The absolute temperature of a system is related to what properties of the molecules that form the system?

1.6 Using the van der Waals equation and assuming a temperature of 298 K and a volume of 25 L calculate the pressure of 1 mol of carbon dioxide gas (use a and b values in Table 1.3).

1.7 What is the pressure exerted by 0.1 mol of an ideal gas in a 2-L chamber at 25°C?

1.8 Calculate the number of molecules in 2 L of air at a pressure of 1 atm and a temperature of 10°C, assuming ideal gas behavior.

1.9 If a helium balloon has an initial volume of 1 L at 25°C but a final volume of 0.95 L after a temperature change, what is the final temperature?

1.10 The mass percentage of dry air is approximately 75.5% for nitrogen, 23.2% for oxygen, and 1.3% for argon. What is the partial pressure of each gas when the total pressure is 1 atm? To calculate the mole fractions use densities of 28.02, 32.0, and 39.9 g mol^{-1} for nitrogen, oxygen, and argon respectively.

1.11 What is the composition of a cell membrane?

1.12 How are the lipids arranged in a cell membrane?

1.13 What is the fluid-mosaic model of a cell membrane?

1.14 What is the structure of an amino acid?

1.15 How many different types of amino acid are commonly found in proteins?

1.16 How are the atoms of an amino acid designated?

1.17 What isomer form of the amino acid is found in proteins?

1.18 Which amino acid residues can be (a) positively charged, (b) negatively charged?

1.19 Which amino acid has a cyclic structure?

1.20 Which amino acid residues have aromatic side chains?

1.21 What reaction can cysteine undergo?

1.22 What bases are present in DNA?

1.23 What are the characteristics of a double helix of DNA?

1.24 What is the base-pairing arrangement in the double helix?

1.25 How does RNA differ from DNA?

Part I

Thermodynamics and kinetics

2

First law of thermodynamics

Classical thermodynamics was developed during the nineteenth century and remains a cornerstone for understanding the properties of matter. This chapter presents the first law of thermodynamics, the conservation of energy. The law may seem to be intuitive, as we know that a reaction cannot produce more energy than was initially present and that hot objects will always cool off due to loss of energy. However, the concept was not always accepted as for many years people sought means to create energy from nothing or from perpetual-energy machines. The failure of such efforts coupled with the development of thermodynamics led to the concept that energy cannot be created or destroyed but only converted from one form to another. The conversion and storage of energy plays a critical role in biological organisms. In mammals, energy is used to contract muscles and released as work is performed in the form of walking or other motions. In all organisms, energy is required to create and transport nutrients and other cellular components such as proteins. The ability to perform such processes is determined by the energy of the reaction, with many key biological pathways making use of energy-rich compounds, such as *adenosine triphosphate*, ATP, to drive a reaction. An understanding of these fundamental biological processes requires knowledge of the basic concepts of thermodynamics. The focus of this chapter is the introduction of the terminology used to define the concepts followed by a description of the different forms of energy, such as heat and enthalpy, and how these terms are related in a thermodynamic setting.

SYSTEMS

Energy can readily flow between objects; for example, if an ice cube is placed into a cup of warm water, the ice cube is heated up and melted by the water. While the energy of the ice cube increases during this process, the conservation of energy requires that the temperature of the water decrease such that there is no net change in energy for both the water and ice cube. This simple example shows that in applying the law it is necessary to think carefully about which objects must be considered in applying energy conservation. Therefore, first we will look at some definitions.

Objects are considered to be either part of a *system* or the *surroundings*. The system is the set of objects under consideration and the surroundings are everything else. In some cases, the surroundings may be considered as being much larger than the system and so not changing as a result of an energy exchange with the system. For example, the surroundings may be represented as a large water bath that remains at constant temperature regardless of energy flow into or out of the system.

The energy of the system is defined as the capacity of the system to perform work. Heat and work are forms of energy that can transfer into a system, or out of it into the surroundings. Energy can be exchanged between the system and surroundings by either performing work or heating. A system does work when the energy transfer is coupled with a motion against an opposing force. For example, gas molecules in a chamber do work when they push against a piston and change its position. Heating occurs when the energy transfer is associated with a temperature difference between the system and surroundings.

The flow of matter and energy in and out of systems can be divided into three categories: open systems, closed systems, and isolated systems (Figure 2.1). Each of these systems is defined in terms of the allowed flow of matter and energy into and out of the system.

- Open systems: a system is described as *open* when both energy and matter of the system can be exchanged with the surroundings. An example of this is a cup of hot water in which matter can leave in the form of water vapor and water can also heat the surroundings.

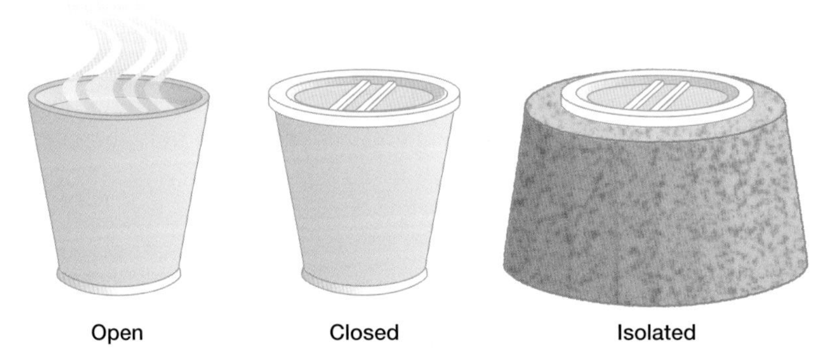

Figure 2.1 A system is described as open if it can exchange both energy and matter with its surroundings, closed if it can exchange only energy, and isolated if neither type of exchange is possible.

- Closed systems: a system is described as *closed* when energy can be exchanged between the system and surroundings but not matter. An example of this is a cup of hot water that is sealed but not insulated. The water vapor cannot escape but heat can still leave the cup.
- Isolated systems: a system can be considered to be *isolated* when neither matter nor energy can leave the system for the surroundings. An example of this is a cup of hot water that is not only sealed to prevent water vapor from escaping but is also insulated, so preventing heat from leaving the cup.

In considering energy flow, two terms must be defined that are commonly used in thermodynamics, adiabatic and diathermic (Figure 2.2). An isolated system where energy cannot flow from the system to the surroundings is called an *adiabatic* system, as occurs when the walls of the cup containing hot water are insulated. When a system is either open or closed, energy can exchange with the surroundings and such a system is called a *diathermic* system. Such a case corresponds to a container of hot water with the walls of the container being thin and allowing heat to pass through the wall. A diathermic process that results in the release of heat into the surroundings is called *exothermic* whereas a process that absorbs heat in called *endothermic*. Of these two types of process, exothermic reactions are much more common with combustible reactions; for example, an organic compound being oxidized by molecular oxygen.

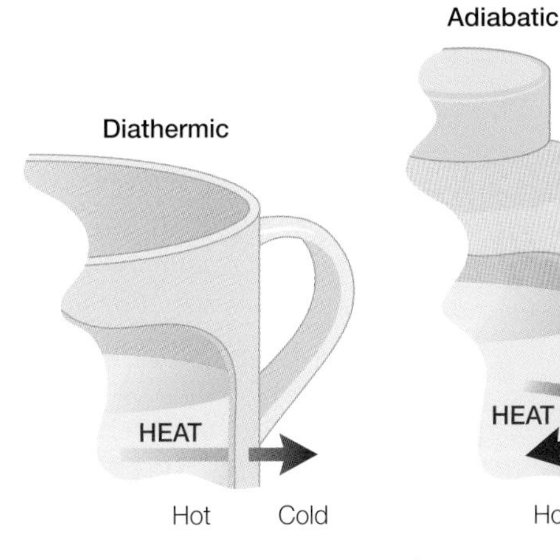

Figure 2.2 The passage of heat from a system is allowed only when the walls are diathermic. No heat transfer is allowed for adiabatic walls even when there is a difference in temperature.

STATE FUNCTIONS

The power of thermodynamics is that relationships can be established among the different properties of the system. An important aspect is that the process by which a system was established does not determine the properties. For example, the heat output of a system due to a temperature change is only dependent upon the initial and final temperature values and not the rate at which the temperature changed. These types of properties are called *state functions*; that is, they are dependent only upon the state of the system and not the path that was used to prepare the system. With state functions, conclusions can be made about reactions that are very general and dependent only upon a few specific parameters. Changes in a state function between any two states of a system are not dependent upon the path; changes are dependent only upon the initial and final conditions. In addition to the state functions already introduced, namely the volume, pressure, and internal energy, systems can be characterized by additional state functions, including the enthalpy, H, which is described below.

FIRST LAW OF THERMODYNAMICS

In considering the first law, we will assign to a system an internal energy, U, which is the total energy available in the system at any given time. Since we have said that energy leaves or enters a system in the form of either heat or work, it follows that any change in the internal energy must be due to either a heat flow or work being performed. Work and heat are defined such that they are positive if they result in a net increase in the internal energy of the system and negative if they result in a decrease in the internal energy. To clarify the signs of these terms, consider the following example. If a diathermic system is placed in contact with surroundings that are colder than the system, then heat will flow from the system to the surroundings. For a warm cup of water in an open cup, heat will flow out of the cup into the air. Since the heat flow is out of the system the sign is negative. If the diathermic system is placed in surroundings that are warmer than the system, then heat will flow into the system and the sign is positive. For example, heat is positive when an open cup of cold water warms up. If work is done on a system, for example a gas is compressed, the work is positive and the internal energy of the system increases. If the system performs work, for example a gas is allowed to expand, the work is negative.

> The first law of thermodynamics states that, for an isolated, system the internal energy of a system is constant.

Isolated systems cannot perform work or exchange heat with the surroundings. Any change in the internal energy, ΔU, must be due to the sum of the work done on (or done by) the system, w, and the heat transferred into or from the system, q:

$$\Delta U = q + w \tag{2.1}$$

Note that this equation deals with the change in the internal energy, ΔU, that can be either positive, negative, or zero, and not the absolute energy. Based upon this law, a perpetual-motion machine, which produces work without consuming an equivalent amount of energy, is an impossible device to construct. Whenever work is performed, the internal energy must decrease. In biological organisms, energy in the form of food and nutrients is required for sustenance in order to perform the work required to live.

RESEACH DIRECTION: DRUG DESIGN I

One common strategy for developing new drugs is large-scale combinatory screening in which thousands of different molecules are tested for activity changes. In addition, the availability of the human genome has led scientists to investigate how to screen large numbers of proteins as possible drug targets. Such high-throughput screening requires methodologies that are readily performed in the laboratory and produce measurable parameters. In addition, once possible drug candidates are identified, the methodology must be amenable to distinguishing high-affinity, high-sensitivity drugs from others that are less effective. For detailed biochemical studies, the ability to perform the measurements over a versatile set of conditions would also be required.

When a drug or ligand binds to a protein, heat is either released or absorbed. Although these changes are small, the heat change can be measured accurately using calorimetry (calorimetry is a Latin-based word that literally means the measurement of heat). There are several methods for measuring heat in biochemical systems, with microcalorimeters being able to measure thermodynamic parameters using volumes as small as 0.1 mL. Differential scanning calorimetery is typically used in biochemistry to monitor folding and unfolding of proteins. Measurements of interactions between molecules and proteins are typically performed using a technique termed isothermal titration calorimetry (ITC), which is performed at constant temperature and involves the systematic titration of the protein with the drug. Modern instruments are able to determine the energetics of drug binding with high reliability and accuracy.

ITC measures the heat absorbed or released when a drug is introduced into a sample cell that contains the protein and is allowed to react (Figure 2.3). In a typical test for a drug, a syringe contains a concentrated solution of the drug and small amounts of the drug are injected in a stepwise manner into a chamber containing the protein.

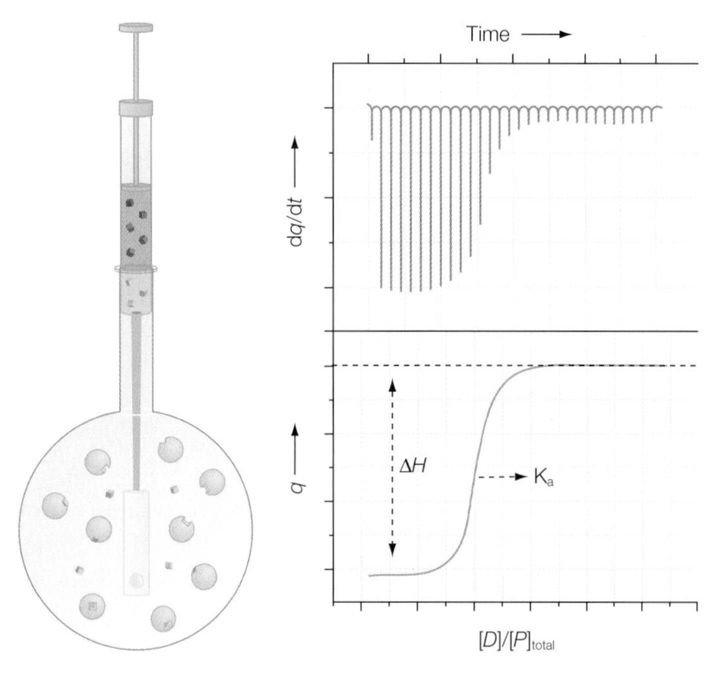

Figure 2.3 A schematic diagram of an ITC sample chamber and the results of a typical binding experiment. The sample chamber contains the protein and the syringe has a concentrated solution of the drug. At stepwise time intervals a small volume of the drug is injected, and subsequent binding leads to the release of heat, q, at the rate dq/dt. Modified from Freire (2004).

After each injection, any heat absorbed or released in the sample cell compared to a reference cell is measured. A feedback circuit continuously supplies thermal power to maintain a constant temperature in both the sample and reference cells, with any release or absorption of heat in the sample cell compensated for by a change in the thermal power. As more of the drug is injected, the number of proteins available for binding decreases systematically, and so the heat release subsequently decreases until the sites are saturated and no additional molecules of the drug can bind to the protein.

The resulting heat profile can be characterized in terms of a traditional binding curve, yielding the number of drug molecules bound to each protein as well as the binding constants. The presence of both the drug D and the protein P in the same chamber will result in binding and formation of a protein–drug complex, PD. For reversible binding in equilibrium, the relative amounts of the unbound states and the complex will be determined by the concentration of the free protein [P], the concentration of the free drug [D], the concentration of the drug–protein complex [PD], and the association constant K_a according to:

$$\mathrm{P} + \mathrm{D} \xleftrightarrow{K_a} \mathrm{PD} \quad K_a = \frac{[\mathrm{PD}]}{[\mathrm{P}][\mathrm{D}]} \tag{2.2}$$

The association constant provides a measure for the binding affinity of the drug to the protein and has units of M^{-1}, with higher values corresponding to higher affinities. The binding constant can be used to determine the amount of bound drug at a specified concentration of the drug and protein. A rearrangement of eqn 2.2 shows that the ratio of the drug–protein complex to free protein, [PD]/[P], is directly proportional to the concentration of the free drug:

$$K_a[\mathrm{D}] = \frac{[\mathrm{PD}]}{[\mathrm{P}]} \tag{2.3}$$

Notice that when the protein sites are half occupied, and [PD] = [P], then the value of K_a can be written simply as 1/[D]. Thus, the concentration of the drug at which half of the binding sites are occupied corresponds to $1/K_a$. Consider a new term, θ, the fraction of binding sites occupied by the drug compared to the total number of binding sites. This fraction is given by the relative number of binding sites occupied to the total number of binding sites, which is the sum of the proteins with and without the drug bound. This ratio can be expressed in terms of the concentrations of the protein without drug bound, [P], the concentration of the total protein, $[\mathrm{P}]_{total}$, and the concentration of protein with the drug bound, [PD]:

$$\theta = \frac{\text{Binding sites occupied}}{\text{Total binding sites}} = \frac{[\mathrm{PD}]}{[\mathrm{P}]_{total}} = \frac{[\mathrm{PD}]}{[\mathrm{P}] + [\mathrm{PD}]} \tag{2.4}$$

Substituting $[PD] = K_a[D][P]$ from eqn 2.3 into the fraction and rearranging the terms yields the expression:

$$\theta = \frac{[PD]}{[P] + [PD]} = \frac{K_a[D][P]}{[P] + K_a[D][P]} = \frac{K_a[D]}{1 + K_a[D]} \tag{2.5}$$

For the ITC measurements, the heat, q, released or absorbed for any given injection is proportional to the number of moles of the PD complex formed upon injection, which can be written in terms of the total volume, V, and concentration [PD]. Making use of the expression for the fraction θ allows the heat to be written in terms of the association constant:

$$q \propto V[PD] = V[P]_{total}\theta = V[P]_{total}\left(\frac{K_a[D]}{1 + K_a[D]}\right) \tag{2.6}$$

Thus, the heat profiles from the ITC experiments can be used to determine the binding affinity for the drug to the protein. The proportionality constant is given by the amount of heat released (or absorbed) for the binding of each molecule.

This approach has been successfully used to probe drug targets. For example, in human leukemia, the mixed-lineage leukemia (MLL) gene is associated with chromosomal translocations and is critical for the regulation of chromatin structure and gene activity (Allen et al. 2006). By using ITC and other structural measurements, the DNA-binding properties of the protein expressed by this gene have been characterized, providing researchers with a platform for developing therapeutics for leukemias that target the DNA-binding site. ITC has also been used with structural studies to understand the mechanism of conotoxins, which are a large family of toxins (Celie et al. 2005). Such toxins are under investigation for neurological diseases, including epilepsy. Knowledge concerning the binding of these toxins to receptors in the cell membrane enhances our understanding of the action of these toxins and the potential for drug design. The determination by ITC of the energetics of cAMP to a transcriptional activator termed CAP, for catabolite-activator protein, showed a biphasic isotherm which pointed to the existence of two nonequivalent binding sites with different characteristics (Popovych et al. 2006).

WORK

In classical mechanics, work is performed when a force, F, is used to move an object through a distance, Δx, according to the relationship:

$$w = -F\Delta x \tag{2.7}$$

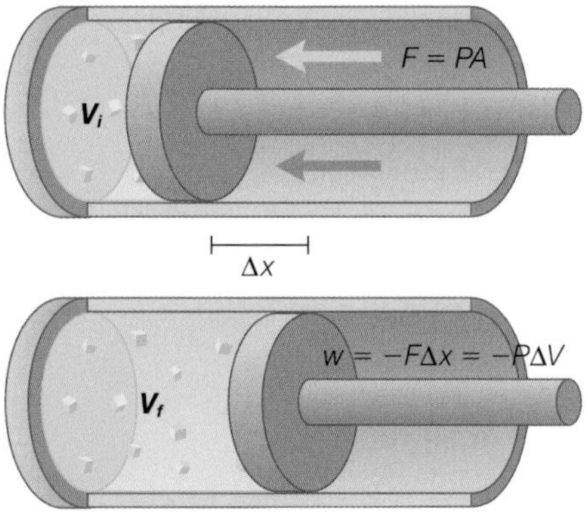

Figure 2.4 When the plunger of the piston slides during expansion, it pushes against a force F that arises from the atmospheric pressure due to gas molecules hitting against the piston. In this case, the work performed can be written as the product of the pressure P and volume change ΔV.

As an example of work, consider a gas inside a cylinder pushing against a piston (Figure 2.4). As the gas expands it must move the piston by pushing with a force that competes with the force exerted by gases in the atmosphere. Work performed by the expanding gas is the product of the force and change in position of the piston. The pressure that gas molecules push against is the force per unit area, or equivalently the force is the product of the pressure and the area. This gives work as being the product of pressure, area, and displacement. However, the area times the displacement is the change in volume, so work can be written as the product of pressure and volume change, with the sign being negative as the work is defined from the system's perspective:

$$w = -F\Delta x = -(PA)\Delta x = -P\Delta V \qquad (2.8)$$

If the force, or equivalently the external pressure, is not constant, we can sum all of the products for the different possible volumes in the form of an integral:

$$w = -\int_{V_1}^{V_2} P\,\mathrm{d}V \qquad (2.9)$$

When work occurs in a system with the opposing forces essentially equal, the work is called *reversible*. A reversible change is a change that can be reversed by an infinitesimal alteration of a variable. In the case of the piston, this happens when the inside and outside pressures are always equal. For example, heating the gas inside the piston will cause the gas inside to expand but the piston is continually sliding, keeping the pressure equal on the two sides. In this case, work can be calculated using the internal pressure (eqn 2.9). For reversible expansion of an ideal gas, when the temperature is held constant, the ideal gas law can be substituted, yielding:

$$w = -\int_{V_i}^{V_f} P\,\mathrm{d}V = -\int_{V_i}^{V_f}\left(\frac{nRT}{V}\right)\mathrm{d}V = -nRT\int_{V_i}^{V_f}\frac{\mathrm{d}V}{V} = -nRT\ln\frac{V_f}{V_i} \qquad (2.10)$$

$P = nRT/V$ for an ideal gas.

$$\int \frac{\mathrm{d}x}{x} = \ln x$$

SPECIFIC HEAT

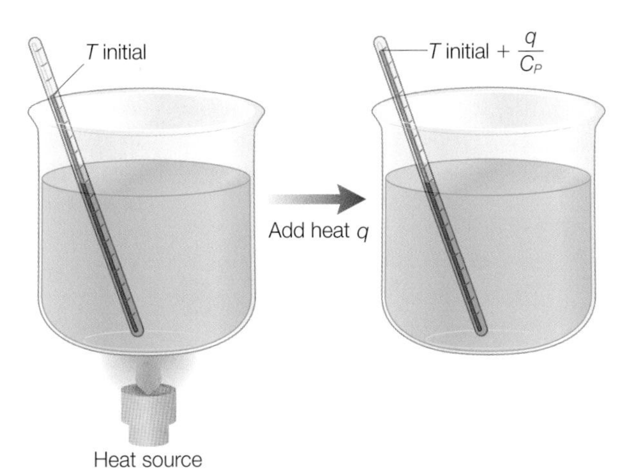

Figure 2.5 Specific heat at constant pressure, C_P, is a thermodynamic parameter that describes how much a substance will change in temperature in response to the addition of heat, q.

The application of heat to an object will cause the object's temperature to increase (Figure 2.5). Since heat represents the transfer of energy and temperature is a measure of the kinetic energy of molecules, the *change in temperature*, ΔT, is directly proportional to the heat applied (provided that heat does not result in a chemical change):

$$q = C\Delta T \tag{2.11}$$

The proportionality constant C is termed the *heat capacity* because it is a measure of how much heat is absorbed or emitted with a given temperature change. The value of C depends upon the properties of the object; for example, the heat capacity of water is very different to that of a metal rod. The heat capacity also is dependent upon the conditions under which heating and cooling are performed. The simplest case is when the volume of the system is held constant, as denoted by the inclusion of a subscript V with the heat capacity, C_V. The other common condition is when pressure is held constant, which is denoted by the subscript P, C_P.

As an example of the use of heat capacity, consider the effect of heating water compared to a piece of copper. Under constant pressure, at 4.18 J K^{-1} g^{-1}, the heat capacity of liquid water is much larger than that of copper (0.38 J K^{-1} g^{-1}). The smaller value of copper's heat capacity means it will become much hotter than an equivalent amount of water when a given amount of heat is applied. For example, applying 1 J of energy to 1 g of water will produce a temperature change of 0.24 K:

$$\Delta T = \frac{q}{C_P} = \frac{1\,\text{J}}{4.18\,\text{J K}^{-1}\,\text{g}^{-1}} = 0.24\,\text{K} \tag{2.12}$$

The same amount of heat will result in a temperature change of 2.63 K for 1 g of copper. Due to its relatively small specific heat, copper is a good heat conductor and so samples are often embedded in a copper block (or one made of another metal) to set a sample at a specific temperature in experiments.

INTERNAL ENERGY FOR AN IDEAL GAS

Energy can be transferred into a system by either a heat flow or by work. Regardless of how the energy was supplied, it can be released in either

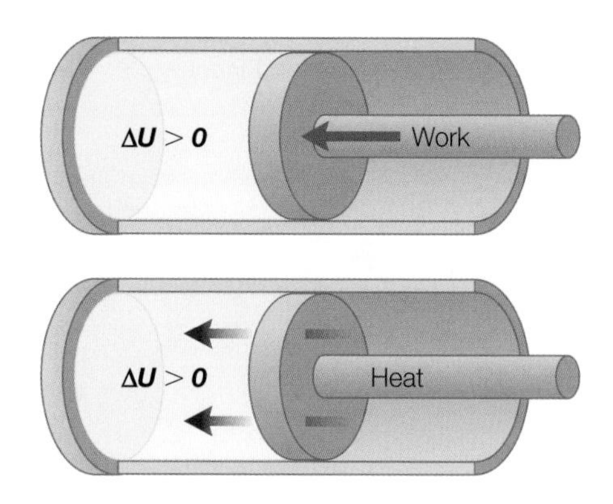

Figure 2.6 For an ideal gas the work performed in moving a piston is equal to the change in heat.

form. The internal energy, U, is a measure of the total capacity of the system to release heat or perform work. When work is done on a system the internal energy will increase; likewise when heat is applied (Figure 2.6). In most cases, the absolute value of the internal energy cannot be determined, but the change in energy can be measured. Therefore, we can define the *change in the internal energy*, ΔU, as the sum of the energy transferred into the system as work, w, and the energy transferred into the system as heat, q:

$$\Delta U = q + w \tag{2.13}$$

When pressure is held constant, work contribution is given by the volume change. The change in internal energy can then be related to the volume change:

$$\Delta U = q + w = q - P\Delta V \tag{2.14}$$

Consider how work, internal energy, and enthalpy are related for an ideal gas that for simplicity is placed in a chamber with a piston, allowing for volume changes (Figure 2.6). The volume change can be related to the change in the pressure, temperature, and number of molecules:

$$\Delta V = \Delta\left(\frac{nRT}{P}\right) \tag{2.15}$$

Since the pressure and the amount of gas are assumed to be held constant, only the temperature can change and the change in volume is proportional to the change in temperature:

$$\Delta V = \Delta\left(\frac{nRT}{P}\right) = \frac{nR}{P}\Delta T \tag{2.16}$$

Substituting this expression into the relationship for ΔU (eqn 2.9) yields:

$$\Delta U = q - P\Delta V = q - P\left(\frac{nR}{P}\Delta T\right) = q - nR\Delta T \tag{2.17}$$

The heat flow can be related to the product of the specific heat at constant pressure and temperature change, leading to the expression:

$$\Delta U = q - nR\Delta T = C_P\Delta T - nR\Delta T = (C_P - nR)\Delta T \tag{2.18}$$

Thus, the change in internal energy is always proportional to the change in temperature, and must be zero if there is no change in temperature. If there is no temperature change the heat flow must exactly balance the work:

$$\text{For } \Delta T = 0,\ \Delta U = q + w = 0 \text{ and } q = -w \tag{2.19}$$

ENTHALPY

Most biological systems are open to the atmosphere, allowing the volume to change in response to a change in energy. For this situation, heat supplied to the system can change not only the internal energy of a system but also result in a concurrent volume change. For example, if some of the energy supplied as heat increases the volume of the system then the change in internal energy will be smaller than found if the volume did not change. For this reason, the heat supplied is equated to another thermodynamic state function, called enthalpy. Formally, *enthalpy*, H, is defined in terms of internal energy, U, and the product of pressure P and volume V according to:

$$H = U + PV \tag{2.20}$$

Biological systems are usually open to the atmosphere and so any processes occur at a constant atmospheric pressure. In addition, the enthalpy in most cases does not need to be considered as an absolute value but only as a relative value, ΔH. Therefore, for most situations only the change in enthalpy at constant pressure will be needed:

$$\Delta H = \Delta U + \Delta(PV) = \Delta U + P\Delta V \tag{2.21}$$

Because enthalpy is defined in terms of state functions (internal energy, pressure, and volume), enthalpy itself is also a state function. Therefore, the change in enthalpy will be independent of the path between two states and dependent only upon the initial and final states. The enthalpy term is used because it alleviates the difficulty in relating the energy change to the changes in other state functions. The change in internal energy is given by heat minus work contribution, provided no other type of work is performed. Substituting this into the expression for the enthalpy change yields:

$$\Delta H = \Delta U + P\Delta V = (q - P\Delta V) + P\Delta V = q \tag{2.22}$$

$\Delta U = q - p\Delta V$ (eqn 2.14)

Thus, at constant pressure, the change in enthalpy is equal to the heat transferred. When heat enters the system maintained at constant pressure the enthalpy must increase and in an exothermic process the enthalpy must decrease. For example, if a gas is placed inside a chamber with a piston that can move so that the pressure is fixed, any heat flow will directly be equal to the change in enthalpy.

DEPENDENCE OF SPECIFIC HEAT ON INTERNAL ENERGY AND ENTHALPY

Derivation box 2.1

State functions described using partial derivatives

A state function describing a system is one that is dependent only upon the state and not the path that was used to prepare the system. Hence state functions are very useful in describing the thermodynamic properties of systems. State functions can be described in formal terms with the use of partial derivatives. As an example of the use of partial derivatives, suppose you wish to describe the trajectory of a boy walking up a hill (Figure 2.7). In addition to knowing the two horizontal coordinates, x and y, it is necessary to also know the altitude at each coordinate, $A(x,y)$. The motion of the boy is described by consideration of how changes in his coordinates lead to changes in his position on the hill; that is, changes in his altitude. Consider the case of the boy moving only along the x direction. For small changes in his position, Δx, there will be a change in his altitude, ΔA, that is determined by the product of the change in coordinate and the slope of the hill:

$$\Delta A = \left(\frac{\partial A(x,y)}{\partial x}\right)_y \Delta x \qquad \text{(db2.1)}$$

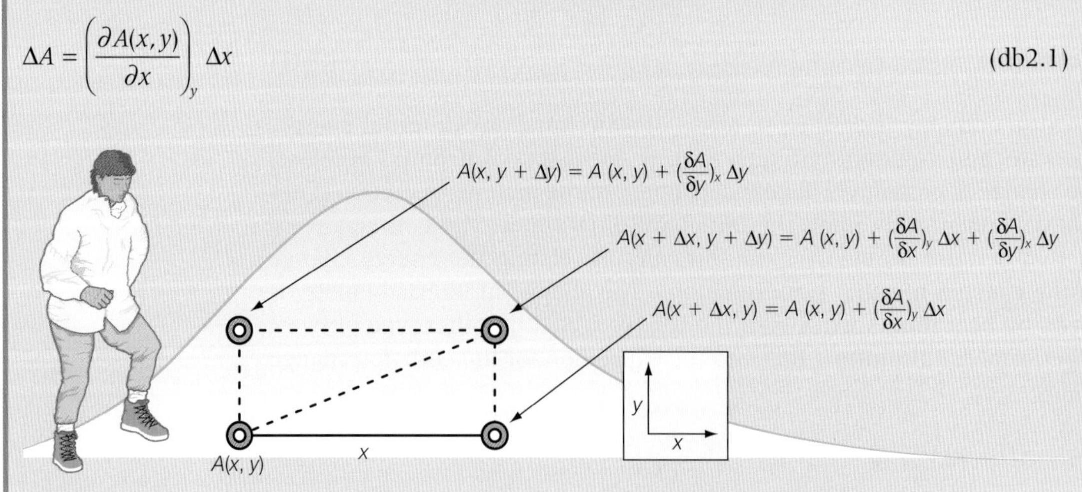

Figure 2.7 The altitude $A(x,y)$ changes due to walking along a hill along two perpendicular directions, x and y.

where the subscript denotes that the process occurs for fixed y. Conversely, if the boy moves along the y direction only by an amount Δy, the change in altitude would be:

$$\Delta A = \left(\frac{\partial A(x,y)}{\partial y}\right)_x \Delta y \qquad \text{(db2.2)}$$

The use of partial derivatives provides the means to determine how the altitude will change in response to a more general move that involves changes in both coordinates. Since altitude is a state function, the change is independent of the path. Regardless of the actual path taken, the change in altitude can be equivalently regarded as being a combination of walking along the x direction followed by walking in the y direction:

$$\Delta A = \left(\frac{\partial A(x,y)}{\partial x}\right)_y \Delta x + \left(\frac{\partial A(x,y)}{\partial y}\right)_x \Delta y \qquad \text{(db2.3)}$$

For any complicated altitude function, this would only be strictly true for *infinitesimal changes* in x and y, written as $\mathrm{d}x$ and $\mathrm{d}y$, respectively, and so the resulting *change in the altitude*, $\mathrm{d}A$, should be written as:

$$\mathrm{d}A = \left(\frac{\partial A(x,y)}{\partial x}\right)_y \mathrm{d}x + \left(\frac{\partial A(x,y)}{\partial y}\right)_x \mathrm{d}y \qquad \text{(db2.4)}$$

Following the same approach, it is possible to write expressions for the changes in all of the state functions due to changes in the parameters that describe them. For example, the internal energy depends upon any two of the three parameters, volume, pressure, and temperature, since these parameters can be related to each other by an equation of state such as the perfect gas law. Consider the *change in the internal energy*, $\mathrm{d}U$, due to changes in volume and temperature. The change $\mathrm{d}U$ can be written in terms of the *changes in the volume and temperature*, $\mathrm{d}V$ and $\mathrm{d}T$, multiplied by their respective slopes:

$$\mathrm{d}U = \left(\frac{\partial U(V,T)}{\partial V}\right)_T \mathrm{d}V + \left(\frac{\partial U(V,T)}{\partial T}\right)_x \mathrm{d}T \qquad \text{(db2.5)}$$

For thermodynamic state functions, the slopes as expressed by the partial derivatives have physical meanings. The second term, namely the change in the internal energy with respect to a change in temperature at fixed volume, is the specific heat at constant volume (eqn 2.25). The nature of the second term can be understood from a consideration of the nature of the internal energy. For an ideal gas, the energy is independent of the volume as long as the temperature is constant, hence this partial derivative has a value of zero. A non-zero value arises due to nonideal properties of real gases. The partial derivative is non-zero when the energy is dependent upon the volume. Equivalently, the energy is dependent upon the separation of the molecules due to an interaction between molecules. Energy divided by volume has units of pressure. This term refers to the internal pressure of the system, because attractive or repulsive interactions between molecules will alter the properties of the molecules, including the pressure of the system. For the van der Waals model of n mol of a gas, the partial derivative is related to the square of the volume, and in general the *internal pressure* is denoted by $P_{internal}$:

$$\left(\frac{\partial U}{\partial V}\right)_T = a\left(\frac{V}{n}\right)^2 = P_{internal} \tag{db2.6}$$

The change in internal energy can then be written in terms of the internal pressure $P_{internal}$ and specific heat at constant volume:

$$dU = P_{internal}dV + C_V dT \tag{db2.7}$$

For an ideal gas, since $P_{internal}$ is zero, the specific heat at constant volume represents the change in the internal energy in response to a change in temperature. Although C_V is determined at constant volume, it can be considered to be a coefficient that determines how much effect a change in temperature will have on the internal energy. When nonideal conditions are included, the change dU is modified according to $P_{internal}$. For real gases, if attractions among molecules are found to dominate, then $P_{internal}$ will be positive and if repulsion interactions dominate then this term will be negative.

Similar expressions can be written for the other thermodynamic parameters. For example, consider the change in enthalpy at constant volume. In this case the remaining parameters are temperature and pressure, and the change in enthalpy can be written in terms of their partial derivatives:

$$dH = \left(\frac{\partial H(P,T)}{\partial T}\right)_P dT + \left(\frac{\partial H(P,T)}{\partial P}\right)_T dP \tag{db2.8}$$

Of these two partial derivatives, the first is the change in enthalpy due to a temperature change at constant pressure, which was shown to be equal to the specific heat at constant pressure (eqn db2.3). The second term, the change in enthalpy in response to a pressure change, can be written in terms of the product of two derivatives involving the temperature following the guidelines for the chain relationship of partial derivatives. Of these two derivatives, the second is just the specific heat at constant pressure and the first is called the *Joule–Thompson coefficient* and denoted by μ:

$$\left(\frac{\partial H}{\partial P}\right)_T = -\left(\frac{\partial T}{\partial P}\right)_H \left(\frac{\partial H}{\partial T}\right)_P = -\mu C_P \tag{db2.9}$$

The Joule–Thompson coefficient is commonly used in understanding how refrigerators work and the process of liquefaction of gases. Refrigerators work with non-ideal gases and take advantage of attractive forces between molecules (and the negative value of μ). The gases used in refrigerators release heat when gas molecules come together, and take up heat when gas molecules are pulled apart. Thus, when a gas is allowed to expand at constant pressure, the temperature of the gas will drop according to the coefficient. In a refrigerator, a gas is cyclically allowed to expand and contract to move heat from inside to outside a system. The Joule–Thompson effect is also used to liquefy gases, as a gas is initially poised at high pressure and is then allowed to expand and hence cool. By performing the volume change cyclically the temperature will systematically drop until the gas condenses to a liquid.

The change in the internal energy is given by the sum of the work and heat contributions. At constant volume, work contribution is zero and so the change in internal energy is given by heat, which is the product of the specific heat and the temperature change. Thus, the change in internal energy divided by the change in temperature is the specific heat:

$$\Delta U = C_V \Delta T \quad \rightarrow \quad C_V = \frac{\Delta U}{\Delta T} \tag{2.23}$$

A more precise statement is to write the specific heat at constant volume in terms of incremental changes in internal energy, or equivalently as a partial derivative, with the subscript denoting that the changes occur with the volume fixed:

$$C_V = \left(\frac{\partial U}{\partial T}\right)_V \tag{2.24}$$

For an ideal gas at constant volume, we can relate the change in enthalpy to the temperature change and specific heat, just as was done for the change in internal energy. The change in enthalpy is directly proportional to the change in temperature:

$$\Delta H = \Delta U + \Delta(PV) = C_V \Delta T + \Delta(nRT) = C_V \Delta T + nR\Delta T = (C_V + nR)\Delta T \tag{2.25}$$

Notice that when there is no change in the temperature then both the internal energy and the enthalpy remain constant for an ideal gas. This is not necessarily true for other systems.

The specific heat at constant pressure has a different dependence. Heat results in not only a temperature increase of the system but also in work that can be performed. As the system heats up it will expand in order to keep the pressure constant, resulting in a $-P\Delta V$ contribution of work. The change in the internal energy is then given by:

$$\Delta U = q - P\Delta V \tag{2.26}$$

The change in enthalpy at constant pressure (eqn 2.27) is given by the contributions of the change in internal energy and change in volume since the change in pressure is zero:

$$\Delta H = \Delta U + \Delta(PV) = \Delta U + P\Delta V \tag{2.27}$$

Substituting the expression for the change in internal energy (eqn 2.26) gives the simple result that the change in enthalpy is equal to the heat:

$$\Delta H = \Delta U + P\Delta V = (q - P\Delta V) + P\Delta V = q \tag{2.28}$$

Following the ideas above for the specific heat at constant volume, the specific heat at constant pressure can be written in terms of the heat divided by the temperature change, with the heat being equal to the change in enthalpy. Again, assuming that the changes in enthalpy are incremental, the expression can be written in terms of a partial derivative, with the specific heat at constant pressure being equal to the derivative of the enthalpy with respect to temperature at constant pressure:

$$C_P = \frac{q}{\Delta T} = \frac{\Delta H}{\Delta T} \rightarrow C_P = \left(\frac{\partial H}{\partial T}\right)_P \tag{2.29}$$

ENTHALPY CHANGES OF BIOCHEMICAL REACTIONS

Many biochemical reactions occur at a constant pressure and the enthalpy change is equal to heat change during the reaction. The enthalpy change is determined by the difference in enthalpy between the initial and final states. By convention, enthalpy changes are reported under what are termed standard ambient temperature and pressure conditions. For most of chemistry, the standard state is defined as 1 mol of an object at 1 barof pressure and at a temperature of 298.15 K. Enthalpy changes in reactions are generally additive, which allows the enthalpy to be determined for a complex process by considering the individual steps. For example, the enthalpy change for a reaction $A \rightarrow C$ is simply the sum of the enthalpy changes for the two stepwise reactions $A \rightarrow B$ and $B \rightarrow C$. Likewise the enthalpy change for a reverse reaction, for example $B \rightarrow A$, has the same value, but the opposite sign, as the change for the forward reaction $A \rightarrow B$. That is, a reaction that releases heat will absorb heat if reversed.

When systems undergo a reaction, the components may change their phase. For example, when water is heated at its boiling point it will become a vapor. The conversion of a substance from one phase to another is called a phase transition. The phase transition often represents a state change between a gas, liquid, and solid. Alternatively, the transition can involve different solid forms; for example, carbon can exist as graphite or diamond solids.

The energy that must be supplied to the system to cause a phase change is called the *enthalpy of formation*. The energy required would depend upon the conditions, such as temperature and pressure, at which the reaction occurs. In order to associate a specific enthalpy change with a given reaction, changes are always reported under standard conditions; that is, at a pressure of 1 bar. For example, when ice melts at 0°C the enthalpy change is +6.01 kJ mol^{-1}. The enthalpies associated with the reverse change will always have the same value but opposite sign. So for water, the enthalpy change is +6.01 kJ mol^{-1} for the ice-to-liquid transition and −6.01 kJ mol^{-1} for the liquid-to-ice transition. The value of the enthalpy change is different for each type of transition for a given molecule. For water, the transition from water to vapor – that is, the boiling transition – has a much larger enthalpy of 40.7 kJ mol^{-1} than found for the freezing transition.

The enthalpy change that is often encountered in biological reactions is a chemical change. For example, an electron may be removed from a molecule, causing a change in the ionization state of a molecule. Bonds can also be broken or formed, resulting in corresponding enthalpy changes that are termed *bond enthalpies*. In any given reaction, there may be several types of change occurring simultaneously. The usefulness of the enthalpy concept is that the overall enthalpy change for the reaction is additive and is given simply by the sum of the changes of the products minus the changes of the reactants. By convention, the enthalpy of elements is equal to zero when they are in a stable state under standard conditions.

As an example, the thermodynamic properties of foods can be discussed in terms of the enthalpy of combustion per gram of food. Humans need about 10^7 J of food every day. If the energy is provided entirely by glucose (Figure 2.8), a total of about 600 g per day is required because glucose has a specific enthalpy of 1.7×10^4 J g^{-1}. By comparison, digestible carbohydrates have a slightly higher enthalpy of 1.7×10^4 J g^{-1} and fats, which are commonly used as an energy-storage material, have values of 3.8×10^4 J g^{-1}.

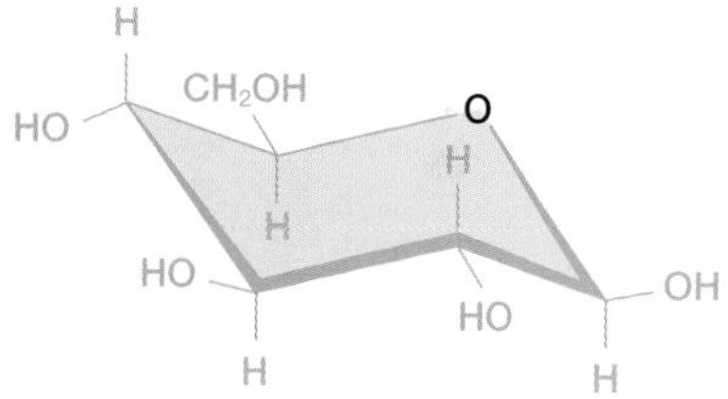

Figure 2.8 The structure of glucose.

These enthalpies of combustion can be calculated by summing the contributions from the combustion process (Table 2.1). The oxidation of glucose results in the production of carbon dioxide and water:

$$C_6H_{12}O_6 \text{ (solid)} + 6O_2 \text{ (gas)} \rightarrow 6CO_2 \text{ (gas)} + 6H_2O \text{ (liquid)} \quad (2.30)$$

The enthalpy contribution from the products is calculated by using the enthalpies of formation for carbon dioxide multiplied by the relative numbers from the balanced equation above:

$$\begin{aligned} \Delta H^{\circ}_{products} &= 6(-393.5 \text{ kJ mol}^{-1}) + 6(-285.8 \text{ kJ mol}^{-1}) \\ &= -4076 \text{ kJ mol}^{-1} \end{aligned} \quad (2.31)$$

There are two reactants that could contribute to the enthalpy. The enthalpy of formation of glucose is −1273.2 kJ mol^{-1} (Table 2.1). Since molecular oxygen by convention is considered to be stable it has zero contribution. The subtraction of the reactants from the products yields:

$$\begin{aligned} (\Delta H^{\circ})_{oxidation} &= -4076 \text{ kJ mol}^{-1} - (-1273.2 \text{ kJ mol}^{-1}) \\ &= -2802 \text{ kJ mol}^{-1} \end{aligned} \quad (2.32)$$

Thus 2.8×10^6 J mol^{-1} of heat is released upon oxidation of glucose during metabolic pathways, or equivalently 1.6×10^4 J g^{-1} as used above. Due to

Table 2.1
Standard enthalpies of formation (1 atm, 25°C).

Substance	$\Delta H°$ (kJ mol^{-1})
Simple molecules	
CO (gas)	−110.5
CO_2 (gas)	−393.5
O_2 (gas)	142.3
H_2O (gas)	−241.8
H_2O (liquid)	−285.84
Carbohydrates	
Glucose	−1273.2
Sucrose	−2225.9
Alcohols	
Acetic acid	−484.5
Ethanol	−277.8
Glycerol	−668.6
Amino acids	
Alanine	−563
Cysteine	−534
Glutamic acid	−1010
Leucine	−647
Serine	−726
Tyrosine	−672
Fatty acids	
Palmitic acid	−891.6
Stearic acid	−947.7

this heat release about 50% of the energy available can be used to perform work when organisms metabolize glucose.

RESEARCH DIRECTION: GLOBAL CLIMATE CHANGE

Consider the Earth as a system that can be characterized with certain thermodynamic parameters. The Earth is a closed system since matter does not leave or enter, neglecting satellites, meteorites, and the escape of light gases such as hydrogen and helium. The Earth does experience an energy flow from the surroundings, namely heating from the Sun. Due to its size and complex composition, the Earth does not have simple, unique properties such as a uniform temperature. However, we can still apply basic thermodynamic ideas using average values.

Let us examine the heat flow into the system. The Earth experiences a certain amount of incoming solar energy that does vary with the time

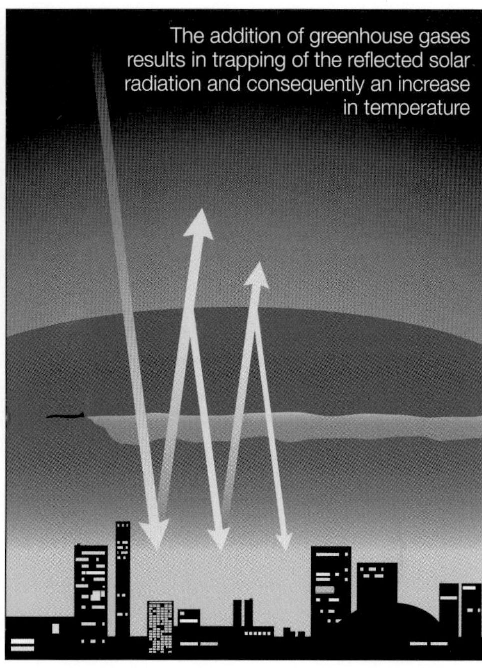

Figure 2.9 Global warming arises from the trapping of solar radiation by gases in the atmosphere. After the radiation from the sun enters the atmosphere, some is reflected back into space. Greenhouse gases limit how much of the reflected light leaves the earth.

of year, but we will assume that this energy is constant over time. In the absence of any atmosphere, the heat from the solar energy required to set the Earth's temperature at a specific temperature is based on how much solar radiation is required to reach the Earth and what fraction is absorbed. The presence of the atmosphere alters how much solar radiation heats the Earth (Figure 2.9). Upon reaching the atmosphere, part of the incident solar energy is reflected whereas the remainder penetrates the atmosphere and reaches the Earth's surface. The fraction of sunlight that reaches the surface heats the surface while another part is reflected back to space. While returning to space, a fraction will escape while part is reflected back. The trapping of a certain fraction of the solar radiation by the atmosphere, termed the greenhouse effect, results in an increase in the average temperature.

The efficiency of energy trapping by the atmosphere depends upon the composition of gases. The atmosphere is composed of primarily nitrogen (78%) and oxygen (21%), with the remaining portion contributed by argon (0.9%), carbon dioxide (0.03%), and several trace gases. The atmosphere contains other naturally occurring components including water vapor, which contributes about 60% to the natural greenhouse effect. There are

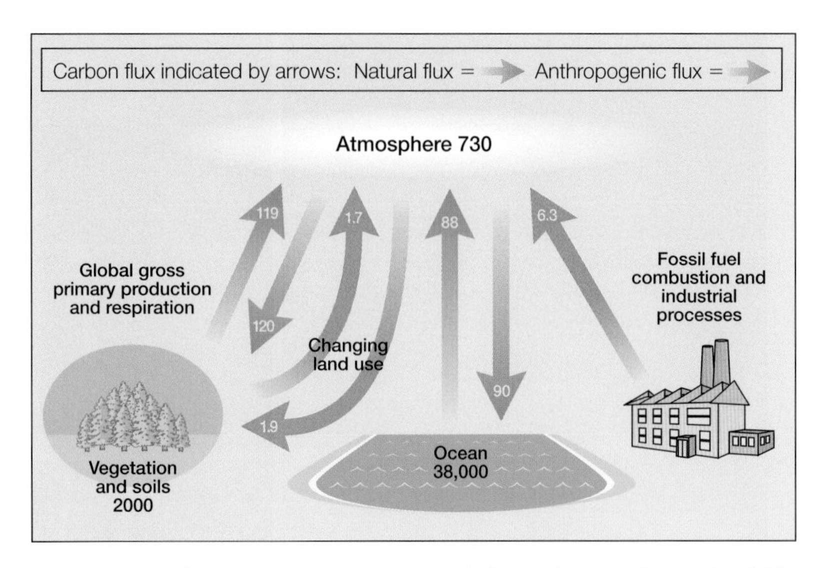

Figure 2.10 A schematic representation of the carbon cycle. Units: billion metric tons measured in carbon equivalent terms. Modified from International Panel on Climate Change (2001).

other components that can influence the energy flow, with the amount determined by human activity. Carbon dioxide accounts for over half of human-induced warming. Two components that are relatively minor in amount but significant due to their ability to absorb heat, are nitrous oxide and sulphur hexafluoride. Some contributors occur exclusively due to human production, such as chloroflurocarbons and halocarbons, which were once used primarily as refrigerants. An international treaty banned these chemicals in 1987 as their presence in the atmosphere led to a decrease in the amount of ozone.

Numerous processes involving the soil, atmosphere, and living organisms regulate the concentrations of these gases, with the flow of carbon dioxide termed the carbon cycle (Figure 2.10). In the cycle there is a balance between two processes. Respiration results in the uptake of oxygen and output of carbon dioxide. Photosynthesis results in the uptake of carbon dioxide and production of oxygen. These natural processes absorb large amounts of carbon dioxide but the additional contributions from the burning of fossil fuels result in a net increase in the amount of carbon dioxide in the atmosphere; hence the amounts of greenhouse gases in the atmosphere increase. Since the Industrial Revolution, the atmospheric concentration of carbon dioxide is estimated to have increased by about 30%, methane concentrations have doubled, and nitrous oxide concentrations have gone up by 15%. Sulfur aerosols result in a cooling of the atmosphere due to their ability to reflect sunlight back into space but these gases are short-lived in the atmosphere.

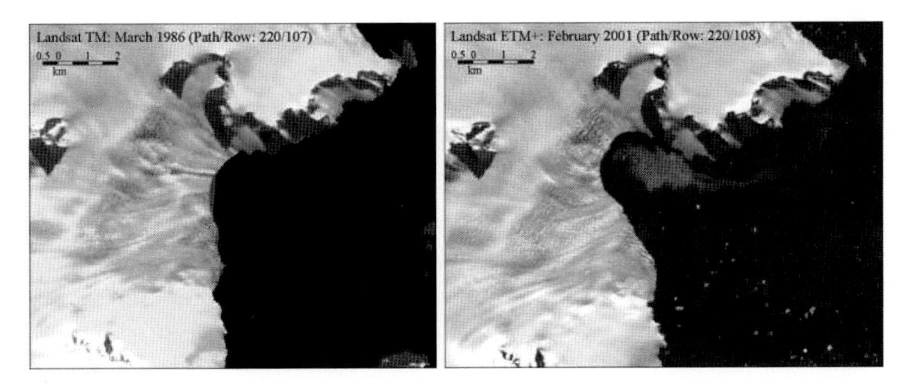

Figure 2.11 Glacial retreat shown by comparison of satellite images of Sheldon Glacier, Adelaide Island in March 1986 (left) and February 2001 (right). From Cook et al. (2005).

Although the increase in greenhouse gases is clear, the effect of these gases on the energy flow of the Earth is difficult to establish due to the complexity of the Earth (Crowley 2000; Cook et al. 2005; Oerlemans 2005). The Earth's temperature varies widely so the use of temperature requires averaging over a certain number of locations. Instead of just relying on recorded temperatures, scientists use global features as markers of temperature. One prominent indicator of the long-term increase in the Earth's temperature is the retreat of glacial fronts (Figure 2.11). The position of the ice shelves in the Antarctic Peninsula shows a pattern of loss that is outside of normal cyclic behavior.

What is the actual contribution of the atmosphere to the Earth's temperature? A simple thermodynamic model of energy flow is that Earth is an airless and rapidly rotating planet so that the temperature is uniform (Sagan & Chyba 1997). In this case, the equilibrium temperature of the Earth is determined by the *distance between the Earth and the Sun*, d_{es}, and the *solar radiance of the Sun*, *S*, at the distance d_{es}:

$$T^4 = (\text{constant})\,\frac{SR^2(1-A)}{d_{es}^2} \tag{2.33}$$

According to this equation, the temperature of the Earth would increase due to an increase in the solar radiation (S), an increase in the area of the Earth (R^2), a decrease in the distance to the Sun (d_{es}), or a decrease in the amount of light reflected back as measured by the reflection coefficient (A). When the values for these parameters are inserted, the temperature is calculated to be 255 K, or −18°C. The mean overall global temperature has been measured to be 288 K, or 15°C. The 33°C difference in temperature is ascribed to the contribution of the atmosphere. The temperature of the Earth does vary slowly with time due to factors such as changes in the amount of solar radiation and atmospheric changes due to volcanoes. In the absence of detailed temperature records over many centuries, it is difficult to separate this natural variability of global temperature from the effect due to an increase in greenhouse gases. However, it is now generally accepted that the temperature has increased by 0.5–1.0°C during the past century. Whereas the absolute magnitude of this change on the global temperature is small, it represents a 2–3% increase in the contribution of greenhouse gases and is projected to have a profound influence on the Earth. For example, computer models of the effect of these temperature increases are predicted to result in a significant rise in the ocean's water levels (Meehl et al. 2005). Correlated with the increase in the increasing sea-surface temperature has been a increasing hurricane frequency and intensity (Webster et al. 2005). In particular, a large increase has been seen in the number of hurricanes reaching categories 4 and 5, such as Hurricane Katrina, which caused so much damage in 2006 in the USA. A possible causal relationship arising from the relationship between saturation vapor pressure and sea-surface temperature is under debate. The trend of increasing and more frequent hurricanes is consistent with climate models, and the complex nature of hurricanes continues to be investigated by scientists. Notably, the efforts of Albert Gore and the Intergovernmental Panel on Climate Change were recognized with the Nobel Prize in Peace in 2007.

REFERENCES

Allen, M.D., Grummitt, C.G., Hilcenko, C. et al. (2006) Solution structure of the nonmethyl-CpG-binding CXXC domain of the leukaemia-associated MLL histone methyltransferase. *EMBO Journal* **25**, 4503–12.

Celie, P.H.N., Kasheverov, I.E., Mordvintsev, D.Y. et al. (2005) Crystal structure of nicotinic acetylcholine receptor homolog AchBP in complex with α-conotoxin PnIA variant. *Nature Structural and Molecular Biology* **12**, 582–8.

Cook, A.J., Fox, A.J., Vaughan, D.G., and Ferrigno, J.G. (2005) Retreating glacier fronts on the Antarctic Peninsula over the past half-century. *Science* **308**, 541–5.

Crowley, T.J. (2000) Causes of climate change over the past 1000 years. *Science* **289**, 270–7.

Freire, E. (2004) Isothermal titration calorimetry: controlling binding forces in lead optimization. *Drug Discovery Today: Technologies* **1**, 295–9.

Intergovernmental Panel on Climate Change (2001) *Climate Change 2001: The Scientific Basis*. UN Press.

Meehl, G.A., Washington, W.M., Collins, W.D. et al. (2005) How much more global warming and sea level rise? *Science* **307**, 1769–72.

Oerlemans, J. (2005) Extracting a climate signal from 169 glacier records. *Science* **308**, 675–7.

Popovych, N., Sun, S., Elbright, R.H., and Kalodimos, C.G. (2006) Dynamically driven protein allostery. *Nature Structural and Molecular Biology* **13**, 831–8.

Sagan, C. and Chyba, C. (1997) The early faint sun paradox: organic shielding of ultraviolet-labile greenhouse gases. *Science* **276**, 1217–21.

Webster, P.J., Holland, G.J., Curry, J.A., and Chang, H.R. (2005) Changes in tropical cyclone number, duration, and intensity in a warming environment. *Science* **309**, 1844–6.

PROBLEMS

2.1 Heat is transferred to a system at 1 atm. Will the internal energy increase more if the system is at constant volume or at constant pressure?

2.2 For the isothermal, reversible compression of an ideal gas, what can be said about the heat?

2.3 Burning gasoline inside a closed, adiabatic system at constant volume will cause what change to the internal energy?

2.4 What can be stated about the internal energy of an isolated system?

2.5 Why do drug candidates have different ITC profiles?

2.6 What type of heat dependence would a favorable drug candidate have?

2.7 If 0.818 mol of an ideal gas at 25°C is in a 2-L container that is expanded to a final volume of 8 L at a constant pressure of 1 atm, what are the heat, work, and change in internal energy?

2.8 What is the work involved in an isothermal, reversible expansion of 0.5 mol of an ideal gas at 30°C from 10 to 20 L?

2.9 A 5-g block of dry ice is placed into a 30-mL container and the container is sealed. (a) What is the pressure in the container after the carbon dioxide has sublimated (vaporized directly to gas; Chapter 4)? (b) Assuming that the pressure is immediately released after the pressure fully develops, how much work is done on the surroundings by the expansion of the gas? Neglect any contribution from the cap.

2.10 A sealed 1-L glass container at 1 atm and 25°C is heated and breaks when the internal pressure reaches 2 atm. When it breaks what is the temperature inside and what is the internal energy? Assume that the specific heat of air has the value of $C_V = 25$ J/(Kmol).

2.11 What is the enthalpy change of a system if 1.0 J of heat is transferred into it at constant pressure?

2.12 If an ideal gas expands isothermally, what can be said about the internal energy?

2.13 One mole of an ideal gas is in a piston and the volume changes from 2 to 4 L at 25°C. What are the work, heat, and change in the internal energy?

2.14 Calculate the heat required to raise the temperature of 1.5 mol of an ideal gas from 15 to 50°C when the volume is fixed at 1 L. What is the change in internal energy using a specific heat value of $C_V = 25$ J/(Kmol)?

2.15 If 2 mol of an ideal gas is allowed to expand isothermally and reversibly from 10 to 30 L at a constant temperature of 20°C, what are the work, heat, and change in internal energy?

2.16 What is the enthalpy of combustion for sucrose?

2.17 How reliable is the melting of the glaciers as an indicator of the temperature if some glaciers are increasing in mass?

2.18 If the amount of carbon dioxide generated by several different biological processes is so large, why should we be concerned about the increase generated by industry and transportation during the past 50 years?

2.19 If the distance between the Earth and Sun decreased by 5% due to an extraterrestrial disaster, how much would the temperature of the Earth change?

2.20 How much would the solar radiation need to increase in order to increase the temperature of the Earth by 10% (neglecting the atmosphere)?

3

Second law of thermodynamics

The first law of thermodynamics provides a means to balance work and heat changes and the parameters that are changed when work is done or when heat flows into or out of a system. However, knowing what is energetically allowable is not the same as knowing which processes will actually occur. From a global perspective, energy within the universe is a constant, yet it is constantly changing with solar systems forming and stars evolving from red giants into white dwarfs. Clearly, these changes are being driven by factors other than energy.

Rather than explore the changes in the universe, let's consider why objects change in simple processes. If a ball is dropped from a height, it will drop until it hits the ground and then bounces back (Figure 3.1). Why did this happen? The ball initially is stationary but has potential energy due to gravity. As the ball falls the potential energy is converted into kinetic energy and the ball speeds up. When it hits the ground the direction of velocity is reversed and the ball slows down and the energy is converted into potential energy. The ball will stop when it reaches the original height and all of the energy is in the form of potential energy again.

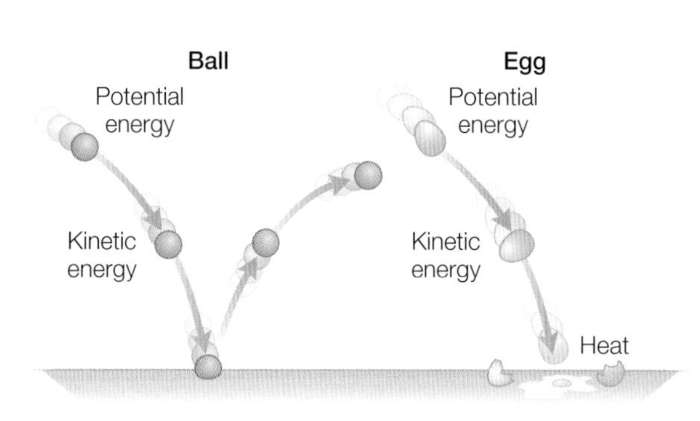

Figure 3.1
A comparison of energy changes for a bouncing ball and an egg.

The ball travels according to the first law, with the energy converting between kinetic and potential energy and the total energy remaining the same throughout the travel. Now consider using an egg instead of the ball. The egg travels downward like the ball but the egg does not bounce up after hitting the ground. Is energy still conserved? For the egg, the kinetic energy is converted upon impact with the ground into heat; that is, random motion of the molecules. Unlike the ball, the egg lost its ability to move upward against the gravitational force and instead became a more disordered state with random motion.

A preference for an increase in the random motion of an object is a simple statement of the second law of thermodynamics. Formally, in addition to energy, objects have another property called *entropy*, which represents *the molecular disorder of a system*. According to the second law, processes will occur spontaneously if the entropy increases. It is the direction of entropy that leads to the direction of processes that occur in the universe. Whether considering the progression of stars through different states as their energy output decreases or the motion of bacteria in search of nutrients, processes are headed away from organized states to random states.

ENTROPY

The increase in disorder is evident in our everyday lives. Each day, food is consumed and largely converted into heat, which is a more disordered state than the original food items. In general, it could be argued that all life forms have devised methods of taking ordered resources, such as molecules or food, and converting them into disordered states in order to generate energy. If all processes are moving to a disordered state, how can highly organized states such as crystals or biological cells be created? The answer lies in considering the overall state and not just one part. For example, your bedroom may tend to become very disordered with daily use until you decide to clean it up. While the clothes and other objects in the room may be more ordered, the process of cleaning the room involves the expenditure of energy in the form of heat. Thermodynamics would say that the disorder associated with the heat is greater than the order associated with the folding and stacking of the clothes.

In a sense, the change in order of an object provides a direction for chemical reactions. When a process results in a change from one state to another that occurs in an irreversible way, the process is called spontaneous. Only irreversible changes are spontaneous; truly reversible processes are not. Truly reversible processes do not occur in nature, as it would require all forces to be perfectly balanced with no driving force for the system to move. However, by moving objects very slowly while keeping forces in nearly perfect balance, processes that are very close to reversible can be created. As an example, gas inside a piston expanding under constant pressure inside and outside does not have any net force to drive the expansion, and the piston does not move. By making the imbalance very small, the piston moves, although at a very slow rate. Such a motion would take a considerable amount of time to be completed, but the rate is not under consideration here, only the direction.

To quantify the concept that matter and energy tend to be dispersed, a new state function is introduced, *entropy*, *S*, which is a measure of the disorder of a system. As energy and matter disperse, entropy is defined to increase. The concept of entropy is explicitly defined in terms of the heat and temperature of a system. For a given reversible process, a small change results in an *entropy change*, dS, which is defined in terms of the *amount of heat produced*, dq, and the temperature:

$$\mathrm{d}S = \frac{\mathrm{d}q}{T} \tag{3.1}$$

For a measurable change of an isothermal process, the *change in entropy*, ΔS, is:

$$\Delta S = \frac{q}{T} \tag{3.2}$$

The change in entropy is equal to the energy transferred as heat divided by the temperature. This definition makes use of heat rather than another energy term, such as work, as heat can be thought of as being associated with the random motion of molecules, while work represents an ordered change of a system. The presence of temperature in the denominator accounts for the effect of temperature on the randomness of motion, as objects which are hot have a larger amount of motion due to thermal energy than cool objects. This definition makes use of the concept of reversible processes, which refers to the ability of infinitesimally small changes in a parameter to result in a change in a process. Thermal reversibility refers to the system having a constant temperature throughout the entire system.

To understand this expression, consider the example of an ideal gas inside a piston that is undergoing an isothermal and reversible expansion (Figure 3.2). In this case, the forces per area on both sides of the piston head are kept closely matched. As was found in the previous chapter, the expansion results in work being performed with a value determined by the volume change:

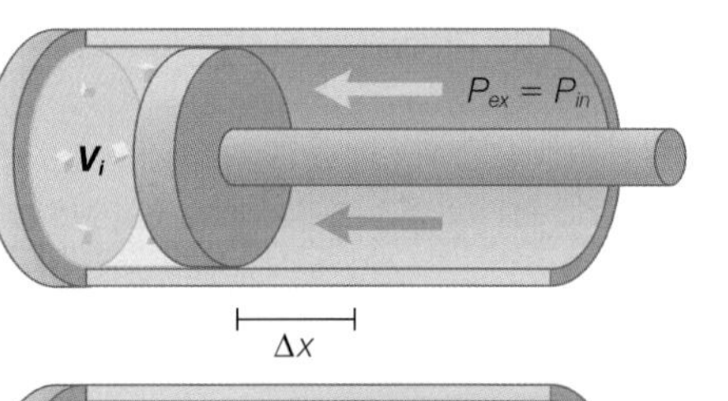

Figure 3.2 The reversible expansion of an ideal gas with the external pressure, P_{ex}, matching the internal pressure, P_{in}.

$$w = -nRT \ln \frac{V_f}{V_i} \tag{3.3}$$

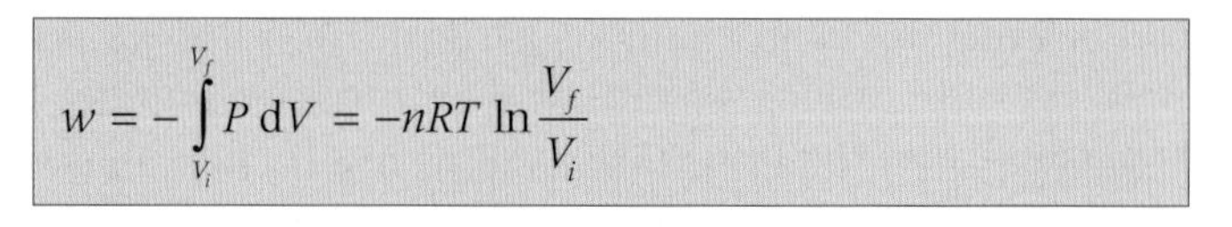

$$w = -\int_{V_i}^{V_f} P\,\mathrm{d}V = -nRT \ln \frac{V_f}{V_i}$$

For an ideal gas, when temperature is fixed, internal energy does not change and the heat flow balances the work, yielding:

$$q = -w = nRT \ln \frac{V_f}{V_i} = T\left(nR \ln \frac{V_f}{V_i}\right) \tag{3.4}$$

For $\Delta T = 0$, $\Delta U = q + w = 0$ and $q = -w$

The entropy change is proportional to the heat (eqn 3.2) and can be written in terms of the volume change:

$$\Delta S = \frac{q}{T} = \left(nR \ln \frac{V_f}{V_i}\right) \tag{3.5}$$

Entropy is the part of the expression for heat flow that represents the change in the volume of the molecules in their final state compared to their initial state. Entropy represents the tendency of molecules to occupy all of the available space. More generally, entropy represents the tendency of a system to explore all of the available states.

ENTROPY CHANGES FOR REVERSIBLE AND IRREVERSIBLE PROCESSES

As the entropy of a system changes, the properties of the surrounding must be addressed. The surroundings are generally considered to be so large that they are isothermal and at constant pressure. Because the surroundings are at constant pressure, the *heat transferred into the surroundings*, q_{sur}, is equal to the *change in the enthalpy of the surroundings*, ΔH_{sur}:

$$dq_{sur} = \Delta H_{sur} \tag{3.6}$$

Since the surroundings are assumed not to change state when the system changes, the transfer of heat to and from the surroundings is effectively reversible, and can be related to the change in entropy using eqn 3.2 regardless of how the heat got to the surroundings:

$$dq_{sur} = TdS_{sur} \tag{3.7}$$

For a reversible change in the system, the heat coming from the system has the same value, but the opposite sign, as the heat going into the surroundings. Then for an isothermal reversible change the *total change in entropy*, dS_{tot}, can be written in terms of the *entropy changes of the system*, dS_{sys}, and the *entropy change of the surroundings*, dS_{sur}, yielding:

$$dS_{tot} = dS_{sys} + dS_{sur} = \left(\frac{q}{T}\right)_{sys} + \left(\frac{q}{T}\right)_{sur} = \frac{q}{T} - \frac{q}{T} = 0 \tag{3.8}$$

Thus, for a reversible change in a system, the total entropy change of the system and surroundings is zero. So any changes in the entropy of

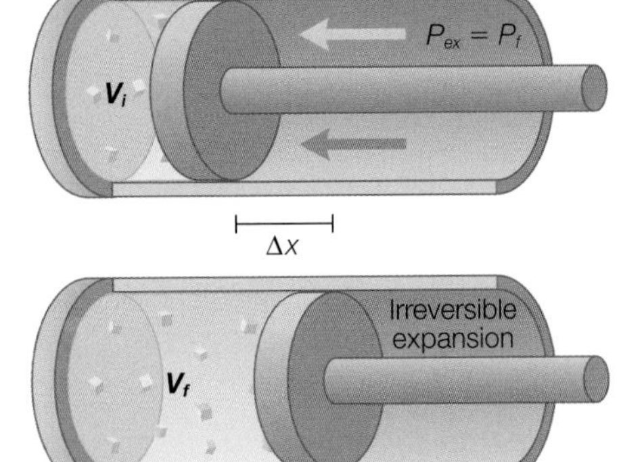

Figure 3.3 The irreversible expansion for an ideal gas with the external pressure, P_{ex}, fixed to the pressure, P_f.

the system are exactly matched by corresponding opposite changes in the entropy of the surroundings.

For a process that occurs at constant pressure, the heat flow is equal to the change in enthalpy (eqn 3.6), and the entropy change of the surroundings can be written in terms of enthalpy:

$$\Delta S_{sur} = -\frac{\Delta H}{T} \tag{3.9}$$

Thus, if the process is *exothermic*, ΔH is *negative* and ΔS_{sur} is *positive*, and the process is *spontaneous*. The entropy of the surroundings increases when heat is released. This conclusion is critical for understanding chemical equilibrium and will be used in the formulation of the Gibbs energy, below.

If the change in the system is irreversible, the entropy change is not zero. To determine the entropy change for an irreversible process, consider as an example the isothermal expansion of an ideal gas against a *fixed pressure*, P_f (Figure 3.3). The value of fixed pressure is poised to be equal to the final pressure of the gas after the expansion is completed. In this case, the work performed is given by the product of the *change in volume*, ΔV, and the pressure, which is expressed using the ideal gas law:

$$w = -P_f\Delta V = -\frac{nRT}{V_f}\Delta V = -nRT\frac{\Delta V}{V_f} \tag{3.10}$$

> Ideal gas law:
>
> $$P = \frac{nRT}{V}$$

The work performed in this irreversible process is different from what was determined for a reversible process (eqn 3.3):

$$w = -nRT\ln\frac{V_f}{V_i} \tag{3.11}$$

For example, if the final volume is twice the initial volume, $V_f = 2V_i$, then the work is calculated to be $-0.5\ nRT$ and $-0.693\ nRT$ for the irreversible and reversible, respectively. Since these are isothermal ideal gases, $q = -w$ and the entropy change can be related to the work:

$$\text{Reversible process: } \Delta S = \frac{q}{T} = -\frac{w}{T} = 0.693\,nRT \quad (3.12)$$

$$\text{Irreversible process: } \Delta S = \frac{q}{T} = -\frac{w}{T} = 0.5\,nRT$$

Thus, the entropy change for the irreversible process is found to be larger than that of the reversible process.

THE SECOND LAW OF THERMODYNAMICS

In general terms, the second law of thermodynamics deals with the concept that energy and matter tend to be dispersed. With the introduction of the concept of entropy, which provides a measure of the order of a system, the second law can be expressed in the following way.

> The second law of thermodynamics states that the entropy of an isolated system tends to increase with time.

This general expression for the second law of thermodynamics can now be stated quantitatively using the expressions for entropy in terms of reversible and irreversible processes. For a reversible process, the entropy change is equal to the heat divided by the temperature:

$$\Delta S = \frac{q}{T} \quad (3.13)$$

The entropic changes for a spontaneous, or irreversible, process can be stated by the *Clausius inequality* to be always greater than the heat divided by the temperature:

$$\Delta S > \frac{q}{T} \quad (3.14)$$

Thus, the second law of thermodynamics can be expressed more precisely as follows.

> The second law states that the entropy of an isolated system increases in the course of a spontaneous change.

Thermodynamically, irreversible processes, such as the free expansion of a gas, will always be accompanied by an increase in the entropy. Thus, the total entropy change can be used to determine whether a process is spontaneous or not. A positive value for the entropy change ΔS means

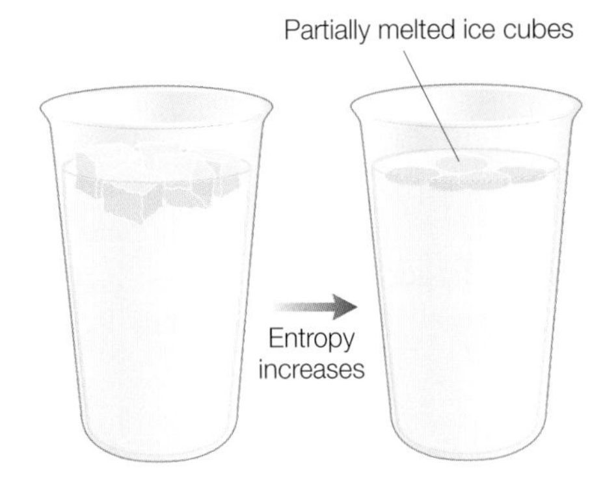

Figure 3.4 The entropy of a cup of ice cubes increases as the ice cubes melt.

that the process will occur, while a negative ΔS means that the process will not happen. If ΔS is zero, then the system is in equilibrium and nothing will change. As an example of entropy changes, consider the spontaneous process of an ice cube melting in a cup of water (Figure 3.4). The change in entropy is given by the transfer of heat, q, from the cup to the ice cube, with the ice cubes and water being at the specific temperatures of T_{cube} and T_{water}:

$$\Delta S = \frac{q}{T_{cube}} - \frac{q}{T_{water}} \tag{3.15}$$

where q is positive for the ice cube since it receives heat and q is negative for the water since the heat leaves the water. The ice cube is at a lower temperature than the water. Thus, the overall value of the entropy change is positive and the ice cube can melt based upon the second law. From a molecular view, the water in the cup is in equilibrium between the frozen and liquid states. In order to melt the ice, the attractive forces that hold the ice molecules together must be overcome. The energy to remove the molecules from the ice comes from the surroundings of the ice; this is called the *enthalpy of fusion*. In general, the enthalpy of fusion provides a measure for the energy required to make a transition from a solid to a liquid state.

INTERPRETATION OF ENTROPY

Entropy, which literally means "a change within," was first coined by Rudolf Clausius in 1851, one of the pioneers in the development of thermodynamics. Whereas entropy can be defined formally in terms of various parameters, its nature can be shown qualitatively with some simple examples. First, consider a cup of boiling water (Figure 3.5). When the water is at 100°C the escaping vapor can perform work. As the cup cools, no work is performed but heat flows from the cup to the surroundings, raising the temperature of the surroundings. The cooling continues until equilibrium is reached and the temperatures of the cup and surroundings are equal. The energy that was once present in the hot cup, and potentially capable of performing work, is no longer available. This irreversible process corresponds to an increase in the entropy through the randomness of the cooling process that will never recover into a cup of boiling water. As a second example, consider the letters on this page. As part of a written text all of the letters are organized in specific patterns. Consider taking the 123 characters from this sentence and randomly distributing them throughout the page in a chaotic pattern that has no meaning. The change into the random distribution corresponds to an increase in entropy. The

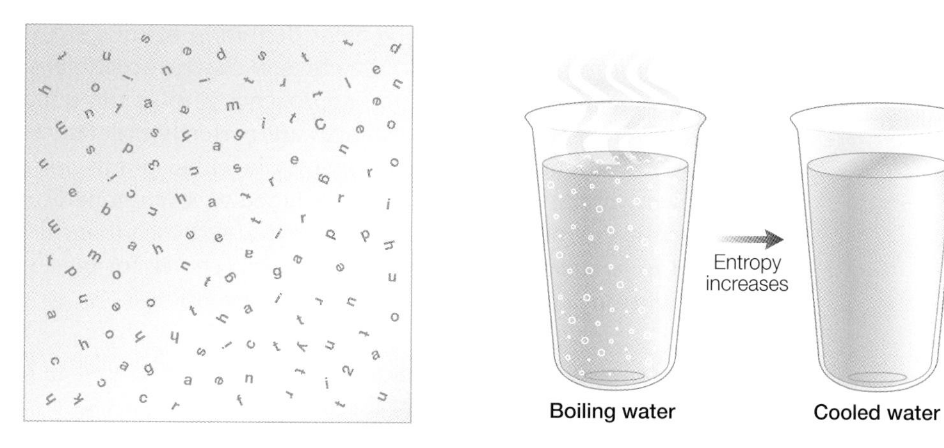

Figure 3.5 Examples of entropy changes: a cup of boiling water that cools and letters from a sentence in the text positioned randomly.

equivalent molecular picture is to replace the letters with gas molecules. Initially, the gas molecules are confined in a balloon and then released by puncturing the balloon. The distribution of the gas molecules is over a larger area and more random than the original pattern, corresponding to an increase in the entropy of the gas molecules. It is interesting to note that living organisms are generally highly organized, with well-defined cellular organizations corresponding to entropy-poor conditions. However, the tendency for entropy to always increase can be seen in considering how living organisms perform basic cellular reactions. For example, the oxidation of glucose with six oxygen molecules leads to the production of six molecules of carbon dioxide and six molecules of water. The increase in the number of molecules, from seven to 12 in this case, allows more molecular movement and disorder and hence a larger entropy.

THIRD LAW OF THERMODYNAMICS

In addition to the two laws that have already been presented, there is a third law of thermodynamics that deals with the question of what happens to objects as the temperature approaches absolute zero. In general, as temperature is decreased, random motion due to thermal motion is quenched. For a crystal, all of the atoms or molecules are located in well-defined, regular arrays and hence spatial disorder is absent. The lack of any thermal motion or spatial disorder suggests that the entropy is zero. From a molecular viewpoint, the entropy can also be viewed as being zero as the arrangement of molecules is uniquely defined. Thus, for any object, the entropy approaches zero as the temperature approaches absolute zero provided that the objects are ordered.

Effectively, this provides a convenient definition of the zero value. In the original formulation by Nerst, the entropy change accompanying any physical or chemical transformation approaches zero as the temperature approaches zero provided the substances are perfectly ordered. With this definition, the absolute value of entropy may be either positive or negative at any given value, with a zero value assigned to any temperature. As a matter of convenience, this law is usually expressed such that the entropy is defined to be always non-negative. With this choice for poising a zero value for entropy, the third law is expressed as follows.

> The third law of thermodynamics states that the entropy of all perfectly crystalline substances is zero at a temperature of zero.

The third law is used for biological applications as it provides a basis for the definition of entropies of materials relative to their crystalline state. However, this law does not normally have a significant impact on biological systems as organisms live at room temperature and their properties at $T = 0$ are not relevant for understanding their cellular processes. The exception arises when cellular components are investigated at low temperatures for spectroscopic studies.

GIBBS ENERGY

A process is spontaneous if the overall entropy change for the system is positive. If the entropy is negative, then the process is not favored and thus *not* spontaneous. However, the calculation of the entropy changes can be difficult as changes in both the system and surroundings must be considered. In the 1800s, the American theoretician Josiah Willard Gibbs established a new state function that is now termed Gibbs energy in his honor. The state function not only provides a means of establishing both positive and negative entropy changes but also is straightforward to establish for a given system. The Gibbs energy can be used to not only determine whether a reaction will proceed but also how much energy is released.

The *overall entropy change*, ΔS_{tot}, is the sum of the entropy changes from the system, ΔS, and surroundings, ΔS_{sur}. The entropy change of the surroundings at constant pressure is just the entropy divided by the temperature (from eqn 3.9):

$$\Delta S_{sur} = -\frac{\Delta H}{T} \tag{3.16}$$

So *the total entropy change*, ΔS_{tot}, at constant temperature and pressure is given by:

$$\Delta S_{tot} = \Delta S - \frac{\Delta H}{T} \quad (3.17)$$

Multiplying both sides by $-T$ yields:

$$-T\Delta S_{tot} = -T\Delta S + \Delta H \quad (3.18)$$

Since the temperature is always a positive number, the reaction is spontaneous if the term $-T\Delta S_{tot}$ is negative. If the process is in equilibrium then this term is equal to zero. The product of temperature and entropy has units of energy and is related to the amount of energy available to do work. This term, $-T\Delta S_{tot}$, is usually called the *Gibbs energy difference*, ΔG, and is written as:

$$\Delta G = \Delta H - T\Delta S \quad (3.19)$$

In summary, the Gibbs energy represents the energy available for the reaction as it includes both enthalpy and entropy contributions. Since biochemical reactions operate at constant temperature and pressure, the Gibbs energy difference is the energy term that will be calculated to determine how a reaction will proceed:

- if ΔG is a positive then the reaction is unfavorable and the initial state is favored,
- if ΔG is zero the reaction is in equilibrium, and
- only if ΔG is negative will the reaction occur spontaneously.

RELATIONSHIP BETWEEN THE GIBBS ENERGY AND THE EQUILIBRIUM CONSTANT

For any given reaction $A \leftrightarrow B$ with an *equilibrium constant* K, the value of the equilibrium constant can be written in terms of the change in the Gibbs energy:

$$K = e^{-\Delta G/kT} \quad (3.20)$$

Thus, the equilibrium constant for a reaction is simply an alternative representation of the Gibbs energy change. This relationship can be divided into three regions (Table 3.1). First, spontaneous reactions occur when the Gibbs energy change is negative; in this case, the association constant is a positive number greater than one. Second, at equilibrium the Gibbs energy is equal to zero, corresponding to a value of one for the equilibrium constant. Third, reactions that are favored to proceed in the reverse direction rather than moving forward correspond to a positive value for the Gibbs energy change, or correspondingly, a value less than one for the equilibrium constant.

Table 3.1

Relationships among the equilibrium constant, K, Gibbs energy change, ΔG, and direction of a chemical reaction*.

K	ΔG	Direction
>1.0	Negative	Proceeds forward
1.0	Zero	At equilibrium
<1.0	Positive	Proceeds in reverse

*Normally the change in the Gibbs energy is the standard value, as discussed in Chapter 6.

As an example, consider how the Gibbs energy can be calculated from the equilibrium constant for the reaction catalyzed by the enzyme phosphoglucomutase:

$$\text{glucose 1} - \text{phosphate} \leftrightarrow \text{glucose 6} - \text{phosphate}$$

Biochemical analysis of the reaction at 25°C and pH equal to 7 shows that the final equilibrium mixture will contain 1 mM glucose 1-phosphate for every 19 mM glucose 6-phosphate. The equilibrium constant is given by the ratio of the relative concentrations:

$$K = \frac{[\text{Glucose 6} - \text{phosphate}]}{[\text{Glucose 1} - \text{phosphate}]} = \frac{19\text{ mM}}{1\text{ mM}} = 19 \qquad (3.21)$$

Using eqn 3.20, the change in the Gibbs energy can be calculated:

$$\begin{aligned}\Delta G &= -RT \ln K = -(8.315\text{ J/(mol K)})(298\text{ K})(\ln 19)\\ &= -7{,}296\text{ J mol}^{-1}\end{aligned} \qquad (3.22)$$

Thus, the reaction proceeds spontaneously, producing a large decrease in the Gibbs energy, as expected for a large positive value for the equilibrium constant.

RESEARCH DIRECTION: DRUG DESIGN II

In Chapter 2, calorimetry was described as a technique that could accurately measure the heat change in response to drug binding. The binding affinity can now be related to the Gibbs energy according to eqn 3.20 and consequently also related to the changes in enthalpy and entropy associated with the binding:

$$K = e^{-\Delta G/kT} = e^{-(\Delta H - T\Delta S)/kT} \qquad (3.23)$$

Consequently, many different combinations of enthalpy and entropy changes can result in the same change in Gibbs energy and elicit the same binding affinity. Most drug-design strategies optimize the binding affinity to minimize the amount of a drug needed while maintaining potency. Achievement of tight binding requires optimization of both the enthalpic and entropic components that arise from different interactions between the drug and protein (Freire 2004).

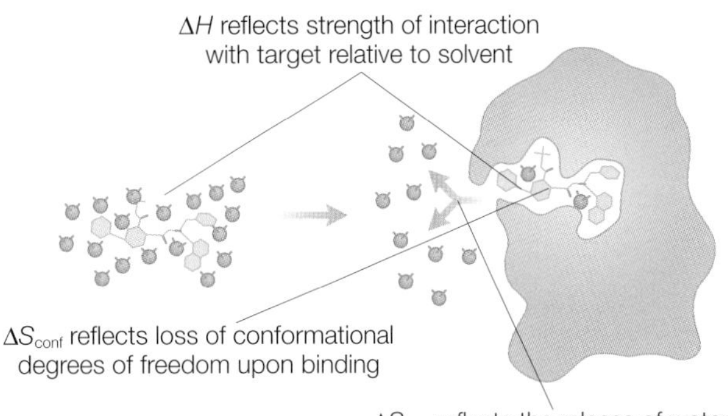

Figure 3.6 Binding-affinity contributions of enthalpy and entropy. Modified from Freire (2004).

Let us consider the different possible interactions that influence the binding of a drug to a targeted protein (Figure 3.6). Upon binding to the protein, the protein may change either its conformation or the protonation state of the amino acid residues located at the binding site. Such protein-related changes are an important aspect of binding and contribute to the resulting change in enthalpy. However, the preferred approach for drug design is to target favorable enthalpy changes that are specific to the drug, such as formation of hydrogen bonds and van der Waals' contacts. Desolvation of polar groups upon binding represents an unfavorable enthalpic contribution that should be compensated for by the favorable interactions. The entropy will change due to two contributions. Before binding, the ligand is solvated and the empty binding site will contain water. Upon binding, water is released resulting in a favorable entropic contribution. However, binding may also result in the loss of the conformations available and so decrease the number of degrees of freedom and hence an yield an unfavorable entropic change. This unfavorable contribution can be minimized by considerations of the drug's flexibility.

In optimization studies of drugs, the initial studies usually focus on optimization of the binding affinity. By measuring the temperature dependence of the binding affinity, the enthalpic and entropic contributions can be identified individually. Drugs that are selected based upon enthalpic optimization are usually more selective, with a higher binding affinity than that obtained using entropic optimization. The higher affinity is due to the less-specific nature of hydrophobic interactions that strongly influence entropy changes. Therefore, enthalpically driven drugs have in general a better potential and should be preferred for optimization.

Among the questions concerning drug design is how to design a molecule that recognizes one enzyme from among hundreds, all of which share a common substrate. Such a challenge confronted biochemists as they developed inhibitors of protein kinases, which are enzymes that regulate many cellular processes. Kinases are inviting drug targets as a number of diseases, including cancer, diabetes, and inflammation, are linked to cell signaling pathways mediated by protein kinases. The human genome encodes 518 protein kinases that share a conserved sequence but which

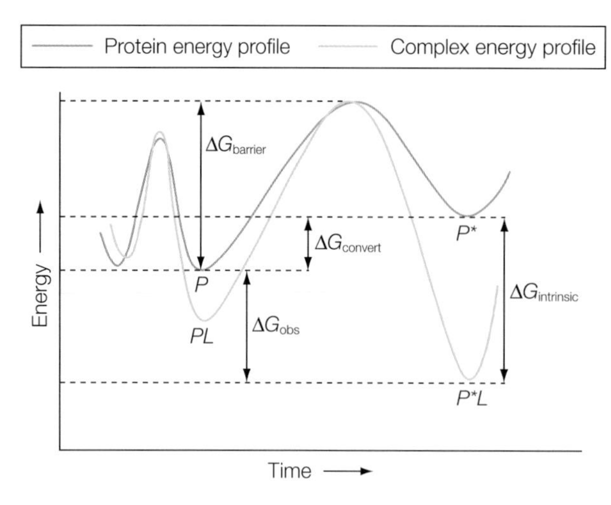

Figure 3.7 Dynamics result in a protein having several possible conformations (*P* and *P**) that each has a different overall energy. The binding of a drug can have different interactions with these conformations, resulting in significantly altered energetics for the drug–protein complex yielding an observed energy change of ΔG_{obs} for ligand binding.

differ in how they are regulated. Whereas inhibitors for kinases have been identified by random compound screening, few successful drugs have been developed using design principles. In recent design studies (Noble et al. 2004; Ahn & Resing 2005; Cohen et al. 2005), the ATP-binding site of protein kinases has been targeted. At the site is a gatekeeper residue that flanks a highly variable hydrophobic portion of the ATP-binding pocket. Differences in the size of this residue as well as the surrounding amino acid residues for different protein kinases have been used to achieve selectivity. Protein kinases can exist in inactive forms that are also attractive for drugs.

Protein dynamics play a role in the affinity and kinetics of drug binding (Freire 2002; Teague 2003). The drug can be viewed as binding to the lowest-energy state of the protein, which must undergo structural rearrangements in order to accommodate the drug. Due to protein dynamics, multiple conformations of a protein are seen by a drug prior to binding (Figure 3.7). In some cases, the drug binding can induce a substantial conformational change due to the hydrophobic interactions between the protein and drug. Although the overall energy of one conformation of the protein may be at a minimum for the unbound complex, it may not be the most favorable for drug binding, as the conformational changes associated with drug binding can lead to a more stable complex for the unbound conformation that has a larger energy.

GIBBS ENERGY FOR AN IDEAL GAS

For an ideal gas, the change in the Gibbs energy can be directly related to its thermodynamic parameters, such as the change in pressure. An *infinitesimal change in the Gibbs energy*, dG, can be related to the enthalpy and entropy according to $\mathrm{d}G = \mathrm{d}H - T\mathrm{d}S$ (eqn 3.19). Since the change in enthalpy can be related to the total energy and the product of pressure and volume (eqn 2.20), an *infinitesimal change in the enthalpy*, dH, is related to the change in volume, dV, and the change in pressure, dP:

$$H = U + PV \tag{3.24}$$

$$\mathrm{d}H = \mathrm{d}U + P\mathrm{d}V + V\mathrm{d}P$$

$$\mathrm{d}(f(x)g(x)) = g(x)\mathrm{d}(f(x)) + f(x)\mathrm{d}(g(x))$$

The first part of this sum, the *total energy change*, dU, can be related to the change in entropy and volume:

$$dU = TdS - PdV \tag{3.25}$$

Inserting this relationship into eqn 3.19 yields the *change in Gibbs energy*, dG:

$$\begin{aligned} dG &= TdS - PdV + PdV + VdP - TdS - SdT \\ dG &= VdP - SdT \end{aligned} \tag{3.26}$$

At constant temperature, the change in temperature, dT, is exactly equal to zero, which simplifies this expression for the change in the Gibbs energy to:

$$dG_{constant\ T} = VdP \tag{3.27}$$

In order to determine the change in the Gibbs energy for a large change, this expression is integrated from the initial pressure to the final pressure. This integration shows that the change in the Gibbs energy is directly given by the thermodynamic properties of the ideal gas, the number of moles, the temperature, and the change in pressure:

$$\Delta G = \int_{P_i}^{P_f} \frac{nRT}{P} = nRT \ln \frac{P_f}{P_i} \tag{3.28}$$

$$\int \frac{dx}{x} = \ln x$$

USING THE GIBBS ENERGY

The Gibbs energy can be used to determine whether a process will occur spontaneously. As an example for the use of entropy and Gibbs energy, consider the values of these parameters for water. The melting of ice results in an increase of enthalpy as heat is absorbed by the surroundings in order to melt the ice:

$$H_2O \text{ (solid)} \rightarrow H_2O \text{ (liquid)} \tag{3.29}$$
$$\Delta H° = 6 \text{ kJ mol}^{-1}$$

The reaction of water to hydrogen gas and oxygen has a positive entropy change as expected for the release of gases from a liquid:

$$2H_2O\ (\text{liquid}) \rightarrow 2H_2\ (\text{gas}) + O_2\ (\text{gas}) \tag{3.30}$$

$$\Delta S^\circ = S^\circ(O_2) + 2S^\circ(H_2) - 2S^\circ(H_2O)$$

$$\Delta S^\circ = (205.1\ \text{J K}^{-1}) + 2(130.7\ \text{J K}^{-1}) - 2(69.9\ \text{J K}^{-1}) = 326.7\ \text{J K}^{-1}$$

If we calculate the change in the Gibbs energy for the same reaction, the Gibbs energy change is positive and so the reaction does not occur spontaneously:

$$\Delta G^\circ = G^\circ(O_2) + 2G^\circ(H_2) - 2G^\circ(H_2O) \tag{3.31}$$

$$\Delta G^\circ = (0\ \text{kJ mol}^{-1}) + 2(0\ \text{kJ mol}^{-1}) - 2(-237.2\ \text{kJ mol}^{-1})$$
$$= 474.4\ \text{kJ mol}^{-1}$$

Whereas liquid water is stable and does not spontaneously dissociate into hydrogen and oxygen gases, we should consider the reverse reaction of hydrogen and oxygen gases forming water:

$$2H_2\ (\text{gas}) + O_2\ (\text{gas}) \rightarrow 2H_2O\ (\text{gas}) \tag{3.32}$$

The enthalpy and entropy changes for this reaction can be calculated by considering each component. For this reaction, the entropy should decrease due to the ordering of the gases into a liquid:

$$\Delta S^\circ = 2S^\circ(H_2O) - S^\circ(O_2) - 2S^\circ(H_2) \tag{3.33}$$

$$\Delta S^\circ = 2(188.7\ \text{J K}^{-1}) - (205.0\ \text{J K}^{-1}) - 2(130.6\ \text{J K}^{-1}) = -88.7\ \text{J K}^{-1}$$

$$\Delta G^\circ = 2G^\circ(H_2O) - G^\circ(O_2) - 2G^\circ(H_2)$$

$$\Delta G^\circ = 2(-237.2\ \text{kJ mol}^{-1}) - (0\ \text{kJ mol}^{-1}) - 2(0\ \text{kJ mol}^{-1})$$
$$= -474.4\ \text{kJ mol}^{-1}$$

This reaction has a large negative value and so is spontaneous. When hydrogen and oxygen gases are mixed an explosion will occur, as happened in the disaster of the hydrogen-filled airship, the Hindenburg, in 1937. A spark is required for the reaction to occur quickly. A spark is needed because the mechanism has many steps involving different intermediates that take place at a fast rate only after energy absorption.

CARNOT CYCLE AND HYBRID CARS

Hybrid cars have significantly increased fuel efficiency. Some predictions suggest that most cars will be hybrids in the not so distant future (see Chapter 12). Why do hybrids get about 20% better fuel efficiency than regular cars? To understand this improvement the thermodynamic principles that govern the efficiency of any engine need to be discussed. Car engines operate in a cyclic process of fuel intake, compression, ignition, expansion, and exhaust. This cycle occurs several thousand times each minute and is used to perform work in the form of moving the car. This traditional type of engine is termed a heat engine since the work is a result of heat released in the combustion process. The ideal gas is trapped inside a piston that expands due to changes in temperature. The expansion drives the piston and the motion of the piston is used to perform work. The efficiency of any heat engine (Figure 3.8) is defined as the ratio of work performed to the total amount of energy available in the form of heat:

$$\text{Efficiency} = \frac{w}{q} \tag{3.34}$$

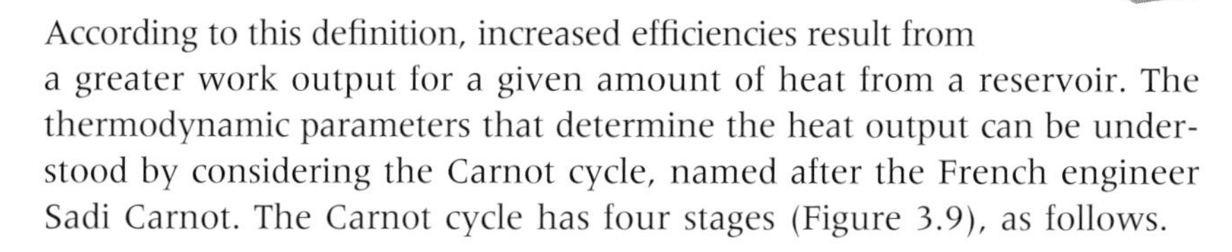

Figure 3.8 A model of heat flow in an engine operating in a reversible Carnot cycle.

According to this definition, increased efficiencies result from a greater work output for a given amount of heat from a reservoir. The thermodynamic parameters that determine the heat output can be understood by considering the Carnot cycle, named after the French engineer Sadi Carnot. The Carnot cycle has four stages (Figure 3.9), as follows.

1. Reversible isothermal expansion from a to b at temperature T_{hot}. The heat supplied to the system is q_1 and the entropy change is q_1/T_{hot}.
2. Reversible adiabatic expansion from b to c. No heat leaves the system so the entropy change is zero. The temperature decreases from T_{hot} to T_{cold}.
3. Reversible isothermal compression from c to d. The heat released to the cold resevoir is q_3, and the entropy change is q_3/T_{cold}.
4. Reversible isothermal compression from d to a. No heat leaves the system and so the entropy change is zero. The temperature changes from T_{cold} to T_{hot}.

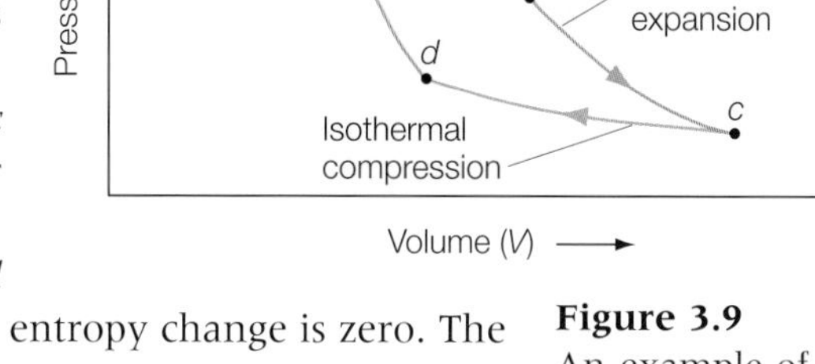

Figure 3.9 An example of a reversible Carnot cycle.

The work per cycle is the sum of the two heat terms, yielding:

$$\text{Efficiency} = \frac{w}{q_{hot}} = \frac{q_{hot} + q_{cold}}{q_{hot}} = 1 + \frac{q_{cold}}{q_{hot}} \text{ where } q_{cold} < 0 \tag{3.35}$$

For an ideal gas, it can be shown that the ratio of the heat terms is the same as the ratio of the temperatures (see Justification below) yielding:

$$\text{Efficiency} = 1 - \frac{T_{cold}}{T_{hot}} \tag{3.36}$$

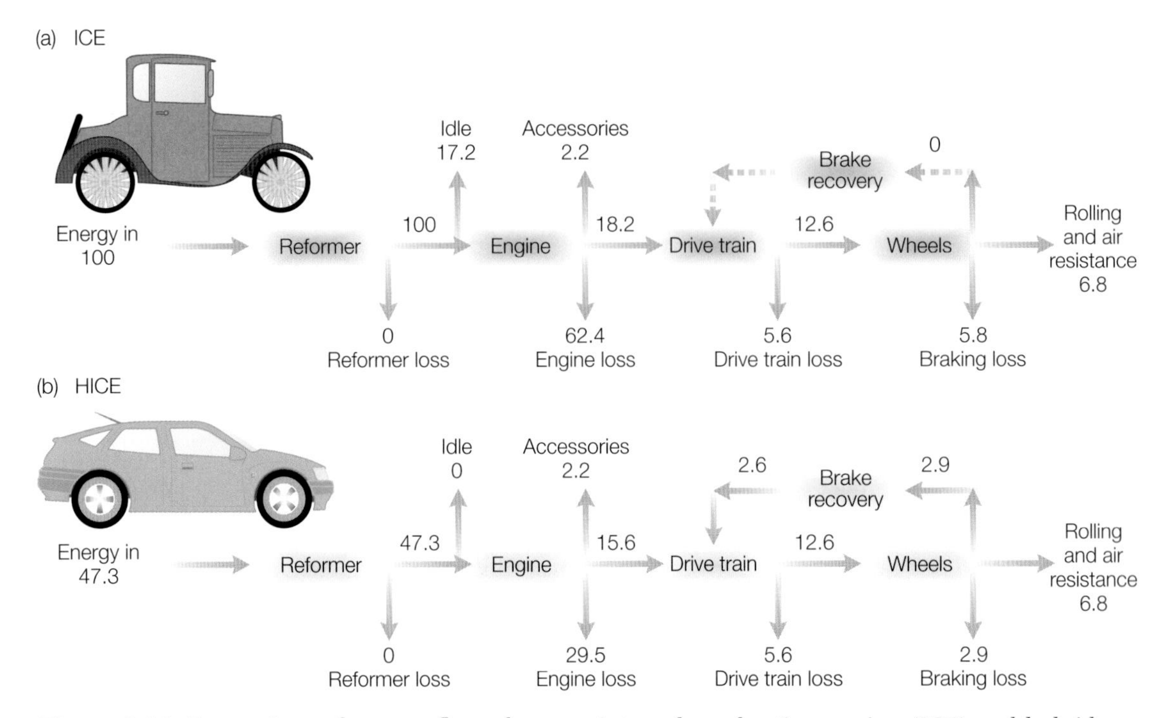

Figure 3.10 Comparison of energy flows from an internal combustion engine (ICE) and hybrid (HICE) car. The ICE makes use of a conventional internal combustion, spark-plug-ignition engine. The hybrid car has an electric motor and parallel drive train to eliminate idling loss and capture some of the energy normally lost during braking. Hybrids are much more efficient, with 12.6/47.3 = 0.266, than conventional cars, with 12.6/100 = 0.126. Modified from Demirdoven and Deutch (2004).

The goal of scientists and politicians is to significantly decrease our dependence on petroleum for transportation (see Chapter 12). This equation tells us that engines have an efficiency that is determined by the difference in temperature (neglecting real-life factors such as friction), which cannot be easily changed. Hybrid cars contain a new technological development that improves efficiency compared to normal internal combustion engines – an electric motor that is immediately adjacent to the gasoline engine (Figure 3.10). The electric motor drives the car primarily during the startup period until the traveling speed is reached because the efficiency of heat engines is best when the engine operates at a constant rate of revolutions. Since the car is started with the electric motor, the gasoline motor is turned off when the car is stationary, rather than burning gasoline while idling, and is only started after the car is moving again. The energy needed to drive the electric motor comes from a battery that is separate from the battery used to drive the starter motor and electrical components. As the electric motor drives the car, the energy available from this battery decreases. Hybrid cars do not need an additional energy source to recharge the batteries because they make use of regenerative braking.

Electric motors draw a current when the motor is turning the transmission (which turns the wheels). With regenerative braking, the electric motor turns backwards when the car is slowing down, generating a current that regenerates the battery. Regenerative braking allows hybrid cars to convert what would normally be simple heat loss into electrical energy that is stored until it is needed to power the car.

Derivation box 3.1

Entropy as a state function

Entropy is defined in terms of state functions, and so entropy should also be a state function. All state functions have the property that the functions are independent of path and are dependent only on the initial and final states. To prove the assertion that entropy is a state function, it is necessary to establish the path independence. One way to establish this property is to demonstrate that the integral of the function around an arbitrary path is zero, as this establishes that the entropy is the same at the initial and final states regardless of path. For the Carnot cycle, the integral of the entropy change is given by the contributions of the first and third steps:

$$\oint dS = \frac{q_{hot}}{T_{hot}} + \frac{q_{cold}}{T_{cold}} \qquad \text{(db3.1)}$$

If entropy is a state function, then this integral is equal to zero. In that case it is necessary to establish that the ratio of heat contributions is equal to the ratio of the temperatures:

$$\text{If} \quad \frac{q_{hot}}{T_{hot}} + \frac{q_{cold}}{T_{cold}} = 0 \quad \text{then} \quad \frac{q_{hot}}{q_{cold}} = \frac{T_{hot}}{T_{cold}} \qquad \text{(db3.2)}$$

To establish this relationship, consider an ideal gas. In this case, the heat has been related to the ratio of the final and initial volumes:

$$q = -w = nRT \ln \frac{V_{final}}{V_{initial}} \qquad \text{(db3.3)}$$

So, the heat terms in the Carnot cycle can be written in terms of the volume changes:

$$q_{hot} = nRT_{hot} \ln \frac{V_b}{V_a}$$

$$q_{cold} = nRT_{cold} \ln \frac{V_d}{V_c} \qquad \text{(db3.4)}$$

The ratio of volumes can be related to a ratio of temperatures for an ideal gas, since the change in internal energy, dU, can be related to the work since the heat flow is zero (dq = 0) for an adiabatic process. In addition, the change in internal energy is proportional to the specific heat and temperature change (eqn 2.23). Combining these two relationships yields:

$$dU = dw + dq = dw = -PdV \quad \text{(db3.5)}$$

$$dU = CdT \quad \text{(db3.6)}$$

$$CdT = -PdV \quad \text{(db3.7)}$$

Using the ideal gas law, the pressure can be substituted, resulting in:

$$CdT = -PdV = -\frac{nRT}{V}dV$$

or (db3.8)

$$C\frac{dT}{T} = -nR\frac{dV}{V}$$

To determine the change for the entire process, these terms are integrated to yield:

$$\int_{T_{initial}}^{T_{final}} C\frac{dT}{T} = \int_{V_{initial}}^{V_{final}} -nR\frac{dV}{V} \quad \text{(db3.9)}$$

$$\int(\text{constant})\frac{dx}{x} = (\text{constant})\int\frac{dx}{x} = (\text{constant})\ln(x)$$

$$C\ln\left(\frac{T_{final}}{T_{initial}}\right) = -nR\ln\left(\frac{V_{final}}{V_{initial}}\right)$$

To simplify this relationship, the variable c is defined as being equal to C/nR and the expression can be revised as:

$$\ln\left(\frac{T_{final}}{T_{initial}}\right)^{c} = \ln\left(\frac{V_{final}}{V_{initial}}\right) \quad \text{(db3.10)}$$

$$a\ln x = \ln(x^{a})$$

$$-\ln(x/y) = \ln(y/x)$$

Since each term has a logarithm, this equation reduces to:

$$\left(\frac{T_{final}}{T_{initial}}\right)^c = \left(\frac{V_{final}}{V_{initial}}\right)$$

or (db3.11)

$$T^c_{final}V_{initial} = T^c_{initial}V_{final}$$

If $\ln x = \ln y$ then $\ln(x/y) = 1$ and $x = y$

For the Carnot cycle, the products are:

$$V_A T^c_{hot} = V_D T^c_{cold} \quad \text{and} \quad V_C T^c_{cold} = V_B T^c_{hot} \qquad \text{(db3.12)}$$

Dividing these two expressions gives:

$$V_A T^c_{hot} = V_D T^c_{cold} \quad \text{and} \quad V_C T^c_{cold} = V_B T^c_{hot}$$

$$\frac{V_A T^c_{hot}}{V_B T^c_{hot}} = \frac{V_D T^c_{cold}}{V_C T^c_{cold}} \quad \text{or} \quad \frac{V_A}{V_B} = \frac{V_D}{V_C} \qquad \text{(db3.13)}$$

Consequently, we can substitute these volume ratios into the expression for the heat flow (eqn db3.4):

$$q_{hot} = nRT_{hot} \ln \frac{V_b}{V_a}$$

$$q_{cold} = nRT_{cold} \ln \frac{V_d}{V_c} = nRT_{cold} \ln \frac{V_a}{V_b} = -nRT_{cold} \ln \frac{V_b}{V_a}$$

and (db3.14)

$$\frac{q_{hot}}{q_{cold}} = \frac{nRT_{hot} \ln \frac{V_b}{V_a}}{-nRT_{cold} \ln \frac{V_b}{V_a}} = -\frac{T_{hot}}{T_{cold}}$$

or

$$\frac{q_{hot}}{T_{hot}} - \frac{q_{cold}}{T_{cold}} = 0$$

Coming back to the integral (eqn db3.1), the change in entropy over the entire path must be zero:

$$\oint dS = \frac{q_{hot}}{T_{hot}} + \frac{q_{cold}}{T_{cold}} = 0 \qquad \text{(db3.15)}$$

Thus the integral of the entropy over the arbitrary path is equal to zero and thus independent of the path. This establishes that entropy is indeed a state function and that the changes in the entropy for any given system can be determined by consideration of the initial and final states of the system. The second law of thermodynamics can then be established for a system based upon the changes between the final and initial states without consideration of the path that led to those changes.

RESEARCH DIRECTION: NITROGEN FIXATION

The nitrogen cycle is a complex biogeochemical cycle that involves many different organisms as well as the soil and atmosphere (Figure 3.11). Part of the nitrogen cycle involves nitrogen fixation, the conversion of molecular nitrogen into nitrites and other compounds suitable for assimilation by algae and plants (Howard & Rees 1996; Ferguson 1998; Barney et al. 2006).

The formation of ammonia, NH_3, from nitrogen and hydrogen gas is exothermic:

$$\begin{aligned} &N_2 \text{ (gas)} + 3H_2 \text{ (gas)} \rightarrow 2NH_3 \text{ (gas)} \\ &\Delta H^\circ = -92.6 \text{ kJ} \end{aligned} \qquad (3.37)$$

Although the reaction is favorable, the rate is extremely slow under standard conditions. Ammonia is one of the main materials used in the production of fertilizers as well as explosives. Therefore, optimization of this reaction for yield on an industrial scale was pursued. Fritz Haber in 1905 devised a process for producing ammonia that was developed for commercial use by Carl Bosch; who won the Nobel Prizes in Chemistry in 1918 and 1931, respectively. It was realized that the reaction was accelerated when the gases were in the presence of a metal surface (Figure 3.12). The gases bind to the surface and the interactions with the surface facilitate the weakening of covalent bonds holding the molecules together. These molecules are then very reactive and the formation of ammonia is favored. The binding of the gas to the surface and the dissociation process requires both a high temperature (≈500°C) and high pressure (several hundred atmospheres) for a high yield.

The synthesis of ammonia is thermodynamically favorable but complicated by a slow rate that reflects the kinetic stability of the nitrogen

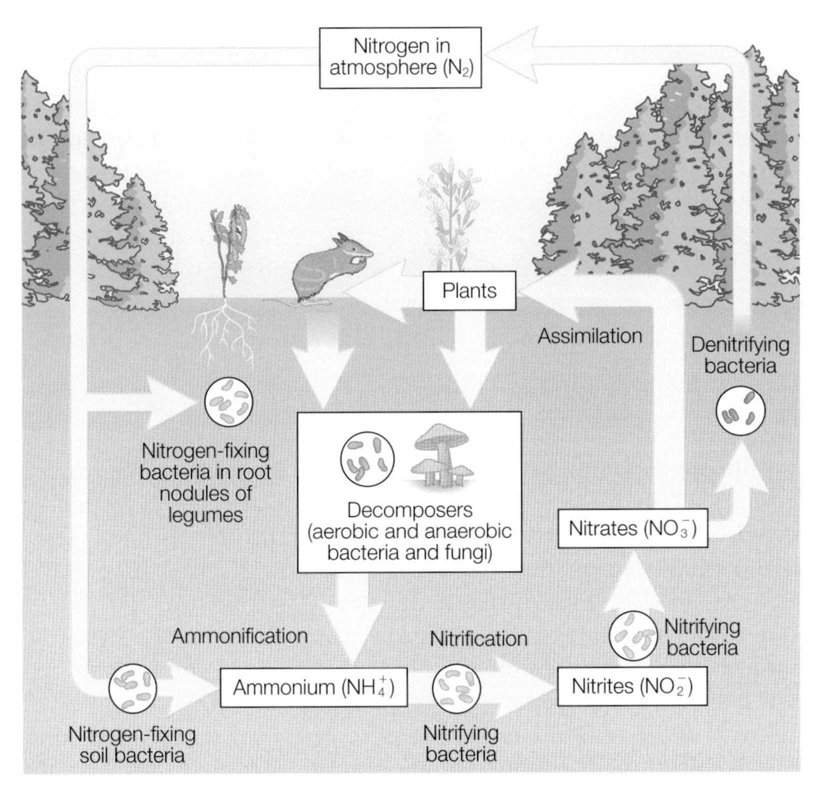

Figure 3.11 Representation of the nitrogen cycle.

molecule. Nitrogen molecules contain a triple bond and, in the gas phase, N_2 does not easily accept or donate electrons. For nitrogen, the triple bond is the most stable, with a bond energy of 225 kcal mol^{-1} compared with 100 and 40 kcal mol^{-1} for the double and single bonds, respectively. The significant greater strength of the triple bond in N_2 is more pronounced in nitrogen compared with other gases; for example the triple-bond energy is less than 3-fold greater than the single bond for oxygen. This stability of the triple bond for N_2 can be seen in an examination of the enthalpies of formation for the different states formed during the reaction:

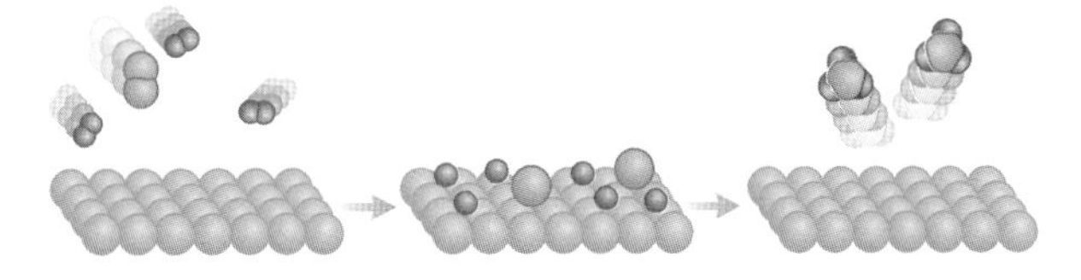

Figure 3.12 The use of a metal surface facilitates the disruption of the bonds in nitrogen molecules.

$$\begin{aligned} &N_2 + H_2 \rightarrow N_2H_2 \quad \Delta H° = +50.9 \text{ kcal mol}^{-1} \\ &N_2H_2 + H_2 \rightarrow N_2H_4 \quad \Delta H° = -27.2 \text{ kcal mol}^{-1} \\ &N_2H_4 + H_2 \rightarrow 2NH_3 \quad \Delta H° = +50.9 \text{ kcal mol}^{-1} \end{aligned} \tag{3.38}$$

The difficulties overcoming the triple bond and creating abundant amounts of ammonia at atmospheric conditions and ambient temperature have been solved by nature. Since nitrogen is a constituent of nearly all biomolecules, nitrogen is essential for life. Although N_2 is abundant in the atmosphere, molecular nitrogen is not reactive and most organisms are unable to directly metabolize this source. Fortunately, certain prokaryote organisms have acquired the ability to reduce dinitrogen to ammonia; these organisms play a critical role in the nitrogen cycle (Figure 3.11). The cellular mechanism that performs this reaction is the nitrogenase enzyme system. Nitrogenase consists of two components, the iron protein and the molybdenum iron protein, named according to the metal cofactors of each protein.

In nitrogenase, the substrate reduction is a multi-electron process:

$$N_2 + 8H^+ + 8e^- + 16MgATP \rightarrow 2NH_3 + H_2 + 16MgADP + 16P_i \quad (3.39)$$

This multi-step reaction involves three types of electron-transfer reaction (Figure 3.13). The iron protein is reduced by cellular electron carriers such as ferredoxin and flavodoxin. The reduced-iron protein forms a complex with the molybdenum protein and transfers an electron in a MgATP-dependent process. Electrons are transferred within the molybdenum protein to the substrate at the active site. Nitrogenase is a relatively slow enzyme, with a turnover time per electron of 5 s^{-1} in comparison with a rate of over 1000 s^{-1} for many other enzymes, The slow rate reflects the complexity of the reaction and the requirement of two different proteins.

As was found for the industrial application, the key to the enzymatic process is the presence of metals. In nitrogenase, the substrate binds to a large metal complex that consists of a molybdenum atom and several iron atoms. The precise mechanism by which the enzyme is able to create ammonia is still not fully understood and research is underway to understand how this protein can perform the same reaction that requires extremely high temperatures and pressures in the human-made protocol. Part of these efforts have centered on the determination and characterization of the metal cofactors using X-ray diffraction (see Chapter 15).

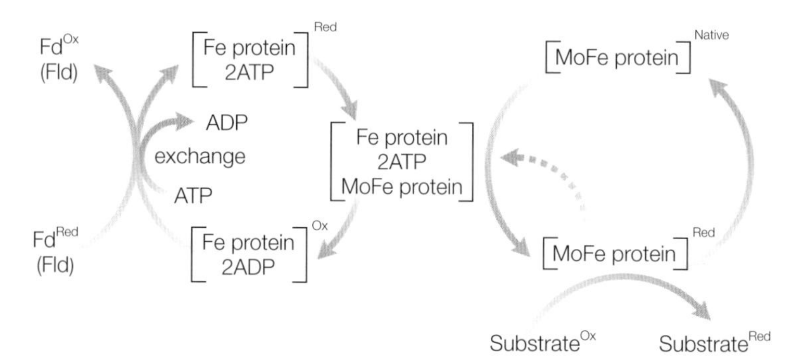

Figure 3.13 A scheme showing the involvement of different molecular complexes in the biological process of nitrogen fixation. Adapted from Rees and Howard (1999).

REFERENCES

Ahn, N.G. and Resing, K.A. (2005) Lessons in rational drug design for protein kinases. *Science* **308**, 1266–7.

Barney, B.M., Lee, H.I., Dos Santos, P.C. et al. (2006) Breaking the N_2 triple bond: insight into the nitrogenase mechanism. *Dalton Transactions* **2006**, 2277–84.

Cohen, M.S., Zhang, C., Shokat, M., and Tauton, J. (2005) Structural bioinformatics-based design of selective, irreversible kinase inhibitors. *Science* **308**, 1318–21.

Demirdoven, N. and Deutsch, J. (2004) Hybrid cars, fuel cell cars and later. *Science* **305**, 974–6.

Ferguson, S.J. (1998) Nitrogen cycle enzymology. *Current Opinion in Chemical Biology* **2**, 182–93.

Freire, E. (2002) Designing drugs against heterogeneous targets. *Nature Biotechnology* **20**, 15–16.

Freire, E. (2004) Isothermal titration calorimetry: controlling binding forces in lead optimization. *Drug Discovery Today: Technologies* **1**, 295–9.

Howard, J.B. and Rees, D.C. (1996) Structural basis of biological nitrogen fixation. *Chemical Reviews* **96**, 2965–82.

Noble, M.E.M., Endicott, J.A., and Johnson, L.N. (2004) Protein kinase inhibitors: insights into drug design from structure. *Science* **303**, 1800–5.

Rees, D.C. and Howard, J.B. (1999). Structural bioenergetics and energy transduction mechanisms. *Journal of Molecular Biology* **293**, 343–50.

Teague, S.J. (2003) Implications of protein flexibility for drug discovery. *Nature Reviews* **3**, 527–41.

PROBLEMS

3.1 If heat is transferred reversibly from the environment into a system with no other changes, what can be said about the entropy change of the system?

3.2 Calculate the change in entropy when 25 kJ of energy is transferred reversibly and isothermally as heat to a large block of iron at (a) 20°C and (b) 200°C.

3.3 For a spontaneous process at constant pressure, how will the Gibbs energy change?

3.4 If a process results in the entropy change for the surroundings being negative, and the pressure is held constant, what can be said about the heat change?

3.5 Calculate the change in molar entropy when a monatomic perfect gas is compressed to half its volume.

3.6 Calculate the change in molar entropy when a monatomic perfect gas is initially compressed to half its volume and then expanded back to the original volume.

3.7 Calculate the Carnot efficiency of a primitive steam engine operating on steam at 100°C and discharging at 20°C. How does the efficiency change if the discharge temperature is 60°C?

3.8 Determine the value of the entropy and enthalpy change for the process if the temperature dependence for the Gibbs energy at constant pressure is given by $\Delta G = -85\text{ J} + (30\text{ J K}^{-1})T$.

3.9 When heat is transferred from a hot block to a cold block, what can be said about the entropy change of each block?

3.10 If the equilibrium constant is greater than one, what can be said about the change in the Gibbs energy?

3.11 For a spontaneous reaction, what can be stated about the entropy change?

3.12 What can be said about the total entropy change for the reversible expansion of an ideal gas?

3.13 Consider a piece of paper burning in a closed, adiabatic, constant-volume container that has an oxygen atmosphere. Are the following changes positive, zero, or negative? (a) The change in internal energy; (b) the change in entropy of the system; (c) the change in entropy of the surroundings.

3.14 Calculate the standard enthalpy of formation for the oxidation of sucrose, using standard enthalpies of formation of −393.5, −285.8, and −2222 kJ mol^{-1} for carbon dioxide, water, and sucrose, respectively.

3.15 If a protein unfolds at 80°C with a standard enthalpy of transition of 500 kJ mol^{-1}, what is the associated entropic change?

3.16 Give some technical reasons for the improved gas mileage of a hybrid car.

3.17 How is a thermodynamic analysis of entropy and enthalpy changes used to indicate a favorable drug candidate?

3.18 Why is nitrogen fixation a slow process?

3.19 How do metal surfaces contribute to the Haber–Bosch process?

3.20 What is the role of nitrogenase in the nitrogen cycle?

4

Phase diagrams, mixtures, and chemical potential

Molecules can exist in three different states, as a solid, liquid, or gas. Changes in these states, or phases, are commonplace and so a description of the factors influencing the transitions plays a critical role in helping us to understand the properties of molecules. In this chapter, we examine the thermodynamic conditions that determine which phase is present and when a phase change occurs. For example, water can exist as ice, liquid, or a gas, and temperature is a key determinant of which phase is present, with water forming ice in the freezer and water changing into vapor when heated in the oven. Once the thermodynamic factors that determine the phase of pure, simple molecules have been discussed, we will turn our attention to the more complex questions such as what factors cause lipids to form cell membranes and proteins to crystallize.

SUBSTANCES MAY EXIST IN DIFFERENT PHASES

A substance is described as having a form of matter termed a *phase* when the substance is *uniform in chemical composition and physical state.* For example, a substance may exist in solid, liquid, or gases phases. Phases may co-exist, as a cup of ice water consists of two distinct phases, each of which is uniform. A substance may exist in many different solid phases depending upon the conditions. For example, as the pressure is increased on water below 0°C, different structural forms of ice are stable as the hydrogen bonds between water molecules are modified by the stress arising from the pressure. Most substances have only a single liquid state, although there are exceptions, and all substances exist in a single gaseous phase.

Substances may make a *phase transition,* which is a *spontaneous conversion from one phase into another phase.* Phase transitions occur at characteristic temperatures and pressures. At 1 atm, ice is a stable phase of water below 0°C but above that temperature the liquid phase of water is more stable. The change in phase indicates that the Gibbs energy of water decreases as ice changes into liquid water above 0°C. At the transition temperature, the two phases are in *equilibrium* as the Gibbs energy is at a minimum value.

If the temperature of a substance passes through the transition temperature, the substance may not be observed to change phase as the energetics may be favorable but the kinetics may be too slow to be significant in practice. Thermodynamically unfavorable phases that are present due to kinetic limitations are referred to as *metastable phases*. Whereas most organisms live only at 1 atm and above 4°C, the isolated components of cells, such as proteins and lipids, can be found in different phases that are revealing about their physical natures.

PHASE DIAGRAMS AND TRANSITIONS

In principle, transitions can occur between two phases. For example, if a solid block of carbon dioxide is placed in an open container at 1 atm, the solid will transform into gaseous carbon dioxide without passing through the liquid phase. The effect of factors such as pressure or temperature on the phase of a molecule is typically mapped out by the use of phase diagrams. For example, water will go from ice to liquid and then to vapor as the temperature increases at a given pressure (Figure 4.1). On the phase diagram this corresponds to moving horizontally from point A to B and then C. The melting and boiling temperatures are dependent upon the pressure. In general, the vapor phase is favored as the temperature increases and the pressure decreases. For example, water boils at a lower temperature at higher elevations due to lower pressure.

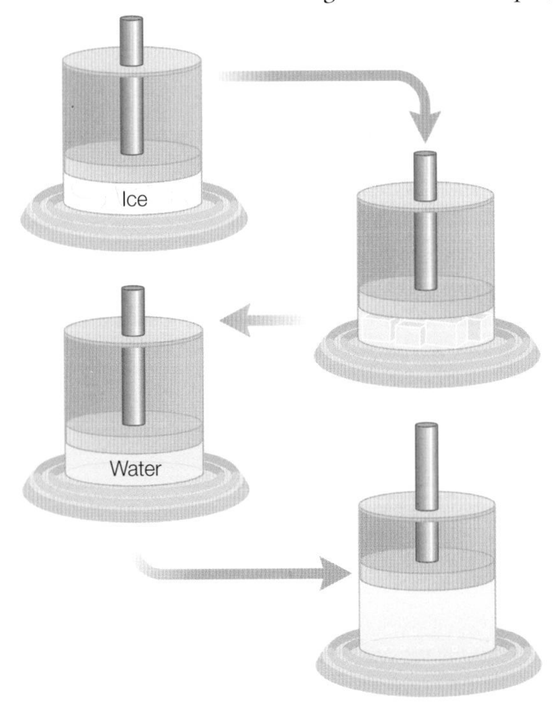

Figure 4.1 In response to heating, the temperature of ice increases, leading to a transition to the liquid phase, followed by a transition to the vapor phase.

On a phase diagram, the dependence of the state upon these parameters is shown by identifying what state is present for every temperature and pressure. A line between two states represents conditions at which both states can exist in equilibrium and is called a *phase boundary*. Notice that in addition to the vapor/liquid and liquid/solid phase boundaries there is also a phase boundary between the solid and vapor states at low temperatures and pressures. This latter boundary represents a direct transition from a solid to a gas without the presence of a liquid intermediate, as occurs when dry ice is warmed and evaporates into carbon dioxide gas.

The phase diagram has a special point called the *triple point* which is located at the *intersection* of the three phase boundaries (Figure 4.2). At the pressure and temperature of the triple point, all three phases exist simultaneously. At high temperatures and pressures there is a point termed the critical point at which the phase diagram stops. When the temperature is at or above the critical point, the liquid and vapor states are no longer distinctive, with both states having the same properties such as density.

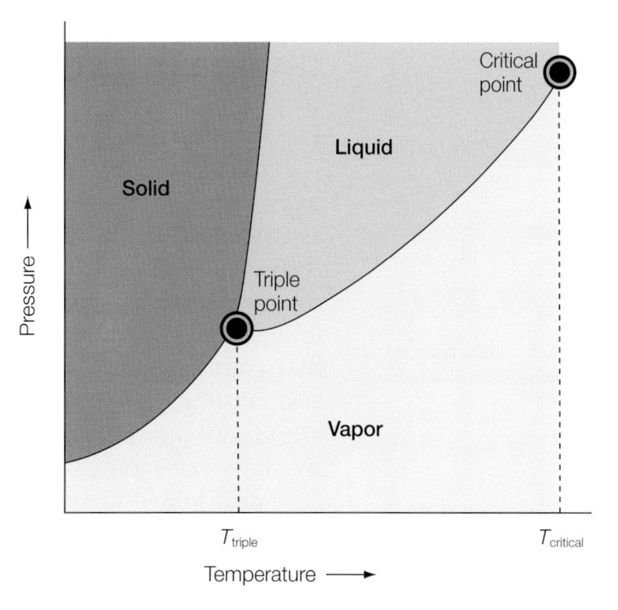

Figure 4.2 A phase diagram of pressure against temperature showing the presence of a triple point and critical point.

CHEMICAL POTENTIAL

The final state function that we need to consider is the chemical potential. The chemical potential provides a measure of Gibbs energy for every component of a mixture. The chemical potential is equal to the Gibbs energy per mole of substance, or equivalently the molar Gibbs energy for a pure substance. If we consider a pure substance with n moles, the *chemical potential*, μ, is defined to be equal to the Gibbs energy, G, divided by the number of moles, n:

$$\mu = \frac{G}{n} \text{ or equivalently } G = n\mu \tag{4.1}$$

When there are multiple substances these contributions are additive. For example, with two substances, A and B, the Gibbs energy for the mixture is just the sum of the contributions from each component:

$$G = n_A\mu_A + n_B\mu_B \tag{4.2}$$

The total Gibbs energy for a mixture is equal to the sum of the individual partial Gibbs energies for each component of the mixture. When using this expression, the partial Gibbs energies in the mixture are distinct from those of pure substances. For example, when oil and water are mixed, the hydrogen bonding of the water surrounding the oil is disrupted. Due to the interactions between the oil and water, the Gibbs energy associated with the oil is then different to that if the oil was present in a pure solution of oil.

PROPERTIES OF LIPIDS DESCRIBED USING THE CHEMICAL POTENTIAL

The lipids forming cell membranes represent an example of a biological system that can be divided into different phases. The membranes serve to establish distinct compartments in the cells. Membranes are complex assemblies of lipids, proteins, and other components organized in asymmetric bilayers with the inner and outer components having different compositions of lipids. The formation of the lipids into bilayers reflects fundamental thermodynamic properties of the lipids. Lipids are molecules that contain both a hydrophilic head group and a hydrophobic fatty acid tail (Figure 4.3). In general, the types of lipids found in membranes are glycerophospholipids, in which the hydrophobic portion contains two fatty acids joined to glycerol, and sphingolipids, which have a single fatty acid joined to sphingosine. In addition, membranes possess sterols that have a rigid set of hydrocarbon rings. In glycerophospholipids and some sphingolipids, a polar head group is joined to the hydrophobic portion by a phosphodiester linkage. The polar head group of phospholipids is usually a common alcohol such as serine, ethanolamine, choline, or glycerol. The fatty acid chains present in cells commonly are composed of 16–18 carbons and may be either saturated or unsaturated.

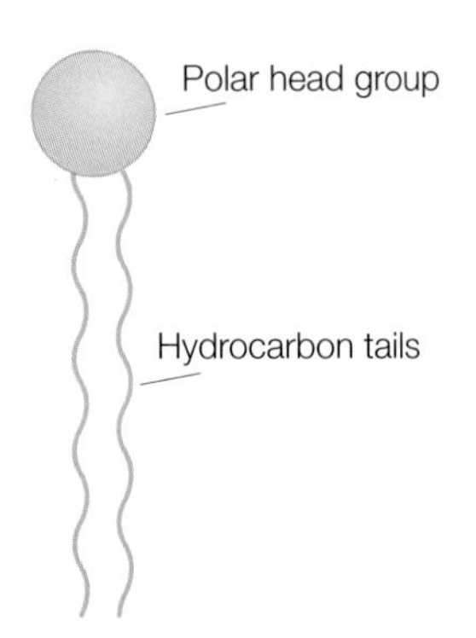

Figure 4.3 A schematic diagram of a phospholipid.

At low concentrations, the lipids will exist as monomers, but as the concentration increases the formation of more complex structures is favored as a result of hydrophobic interactions. This behavior can be modeled by considering the chemical potentials of solutions. The amount of each component in a solution can be described by the *mole fraction*, X_i, which is defined as the amount of the ith component, n_i, divided by the total number of molecules, n:

$$X_i = \frac{\text{Amount of molecule } i}{\text{Total number of molecules}} = \frac{n_i}{n} \quad (4.3)$$

For an ideal solution, the *chemical potential*, μ_i, of the ith component is related to the mole fraction X_i according to:

$$\mu_i = \mu_i^0 + RT \ln X_i \quad (4.4)$$

where the quantity μ_i^0 is the standard-state chemical potential which equals the molar energy of a pure compound:

$$\mu_i = \mu_i^0 \quad \text{for} \quad X_i = 1 \quad (4.5)$$

As an example, consider a two-component system, A and B. The chemical potentials for each component can be expressed in terms of their standard-state chemical potentials and mole fractions:

$$\mu_A = \mu_A^0 + RT \ln X_A \quad \text{and} \quad \mu_B = \mu_B^0 + RT \ln X_B \tag{4.6}$$

Since there are only two components, the sum of the mole fractions must be equal to one ($X_A + X_B = 1$). Hence the chemical potential of A can be expressed in terms of the mole fraction of B:

$$\text{Since } X_A + X_B = 1,\ \mu_A = \mu_A^0 + RT \ln X_A = \mu_A^0 + RT \ln(1 - X_B)$$

$$\approx \mu_A^0 + RT\left(X_B + \frac{X_B^2}{2}\right) \tag{4.7}$$

$\ln(1 - X) \approx X + X^2/2$

Expressing the mole fraction X_B in terms of the *solute concentration* c_B for a dilute solution gives:

$$X_B = \frac{\text{Moles B}}{\text{Moles A}} = \left(\frac{c_B \text{ g ml}^{-1}}{M_B \text{ g mol}^{-1}}\right) V_A \text{ ml mol}^{-1} = \frac{c_B V_A}{M_B} \tag{4.8}$$

and the standard free energy becomes:

$$\mu_A = \mu_A^0 - RTV_A\left(\frac{c_B}{M_B} + Bc_B^2\right) \tag{4.9}$$

where B is the *second viral coefficient*. For an ideal solution, the second viral coefficient can be written as:

$$B_{ideal} = \frac{V_A}{2M_B^2} \tag{4.10}$$

When $B > B_{ideal}$ the chemical potential decreases more rapidly than the ideal case and the system behaves as a good solvent. When $B \ll 0$, the chemical potential decreases less rapidly and the system behaves as a poor solvent and phase separation occurs. For lipids and detergents, this phase separation can be in the form of the micelles, bilayers, or other aggregates.

LIPID AND DETERGENT FORMATION INTO MICELLES AND BILAYERS

The actual packing of the lipids will be determined by the geometric aspects and interactions between head groups. For lipids, the formation of the bilayers (Figure 4.4) is a rapid and spontaneous process once the concentration reaches a critical point. Other types of lipid arrangement are

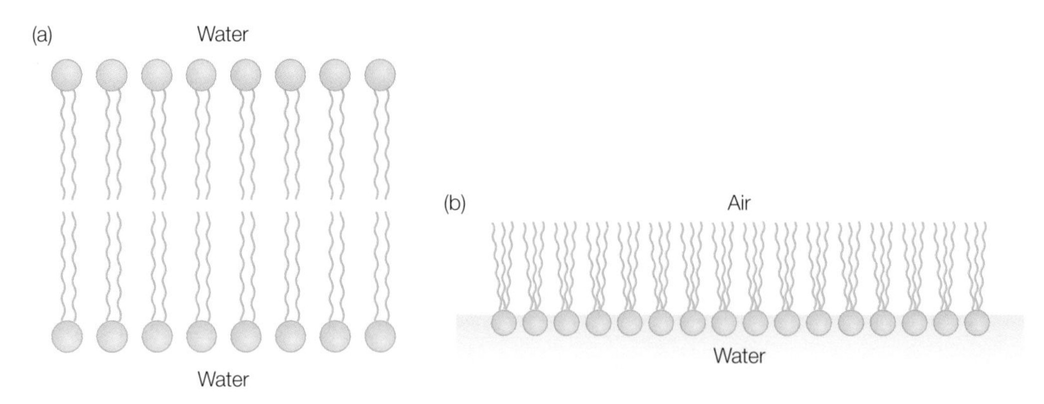

Figure 4.4 Lipids can form both (a) bilayers and (b) monolayers.

also possible. At an air/water interface the lipids will form monolayers with the polar head groups directed towards the water and the fatty acid chains exposed to air (Figure 4.4).

Detergent molecules are also amphipathic compounds and usually contain a hydrophilic polar head group and a hydrophobic hydrocarbon chain. It should be noted that detergents usually have only a single carbon chain and can pack into other arrays, micelles, and liposomes, as shown below. In a micelle, the hydrophobic chains are in the interior and the polar head groups are exposed to the water. Micelles can be quite large and can contain hundreds of detergent molecules. Liposomes have a bilayer arrangement with a central aqueous compartment lined by the polar groups of the inner lipids (Figure 4.5). Since liposomes and micelles have no hydrophobic edges, both structures are relatively stable in an aqueous environment.

The ability of detergents and lipids to form micelles and liposomes can be described with the use of phase diagrams. The formation of a micelle

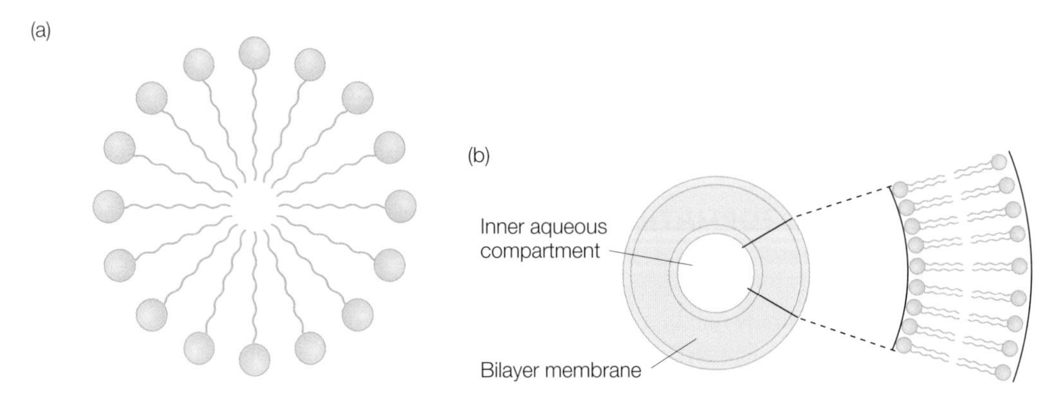

Figure 4.5 Detergent molecules can form (a) micelles or (b) liposomes.

from a detergent solution will depend upon the thermodynamic parameters that are effectively summarized with the phase diagram. For example, whether a detergent exists in solution as a monomer or micelle is dependent upon the detergent concentration and temperature (Figure 4.6). As the detergent concentration increases the micelle is more likely to form. However, the specific concentrations at which this occurs is strongly dependent upon temperature and there is usually a critical temperature below which the formation does not occur. Also, detergents and lipids can form much more complex phases under specific conditions. One such phase, termed the cubic lipid phase, is discussed in Chapter 17.

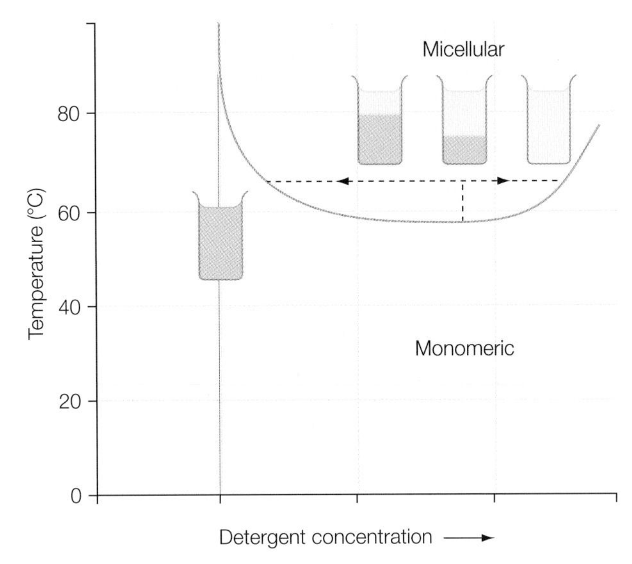

Figure 4.6 A phase diagram for a detergent showing the monomeric and micellular regions. As the detergent concentration is increased at 65°C, the solution undergoes a phase separation.

RESEARCH DIRECTION: LIPID RAFTS

Cell membranes are bilayers composed of a number of different lipids, with the most common shown below. These lipids are not uniformly distributed but rather are found in specific regions of the bilayers. The classic picture of the cell membrane, as developed by S. Jonathan Singer and Garth Nicholson (1972), is the *fluid mosiac concept.* In this model, the bilayer serves as a two-dimensional neutral environment for the membrane proteins, lipids, and proteins that are free to move within the bilayers (see Figure 1.5). In model lipid bilayers, the behavior is more complex, as moving lipids can be highly restricted to motion in gel states that are influenced by the lipid composition.

In cell membranes, the complex composition leads to non-uniform distributions of the membrane components. A well-established example of such heterogeneity is the asymmetry in the lipid composition for the two sides of the bilayer. Another more controversial example is a lipid raft (Simons & Ikonen 1997; Jacobson & Dietrich 1999; Parton & Hancock 2004). The basic concept of a lipid raft is that lipid microenvironments exist in the cell membrane, with the microenvironments having a composition enriched in certain lipids and proteins. The raft concept remains unproven despite the large amount of experimental evidence in its favor (Brown & London 1998; Edidin 2001). As one example of the difficulties of unambiguously establishing their presence in cells, consider the straightforward measurement of the density of membrane fractions that first indicated the presence of lipid rafts. Fractionation studies of cell membranes showed both low-density and high-density components (see Figure 4.7). The low-density

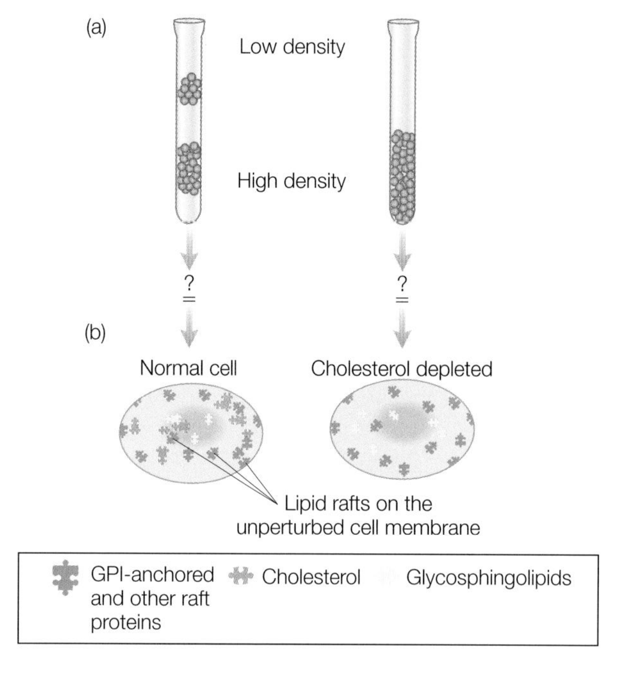

Figure 4.7 Density determinations of cell membrane fractions show the presence of a low-density component assigned to a lipid-raft contribution in (a) normal cells but not (b) cholestrol-depleted cells. Modified from (Edidin 2001).

regions were rich in cholesterol, glycosphingolipids, and lipid-anchored proteins. When the cell cholesterol was depleted, the low-density fraction was not observed. In model membrane systems, mixtures of these lipids have been found to organize domains. The simple interpretation of these experiments was that the low-density region exists in the cell membrane as a so-called lipid raft. However, the complexity of the membrane allows for the possibility that this simple interpretation is not correct due to unforseen artifacts. Support for such localized regions is provided by antibody patching and immunofluorescence microscopy but variation in the results among cells makes quantification difficult. Other experiments, such as tracking the location of single fluorophores to monitor the diffusion of individual raft proteins or lipids, would potentially be very revealing but such measurements are technically very challenging.

Interest in lipid rafts rose when it was recognized that many critical membrane-associated proteins could be part of lipid rafts (Simons & Toomre 2000; Kenworthy 2002). It appears that the rafts are centers for receptor-mediated signaling involving various membrane receptors such as the T-cell, B-cell, epidermal growth factor, and insulin receptors. In many cases, the receptors belong to a class of proteins called glycosylphosphatidylinositol (GPI)-anchored proteins. GPI, is a hydrophobic molecule that is attached to the C-terminus of certain proteins. The GPI molecule allows the protein to be closely associated with the cell membrane through a flexible tether while having the ability to interact with different proteins located on the extracellular side of the membrane. Enrichment of GPI-anchored proteins is observed in the low-density fractions, and treatments that disrupt the proposed rafts also disrupt the receptor function. A simple model of lipid rafts is that the rafts segregate in the lipid bilayer and sequester proteins attached by the GPI anchors. Several proteins are known to depend upon cholesterol that is enriched in the raft. Making analysis of these domains difficult is both the small size of the rafts and the presence of invaginations called caveolae that are rich in specific lipids. Although separate from rafts, the presence of the caveolae may be regulated by lipid rafts. Adding to the complexity of rafts is the observation that the distribution of lipid rafts is not uniform. This is presumably due to the function of the rafts, such as a proposed role in cellular trafficking. For example, the rafts may serve as sorting platforms to which vesicular integral membrane proteins associate as stabilizers for other proteins in

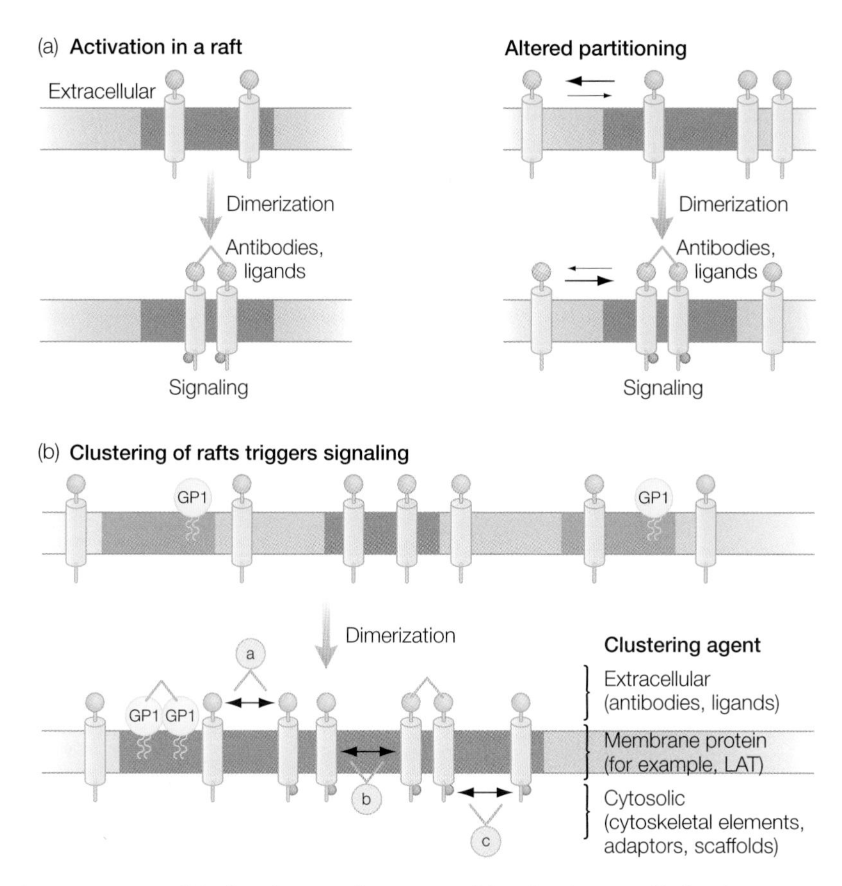

Figure 4.8 Models for the involvement of lipid rafts in cellular functions, with signaling initiated in (a) single rafts or (b) clustered rafts. Modified from Simons and Toomre (2000).

the raft. In these models, signaling could be initiated through the raft (see Figure 4.8). Many of these proteins are active only after dimerization or association with other proteins. The enrichment of specific proteins in the raft may facilitate this process. Two proteins within the raft may become activated upon association with other proteins. Alternatively, different rafts may become clustered, leading to dimerization of the protein components. The interactions that drive raft assembly are dynamical and reversible, providing a mechanism for turning off the activity.

DETERMINATION OF MICELLE FORMATION USING SURFACE TENSION

The formation of these lipid and detergent structures can be measured by a number of physical properties. As an example, the formation of micelles can be determined by measuring the surface tension of the solution. Liquid drops will generally adopt a shape that minimizes the exposed surface area, resulting in spheres due to their small surface area/volume ratio. For a curved surface with a radius r, the pressure on the concave side of the interface, P_{in}, is always greater than the pressure on the convex side, P_{out}. The outward force is given by the product of the pressure and the area, $4\pi r^2 P_{in}$. The inward force is given by two contributions. The first contribution is from the pressure, $4\pi r^2 P_{out}$. The second is due to the surface tension, γ, defined as the proportionality constant relating the work needed to change the surface area, A, of a sample by an infinitesimal amount. For the case of a drop, the work done by changing a drop from a radius r to $r + dr$ is given by:

$$\gamma[A(r + dr) - A(r)] = \gamma[4\pi(r + dr)^2 - 4\pi r^2] = 8\pi\gamma r dr \tag{4.11}$$

$$(r + dr)^2 - r^2 = r^2 + 2rdr + dr^2 - r^2 = 2rdr + dr^2 \approx 2rdr \quad dr \ll 1$$

Since work is equal to force times distance, the force opposing the movement through the distance dr is given by $8\pi\gamma dr$. At equilibrium the inward and outward forces balance and so:

$$4\pi r^2 P_{in} = 4\pi r^2 P_{out} + 8\pi\gamma dr \tag{4.12}$$

$$P_{in} = P_{out} + \frac{2\gamma}{r}$$

This relationship shows that the pressure on the concave surface is always greater than the pressure on the convex surface by an amount given by the surface tension.

The tendency of liquids to move up a capillary is called capillary action and is a consequence of surface tension. When a glass capillary is initially immersed in a liquid, the liquid will adhere to the walls (Figure 4.9). As the film climbs up, the surface curves and the pressure beneath the meniscus is less than the atmospheric pressure by $2\gamma/r$. This pressure difference drives the liquid upward until hydrostatic equilibrium is reached:

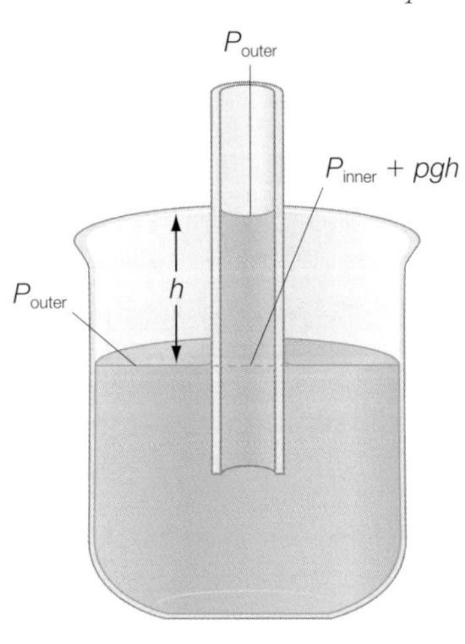

Figure 4.9 Water will travel up a small capillary until the pressure difference is equal to the hydrostatic pressure arising from the water in the capillary.

$$\Delta P = \frac{2\gamma}{r} = \rho g h \quad \text{or} \quad h = \frac{2\gamma}{\rho g r} \tag{4.13}$$

where ρ is the density of the liquid and g is the gravitational force constant. The surface tension of water at 293 K is equal to 72.7 mN m^{-1}. This value is strongly dependent upon the properties of the solutions, as methanol has a much lower value, 28.9 mN m^{-1}, whereas mercury has a much higher value, 472 mN m^{-1}.

At room temperature, a solution of pure water should give a surface tension of 72.8 mN m^{-1}. When a detergent is added to water the presence of the detergent disrupts the interactions between water molecules and the surface tension decreases. Thus, as the concentration of a detergent is increased, the surface tension will usually decrease until it reaches a specific concentration, above which there is no longer any change in surface tension (Figure 4.10). This is interpreted as the concentration at which it is thermodynamically favorable for micelles to form. The detergent concentration at which micelles form is called the *critical micelle concentration*, or *CMC*.

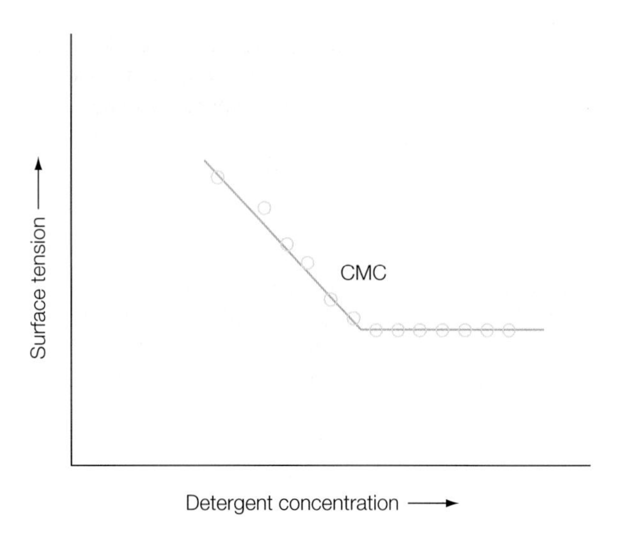

Figure 4.10 The dependence of surface tension on the presence of detergent can be used to determine the CMC, or critical micelle concentration.

Proteins present in membranes can be purified by the use of detergents. The lipid environment surrounding the hydrophobic surface of the protein is replaced by detergents, allowing extraction from the cell as seen in Figure 4.11. The proteins can then be further purified by conventional techniques provided the presence of detergents is maintained in solutions. Once isolated and purified, the action of the protein in the cell membrane can be studied by inserting the proteins back into membranes. One traditional approach was to make synthetic bilayers by placing a small septum between two chambers. Lipids present in the solutions will spontaneously form a bilayer across the septum. If membrane proteins are present in solution they will also be incorporated into the bilayer. Another approach is to introduce detergent-solubilized proteins into a solution that can form liposomes (Figure 4.11). The detergent is removed from the protein by dialysis and the protein is spontaneously inserted into the lipsosome

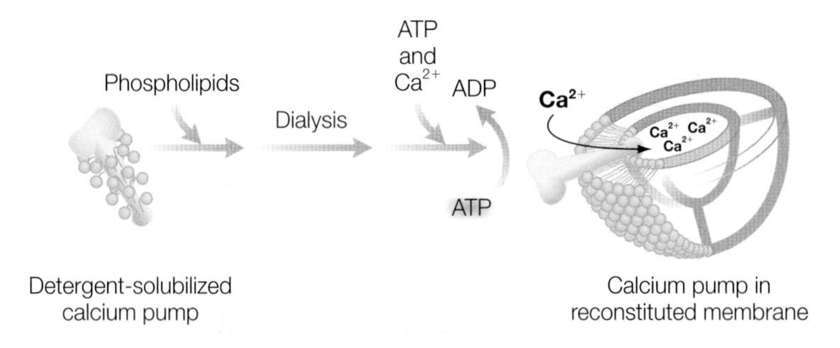

Figure 4.11 Membrane proteins can be embedded in bilayers by insertion into liposomes.

(Figure 4.11). Alternatively, the proteins can be suspended in the bilayer formation on a glass slide, which provides a surface that can be addressed readily with spectroscopic probes to study the motion of proteins and other components in the bilayer.

MIXTURES

An understanding of metabolic pathways and other cellular reactions involves dealing with mixtures of substances. Just as the chemical potential can be used to understand the formation of lipids and detergents into bilayers and liposomes, it can also be used to understand the properties of simple mixtures. For simplicity, we will consider only binary mixtures, although the ideas can be extended easily to more complex mixtures. Consider a mixture of molecules A and B. The Gibbs energy is given by the product of the chemical energy and the number of molecules (eqn 4.2):

$$G = n_A\mu_A + n_B\mu_B \tag{4.14}$$

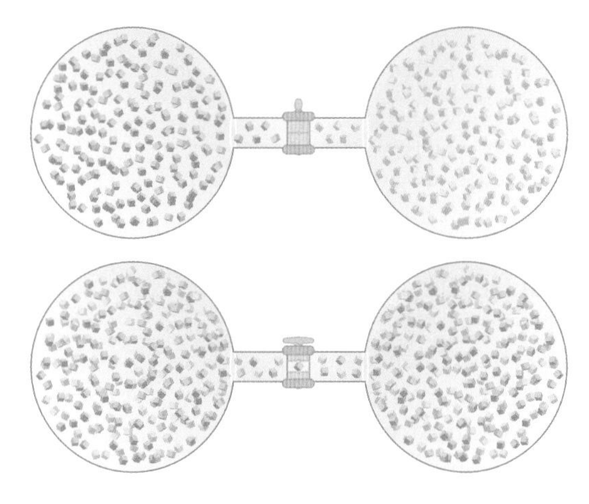

Figure 4.12 Two gases are initially separated but then allowed to mix.

As an example, consider a container that has two different ideal gases, gas A and gas B (Figure 4.12). Initially, the gases are located in separate containers at a certain pressure. Before mixing, the total Gibbs energy is the sum of the individual Gibbs energies, μ_A and μ_B. For an ideal gas undergoing a pressure change, the Gibbs energy is given by a logarithmic ratio of the final and initial pressure. Thus, the chemical potential for each molecule can be related to the ratio of the initial and final pressures:

$$\mu_A - \mu_A^i = \frac{G_A}{n_A} = RT \ln\frac{P_f}{P_i} \tag{4.15}$$

$$\Delta G = nRT \ln\frac{P_f}{P_i}$$

where μ_A^i represents the chemical potential of A under initial conditions. As is done for the other thermodynamic parameters, the use of the chemical potential is enhanced with the definition of a standard state. The chemical potential at a given pressure, $\mu(P)$, is defined relative to the chemical potential at a standard pressure of 1 bar, $\mu(P_0)$:

$$\mu(P) = \mu(P_0) + RT \ln \frac{P}{P_0} \tag{4.16}$$

Substituting this expression for the chemical potential into the relationship for the Gibbs energy yields:

$$G_i = n_A\left(\mu_A(P_0) + RT \ln \frac{P}{P_0}\right) + n_B\left(\mu_B(P_0) + RT \ln \frac{P}{P_0}\right) \tag{4.17}$$

After mixing, the molecules A and B will distribute over the entire volume and the partial pressures will decrease and the Gibbs energy becomes:

$$G_f = n_A\left(\mu_A(P_0) + RT \ln \frac{P_A}{P_0}\right) + n_B\left(\mu_B(P_0) + RT \ln \frac{P_B}{P_0}\right) \tag{4.18}$$

$P = P_A + P_B$

Notice that since the partial pressures are less than the total pressure, the ratio of the pressures has decreased and so the Gibbs energy has consequently decreased. The change in the Gibbs energy due to mixing, ΔG_{mix}, can now be calculated:

$$\Delta G_{mix} = G_f - G_i \tag{4.19}$$

$$\Delta G_{mix} = \left[n_A\mu_A(P_0) + n_A RT \ln \frac{P_A}{P_0} + n_B\mu_B(P_0) + n_B RT \ln \frac{P_B}{P_0}\right]$$
$$- \left[n_A\mu_A(P_0) + n_A RT \ln \frac{P}{P_0} + n_B\mu_B(P_0) + n_B RT \ln \frac{P}{P_0}\right]$$

$$\Delta G_{mix} = n_A RT \ln \frac{P_A}{P_0} + n_B RT \ln \frac{P_B}{P_0} - n_A RT \ln \frac{P}{P_0} + n_B RT \ln \frac{P}{P_0}$$

$$\Delta G_{mix} = RT\left(n_A \ln \frac{P_A}{P_0} - n_A \ln \frac{P}{P_0} + n_B \ln \frac{P_B}{P_0} - n_B \ln \frac{P}{P_0}\right)$$
$$= RT\left(n_A \ln \frac{P_A}{P} + n_B \ln \frac{P_B}{P_0}\right)$$

$$\ln\left(\frac{a}{b}\right) - \ln\left(\frac{c}{b}\right) = \ln\left(\frac{a}{b}\cdot\frac{b}{c}\right) = \ln\left(\frac{a}{c}\right)$$

This last expression can be written in terms of mole fractions, yielding an expression for the Gibbs energy of mixing in terms of the mole fractions:

$$X_A = \frac{n_A}{n} = \frac{P_A}{P}$$

$$X_B = \frac{n_B}{n} = \frac{P_B}{P} \tag{4.20}$$

$$n = n_A + n_B$$

$$\Delta G = nRT(X_A \ln X_A + X_B \ln X_B)$$

This final relationship expresses a simple idea that, by simply allowing different substances to mix together, the Gibbs energy will decrease. Although this was derived for ideal gases, the same idea will hold for real gases provided that the interactions between the gas molecules are negligible.

Thus, for ideal gases, the change in the Gibbs energy is sum of the contributions from each molecule. Since the sum of the mole fractions is equal to one, the change in Gibbs energy is uniquely determined by the mole fraction for any one gas (Figure 4.13). The difference ranges from zero, when only one type of gas is present, to a minimum value when the two gases are present in equal amounts. Because mole fractions are never greater than one, the logarithms are always negative and the change in the Gibbs energy is negative for all ideal gases. The decrease in energy results in a spontaneous mixing of the two gases. For non-ideal gases, the gas molecules will interact and there are additional contributions to the energy difference.

Figure 4.13 The Gibbs energy of mixing two ideal gases.

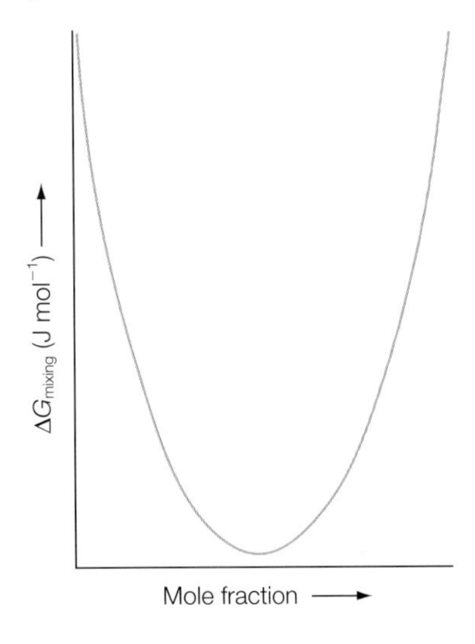

If the two gases in the example are the same gas then the change in the Gibbs energy is zero because the final mole fraction is one. The identity of the gases influences the outcome because the driving force for mixing is actually the entropy change of the system upon mixing. To probe the role of entropy, consider mixing two buckets of balls together. If the balls are identical, then the amount of disorder, or entropy, does not change upon mixing. However, if the balls are distinguishable, the entropy does change. For example, consider two different-sized balls that are initially placed in separate chambers. The content of the two chambers is mixed by a central paddle that allows the small balls to pass but not the large balls. In the course of mixing, work is performed and the entropy changes (Figure 4.14).

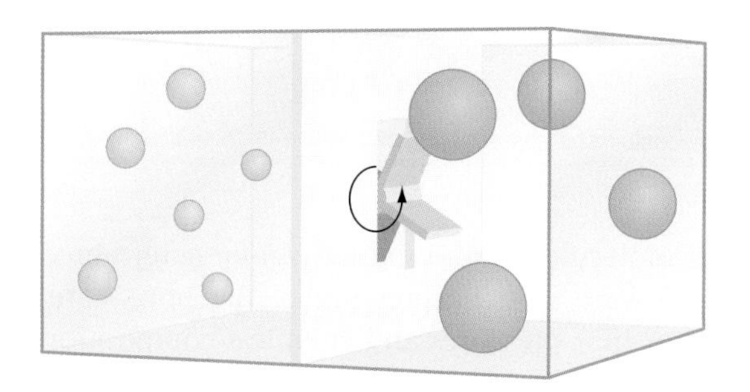
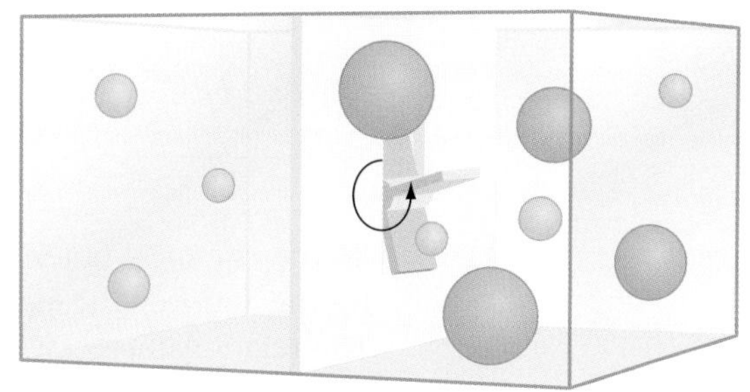

Figure 4.14 Mixing of two sizes of balls results in a change in entropy of the system.

RAOULT'S LAW

Vapor pressure is the pressure due to molecule A that would be generated if the gas and the liquid states of this molecule were allowed to reach equilibrium in a closed system. Consider a system that contains only a pure state of molecule A. At equilibrium, the chemical potential of the gas phase, μ^0 (gas), is equal to the chemical potential of the liquid state, μ^0 (liquid). The chemical potential of the gas state, and hence also the liquid state, can be written in terms of the chemical potential under standard conditions and the ratio of the given and standard pressures:

$$\mu_{pure}\,(\text{gas}) = \mu_A^0 + RT\ln\left(\frac{P_{pure}}{P^0}\right) = \mu_{pure}\,(\text{liquid}) \tag{4.21}$$

If there are two liquids mixed together, the chemical potential of molecule A will depend upon the partial pressure associated with molecule A:

$$\mu_A\,(\text{gas}) = \mu_A^0 + RT\ln\left(\frac{P_A}{P^0}\right) = \mu_A\,(\text{liquid}) \tag{4.22}$$

Now consider the difference in the chemical potentials for the pure and mixed liquid states:

$$\mu_A\,(\text{liquid}) - \mu_{pure}\,(\text{liquid}) = \mu_A^0 + RT\ln\left(\frac{P_A}{P^0}\right) - \mu_A^0 - RT\ln\left(\frac{P_{pure}}{P^0}\right) \tag{4.23}$$

$$\mu_A\,(\text{liquid}) = \mu_{pure}\,(\text{liquid}) + RT\ln\left(\frac{P_A}{P^0}\right) - RT\ln\left(\frac{P_{pure}}{P^0}\right)$$

$$= \mu_{pure}\,(\text{liquid}) + RT\ln\left(\frac{P_A}{P_{pure}}\right)$$

$$\ln(A) - \ln(B) = \ln\left(\frac{A}{B}\right)$$

So the chemical potential of molecule A in a mixture can be determined by the ratio of the partial pressure and the pressure of the pure state. The relationship between the vapor pressures and chemical composition of liquids was established by the French scientist Francois Raoult, who established the following relationship between the partial pressure P_A due to molecule A in a mixture and the mole fraction X_A:

$$P_A = P_{pure}X_A \tag{4.24}$$

This relationship states that the partial pressure of each of the components is directly proportional to the vapor pressure of the corresponding pure substance with the proportionality constant being the mole fraction. The relationship also provides the basis for deriving the chemical potential for the mixture. Substituting this relationship for partial pressure (eqn 4.22) into the relationship for chemical potential (eqn 4.21) yields an expression for the chemical potential of a mixture in terms of the chemical potential of the pure state and the mole fraction of the mixture:

$$\mu_A\,(\text{liquid}) = \mu_{pure}\,(\text{liquid}) + RT\ln\left(\frac{P_{pure}X_A}{P_{pure}}\right) = \mu_{pure}\,(\text{liquid}) + RT\ln X_A \tag{4.25}$$

Now the various thermodynamic parameters are eliminated and the only parameters left are those that can be defined for the liquid phase. The determination of this simple relationship hinges on Raoult's law. The molecular basis for Raoult's law can be understood by considering the surface of a liquid with molecules leaving and returning (Figure 4.15). The rate at which the molecules leave the surface, or equivalently the rate of vaporization, is proportional to the number of molecules on the surface, which is proportional to the number in solution:

Figure 4.15 A molecular view of Raoult's law.

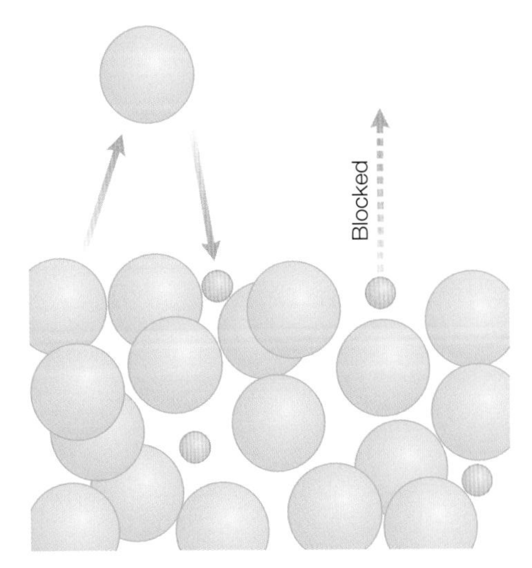

$$\text{Rate of leaving} = c_{leaving}X_A \tag{4.26}$$

where $c_{leaving}$ is the proportionality constant. Assuming that the surface is large and molecules leaving do not interact with molecules returning, the rate at which the molecules return, or the rate of condensation, can be written as:

$$\text{Rate of return} = c_{return} P_A \tag{4.27}$$

since the partial pressure is proportional to the concentration of the molecules in the vapor phase. At equilibrium, the rates of leaving and returning are the same so these two expressions can be equated, allowing the partial pressure to be written in terms of the mole fraction:

$$\text{Rate of leaving} = \text{rate of return} \tag{4.28}$$

$$c_{leaving} X_A = c_{return} P_A$$

$$P_A = \frac{c_{leaving}}{c_{return}} X_A$$

For a pure liquid with only one molecule, $X_A = 1$. Substituting this value into the equation yields:

$$P_{pure} = \frac{c_{leaving}}{c_{return}} \tag{4.29}$$

The ratio of the partial pressure P_A to the pressure for the pure liquid can be found by dividing these two equations, yielding an expression for the partial pressure:

$$\frac{P_A}{P_{pure}} = \frac{\left(\frac{c_{leaving}}{c_{return}} X_A\right)}{\left(\frac{c_{leaving}}{c_{return}}\right)} = X_A \tag{4.30}$$

$$P_A = X_A P_{pure}$$

Thus, the solute molecules in a solution are distributed throughout the solvent. The solution is considered to be in equilibrium with the gas phase. The composition of the gas phase is determined by a balance of a gain of molecules through evaporation and a loss through condensation.

This equation does assume an ideal behavior for molecules in which the total vapor pressure of a mixture is given by the sum of the mole fractions of each molecule weighted by the partial pressures of each molecule:

$$P = X_A P_{A,pure} + X_B P_{B,pure} \tag{4.31}$$

Deviations from this ideal behavior can be accounted for by including empirically determined constants, as described originally by the English scientist William Henry. As a rule, the deviations are small when the solutions are very dilute and appear to be homogeneous.

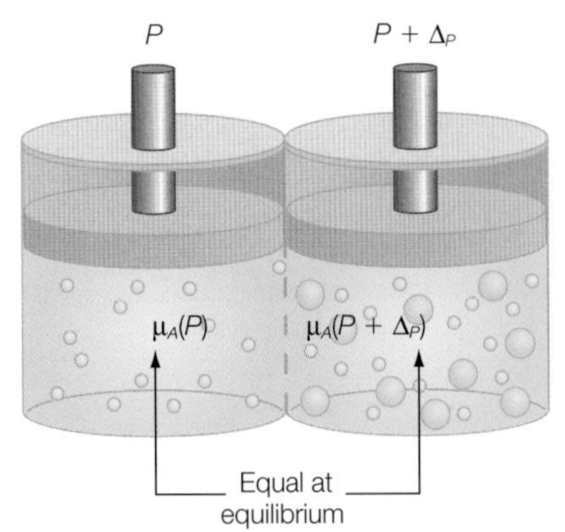

Figure 4.16
Molecules can pass from one solution into another by passage through a semipermeable membrane via osmosis.

OSMOSIS

Osmosis is the spontaneous movement of a pure solvent into a solution across a semipermeable membrane that allows the solvent, but not the solute, to pass through. Consider the simple case of two liquid solutions separated by a membrane in a chamber with a solute added to one side but the other side being initially a pure solvent (Figure 4.16). Since the side with the solute has a lower chemical potential, the solvent will migrate across the membrane towards this side. The osmotic pressure is defined as the pressure that must be applied to the side with the solute to prevent solvent transfer. The *van't Hoff equation* describes the osmotic pressure by comparison of the chemical potentials of the two sides.

The difference in chemical potentials is given by the reduction of the mole fraction of the solvent from 1 to X_A:

$$\mu_A(P + \Delta_P) - \mu_A(P) = RT \ln X_A \tag{4.32}$$

At equilibrium, the difference in chemical pressure is balanced by the difference in pressure multiplied by the volume:

$$(\Delta P)V = -RT \ln X_i \approx -RT(1 - X_A) = RTX_B \tag{4.33}$$

$\ln x \approx 1 - x$ and $X_A + X_B = 1$

Dividing both sides of the equation by the volume and replacing X_B/V with the *molarity* n_B yields the relationship:

$$\Delta P = n_B RT \tag{4.34}$$

Semipermeable membranes are commonly used in biochemistry in dialysis experiments in which a protein solution is enclosed within a membrane and placed into another solution containing a different buffer or salt. While the protein remains in the membrane bag, the solutions equilibrate. The protein is then available for experimentation under the new conditions. Dialysis can also be used to characterize the binding of small molecules to proteins.

RESEARCH DIRECTION: PROTEIN CRYSTALLIZATION

Through advances in molecular biology, every gene has been sequenced for a very large number of organisms. The wealth of genomic information is being studied by many directions in research. For example, the response of every gene of an organism to environmental changes is being investigated. Every gene encodes a protein that has a specific function in the cell. Proteomics is the study of how every protein of an organism responds to changes. Although the main characteristic of a gene is its nucleotide sequence, an understanding of the properties of a protein requires identification of not only the protein sequence, but also of how the polypeptide chain folds as a three-dimensional object. Therefore a major goal of proteomics is to determine the three-dimensional structures of every protein within certain targeted organisms.

The technique of X-ray diffraction is the primary tool used by scientists to determine the three-dimensional structure of proteins. Many aspects of this technique have been optimized as described in Chapter 15. The major limitation in most cases is that the technique requires a crystallized protein. Since proteins can assemble into complexes and undergo post-translational modifications, the number of proteins per organism is not uniquely determined by the genome, but certainly is in the range of tens of thousands. To determine every protein structure within a reasonable time, it is necessary to apply automated, high-throughput methods. Despite these improvements, the crystallization of any given protein may take years, largely because the thermodynamic properties that lead to crystallization remain difficult to apply and crystallization proceeds using a trial-and-error approach involving thousands of conditions (Kam et al. 1978; McPherson et al. 1995).

Proteins need to always be in the proper buffering conditions and ionic strength. Crystallization represents changing the state of the protein by its removal from a solution into a crystalline array. Thus, there is a competition between the protein in liquid solution and in the crystalline state. The protein in solution will have a certain free energy (Figure 4.17). If the protein forms a microcrystal consisting of only a few proteins, the free energy will increase. The favorable introduction of additional bonds between proteins cannot overcome the penalty of the decreased entropy. Thus, the monomer state is favored over the microcrystalline state. Only when the crystal reaches a certain critical size does the free energy decrease with increasing size, hence favoring the crystallization process. In practice, if the energy increase due to microcrystalline formation is low, then the system, after a period of time, can form a crystal of the critical size and the crystallization can proceed. While this may seem straightforward, the problem is that proteins can also form another state, a disordered precipitate, that is energetically favorable to growing large crystals, and this can deplete the supply of

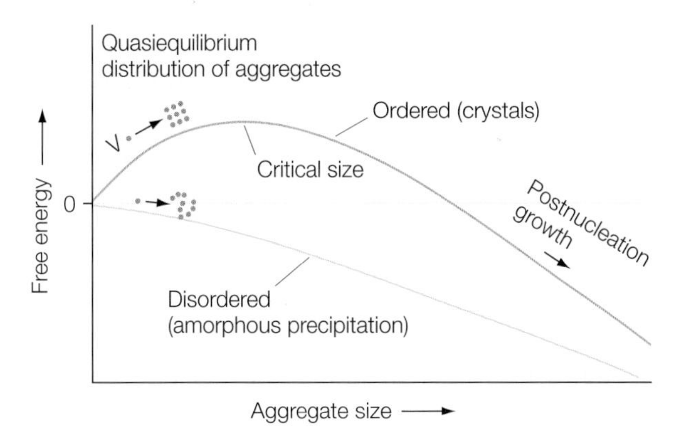

Figure 4.17 Dependence of the Gibbs energy on the size of an aggregate. Modified from Kam et al. (1978).

protein. Thus, in most cases, the protein precipitates and crystals are obtained only by surveying many different solvent conditions in the hope that one favors the slow and ordered growth needed for crystallization.

The preferred approach for producing protein crystals is to use vapor diffusion, in which the process is driven by the removal of water molecules from the protein solution. Two solutions are placed in a closed container: a reservoir and a protein solution. The reservoir contains the precipitating agent, usually a salt or polyethylene glycol, and has a relatively large volume (typically 1 mL). The second solution has the protein, buffer, and a smaller concentration of the precipitating agent, and a much smaller volume (about 1 μL). Because both solutions are in the same closed container, each solution will equilibrate through the vapor exchange between solutions. With time the liquid from the small drop will change into vapor and recondense with the large solution. This process will continue until equilibrium is achieved. At equilibrium, the chemical potentials associated with the precipitant are equal, hence the precipitant concentrations are the same. Thus, the protein drop has been concentrated by a factor given by the ratio of the initial precipitant concentrations of the two solutions. For example, if initially the salt concentrations are 0.5 and 1.0 M for the protein drop and reservoir, respectively, the volume of the protein drop will be reduced by half and the concentrations of the protein and other components will be doubled (so that the final salt concentration in the protein drop will be 1.0 M, matching the reservoir concentration).

Why does this lead to crystallization? Vapor diffusion causes a slow decrease in the amount of water in the small drop and hence a slow increase in the concentrations of the protein and precipitating agent (as well as the buffer). The dependence of the state of the protein, either in solution or in a crystal, on the salt and protein concentration can be mapped out with a phase diagram (Figure 4.18). On these diagrams, the lines represent

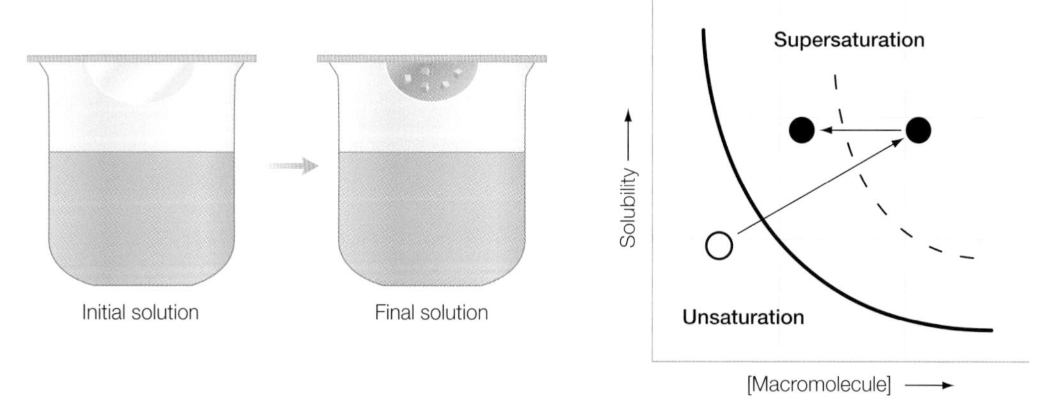

Figure 4.18 The crystallization process using vapor diffusion can be described in terms of a phase diagram.

boundaries between a phase where the protein is in solution to where the protein has become either a precipitant or crystal. With vapor diffusion, the concentration of both the protein and salt increase simultaneously, as shown by the diagonal line. The dark line represents the solubility dependence, so when this line is crossed the protein concentration is above the solubility, corresponding to a supersaturated solution. Since the supersaturated solution is not stable, the protein will leave the solution as either a precipitate or crystals. Phase diagrams such as these are sometimes used to design better conditions for crystallization.

Crystallization can be enhanced when the crystallization trial is poised such that the presence of a critical nucleus is favored. One classical approach is *seeding*, in which a microscopic protein crystal is inserted into the crystallization solution. Since the nucleus is already formed, energetically, the growth of the crystal is favorable. Although this approach can be highly effective, often the precise conditions needed for the crystal growth are highly specific and time-consuming to identify. The best seed is to make use of small protein crystals that one hopes will turn into much larger crystals. An alternative is to make use of other seed material that has the proper surfaces, but this has been successful only for small molecules. More generic approaches have been proposed, such as the use of porous templates (Page & Sear 2006; Figure 4.19). In this case, the base material is a porous surface that has a distribution of sizes on the microscopic level. Among the large number of pores, there will be one with a size that matches the length needed to accommodate a critical nucleus. Crystallization starts at one edge of the pore, and protein molecules will attach to the nucleus until the pore is filled. Once the crystal has grown and filled the pore then the crystallization continues and a large crystal can be formed. Ideally, the development of tools such as porous templates or microfluidic chambers (Hansen et al. 2004; Page & Sear 2006) should help overcome the major bottleneck in the determination of protein structures.

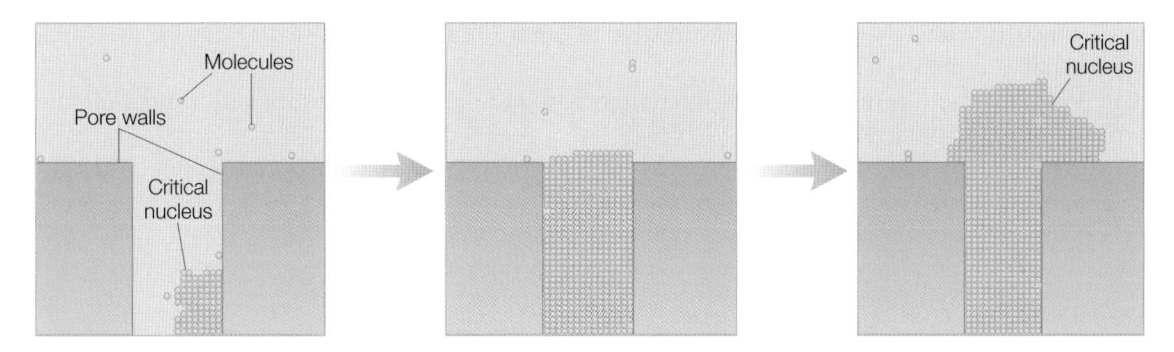

Figure 4.19 Crystal nucleation occurs in the pore of a surface and serves as a seed for growing a large crystal. Modified from Page and Sear (2006).

REFERENCES

Brown, D.A. and London, E. (1998) Functions of lipid rafts in biological membranes. *Annual Review of Cell and Developmental Biology* **14**, 111–36.

Edidin, M. (2001) Shrinking patches and slippery rafts: scales of domains in the plasma membrane. *Trends in Cell Biology* **11**, 492–6.

Hansen, C., Sommer, M.O.A., and Quake, S.R. (2004) Systematic investigation of protein phase behavior with a microfluidic formulator. *Proceedings of the National Academy of Sciences USA* **101**, 14431–6.

Jacobson, K. and Dietrich, C. (1999) Looking at lipid rafts? *Trends in Cell Biology* **9**, 87–91.

Kam, Z., Shore, H.B., and Feher, G. (1978) On the crystallization of proteins. *Journal of Molecular Biology* **123**, 539–55.

Kenworthy, A. (2002) Peering inside lipid rafts and caveolae. *Trends in Biochemical Sciences* **27**, 435–8.

McPherson, A., Malkin, A.J., and Kuznetsov, Y.G. (1995) The science of macromolecular crystallization. *Structure* **3**, 759–68.

Page, A.J. and Sear, R.P. (2006) Heterogeneous nucleation in and out of pores. *Physical Review Letters* **97**, 065701.

Parton, R.G. and Hancock, J.F. (2004) Lipid rafts and plasma membrane microorganization: insights from Ras. *Trends in Cell Biology* **14**, 141–7.

Simons, K. and Ikonen, E. (1997) Functional rafts in cell membranes. *Nature* **387**, 569–72.

Simons, K. and Toomre, D. (2000) Lipid rafts and signal transduction. *Nature Reviews: Molecular Cell Biology* **1**, 31–9.

Singer, S.J. and Nicholson, G.L. (1972) The fluid mosaic model of the structure of cell membranes. *Science* **175**, 720–31.

PROBLEMS

4.1 Does the change from a solid to a liquid represent a phase change?

4.2 Does the expansion of a metal during heating represent a phase change?

4.3 Why does adding salt to a mixture of ice and liquid water cause the temperature to drop?

4.4 What is the general structure of a phospholipid?

4.5 Explain how chemical potentials predict the behavior of detergents and lipids and how the second viral coefficient is used.

4.6 Explain the difference between a liposome and a micelle.

4.7 What is the critical micelle concentration?

4.8 How would a phase diagram be expected to change if the hydrocarbon chain of a detergent became longer?

4.9 Some lipids can form complex structures in what is termed a cubic lipid phase. Estimate where this phase would be found relative to other phases on a phase diagram for this lipid.

4.10 What is a lipid raft?

4.11 Why are these results from cell-fractionation experiments not definitive indicators for the presence of lipid rafts?

4.12 Why is the concentration of glycosylphosphatidylinositol (GPI)-anchored proteins thought to be important to the function of lipid rafts?

4.13 How high does water travel up a capillary with a radius of 2.0 mm?

4.14 If a blue liquid travels up a capillary only half as high as red liquid, what can be concluded about the two liquids?

4.15 Why does the surface tension of a detergent solution stop decreasing once the detergent concentration reaches a certain value?

4.16 What can be said about the enthalpy change associated with the mixing of two different ideal gases at constant temperature?

4.17 Why do reactions reach an equilibrium instead of entirely producing the products?

4.18 If a molecule cannot be mixed with water, it is said to be immiscible. What can be said about the enthalpy of mixing in this case?

4.19 When a spontaneous, endothermic reaction occurs at constant temperature and pressure, what can be said about the entropy change of the surroundings?

4.20 In a dialysis experiment, a 10-mL protein solution initially has a buffer concentration of 0.1 M. How much will the buffer concentration decrease if the protein solution is dialyzed against (a) 100 mL of water and (b) 1 L of water?

4.21 Why does the free-energy dependence for the crystallization of a protein have a maximum value?

4.22 If only the protein concentration were increased in a crystallization experiment, how would this be shown on a crystallization phase diagram?

4.23 What can be said about the total entropy change for the crystallization of a protein from solution if it occurs spontaneously?

5

Equilibria and reactions involving protons

The concepts of Gibbs energy and chemical potential are developed in this chapter in terms of the equilibrium compositions of reactions. The minimum value of Gibbs energy is shown to represent the equilibrium, and it provides a means to express the equilibrium constant in terms of the Gibbs energy. The thermodynamic formulation provides a platform to quantify the changes in pH associated with some acids, bases, and buffers that play a key role in biological systems, such as cardiovascular systems. The involvement of protons at an active site of a protein that undergoes redox reactions is presented, including how protons can strongly influence the energetics of the process, and why the presence of a proton pathway can be required.

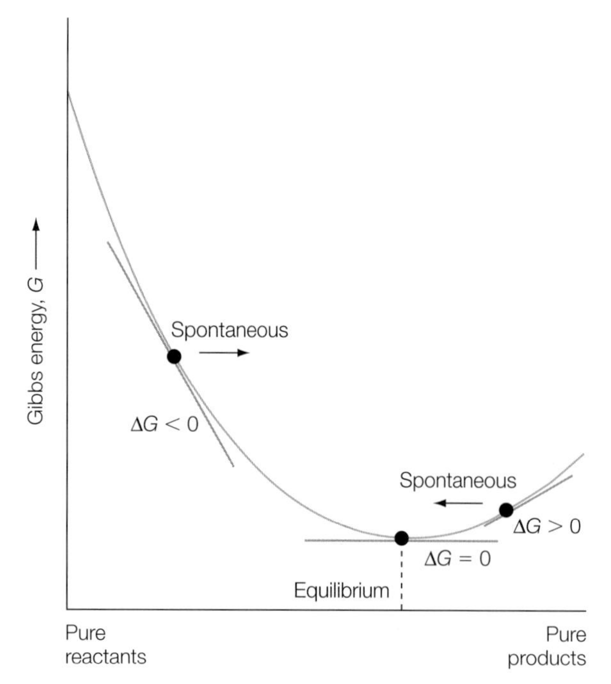

Figure 5.1 As a reaction proceeds, the Gibbs energy of the system changes due to the alteration of the relative amounts of reactants and products.

GIBBS ENERGY MINIMUM

After a reaction has started, it will eventually reach an equilibrium that depends upon the Gibbs energy difference, ΔG. For example, consider a simple reaction such as A $\leftrightarrow$ B. If the system starts with molecule A, with time molecule A will be converted into molecule B. This conversion does not necessarily proceed completely, but rather the reaction will come to an equilibrium that has a mixture of both molecules. In general, as the Gibbs energy for the reactant A compared to product B increases, then the amount of molecule B should increase (Figure 5.1). A reaction is spontaneous when the Gibbs energy of a mixture of A and B decreases as the amount of B increases. For most common reactions, the formation of the absolutely pure state of

B is not favored, as this state will not have the lowest value of G. Rather, the reaction will reach an equilibrium point that has a mixture of both A and B. At the equilibrium, the value of G is at the minimum value and moving away from the equilibrium is energetically unfavorable.

The slope of the free-energy dependence shows the direction in which the reaction will proceed. If the slope is less than zero, it is energetically favorable for the reaction to go forward and the process is spontaneous. Notice that this will cause the reaction to always move towards the equilibrium as, if the mixture has mostly B, the slope is negative for moving towards the equilibrium and not towards a larger amount of B. At the equilibrium point, the slope is zero and the amounts of A and B have no tendency to change.

The equilibrium point plays a critical role in a reaction and so we would like to use thermodynamics to understand what determines the equilibrium position. The Gibbs energy difference can be written as in logarithmic terms of the *equilibrium constant*, K_{eq}:

$$\Delta G^{\mathrm{o}} = -RT \ln K_{eq} \tag{5.1}$$

The value of the equilibrium constant is determined by the relative amount of the products and reactants. Consider a general reaction of A and B converting to C and D with the relative number of moles identified as n_{A} to n_{D}:

$$n_{\mathrm{A}}\mathrm{A} + n_{\mathrm{B}}\mathrm{B} \rightarrow n_{\mathrm{C}}\mathrm{C} + n_{\mathrm{D}}\mathrm{D} \tag{5.2}$$

If compounds A–D have the concentrations c_{A} to c_{D}, the equilibrium constant is given by:

$$K_{eq} = \frac{c^c_{\mathrm{C}} c^d_{\mathrm{D}}}{c^a_{\mathrm{A}} c^b_{\mathrm{B}}} \tag{5.3}$$

Derivation box 5.1

Relationship between the Gibbs energy and equilibrium constant

Consider the case when the pressure is constant and the reaction enthalpy is zero, so there is no heat flow. The determinant of the reaction then must be the entropy. The Gibbs energy is given by the mole fractions of the components (Chapter 4):

$$\Delta G = nRT(X_{\mathrm{A}} \ln X_{\mathrm{A}} + X_{\mathrm{B}} \ln X_{\mathrm{B}}) \tag{db5.1}$$

where the mole fractions have values between 0 and 1, with the sum always equal to 1. At the extreme point, where $X_{\mathrm{A}} = 1$ and $X_{\mathrm{B}} = 0$, the value of ΔG is 0. Likewise, ΔG is zero

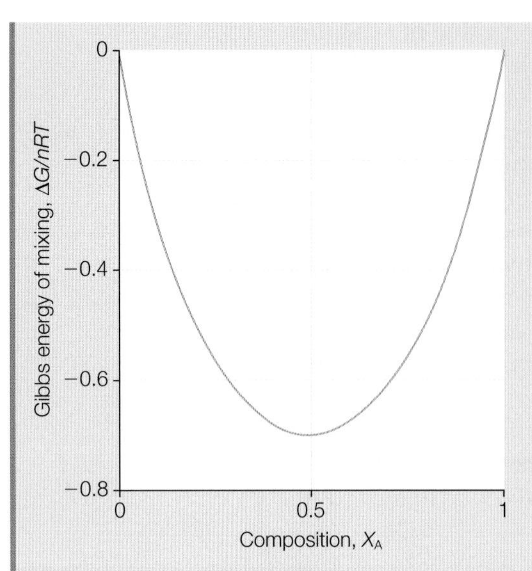

Figure 5.2 The Gibbs energy of mixing for different mole fractions of component A.

when $X_A = 0$ and $X_B = 1$. For intermediate values, the logarithms are always negative and ΔG is always negative. Thus, ΔG will have a parabolic shape with a minimum at $X_A = X_B = 0.5$ (Figure 5.2).

If no other factors are involved, then the distribution of A and B could change with time, with the average value being equal amounts, as this represents the state with the lowest energy. When enthalpy is involved, the situation changes, as seen for an ideal gas. Consider the Gibbs energy for the reaction, $(\Delta G)_{rec}$, which is the difference in the chemical potentials between the reactants and products. For an ideal gas the chemical potentials can be written in terms of the *total pressure*, P, and the *partial pressures*, P_A and P_B:

$$(\Delta G)_{rec} = \mu_B - \mu_A = \mu_B^0 + RT\ln\frac{P_B}{P} - \mu_A^0 - RT\ln\frac{P_A}{P} \quad \text{(db5.2)}$$

$$(\Delta G)_{rec} = \mu_B^0 - \mu_A^0 + RT\ln\frac{P_B}{P_A} \quad \text{(db5.3)}$$

$$\ln\frac{x}{y} - \ln\frac{z}{y} = \ln\frac{xy}{yz} = \ln\frac{x}{z}$$

The difference in the standard chemical potentials is usually referred to as the Gibbs energy of reaction at standard conditions or the standard Gibbs energy of reaction, $(\Delta G)_{rec}^{\circ}$. Using the standard Gibbs energy of reaction in eqn db5.3 yields:

$$(\Delta G)_{rec} = (\Delta G)_{rec}^{\circ} + RT\ln\frac{P_B}{P_A} \quad \text{(db5.4)}$$

The minimum of the Gibbs energy will occur when it is zero and the reaction will neither go forwards nor backwards, because the slope is zero. At this equilibrium point:

$$(\Delta G)_{rec} = 0 = (\Delta G)_{rec}^{\circ} + RT\ln\frac{P_B}{P_A} \quad \text{(db5.5)}$$

Expressing the equilibrium constant in terms of the partial pressures:

$$K_{eq} = \frac{P_B}{P_A} \quad \text{(db5.6)}$$

allows the standard Gibbs energy difference to be written in terms of the equilibrium constant:

$$(\Delta G)_{rec} = 0 = (\Delta G)^{\circ}_{rec} + RT \ln K_{eq} \quad \text{(db5.7)}$$

$$K_{eq} = e^{-\frac{(\Delta G)^{\circ}_{rec}}{RT}}$$

$y = e^x$ then $\ln y = x$

For a more general situation, the chemical potential of the *i*th molecule can be related to what is termed the *activity*, a_i, according to:

$$\mu_i = \mu_i^0 + RT \ln a_i \quad \text{(db5.8)}$$

The activity is a measure of the concentration of a molecule. For an ideal solution, the activity is equal to the mole fraction. For a nonideal solution, the activity of the *i*th molecule is proportional to the mole fraction, x_i, and the *activity coefficient*, γ, according to:

$$a_i = \gamma_i x_i \quad \text{(db5.9)}$$

For the cases under consideration, solutions are considered to be ideal with $\gamma = 1$. For the reaction shown (eqn 5.3), the Gibbs energy of reaction can then be written as:

$$(\Delta G)_{rec} = d\mu_D + c\mu_C - b\mu_B + a\mu_A \quad \text{(db5.10)}$$

Substituting the expression for activity (eqn db5.8) yields:

$$(\Delta G)_{rec} = d(\mu_D^0 + RT \ln a_D) + c(\mu_C^0 + RT \ln a_C) - b(\mu_B^0 + RT \ln a_B) + a(\mu_A^0 + RT \ln a_A)$$

$$(\Delta G)_{rec} = (d\mu_D^0 + c\mu_C^0 - b\mu_B^0 - a\mu_A^0) + RT(d \ln a_D + c \ln a_C - b \ln a_B - a \ln a_A) \quad \text{(db5.11)}$$

The standard terms can be collected and the terms depending on the activities can be rewritten:

$$(\Delta G)^{\circ}_{rec} = (d\mu_D^0 + c\mu_C^0 - b\mu_B^0 - a\mu_A^0) \quad \text{(db5.12)}$$

$$RT(d \ln a_D + c \ln a_C - b \ln a_B - a \ln a_A) = RT \ln \frac{a_C^c a_D^d}{a_A^a a_B^b} \quad \text{(db5.13)}$$

The expression for the reaction Gibbs energy of a mixture becomes:

$$(\Delta G)_{rec} = (\Delta G)^{\circ}_{rec} + RT \ln \frac{a^c_{\mathrm{C}} a^d_{\mathrm{D}}}{a^a_{\mathrm{A}} a^b_{\mathrm{B}}} \qquad \text{(db5.14)}$$

Assuming behavior of an ideal solution, this can be written in terms of the concentrations because the activity coefficients all become 1 and the change in the Gibbs energy can be expressed in terms of an *equilibrium constant*, K_{eq}, according to:

$$(\Delta G)_{rec} = (\Delta G)^{\circ}_{rec} + RT \ln \frac{c^c_{\mathrm{C}} c^d_{\mathrm{D}}}{c^a_{\mathrm{A}} c^b_{\mathrm{B}}} = (\Delta G)^{\circ}_{rec} + RT \ln K_{eq} \qquad \text{(db5.15)}$$

$$K_{eq} = \frac{c^c_{\mathrm{C}} c^d_{\mathrm{D}}}{c^a_{\mathrm{A}} c^b_{\mathrm{B}}} \qquad \text{(db5.16)}$$

RESPONSE OF THE EQUILIBRIUM CONSTANT TO CONDITION CHANGES

Equilibrium conditions for any given reaction will shift in response to changes to a system, such as a change in pressure, temperature, or concentrations of the various reactants and products. The equilibrium conditions are not altered by the presence of an enzyme; this only changes the rate of a process. The response of a reaction due to such a change was originally described by the French chemist Henri Le Chatelier, who proposed what is now known as Le Chatelier's principle:

> Le Chatelier's principle states that when a system at equilibrium is subjected to a disturbance, the system will respond such that the effect of the disturbance is minimized.

Consider a chamber containing gas molecules with the pressure controlled by a piston (Figure 5.3). The gas molecules can be present in equilibrium as both monomers and dimers through a reaction $A \leftrightarrow A_2$. If the piston is pushed into the chamber the gas will be compressed, as the response of the system will be to minimize the increase in pressure. In this case, the pressure increase can be reduced by decreasing the number of particles in the gas phase; that is, shifting the equilibrium towards a larger number of dimers.

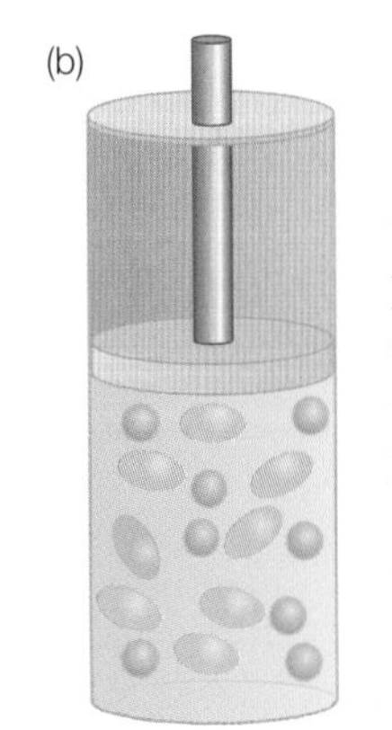

Figure 5.3 When gas molecules are compressed they will respond to minimize the effect, as shown in this case by forming dimers, thus decreasing the number of particles in the gas phase.

Following Le Chatelier's principle, the response of a system to a change in temperature can be predicted. For a system in equilibrium, an endothermic reaction will shift in favor of the products for a temperature increase. Likewise, for an exothermic reaction, an increase in the temperature will favor the reactants. Whereas Le Chatelier's principle provides qualitative guidelines, the effect of a temperature change on the equilibrium constant can be quantified by writing the Gibbs energy in terms of enthalpy and entropy (Chapter 6):

$$(\Delta G)^\circ_{rec} = (\Delta H)^\circ_{rec} - T(\Delta S)^\circ_{rec} \tag{5.4}$$

Using eqn db5.7, this expression can be revised in terms of the equilibrium constant:

$$-RT \ln K_{eq} = (\Delta H)^\circ_{rec} - T(\Delta S)^\circ_{rec} \tag{5.5}$$

Dividing both sides of the equation by RT yields the *van't Hoff equation*:

$$\ln K_{eq} = -\frac{(\Delta H)^\circ_{rec}}{RT} + \frac{(\Delta S)^\circ_{rec}}{R} \tag{5.6}$$

The temperature dependence of the equilibrium constant is a straight line when plotted as a function of $1/T$, with the slope giving the enthalpy and the intercept giving the entropy. Such a graph is commonly used to yield the enthalpy and entropy of a reaction. For an exothermic reaction, the slope will be positive and hence an increase in the temperature will decrease $1/T$, resulting in a decrease in the equilibrium constant as predicted by Le Chatelier's principle.

ACID–BASE EQUILIBRIA

Consider an aqueous solution with acids, which are proton-donating molecules, and bases, which are proton-accepting molecules. Any given molecule may act as either an acid or a base depending upon the conditions. For example, consider a molecule that is initially protonated, HA, which acts as an acid by giving up a proton to water. After the proton transfer, this molecule can act as a base by taking the proton back from A^-, the conjugate base:

$$HA + H_2O \leftrightarrow H_3O^+ + A^- \tag{5.7}$$

In an aqueous solution, the proton is never isolated and always solvated. However, the water molecule does not undergo a change and this equilibrium reaction is often written without explicitly including the involvement of the water molecule:

$$HA \leftrightarrow H^+ + A^- \tag{5.8}$$

Water can also serve directly as an acid by donating a proton to a base, B. Once the water molecule donates the proton, the OH^- can serve as a proton acceptor for the conjugate acid, BH^+, according to the following reaction:

$$B + H_2O \leftrightarrow BH^+ + OH^- \tag{5.9}$$

Since water is capable of serving as both an acid and a base, two water molecules can exchange a proton:

$$H_2O + H_2O \leftrightarrow H_3O^+ + OH^- \tag{5.10}$$

For this reaction, the equilibrium constant, K_W, can be written in terms of the activities:

$$K_W = \frac{a_{OH^-} a_{H_3O^+}}{a_{H_2O} a_{H_2O}} = a_{OH^-} a_{H_3O^+} \tag{5.11}$$

The reason why this expression can be reduced is that the activity is defined such that in its standard state a substance has an activity of one. This can be seen by writing the chemical potential of water in terms of the activity using eqn db5.8:

$$\mu_{H_2O} = \mu^0_{H_2O} + RT \ln a_{H_2O} \tag{5.12}$$

In this equation, the chemical potential is equal to the standard chemical potential when the activity term is zero, or equivalently, the activity is one:

$$\mu_{H_2O} = \mu^0_{H_2O} + RT \ln a_{H_2O} \rightarrow RT \ln a_{H_2O} = 0; \text{ which is true when } a_{H_2O} = 1 \tag{5.13}$$

$\ln(1) = 0$ as $e^0 = 1$

Notice that the activity of an ion is always relative to the standard state. Typically, the standard state of a solvent is defined for the pure solvent for which the activity is one. For a solute, the standard state is defined at 1 molal and activity is always relative to 1 molal.

Since the energy of a reaction is related to the natural logarithm of the equilibrium constant, the equilibrium constant is often referenced in terms of the logarithm and is termed the pK value:

$$\text{p}K = -\log K \qquad (5.14)$$

For water, the $\text{p}K_W$ can be written as:

$$\text{p}K_W = -\log(K_W) = -\log(a_{OH^-}a_{H_3O^+}) = -\log(a_{OH^-}) - \log(a_{H_3O^+}) \qquad (5.15)$$

$\log(xy) = \log x + \log y$

This expression can be revised by introducing two new terms, pH and pOH:

$$\text{pH} = -\log(a_{H_3O^+}) \qquad (5.16)$$
$$\text{pOH} = -\log(a_{OH^-})$$
$$\text{p}K_W = \text{pH} + \text{pOH}$$

For water at 25°C, the equilibrium constant K_W has a value of 1.008×10^{-14}. The $\text{p}K_W$ then has a value of 14 and the two terms pH and pOH will always add up to 14. Since the pOH can be expressed in terms of the pH, the concentrations are usually referenced to the pH. For example, rather than state that the H_3O^+ concentration is 10^{-5} M, the solution is described as having a pH of 5. For pure water, the expression:

$$H_2O + H_2O \leftrightarrow H_3O^+ + OH^- \qquad (5.17)$$

dictates that the concentrations of H_3O^+ and OH^- must be the same, which is true when the pH and pOH have values of 7:

$$\text{pH} + \text{pOH} = 14 \qquad (5.18)$$

$$\text{pH} = \text{pOH} = \frac{14}{2} = 7$$

Water is referred to as a weak acid or base since the equilibrium constant is small and the protons can be titrated, in contrast to a strong acid or base that has an equilibrium constant larger than one. When a weak acid, HA, is added to water, the concentrations of ions in the solution will change according to eqn 5.7, and the equilibrium constant K_A is given by:

$$HA + H_2O \leftrightarrow H_3O^+ + A^- \qquad (5.19)$$

$$K_A = \frac{[H_3O^+][A^-]}{[HA]}$$

Because the concentration of water is changed little by ionization, the expression is written assuming that the water concentration is unchanged.

Since the starting material is only the weak acid HA and water, there must be a balance between the number of ions:

$$[H_3O^+] = [A^-] \tag{5.20}$$

so eqn 5.19 can be written as:

$$K_A = \frac{[H_3O^+]^2}{[HA]} \tag{5.21}$$

Using the definitions of eqns 5.14 and 5.16 yields:

$$\log(K_A) = 2\log[H_3O^+] - \log[HA] \tag{5.22}$$

$$pK_A = 2pH + \log[HA] \text{ or}$$

$$pH = \frac{1}{2}pK_A - \frac{1}{2}\log[HA]$$

$$\log\frac{x^n}{y} = \log x^n - \log y = n\log x - \log y$$

Thus, the pH can be calculated with a certain amount of a weak acid, provided the weak acid is a significantly stronger acid than water. To titrate the proton in a weak acid or base, a strong acid or base is added. For example, as a strong base is added to a weak acid, the proton on the acid will be transferred from the acid to the base, with the location of the proton at any given point determined by the concentration of the added base (Figure 5.4). The shape of the dependence of the pH on the amount of the added base can be understood by examining two points of the titration curve.

The addition of the strong acid will stoichiometrically convert the weak acid to its conjugate base following eqn 5.19. The pH at any given point can be determined by revising this dependence in terms of the pK_A and pH values:

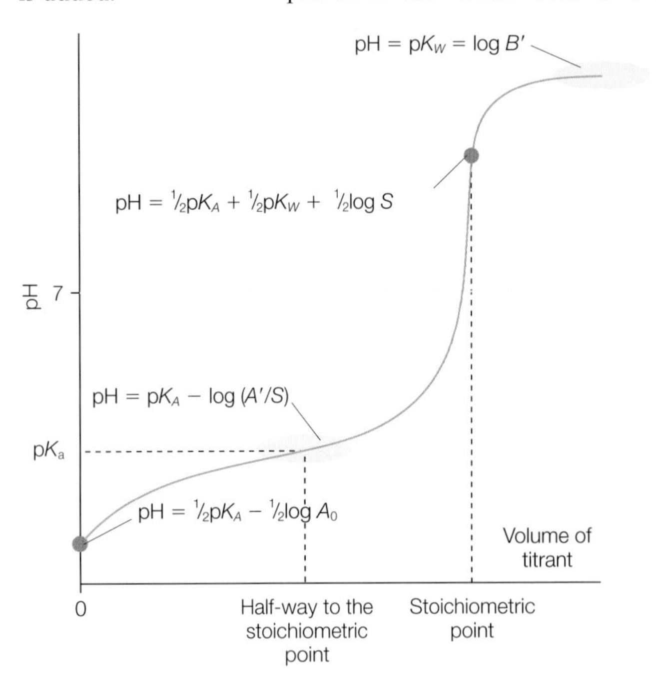

Figure 5.4 The titration curve for removing a proton from a weak acid as a strong base is added.

$$K_A = \frac{[H_3O^+][A^-]}{[HA]} \tag{5.23}$$

$$\log K_A = \log\frac{[H_3O^+][A^-]}{[HA]} = \log[H_3O^+] + \log\frac{[A^-]}{[HA]}$$

$$pK_A = pH - \log\frac{[A^-]}{[HA]}$$

$$pH = pK_A + \log\frac{[A^-]}{[HA]}$$

$pK = -\log K$	(eqn 5.14)

The dependence of pH on the concentrations of A^- and HA is called the *Henderson–Hasselbach* equation. This relationship allows the calculation of the pH from the ratio of the base to acid forms of the weak acid and its pK_A. This dependence forms the basis for pH buffers, as described below.

The stoichiometric point is when enough strong base has been added to convert all of the weak acid HA to the conjugate weak-base form A^-. Since it now behaves as a base, the pH dependence is altered. At this point, A^- may pick up a proton from the water, and the equilibrium constant K_B can be expressed in terms of K_A using eqn 5.21:

$$A^- + H_2O \leftrightarrow HA + OH^- \tag{5.24}$$

$$K_B = \frac{[HA][OH^-]}{[A^-]} = \frac{[HA]}{[H_3O^+][A^-]}[H_3O^+][OH^-] = \frac{1}{K_A}K_W$$

Assuming that the concentration of HA approximately matches the amount of OH^- allows this equation to be rewritten as:

$$[HA] \approx [OH^-] \tag{5.25}$$

$$\frac{K_W}{K_A} = \frac{[HA][OH^-]}{[A^-]} = \frac{[OH^-]^2}{[A^-]} = \frac{1}{[A^-]}\frac{[OH^-]^2[H_3O^+]^2}{[H_3O^+]^2} = \frac{1}{[A^-]}\frac{K_W^2}{[H_3O^+]^2}$$

Now, inserting the definitions of pK_W, pK_A, and pH (eqns 5.14–5.16) yields:

$$\log\frac{K_W}{K_A} = \log\frac{1}{[A^-]}\frac{K_W^2}{[H_3O^+]^2} \tag{5.26}$$

$$\log K_W - \log K_A = -\log[A^-] + \log K_W^2 - \log[H_3O^+]^2$$

$$-pK_W + pK_A = -\log[A^-] - 2pK_W + 2pH \text{ or}$$

$$pH = \frac{1}{2}pK_A + \frac{1}{2}pK_W + \frac{1}{2}\log[A^-]$$

$$\log x^n = n \log x$$

At the stoichiometric point, this relationship leads to a much sharper dependence on the added base than was determined at lower concentrations (Figure 5.4), due to the conversion of the weak acid into its conjugate base. If the strong base is added beyond the stoichiometric point, then the pH behavior will predominately reflect the properties of the strong base alone:

$$\mathrm{pH} = \mathrm{p}K_W + \log[\text{excess base}] \tag{5.27}$$

As an example of this pH behavior, consider a mixture of acetic acid, CH_3COOH, as the weak acid that is titrated with the strong base sodium hydroxide, NaOH. Acetic acid has a $\mathrm{p}K_A$ value of 4.75. For a solution with 0.1 M acetic acid, the pH is given by eqn 5.22:

$$\mathrm{pH} = \frac{1}{2}\mathrm{p}K_A + \log[\mathrm{HA}] = \frac{4.75}{2} + \log(0.1) = 2.88 \tag{5.28}$$

Adding sodium hydroxide at a concentration of 0.05 M, which is half of the concentration of the acetic acid, will convert half of the acetic acid to its conjugate base, yielding from eqn 5.23 a pH of:

$$\mathrm{pH} = \mathrm{p}K_A + \log\frac{[\mathrm{A}^-]}{[\mathrm{HA}]} = 4.75 + \log\frac{0.05}{0.05} = 4.75 \tag{5.29}$$

$$\log 1 = 0 \text{ since } e^0 = 1$$

Notice that the pH equals the $\mathrm{p}K_A$ when the amount of the acid and its conjugate base are equal.

Adding more sodium hydroxide to a concentration of 0.1 M converts the acetic acid to its conjugated base and the pH is determined by eqn 5.26:

$$\begin{aligned} \mathrm{pH} &= \frac{1}{2}\mathrm{p}K_A + \frac{1}{2}\mathrm{p}K_W + \frac{1}{2}\log[\mathrm{A}^-] \\ \mathrm{pH} &= \frac{4.75}{2} + \frac{14}{2} + \frac{1}{2}\log(0.1) = 2.37 + 7.00 - 0.5 = 8.87 \end{aligned} \tag{5.30}$$

Finally, increasing the sodium hydroxide concentration to 0.2 M results in a pH, as determined by eqn 5.27, of:

$$\mathrm{pH} = \mathrm{p}K_W + \log[\text{excess base}] = 14.0 - \log(0.2) = 12.7 \tag{5.31}$$

PROTONATION STATES OF AMINO ACID RESIDUES

When an amino acid is dissolved in water, it can exchange a proton with water, acting as either a weak base or a weak acid (Figure 5.5). The amino acid acts as a weak acid when the protonation state of the N-terminus changes:

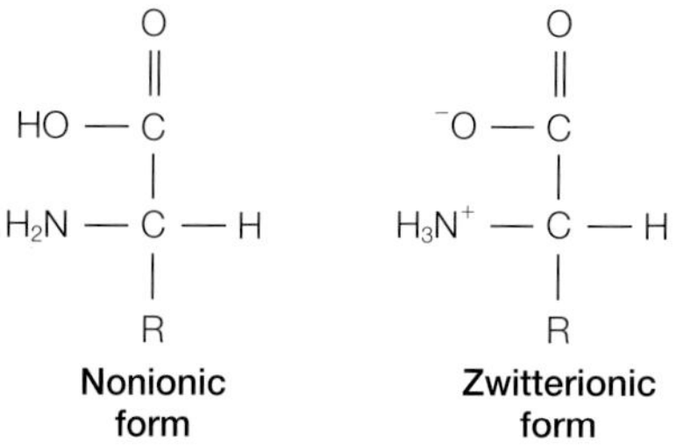

$$NH_3^+ \leftrightarrow NH_2 + H^+ \quad (5.32)$$

Or, it acts as a weak base when the C-terminus changes protonation state:

$$COO^- \leftrightarrow COOH \quad (5.33)$$

Figure 5.5 The nonionic and zwitterionic forms of amino acids.

Thus, there are at least two pK_A values, of around 2.35 and 9.60, associated with every amino acid. For an amino acid that is part of a polypeptide chain, the carboxyl and amino groups do not undergo changes in protonation state because they have formed the peptide bond. In a protein, only the side chains of the amino acids can change protonation with pH (except for the N- and C-termini of the chain). Of the 20 common amino acid residues, only seven have side chains that can be protonated, and the pK_A values range from 4.0 to 12 (Table 5.1). For example, the imidazole group of histidine has a pK_A of 6.5–7.5 and is charged when the pH is less than the pK_A (Figure 5.6).

In proteins, the pK_A values may shift dramatically compared to those seen for the isolated amino acids in solution. If an amino acid residue is buried inside of a protein in a very hydrophobic pocket, then the pK_A will shift to avoid the presence of a buried charge. For example,

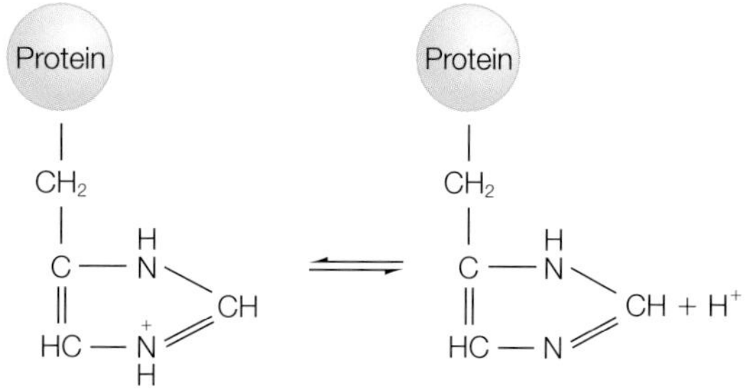

Figure 5.6 For amino acids in a polypeptide chain, only the side chains can undergo protonation changes, as shown for histidine.

Table 5.1
Typical pK_A values of the protonatable amino acid residues.

Amino acid residue	pK_A of side chain
Aspartic acid	4.0–5.0
Glutamic acid	4.0–5.0
Histidine	6.5–7.5
Cysteine	8.5–9.0
Tyrosine	9.5–10.5
Lysine	10–10.5
Arginine	12

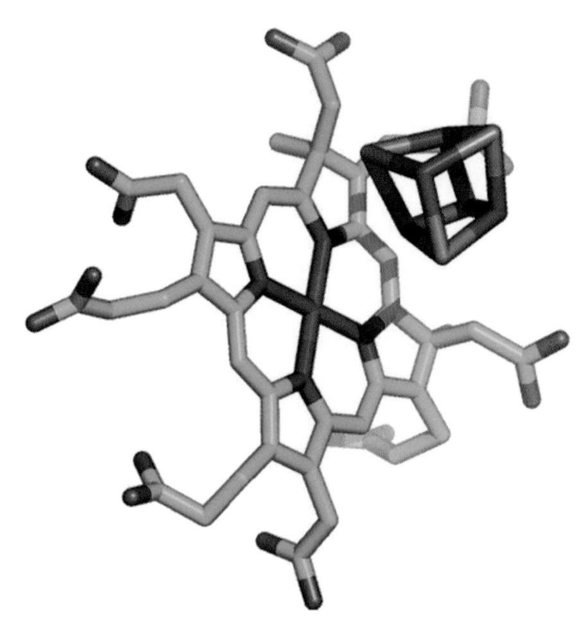

Figure 5.7 Structure of nitrite reductase showing the siroheme, iron–sulfur cluster, and seven ionizable amino acid residues. The substrate (not shown) binds to the iron of the siroheme.

glutamic acid has been found to have pK_A values shifted by several pH units to maintain its neutral form through interactions with its environment. The values will also shift if several ionizable amino acid residues are near each other. For example, inorganic nitrogen assimilation in oxygenic photosynthetic organisms has a key step involving the six-electron reduction of nitrite to ammonia, which is catalyzed by nitrite reductase. Nitrate binds to a special cofactor, a siroheme that is coupled to a four-iron–four-sulfur metal cluster (Figure 5.7). At this site of the enzyme there are seven ionizable amino acid residues, six arginine and one lysine. The site is buried within the protein and so it is likely that some of these amino acid residues are not ionized to avoid the presence of a significant charge in the protein. Although the precise mechanism is not known, as the reaction proceeds it is likely that the protonation states of the residues change in an interactive fashion due to electrostatic interactions involving the substrate, cofactor, and amino acid residues. Experimentally, the determination of the individual pK_A values of amino acid residues is difficult to determine, especially when the electrostatic interactions are complex.

BUFFERS

The titration curve shows a general feature of certain molecules, termed buffers. When the pH is near the pK_A of the buffer, changes in the amount of ions do not significantly alter the pH of the solution. In comparing this region of the curve for different buffers (Figure 5.8), several details are evident. The curves for the different buffers are determined by the Henderson–Hasselbalch equation (eqn 5.23) and have similar shapes that are shifted vertically along the pH axis. The pH corresponding to the midpoint of the curve – that is, when [HA] = [A$^-$] – always occurs at the pK_A of the buffer (as noted above). The slope of the curve near the midpoint is very shallow so that the pH of the solution is relatively insensitive to the addition of a strong base or acid.

Buffers are commonly used with biological samples to maintain the pH of a solution at a specific value. A number of compounds have been identified that can be used as buffers (Table 5.2). The ability of a buffer to maintain the pH is maximal at the pK_A of the buffer, with the useful range being within one pH unit of the pK_A value.

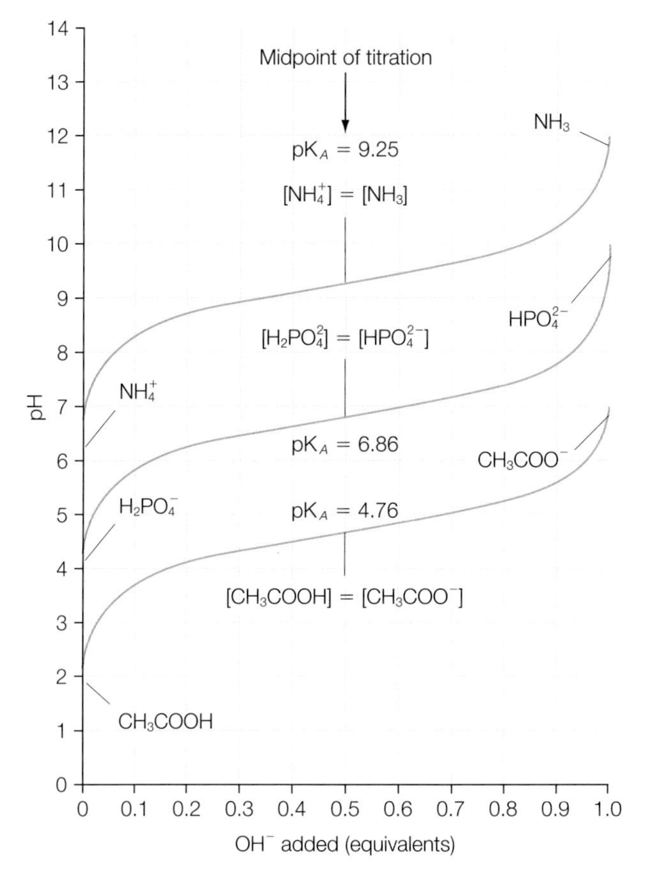

Figure 5.8 The titration curves for acetic acid, phosphate, and ammonia.

Table 5.2

A partial listing of some common buffers and their pK_A values at 20°C.

Buffer	pK_A
2-*N*-(morpholino)Ethanesulfonic acid (MES)	6.15
Piperazine-*N,N'*-bis(2-ethanesulfonic acid) (PIPES)	6.76
Imidazole	7.00
3-(*N*-morpholino)Propanesulfonic acid (MOPS)	7.20
Phosphate*	7.21
N-2-Hydroxyethylpiperazine-*N'*-2-ethanesulfonic acid (HEPES)	7.55
Tris(hydroxymethyl)aminomethane	8.08

*Phosphate buffers are normally prepared from a combination of monobasic and dibasic salts mixed together in proportion to yield the desired pH. Phosphoric acid has three pK_A values: 2.12, 6.86, and 12.32.

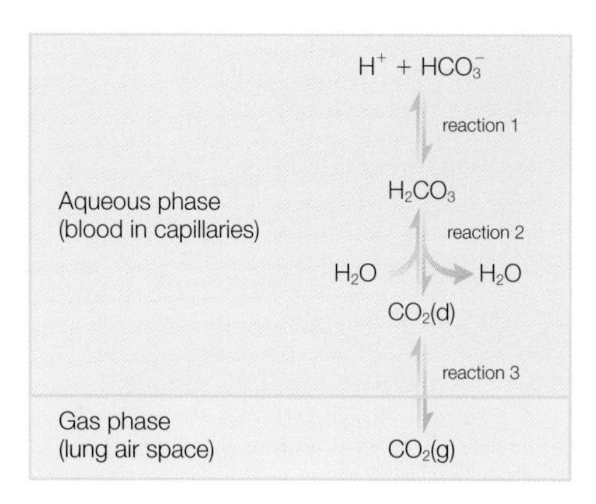

Figure 5.9 Carbon dioxide (CO_2) in the lungs can exchange with carbon dioxide dissolved in the bloodstream, which is in equilibrium with carbonic acid (H_2CO_3) and bicarbonate (HCO_3^-).

Buffering in the cardiovascular system

In animals with lungs, bicarbonate is an effective buffer found in the blood and maintains the pH near 7.4. The pH of blood is dependent upon two coupled reactions. First, the amount of gaseous carbon dioxide dissolved in the blood will equilibrate with water and produce carbonic acid (Figure 5.9). Second, there will be an equilibrium between carbonic acid and bicarbonate. The amount of carbon dioxide in the blood is coupled to the amount present in the lungs.

Vigorous exercise will produce lactic acid and increase the amount of H^+. In turn, the concentration of carbon dioxide in the blood increases, thus increasing the pressure of carbon dioxide in the lungs, causing the extra carbon dioxide to be exhaled. Conversely, when the pH of blood plasma is raised by a metabolic process the equilibrium shifts, causing more carbon dioxide to dissolve in the blood. Since the lungs have a large capacity for carbon dioxide, changes in breathing rapidly lead to compensations for changes in pH. Shifts of the pH of blood can lead to a condition known as acidosis, when the pH drops from the normal pH of 7.4 to 7.1, or alkalosis, when the pH rises to 7.6. Such conditions can arise for different reasons, such as severe diarrhea.

RESEARCH DIRECTION: PROTON-COUPLED ELECTRON TRANSFER AND PATHWAYS

Proton and electron transfer are two fundamental chemical processes in biological systems. In many cases, proton transfer is often coupled with electron transfer to maintain an overall balance in the charge. Three possible pathways are possible for proton-coupled electron transfer. First, the electron transfer can precede the proton transfer; second, the proton transfer can precede the electron transfer; or third, the two transfers can occur simultaneously or in a concerted fashion. Although the outcomes of the three pathways are identical, the dependencies of these three pathways on biochemical factors vary tremendously.

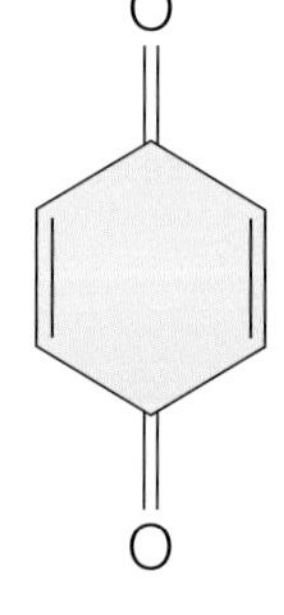

Figure 5.10 Structure of benzoquinone.

As an example, consider the properties of quinones as electron acceptors in proteins. Quinone molecules are six-membered carbon rings with different possible substitutents attached, the simplest molecule being benzoquinone (Figure 5.10). In photosynthetic organisms (see Chapter 20), quinone serves as an electron carrier between protein complexes, with light driving the reduction to quinol by coupling electron transfer with proton transfer:

$$Q + 2e^- + 2H^+ \leftrightarrow QH_2 \tag{5.34}$$

In photosynthetic complexes, this process involves the sequential transfer of one electron to a quinone acceptor, Q_B, forming a semiquinone. Subsequently, another electron is transferred in a process that leads to the transfer of two protons from the solution to the quinone. In both cases, the electron donor to Q_B is another quinone, Q_A. If the proton transfer precedes electron transfer, a protonated semiquinone intermediate state is involved (Figure 5.11). If electron transfer precedes proton transfer, the intermediate state is a doubly reduced quinone. If the transfer is concerted, then only one step takes place and no intermediate states are formed.

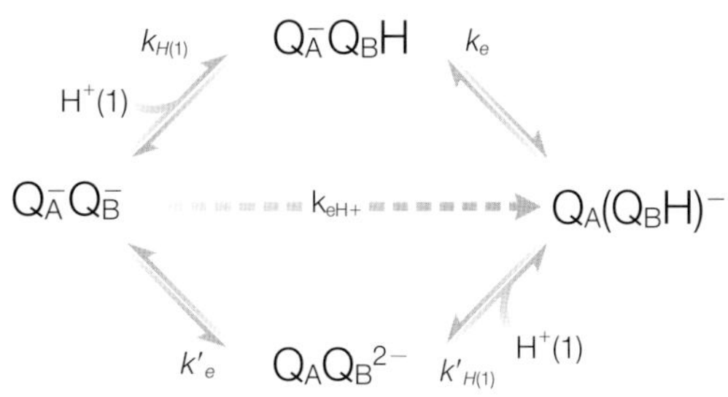

Figure 5.11 Possible reaction paths for the coupling of protons to the transfer of an electron to the quinone Q_B, which has previously been reduced to the state Q_B^-. The electron donor is a reduced quinone, Q_A^-, and the transfer of the electron may occur either along the upper path, in which case proton transfer precedes electron transfer, or along the lower path, in which case electron transfer precedes proton transfer. For a concerted mechanism, no intermediate is formed and the final state is formed directly. Modified from Graige et al. (1996).

In both of the two-step processes, the overall reaction rate will be limited by either proton or electron transfer. If proton transfer is first and rate-limiting, then the overall rate is given by the rate of proton transfer. However, if proton is rate-limiting but second, then the overall rate is given by the product of the proton-transfer rate and the fraction of the intermediate, namely the doubly reduced quinone. If electron transfer is first and rate-limiting, then the overall rate reflects the electron-transfer rate. However, if the electron transfer is second and rate-limiting, then the rate is proportional to the product of the electron-transfer rate and the amount of the intermediate, the protonated quinone. For the concerted transfer, the modeling involves a proton-dependent electron transfer with no intermediate.

Each of these possible combinations of rate-limiting steps and intermediates predicts a different rate dependence on the pH and free-energy dependence for the reaction. For the bacterial reaction center, data were consistent with a rate-limiting electron-transfer step followed by a rapid proton transfer, although the concerted model was also possible (Graige et al. 1996). The coupling of proton and electron transfer occurs for many other biological electron-transfer cofactors. For example, the oxidation of tyrosine is always coupled with release of the phenolic proton, resulting in a neutral tyrosyl radical. Even in complex multi-electron enzymatic reactions, reductive or oxidative steps are usually found to have associated proton-transfer steps. For example, proton and electron transport are coupled in cytochrome *c* oxidase, a protein complex involved in respiration (see Chapter 6), and energetic calculations are most consistent with a proton-transport process that acts in concert with electron transport and is controlled by a protonatable amino acid residue (Glu-286; Olsson et al. 2005).

The coupling of electron and proton transfer requires that proteins have a means to store and donate, or accept, protons as a reaction occurs.

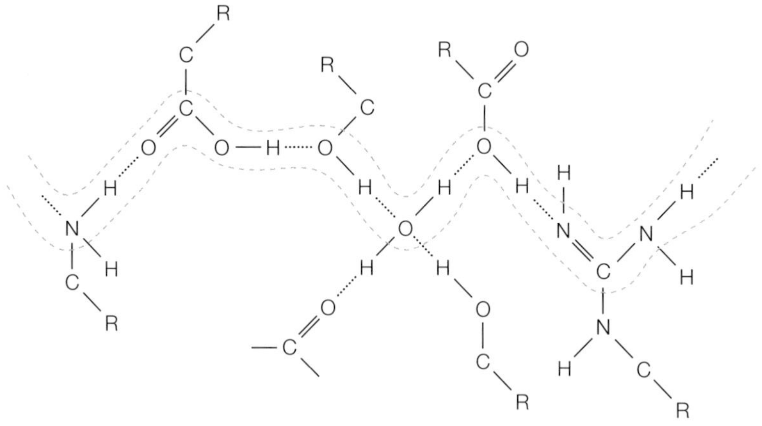

Figure 5.12 Proton pathway formed by a hydrogen-bonded chain of amino acid residues and bound water molecules.

In such cases a protonatable amino acid residue is always found near the active site to serve as the donor or acceptor of the proton. Enzymatic reactions require repeated reactions so proteins must also have a means of transporting protons from the bulk solution to the amino acids at the active site. Proton conduction usually occurs by the transfer of the proton through a pathway formed by a series of hydrogen-bonded amino acid residues (Figure 5.12). By such pathways, protons are carried rapidly through proteins up to distances of 20 Å. The amino acid side chains participating in the pathway must serve as first a proton acceptor and then donor. The pK_A values must not be too high, and residues with moderate or low pK_A values are generally favored. The protons are generally thought to travel in only very short distances (see Chapter 13), so the pathway must be unbroken. Any breaks may be compensated for by rotations of the side chains or the presence of bound water molecules, as found in bacteriorhodopsin (Chapter 17). The precise proton pathway in a protein can be identified by systematic mutation of possible participants in such a pathway (Paddock et al. 2003). For example, in the bacterial reaction center, the protonation of quinone was found to occur by the transfer of protons through pathways involving two different sets of amino acid residues (Figure 5.13).

Whereas the traditional view of proton transfer involving a well-defined pathway is appropriate for many enzymes, some research groups have suggested that in some enzymes additional mechanisms must be considered. Consideration of potential proton-conducting pathways in a static model of a protein may not be sufficient to detect possible pathways. Dynamics must also be considered because random fluctuations of amino acid residues found on the surface of an enzyme can result in carboxylates either sharing a common proton through a hydrogen bond or being connected through water molecules, thus forming the desired proton pathway, as has been found in cytochrome P450cam (Benkovic & Hammes-Schiffer 2003, 2006;

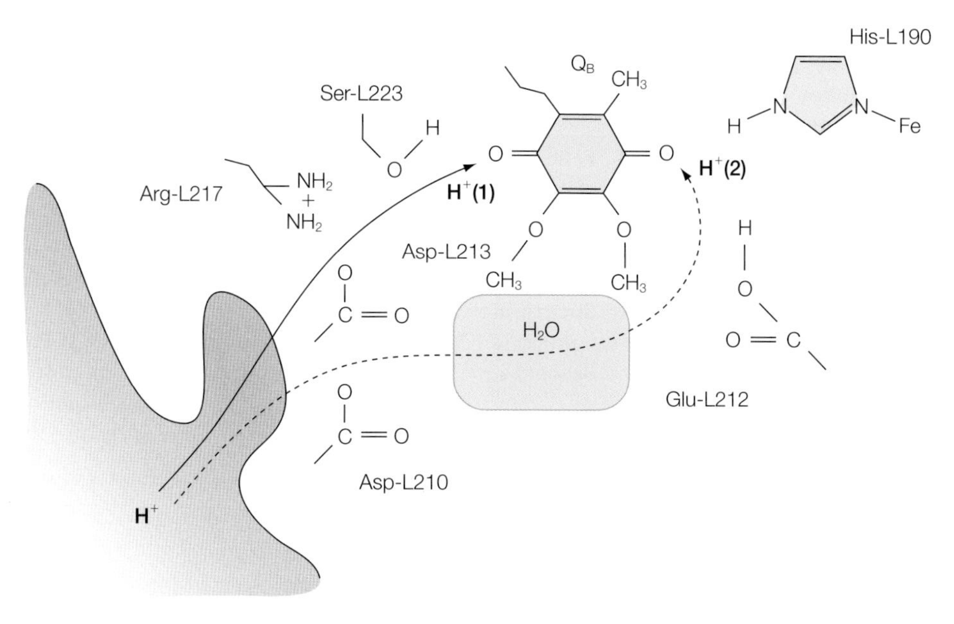

Figure 5.13 Proton-transfer pathways to the quinone in bacterial reaction centers. Modified from Paddock et al. (2003).

Taraphder & Hummer 2003; Friedman et al. 2005; see Chapter 7). The traditional pathways are viewed as requiring proton donors and acceptors to be within 2 Å for the proton to be transferred. However, hydrogen has been proposed to be capable of tunneling (see Chapter 10), which would enable the protons to transverse longer distances (Cha et al. 1989; Liang & Klinman 2004). Electrons have long been known to travel over long distances in proteins (Chapter 10) but protons and hydrogen had been thought to be not capable of such travel due to their much larger size. Efforts are underway to delineate the combined efforts of theoreticians and experimentalists to elucidate the factors that would control proton tunneling and to probe the coupling of tunneling with dynamics (Francisco et al. 2002; Tanner et al. 2003; Garcia-Viloca et al. 2004).

REFERENCES

Benkovic, S.J. and Hammes-Schiffer, S. (2003) A perspective on enzyme catalysis. *Science* **301**, 1196–1202.

Benkovic, S.J. and Hammes-Schiffer, S. (2006) Enzyme motions inside and out. *Science* **312**, 208–9.

Cha, Y., Murray, C.J., and Klinman, J.P. (1989) Hydrogen tunneling in enzyme reactions. *Science* **243**, 1325–30.

Francisco, W.A., Knapp, M.J., Blackburn, N.J., and Klinman, J.P. (2002) Hydrogen tunneling in peptidlyglycine alpha-hydroxylating monooxygenase. *Journal of the American Chemical Society* **124**, 8194–5.

Friedman, R., Nachliel, E., and Gutman, M. (2005) Application of classical molecular dynamics for evaluation of proton transfer mechanism on a protein. *Biochimica Biophysica Acta* **1710**, 67–77.

Garcia-Viloca, M., Gao, J., Karplus, M., and Truhlar, D.G. (2004) How enzymes work: analysis by modern rate theory and computer simulations. *Science* **303**, 186–95.

Graige, M.S., Paddock, M.L., Bruce, J.M., Feher, G., and Okamura, M.Y. (1996) Mechanism of proton-coupled electron transfer for quinone (Q_B) reduction in reaction centers of *Rb. sphaeroides*. *Journal of the American Chemical Society* **118**, 9005–16.

Liang, Z.X. and Klinman, J.P. (2004) Structural bases of hydrogen tunneling in enzymes: progress and puzzles. *Current Opinion in Structural Biology* **14**, 648–55.

Olsson, M.H.M., Sharma, P.K., and Warshel, A. (2005) Simulating redox coupled proteon transfer in cytochrome c oxidase: looking for the proton bottleneck. *FEBS Letters* **579**, 2026–34.

Paddock, M.L., Feher, G., and Okamura, M.Y. (2003) Proton transfer pathways and mechanism in bacterial reaction centers. *FEBS Letters* **555**, 45–50.

Tanner, C., Manca, C., and Leutwyler, S. (2003) Probing the threshold to H atom transfer along a hydrogen bonded ammonia wire. *Science* **302**, 1736–9.

Taraphder, S. and Hummer, G. (2003) Protein side-chain motion and hydration in proton-transfer pathways. Results for cytochrome P450cam. *Journal of the American Chemical Society* **125**, 3931–40.

PROBLEMS

5.1 At equilibrium, what can be said about the Gibbs energy?

5.2 For a spontaneous process at constant pressure, will the change in Gibbs energy be positive, negative, or zero?

5.3 For a process with two components, what can be said about the Gibbs energy when the mole fractions are equal?

5.4 If the equilibrium constant is strongly temperature-dependent, what can be said about the relative entropic and enthalpic contributions to the Gibbs energy?

5.5 If the equilibrium constant is independent of temperature, what can be said about the relative entropic and enthalpic contributions to the Gibbs energy?

5.6 In the reaction $CH_3COO^- + H_2O \leftrightarrow CH_3COOH + OH^-$ state whether each species is acting as a base, acid, neither, or both.

5.7 Calculate the pH for a solution with an H_3O^+ molar concentration of (a) 10^{-5} M and (b) 10^{-9} M.

5.8 The equilibrium constant for a particular electrochemical reaction is 0.5. Is the standard-reaction Gibbs energy greater than zero, less than zero, zero, or undetermined?

5.9 At equilibrium, for the reaction A $\leftrightarrow$ B, the final concentration of A is twice that of B. What can be said about the Gibbs energy for this reaction at equilibrium?

5.10 After 100 mL of 0.1 M NaOH is added to 900 mL of a solution containing 0.03 mol of acetic acid (CH_3COOH, $pK_A = 4.75$) and 0.02 mol of NaCl, what is the pH of the final solution?

5.11 Acetic acid is a weak acid with a pK_A of 4.75. What is the pH of a solution prepared by adding 0.1 mol of acetic acid and 0.03 mol of NaOH to water at a final volume of 1 L?

5.12 After adding 0.01 M HCl to distilled water that is initially at pH 7, what is the final pH?

5.13 An isoelectric pH is the pH at which the molecule has no net average charge. What is the isoelectric pH of glycine?

5.14 A 25-mL solution has a strong base at 0.15 M. What is the initial pH?

5.15 The buffer Tris, or tris(hydroxymethyl)aminomethane, has a pK_A of 8.3. At what pH would this buffer have equal molar concentrations of Tris and its conjugate base?

5.16 Why is electron transfer often coupled to proton transfer?

5.17 If formation of quinol is independent of the Gibbs energy difference, what can be said about the rates of proton and electron transfer?
5.18 What are the different protonation states of tyrosine?
5.19 Why is release of the phenolic proton coupled to tyrosyl formation?
5.20 How are protons traditionally thought to be transferred long distances in proteins?
5.21 Experimental data support what type of proton-transfer reactions?

6

Oxidation/reduction reactions and bioenergetics

The previous chapters have dealt with the principles of thermodynamics and the involvement of protons in acid/base reactions. In this chapter the focus is on reduction/oxidation reactions, or redox reactions. Redox reactions are significant for many fundamental biological processes involving the transfer of electrons from one species to another. The Nernst equation provides a thermodynamic description of these reactions with the oxidation/reduction midpoint potential being a useful parameter, characteristic of a redox-active species. Redox reactions lead to the presence of charged species. The properties of ions in solution can be highly dependent upon the properties of the solvent molecules. The interactions of ions with the molecules of the solution are described in terms of their Gibbs energy of formation and a new parameter, the activity coefficient. The critical role of such processes in biological systems is illustrated for two different biological applications, the chemiosmotic hypothesis and respiration.

OXIDATION/REDUCTION REACTIONS

The transfer of an electron from one charged species to another species is central to many biological reactions. Proteins can carry electrons and usually possess cofactors that can be oxidized or reduced. The process of electron transfer is usually broken into two reactions, called half reactions, that are not real reactions but are used for bookkeeping. For example, consider a reaction in which a molecule X accepts an electron, producing a reduced species, X^-:

$$X + e^- \rightarrow X^- \tag{6.1}$$

Since the electron must come from somewhere, there must be an electron donor, Y, producing an oxidized species, Y^+:

$$Y \rightarrow Y^+ + e^- \quad (6.2)$$

The actual reaction is written by summing these two equations:

$$X + e^- + Y \rightarrow X^- + Y^+ + e^- \quad \text{or} \quad X + Y \rightarrow X^- + Y^+ \quad (6.3)$$

Reduction/oxidation reactions are similar to acid/base reactions. In both cases a charge, namely an electron or proton, is transferred. For acids and bases the parameter pK_A describes the affinity of a molecule for a proton. The equivalent parameter for oxidation/reduction reactions is the *midpoint potential*; that is, the potential at which half of a molecule is oxidized.

ELECTROCHEMICAL CELLS

Electrochemical reactions are often driven in electrochemical cells, such as a battery. An electrochemical cell is a system that has two electrodes (Figure 6.1). These electrodes are connected electrically with an electrolyte – the solution where the oxidation/reduction reaction occurs. As the reaction proceeds, electrons are released and enter the anode. Electrons re-enter the cell through the second electrode, termed the cathode. The electrodes are either in the same solution or connected by a solution that is termed a salt bridge. In a Galvanic cell, a spontaneous reaction generates electricity, whereas an electrolytic cell is one that needs an external voltage to drive a non-spontaneous reaction.

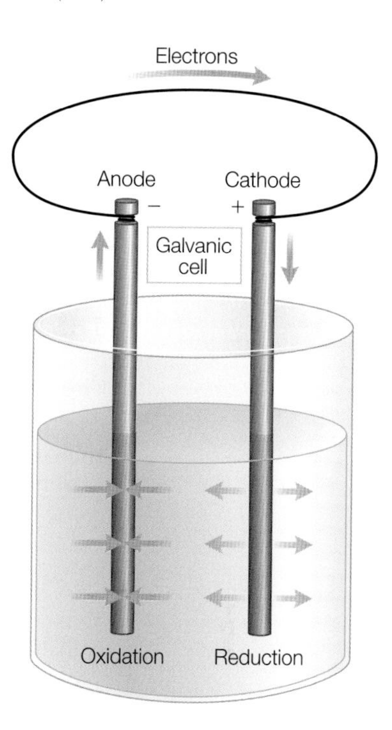

Figure 6.1
A galvanic cell with electrons being transferred from solution to the anode and traveling from the anode to the cathode, where the reduction occurs.

In order to calculate the energy of an electrochemical cell, a standard must be chosen in order to address the pairing of ions. The standard chosen may vary, but the most common one is the hydrogen electrode. This reaction involves the transfer of two electrons to two protons in solution, forming hydrogen gas:

$$2H^+ \text{ (aq)} + 2e^- \rightarrow H_2 \text{ (gas)} \quad (\Delta G)^\circ_{rec} = 0 \quad (6.4)$$

As a standard, the *Gibbs energy of reaction*, $\Delta G^\circ_{reaction}$, is defined as zero. With this definition, it is possible to determine relative reaction Gibbs energies for any half reaction.

In electrical circuits, energies are usually described in terms of voltages rather than Gibbs energies. The unit of voltage, namely a Volt (V), is a Joule per Coulomb ($J\ C^{-1}$) and so voltage is a measure of the energy required to move the charge. A Coulomb is a certain number of electron charges, 6.24×10^{18}. Voltage has similarities to a molar Gibbs energy (the amount of free energy it takes to convert 1 mol of a substance from one form or state to another). Therefore, it should follow that the *potential, E,* is proportional to the Gibbs energy of reaction according to:

$$(\Delta G)_{rec} = -nFE \tag{6.5}$$

where n is the number of electrons involved in the oxidation/reduction reaction and F is a proportionality constant called the *Faraday constant.* The Faraday constant provides the conversion from the Gibbs energy, which is a value on a per-mole basis, to voltage, which is on a per-Coulomb basis. The value of the Faraday constant is then given by:

$$F = \frac{6.02 \times 10^{23}\ \text{mol}^{-1}}{6.24 \times 10^{18}\ \text{C}^{-1}} = 9.65 \times 10^{4}\ \text{C mol}^{-1} \tag{6.6}$$

THE NERNST EQUATION

The Gibbs energy difference can be related to the equilibrium constant, or equivalently to the ratio of the product, namely the concentration of the oxidized species, A_{oxidized}, and the reactant, or reduced species, A_{reduced} (see Chapter 5):

$$(\Delta G)_{rec} = (\Delta G)^{\circ}_{rec} + RT \ln \frac{[A_{\text{oxidized}}]}{[A_{\text{reduced}}]} \tag{6.7}$$

Substituting the relationship between the Gibbs energy and the voltage (eqn 6.5) yields:

$$-nFE = -nFE^{0} + RT \ln \frac{[A_{\text{oxidized}}]}{[A_{\text{reduced}}]} \tag{6.8}$$

$$E = E^{0} - \frac{RT}{nF} \ln \frac{[A_{\text{oxidized}}]}{[A_{\text{reduced}}]}$$

This final equation is called the Nernst equation, named after Walther Nernst who won the Nobel Prize in Chemistry in 1920. In a sense, it proves a measure for whether a molecule has an electron, just as the pK_A provides a measure for a proton. The equation can be used to calculate the potential, given the midpoint potential and the concentrations of the reactants and products. A useful number to remember is that when the equilibrium constant is increased 10-fold, the potential changes by:

$$E - E^{0} = -\frac{(8.314\ \text{J/(K mol)} \times 298\ \text{K}}{(1)(9.65 \times 10^{4}\ \text{C mol}^{-1})} \ln(10) = 0.0592\ \text{J C}^{-1} = 59.2\ \text{mV} \tag{6.9}$$

Table 6.1
Oxidation/reduction midpoint potentials for some biological reactions.

Redox reaction	E_m (V)
$P680^+ + e^- \leftrightarrow P680$	~1.1
$O_2 + 4H^+ + 4e^- \leftrightarrow 2H_2O$	0.82
$Chl^+ + e^- \leftrightarrow Chl$	0.78
$Bchl^+ + e^- \leftrightarrow Bchl$	0.64
$P870^+ + e^- \leftrightarrow P870$	0.50
Cyt *f* (Fe^{3+}) + e^- $\leftrightarrow$ Cyt *f* (Fe^{2+})	0.37
Cyt *c* (Fe^{3+}) + e^- $\leftrightarrow$ Cyt *c* (Fe^{2+})	0.25
$Q + 2e^- + 2H^+ \leftrightarrow QH_2$	0.06
$2H^+$ (aq) + $2e^-$ $\rightarrow$ H_2 (gas) (standard conditions, pH 0)	0.00
$FMN_{oxidized} + 2e^- + 2H^+ \leftrightarrow FMNH_2$	−0.21
$FAD + 2e^- + 2H^+ \leftrightarrow FADH_2$	−0.22
Glutathione + $2e^-$ + $2H^+$ $\leftrightarrow$ 2 $glutathione_{reduced}$	−0.23
$NAD^+ + 2e^- + H^+ \leftrightarrow NADH$	−0.32
$2H^+$ (aq) + $2e^-$ $\rightarrow$ H_2 (gas) (pH 7)	−0.42
Ferredoxin (Fe^{3+}) + e^- $\leftrightarrow$ ferredoxin (Fe^{2+})	−0.43

Bchl, bacteriochlorophyll; Chl, chlorophyll; Cyt, cytochrome; FAD, flavin adenine dinucleotide; FMN, flavin mononucleotide; NAD, nicotiamide adenine dinucleotide; P680, primary donor of photosystem II; P870, primary donor of bacterial reaction centers; Q, ubiquinone.

MIDPOINT POTENTIALS

Notice that when the concentrations of the oxidized and reduced species are equal, the potential E is equal to E^0, so this constant is usually termed the oxidation/reduction midpoint potential, E_m. Many biological half reactions involve protons so the midpoint potential is defined at a pH of 7. The midpoint potentials for biological molecules have been tabulated (Table 6.1) and provide the basis for electron-transfer reactions, as discussed in the next section. In general, increasing midpoint potentials correspond to a greater affinity for an electron and hence increasing oxidation capability.

The potentials measured for biologically relevant oxidation/reduction reactions cover a very large range. The primary donor of photosystem II, P680, is the most oxidizing cofactor found in biology. The potential of P680 is high enough even to oxidize water. Notice that the midpoint potential of P680 is larger than for its chemical component, chlorophyll *a*, in solution and much larger than the potential of the corresponding bacterial electron donor, P870 (see Chapter 20). Cytochromes are proteins with hemes as cofactors that serve as electron carriers, as for cytochrome *c*, or as membrane proteins that are part of electron-transfer chains, such as cytochrome *f*, which is part of the cytochrome b_6f complex. Ubiquinones serve as electron acceptors in different protein complexes, including the bacterial reaction center (Chapter 5). Notice that hydrogen has a zero midpoint potential since it serves as a standard, but only under standard conditions. At pH 7, the midpoint potential is decreased to 0.42 V due to the 0.059 decrease per pH unit expected for reactions coupled to proton transfer. Ferredoxin is a small protein that contains an iron–sulfur cluster that becomes oxidized or reduced during different metabolic processes. In some cases, enzymes that catalyze oxidation/reduction reactions transfer electrons into universal electron carriers. Several compounds have low midpoint potentials and so serve as good electron carriers, including flavin mononucleotide (FMN), flavin adenine dinucleotide (FAD), and glutathione. In some cases the electron carrier readily moves between enzymes, as found for NAD^+ and $NADP^+$, whereas in other cases the cofactor is tightly bound, as usually found for FMN and FAD. As Table 6.1 indicates, many but not all proteins that participate in oxidation/reduction reactions contain metals that serve as the electron donor or acceptor.

In many biological reactions, the oxidation/reduction reaction involves the transfer of two electrons and two protons. Such reactions are termed dehydrogenations and the enzymes that catalyze them are called dehydrogenases. For example, the conversion of lactate to pyruvate involves the removal of two protons from the ketone group at the second carbon position in addition to the removal of two electrons (Figure 6.2). The net transfer of two protons with two electrons is common but not required. For example, the oxidation of NAD involves the release of two protons in the dehydrogenation reaction (Figure 6.3). One of these protons is released into the aqueous solution but the oxidized form of the molecule accepts a hydride ion, yielding a net release of one proton.

The values reported in Table 6.1 have been determined experimentally by one of two means. One approach is to poise the ambient potential at a series of values with the use of chemical reductants and oxidants (Figure 6.4). Alternatively, the potential can be established using an electrochemical cell. For each potential, the oxidation state of a particular cofactor must then be measured by spectroscopic means, such as by monitoring the changes in an optical absorption spectrum. From the spectra, the fraction reduced at each potential is fitted using the Nernst equation (eqn 6.8) and the midpoint potential is determined. Since the cofactors in proteins are usually buried inside the protein, special mediator compounds for these measurements may be used to facilitate the transfer of electrons between the electrodes and the cofactor.

In addition to factors such the pH and ionic strength of the solution surrounding the protein, the midpoint potential of a cofactor in a protein can differ by up to 0.5 V compared to the value in solution due to cofactor–protein interactions. The most critical factor is the ligation of the cofactor that will preferentially stabilize the reduced or

Figure 6.2
The oxidation of lactate to pyruvate requires the loss of two protons.

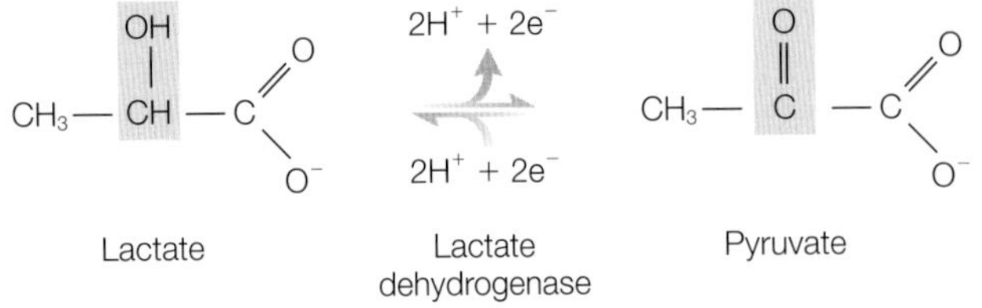

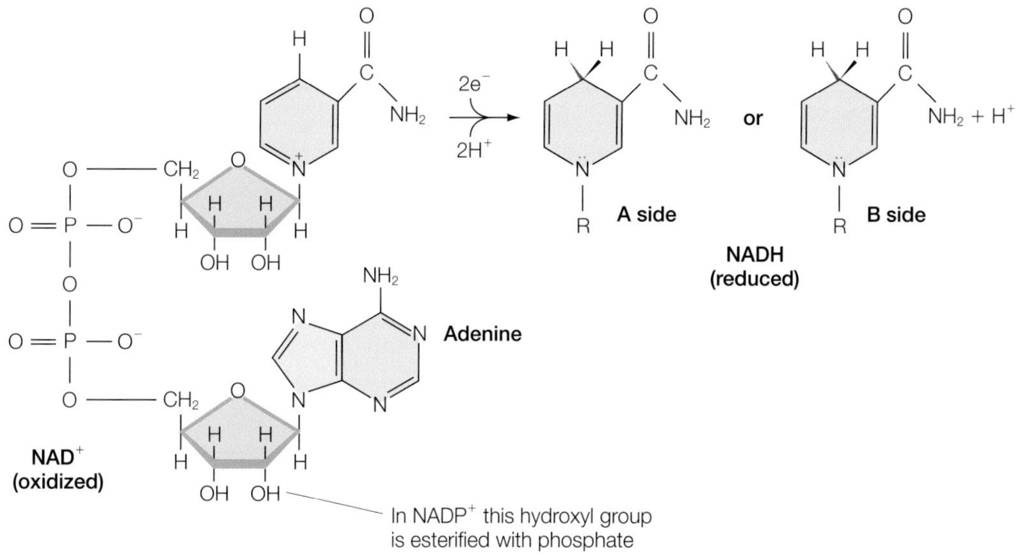

$2e^-$, $2H^+$

A side

or

B side

$NH_2 + H^+$

NADH (reduced)

Adenine

NAD^+ (oxidized)

In $NADP^+$ this hydroxyl group is esterified with phosphate

Figure 6.3 The oxidation of NAD^+ to NADH is a two-electron process with the release of only one proton.

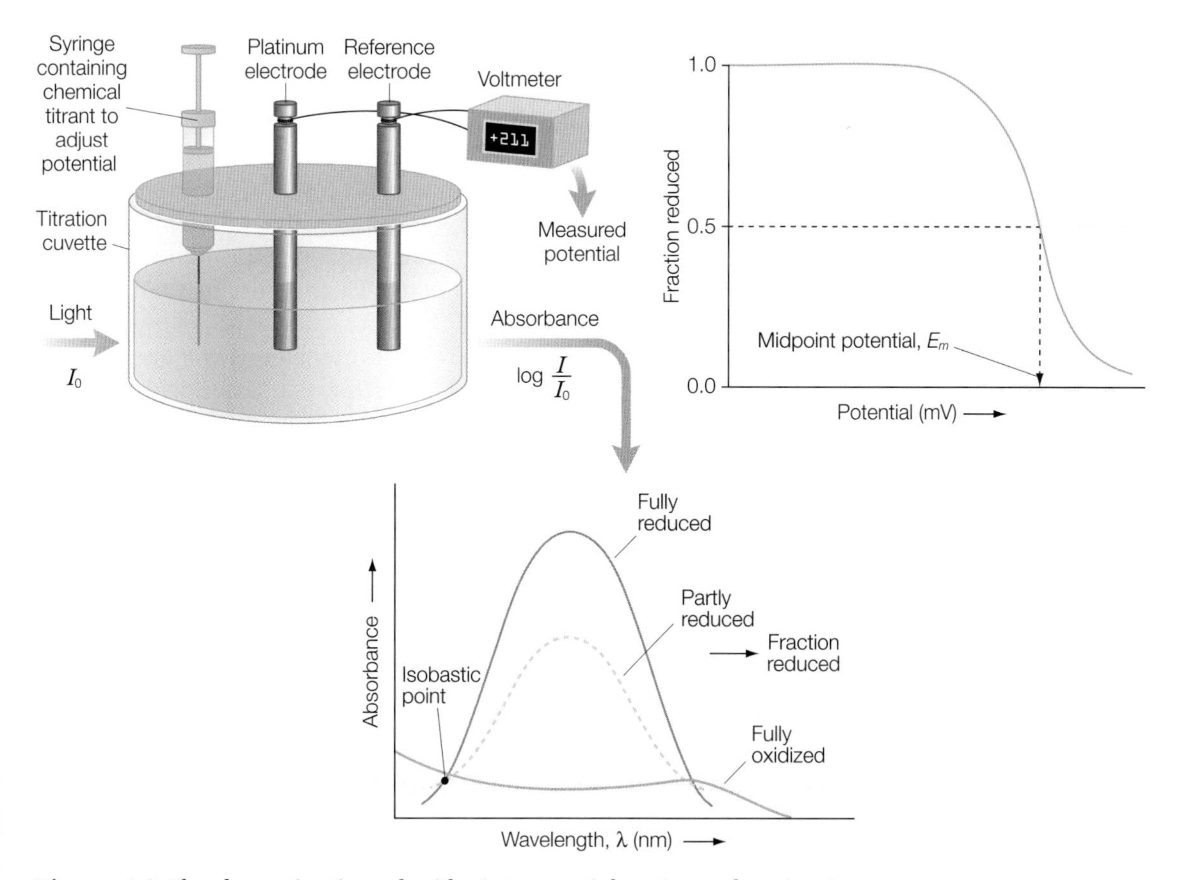

Figure 6.4 The determination of midpoint potentials using redox titrations.

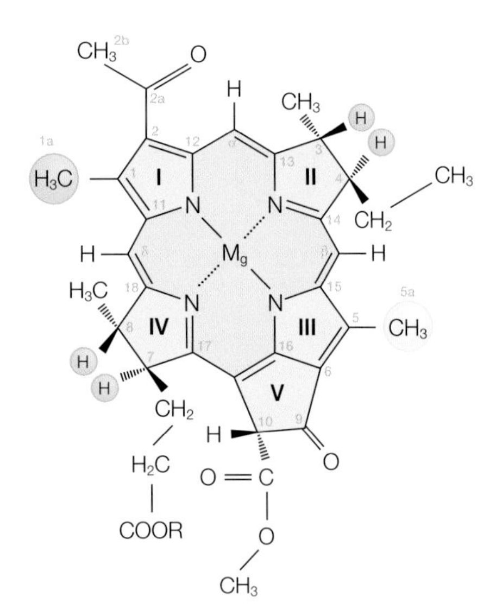

Figure 6.5 The structure of bacteriochlorophyll *a* found in the bacterial reaction center.

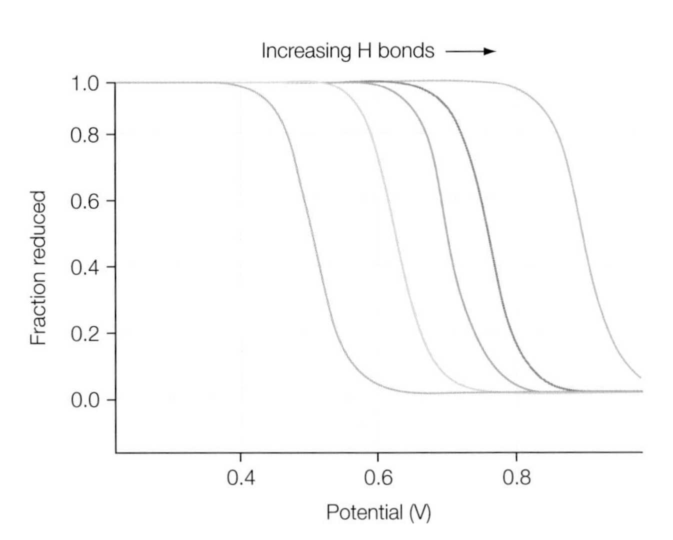

Figure 6.6 Redox titrations of the bacterial reaction center showing systematic increases in the midpoint potential due to the addition of hydrogen bonds.

oxidized state. For example, the iron in heme has two axial ligands (Chapters 15 and 16). A more basic axial ligand is a better electron donor with a greater attraction for Fe^{3+}. Thus a more basic ligand will stabilize the oxidized state and lower the midpoint potential. Cytochrome hemes with two axial ligands usually have a more negative midpoint potential than hemes with one methionine ligand and one histidine since the imidazole side chain of histidine is a better electron donor than the side chain of methonine.

Hydrogen bonding and other electrostatic interactions will also systematically alter the midpoint potential of a cofactor. For bacteriochlorophylls, there are two carbonyl oxygens that are part of the conjugated ring and serve as acceptors of hydrogen bonds from the surrounding protein (Figure 6.5). As proton donors were introduced in the hydrogen-bonding position, the midpoint potential was found to increase (Figure 6.6). By performing electron nuclear double resonance (ENDOR) measurements (Chapter 16), the distribution of electrons over macrocycles was determined and the increase in midpoint potential could be explained by a Hückel model (Chapter 13), with changes in midpoint as a result of stabilization of the reduced state due to the hydrogen-bonding interactions.

GIBBS ENERGY OF FORMATION AND ACTIVITY

When a salt such as sodium chloride is present in solution, the salt dissolves and the sodium and chloride ions separate due to the resulting increase in entropy. Although salt will dissolve in polar solutions such as water, it will not dissolve in other liquids such as oil (a nonpolar solution), because there are interactions that stabilize the ions in a soluble form. For example, water is very polar and can always align itself in such a way that the more positive hydrogen ions are pointing at the chloride ions and the more negative oxygen ions are pointing at the sodium ions (Figure 6.7). This realignment of the water dipoles lowers the Gibbs energy of the system, making it possible for the chloride anions and sodium cations to separate in a stable configuration in solution. Due to the electrostatic interactions among the ions, the solutions have nonideal properties, as ions of opposite charge will be attracted to each other, resulting in clusters of an ion surrounded by ions of the opposite charge. These interactions were modeled by Peter Debye and Erich Hückel in 1923 in what is now described as the *Debye–Hückel theory* of ionic solutions. These interactions are sometimes parameterized by the use of activity coefficients that decrease from the ideal value of one due to nonideal behavior.

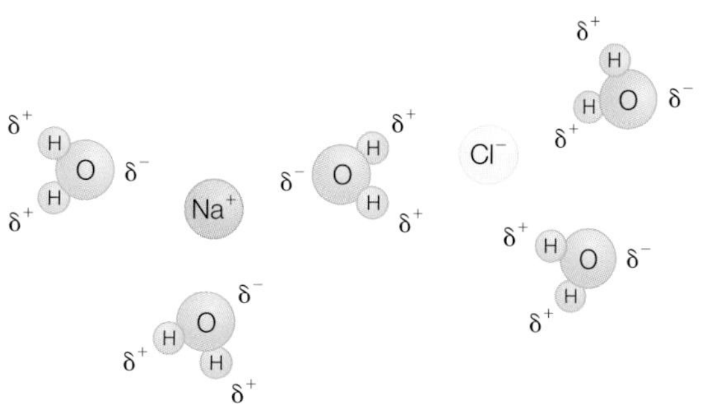

Figure 6.7 In water, salt will dissolve and the resulting Na^+ and Cl^- ions will be stabilized by the rearrangements of the orientations of the dipoles of the surrounding water molecules.

Any experimental evaluation of the thermodynamic properties of an ion will necessarily involve consideration that the solution contains both the ion and the accompanying cation or anion in solution. Although a property, such as the Gibbs energy, for a pair of ions can be determined experimentally, application of the observed results usually requires determination of the property for only one of the ions. To separate the two contributions, the convention is adopted of using the hydrogen ion as a standard with zero standard enthalpy and Gibbs energy at all temperatures. The enthalpy and Gibbs energy for other ions are then defined relative to the hydrogen ion. For example, when hydrogen chloride is dissolved in water, hydrogen and chloride ions are formed with a change ΔG_f° in the *Gibbs energy of formation*:

$$\frac{1}{2}H_2 \text{ (gas)} + \frac{1}{2}Cl_2 \text{ (gas)} \leftrightarrow H^+ \text{ (aqueous)} + Cl^- \text{ (aqueous)} \tag{6.10}$$

$$\Delta G^\circ(HCl) = -131.23 \text{ kJ mol}^{-1}$$

By convention, the Gibbs energy of formation for the hydrogen ion is zero; hence the Gibbs energy for hydrogen chloride is equal to that for the chloride alone:

$$\Delta G^\circ(HCl) = \Delta G_f^\circ(H^+) + \Delta G_f^\circ(Cl^-) = \Delta G_f^\circ(Cl^-) = -131.23 \text{ kJ mol}^{-1} \tag{6.11}$$

$$\text{since} \quad \Delta G_f^\circ(H^+) = 0$$

Knowing $\Delta G_f^\circ(Cl^-)$, the Gibbs energy of formation of sodium, $\Delta G_f^\circ(Na^+)$, can be determined by dissolving sodium chloride in solution, forming Na^+

and Cl^-. The contribution of the chloride to the measured value for sodium chloride can be subtracted resulting in the Gibbs energy of formation of sodium alone:

$$\text{Na (solid)} + \frac{1}{2}\text{Cl}_2\text{ (gas)} \leftrightarrow \text{Na}^+\text{ (aqueous)} + \text{Cl}^-\text{ (aqueous)}$$

$$\Delta G_f^\circ(\text{NaCl}) = -393.1\text{ kJ mol}^{-1} \tag{6.12}$$

$$\begin{aligned}\Delta G_f^\circ(\text{Na}^+) &= \Delta G_f^\circ(\text{NaCl}) - \Delta G_f^\circ(\text{Cl}^-)\\ &= -393.1\text{ kJ mol}^{-1} - (-131.23\text{ kJ mol}^{-1}) = -261.9\text{ kJ mol}^{-1}\end{aligned}$$

By this type of analysis, the Gibbs energy of formulation can be determined for different ions. Likewise, the enthalpies and entropies of formation can be determined using hydrogen as the standard.

IONIC STRENGTH

The properties of solutions containing ions are influenced by both the concentrations of the ions and the charges of the ions. A solution of sodium chloride will contain monovalent sodium and chloride ions, while a solution of calcium chloride will have the divalent calcium ion as well as the monovalent chloride ion. The ionic strength is a widely used dimensionless parameter that provides a measure of both of these aspects. The *ionic strength*, I, is determined by all of the ions in solutions and so is given by the sum of the product of the *molarity*, m_i, and the *charge*, z_i, of all ith ions:

$$I = \frac{1}{2}\sum_i m_i z_i^2 \tag{6.13}$$

As an example, the ionic strength of a 100 mM solution of sodium chloride, NaCl, is determined by combining the contributions of the individual ions that are each at a 100 mM concentration, or equivalently a 0.1 molarity:

$$I = \frac{1}{2}m_{\text{Na}^+}z_{\text{Na}^+}^2 + \frac{1}{2}m_{\text{Cl}^-}z_{\text{Cl}^-}^2 = \frac{1}{2}(0.1)(1)^2 + \frac{1}{2}(0.1)(-1)^2 = 0.1 \tag{6.14}$$

The ionic strength of 100 mM magnesium chloride, $MgCl_2$, is different than 100 mM NaCl as the magnesium chloride will dissolve into divalent magnesium ions, with monovalent chloride ions being at twice the molarity of the magnesium:

$$I = \frac{1}{2}m_{\text{Mg}^{2+}}z_{\text{Mg}^{2+}}^2 + \frac{1}{2}m_{\text{Cl}^-}z_{\text{Cl}^-}^2 = \frac{1}{2}(0.1)(2)^2 + \frac{1}{2}(0.2)(-1)^2 = 0.3 \tag{6.15}$$

When ions from different salts are dissolved together, the ionic strength is calculated by combining all of the contributions of the ions.

ADENOSINE TRIPHOSPHATE

In living organisms, the energy released by the oxidation of nutrients is stored as *adenosine triphosphate* (ATP; Figure 6.8). The history of ATP and its utilization in living cells has been fairly controversial at times. The ATP molecule was first isolated in 1929 and biochemical studies in the 1940s showed that ATP was utilized in oxidative phosphorylation. The usefulness of ATP lies in its ability to convert to *adenosine diphosphate* (ADP), with the loss of the terminal phosphate through hydrolysis, producing *inorganic phosphate* (P_i):

$$ATP + H_2O \leftrightarrow ADP + P_i + H_3O^+ \qquad (6.16)$$

The reaction is strongly exothermic, with a 30.5 kJ mol^{-1} release of energy under normal biological conditions. The ADP–phosphate bond is sometimes termed a high-energy bond to identify the strong tendency to undergo the reaction. This term is not meant to imply that the bonds are unstable, as they represent normal chemical bonds with a bond dissociation energy of about 500 kJ mol^{-1}. Rather, the bond is described as a high-energy bond because chemical reactions in which the bonds are hydrolyzed by water have a negative energy change, making the process spontaneous. As an example, the breakdown of foods begins with the oxidation of glucose into glucose 6-phosphate in an endothermic process (see Chapter 2). This endothermic process, with $\Delta G° = +13.8$ kJ mol^{-1}, can be driven by coupling it to the highly exergonic ATP reaction, resulting in a favorable overall Gibbs energy change:

$$\text{Glucose} + ATP \leftrightarrow \text{glucose 6-phosphate} + ATP \qquad (6.17)$$
$$\Delta G° = 13.8 - 30.5 \text{ kJ mol}^{-1} = -16.7 \text{ kJ mol}^{-1}$$

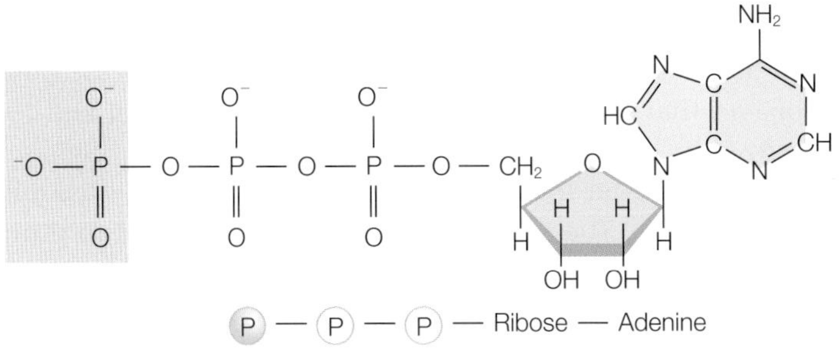

Figure 6.8 The chemical structure of adenosine triphosphate, ATP, with the terminal phosphate group shaded.

The value of -30.5 kJ mol^{-1} for ATP hydrolysis represents the standard-state Gibbs energy change when all reactants are at a concentration of 1 M and pH 7. In a cell, the concentrations and pH are not at the standard-state values and therefore the actual energy change will differ from that calculated using the standard-state conditions. The actual Gibbs energy change is given by considering the actual concentrations of the ADP, ATP, and inorganic phosphate:

$$\Delta G = \Delta G^{\circ} + RT \ln \frac{[\mathrm{ADP}][\mathrm{P_i}]}{[\mathrm{ATP}]} \tag{6.18}$$

Most cells maintain the concentrations of ATP, ADP, and inorganic phosphate within a very narrow range. Typical concentrations of ATP and inorganic phosphate are 2.5 and 2.0 mM, respectively, with the ADP concentration being at a much lower value of 0.25 mM. Substituting these concentrations into eqn 6.18 yields a more negative energy change, of -52 kJ mol^{-1} at 298 K and pH 7, than the standard value.

CHEMIOSMOTIC HYPOTHESIS

Whereas the general properties of ATP were established, an understanding of the mechanism was elusive. In 1961, Peter Mitchell proposed a mechanism in which the Gibbs energy is stored in the form of a pH gradient and electrical potential across the cell membrane. This became known as the chemiosmotic hypothesis. The proposal was initially poorly received but eventually gained acceptance as experimental studies proved the correctness of the ideals, and now the hypothesis is the cornerstone for understanding energy usage in cells.

> The essence of the Mitchell chemiosmotic hypothesis is that electron transfer occurs in a vectorial fashion across biological membranes.

As electrons are transferred through a series of carriers across the membrane, protons are also transported, generating a pH difference between the interior and exterior sides of the cell membrane. Since the membrane is a lipid bilayer that is impermeable to charges (Chapters 4 and 19), after transfer across the membrane the protons can be stored. Since protons are charged, the transfer leads to the generation of a potential difference across the membrane. The sum of these two effects is used to provide the energy for the synthesis of ATP.

The hypothesis provided a natural explanation for a number of experimental observations. For example, it was known that a class of compounds called uncouplers could inhibit ATP synthesis in intact systems. These compounds are lipophilic weak acids such as dinitrophenol. Mitchell proposed

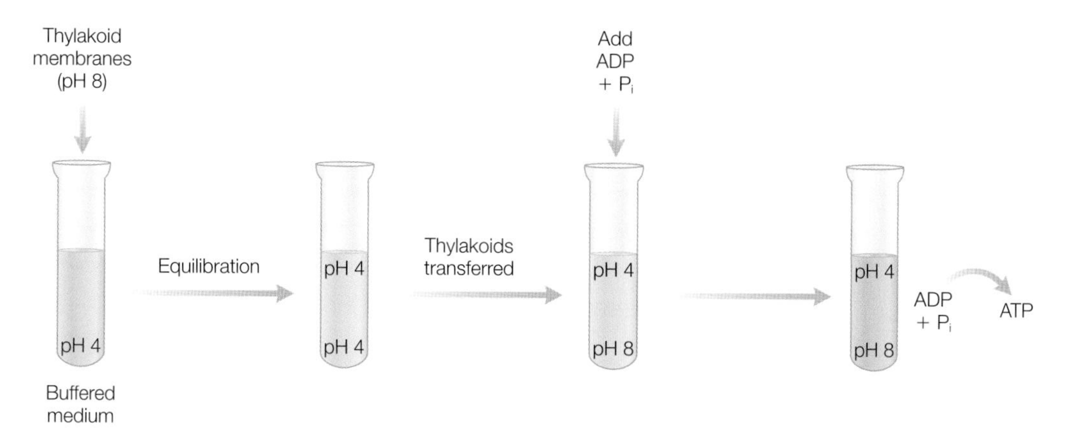

Figure 6.9 Support for the chemiosmotic hypothesis was provided by the formation of ATP in the experiments of Jagendorf and coworkers.

that these compounds could dissipate the proton gradient due to their ability to diffuse through the membrane in both protonated and deprotonated forms. In 1966, the hypothesis was dramatically supported by experiments by Andre Jagendorf and coworkers (see Jagendorf, 1998). Thylakoids were suspended in a buffer of pH 4 that caused both the interior and exterior of the cells to equilibrate at that pH (Figure 6.9). A buffer of pH 8 was injected rapidly into the solution, creating a pH difference of approximately four units across the thylakoid membrane. The pH difference resulted in large amounts of ATP being formed from ADP and inorganic phosphate, supporting the chemiosmotic hypothesis. Ultimately most skeptics were convinced and finally in 1978 Mitchell received a Nobel Prize for Chemistry.

The transport of protons across the membrane creates both a concentration difference and a charge difference, with both effects influencing the energetics. One contribution to the Gibbs energy difference arises from the difference in proton concentration for the two sides of the cell membrane. The proton difference arises from different metabolic processes (Figure 6.10) or from the action of proton pumps (Chapter 18). For an interior proton concentration $[H^+]_{in}$ and proton concentration outside the membrane $[H^+]_{out}$, the Gibbs energy difference is given by the ratio of these two concentrations:

$$\delta G = \Delta G - \Delta G^\circ = +RT \ln \frac{[H^+]_{in}}{[H^+]_{out}} \tag{6.19}$$

This expression for the Gibbs energy difference can be rewritten in terms of the pH difference:

$$\ln \frac{[H^+]_{in}}{[H^+]_{out}} = \ln[H^+]_{in} - \ln[H^+]_{out} = 2.3(\log[H^+]_{in} - \log[H^+]_{out}) \tag{6.20}$$

$$= -2.3(pH_{in} - pH_{out}) = -2.3\ \delta pH \quad \Delta G - \Delta G^\circ = +RT(-2.3\ \Delta pH)$$

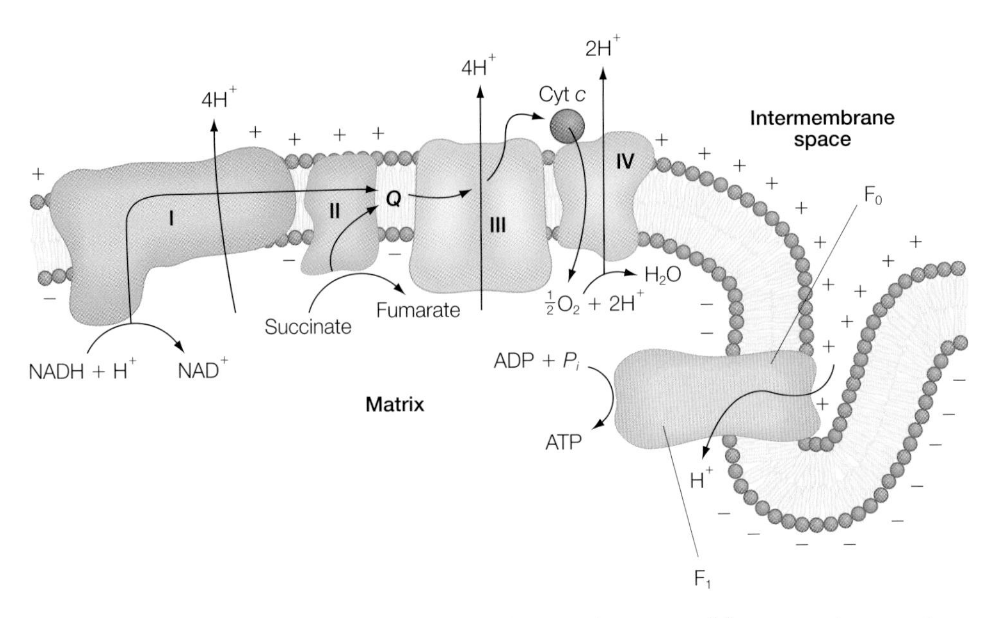

Figure 6.10 A schematic representation showing the involvement of four protein complexes, identified as complexes I–IV, and ATP synthase in the chemiosmotic hypothesis.

The second contribution arises from the difference in charge for the two sides of the membrane. The difference in the Gibbs energy for that contribution is given by the voltage difference ΔV across the cell membrane (eqn 6.5) using $n = 1$ for a proton charge:

$$(\Delta G)_{rec} = -nF\Delta V = -F\Delta V \tag{6.21}$$

Together, the total Gibbs energy available due to the difference in the concentration of protons on the two sides of the cell membrane is termed the *protonmotive force*, Δp, and can be written as:

$$\Delta p = -FV - 2.3RT(\Delta \text{pH}) \tag{6.22}$$

Experiments have established that only the total value of Δp is critical for the synthesis of ATP, as different relative fractions of the two components result in the same outcome. In thylakoids, the membrane potential is small and so Δp is due primarily to the pH difference across the membrane, although in intact plants the membrane potential may be larger.

RESEARCH DIRECTION: RESPIRATORY CHAIN

The cellular pathway for the development of the protonmotive force is carried out by four integral membrane proteins, identified as complexes

I–IV (Figure 6.10). Complex I, also called NADH:ubiquinone oxidoreductase, is an 850-kDa enzyme composed of over 40 protein subunits, including a FMN-containing flavoprotein and several iron–sulfur cofactors. Complex I catalyzes the conversion of NADH to NAD^+, which is coupled to electron transfer to ubiquinone and the pumping of protons from the matrix to the intermembrane space:

$$NADH + 5H^+_{matrix} + Q \rightarrow NAD^+ + QH_2 + 4H^+_{intermembrane} \qquad (6.23)$$

Complex II, or succinate dehydrogenase, is a 140-kDa enzyme that contains a number of cofactors. This enzyme couples the electron transfer of succinate to fumarate with the conversion of FAD to $FADH_2$ (Figure 6.11). In this reaction, electrons pass from succinate through FAD and the iron–sulfur cofactors to ubiquinone. Complexes I and II, together with the proteins acyl-CoA dehydrogenase, ETF:ubiquinone oxidoreductase, and glycerol 3-phosphate dehydrogenase, produce a pool of reduced ubiquinone, QH_2, that is re-oxidized by complex III.

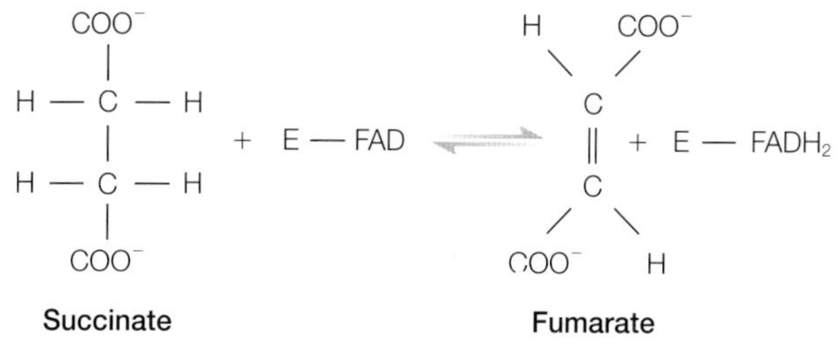

Succinate + E — FAD ⇌ Fumarate + E — $FADH_2$

Figure 6.11 In respiration, succinate is converted into fumarate with the involvement of FAD.

Complex III, also called the cytochrome bc_1 complex, is a 250-kDa protein with 11 protein subunits and a number of hemes and iron–sulfur centers. Complex III couples the transfer of electrons from the ubiquinones to cytochrome *c* with the accompanying transfer of protons from the matrix across the membrane to the intermembrane space. The net oxidation/reduction reaction, often termed the Q cycle, couples the transfer of electrons from the ubiquinones to the transfer of protons across the cell membrane:

$$QH_2 + 2\text{cyt } c_1(\text{oxidized}) + 2H^+_{matrix} \rightarrow Q + 2\text{cyt } c_1(\text{reduced}) + 4H^+_{intermembrane} \qquad (6.24)$$

Complex IV, also named cytochrome oxidase, completes the respiratory chain. The size of complex IV varies in different organisms, from three or four small protein subunits in bacteria to 13 in eukaryotic cells. Heme and copper cofactors perform the overall four-electron reduction of oxygen by a mechanism that is sequential without the release of intermediates:

$$4\text{Cyt } c(\text{reduced}) + 8H^+_{matrix} + O_2 \rightarrow 4\text{cyt } c(\text{oxidized}) + 4H^+_{intermembrane} + 2H_2O \qquad (6.25)$$

In considering the net flow of electrons through the respiratory chain, electrons from NADH reduce molecular oxygen according to:

$$NADH + H^+ + \frac{1}{2}O_2 \rightarrow NAD^+ + H_2O \qquad (6.26)$$

This reaction is highly exergonic, with a Gibbs energy difference of -220 kJ mol^{-1} under standard conditions. Much of the energy is used to transfer protons out of the matrix into the intermembrane space. In considering the proton flow, for every pair of electrons, there are four protons transferred by complex I, four by complex III, and two by complex IV. Including the proton transfer with the electron transfer yields:

$$\mathrm{NADH} + 11\mathrm{H}^+_{\mathrm{matrix}} + \frac{1}{2}\mathrm{O}_2 \rightarrow \mathrm{NAD}^+ + 10\mathrm{H}^+_{\mathrm{intermembrane}} + \mathrm{H_2O} \quad (6.27)$$

The molecular mechanisms by which each step of this net reaction proceeds is under active investigation by a number of laboratories. For each complex, a number of cofactors are involved in multi-electron transfers that are coupled with proton transfer. The determination of the structures of complexes II, III, and IV and the water-soluble domain of complex I has provided the foundation for such investigations (Xia et al. 1997; Iwata et al. 1998; Iverson et al. 1999; Lancaster et al. 1999; Sarasta 1999; Lange & Hunte 2002; Yankovskaya et al. 2003; Sazanov & Hinchiffe 2006). Among the revelations from these structures was the realization that the protein surrounding the cofactors is not simply scaffolding. Motion of the protein plays a critical role, with the electron transfer between the iron–sulfur cluster and cytochrome c_1 heme being driven by a rocking motion of the Rieske iron–sulfur protein (Figure 6.12; Zhang et al. 1998; Darrouzet et al. 2001; Berry & Huang 2003).

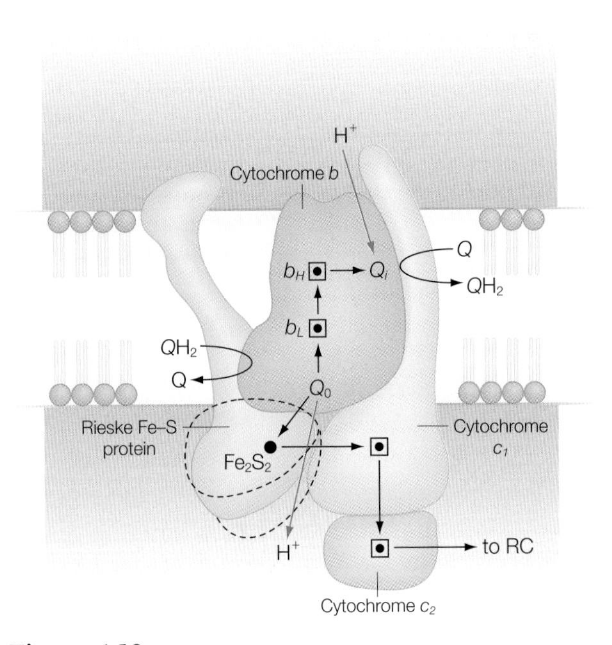

Figure 6.12 Schematic representation of the cytochrome bc_1 complex showing the motion of the Rieske iron–sulfur protein that plays a key role in driving the electron-transfer pathway. Modified from Crofts and Berry (1998).

RESEARCH DIRECTION: ATP SYNTHASE

This strategy of overcoming endothermic reactions by coupling the reactions with ATP hydrolysis is used in all living cells for the synthesis of metabolic intermediates and cellular components. To be practical, ATP must be available to drive these reactions. Towards this end, the transfer of protons across the cell membrane is used to drive the synthesis of ATP from ADP through the transfer of protons from the intermembrane space to the matrix:

$$\mathrm{ADP} + \mathrm{P_i} + n\mathrm{H}^+_{\mathrm{intermembrane}} \rightarrow \mathrm{ATP} + \mathrm{H_2O} + n\mathrm{H}^+_{\mathrm{matrix}} \quad (6.28)$$

The number of protons involved in ATP synthesis has been actively discussed and still is not resolved. Early measurements yielded a ratio of protons transferred for ATP synthesized of $3H^+$/ATP, but other measurements have yielded $4H^+$/ATP and $4.67H^+$/ATP.

The coupling of proton transfer and ATP synthesis occurs in the enzyme ATP synthase. This enzyme is a large multisubunit complex that has been studied by many groups, most notably John Walker and Paul Boyer who received the Nobel Prize for Chemistry in 1997 for their efforts (Abrahams et al. 1994; Boyer 2000). The complex has two domains, identified as F_0 and F_1, with the chloroplast enzyme domains denoted by CF_0 and CF_1. ATP synthase enzymes from different cells have a similar composition and structure. The F_1 domain has three copies of the α and β subunits and one copy of the other F_1 subunits δ, γ, and ε. The composition of the F_0 domain differs in different organisms, with the bacterial and mitochondrial enzymes having one copy of the *a* subunit, two copies of the *b* subunit (or an analogous subunit), and 10–14 copies of the *c* subunit.

The three-dimensional structure of the F_1 domain showed that the α and β subunits are in an approximate hexameric arrangement but with each subunit exhibiting a different conformation that reflects the three different functional states: with ATP bound, with ADP bound, and with the binding site empty (Figure 6.13). At the center is the single γ subunit that forms a long, bent helical structure in the center of the structure. The γ subunit is asymmetrically placed in the structure and interacts with only one of the three β subunits.

The γ subunit extends below the F_1 domain into the region of the F_0 domain. The F_0 domain is composed of three protein subunits, *a*, *b*, and *c*. The *c* subunit is very hydrophobic and consists of two transmembrane helices with small loops. The *c* subunits are arranged symmetrically around the F_1 symmetry axis with the two sets of helices forming two concentric circles (Figure 6.14; Meier et al. 2005; Murata et al. 2005). The rings of *c* subunits are thought to be tightly associated with the γ subunit of the F_1 domain but not with the other subunits. In a motor, the stator is stationary and the rotor spins in the center. The *c* subunits can be considered to form a rotor that can move independently of the remaining portions of the protein that would be the stator.

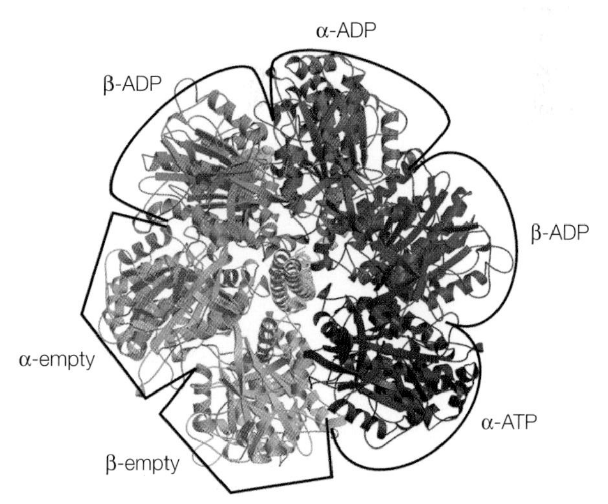

Figure 6.13 Structure of the mitochondrial ATP synthase F_1 domain viewed down the approximate 6-fold symmetry axis of the α and β subunits with structural differences due to the ATP/ADP binding, as indicated. At the center is the γ subunit. Based upon Abrahams et al. (1994).

The rotation of the subunits was demonstrated by experiments performed by Masamiysu Yoshida and Kazuhiko Kinosita

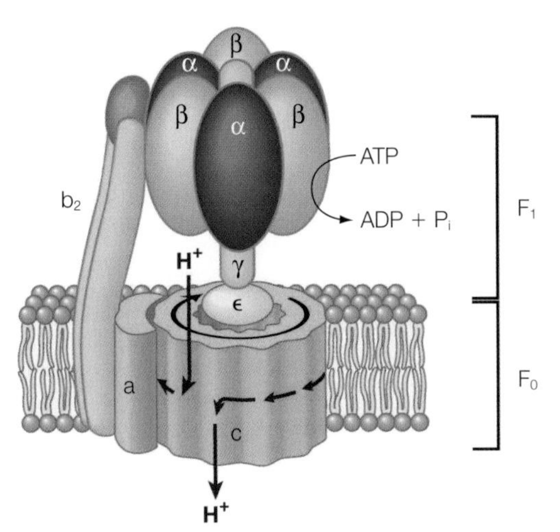

Figure 6.14 The structure of the F_0F_1 complex and a model of how rotation of the *c* subunits in the cell membrane relative to the F_1 domain and *a* and *b* subunits can couple ATP synthesis to proton transport. Based upon Murata et al. (2005).

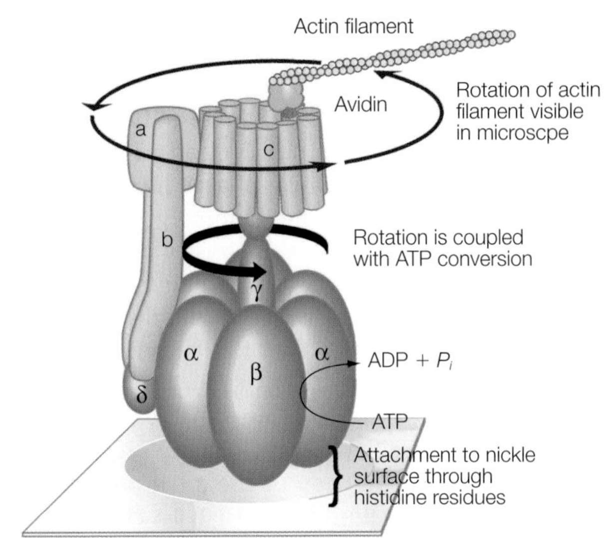

Figure 6.15 An experimental demonstration of the rotation of the ATP synthase by use of fluorescently labeled actin filament. Based upon Kinosita et al. (2004).

(Figure 6.15; Kinosita et al. 2004; Junge & Nelson 2005). The F_1 domain was attached to a glass slide with a nickel coating through histidine residues that were on extensions of the protein. A long, thin actin filament with a fluorescent label was attached through a streptavidin/avidin connector to the *c* subunits. When ATP was added, the enzyme ran in reverse and the filament was found to spin rapidly. In single-molecule experiments (Chapter 14) the filament was found to be positioned only along certain discrete angles.

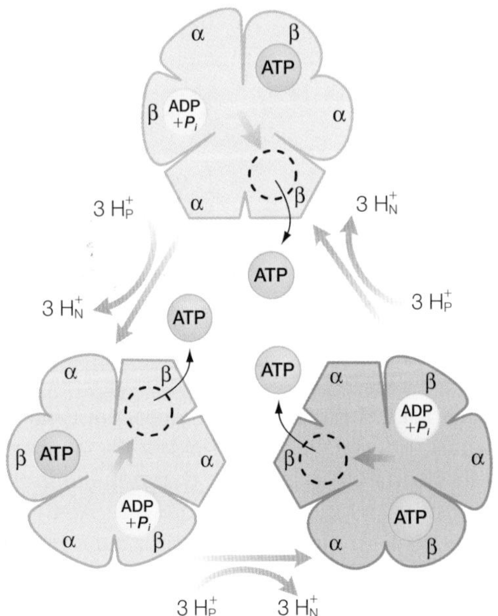

Figure 6.16 The binding-change mode for ATP synthesis. Modified from Boyer (2000).

Based upon a long series of experiments, the binding-change mechanism for ATP synthesis has been elucidated. The mechanism involves the presence of three sites, one of which binds ATP tightly, the second of which binds weakly, and the third of which is empty (Figure 6.16). Energy is required to release ATP, not to form it. The position of the three sites in the subunits is not fixed, but varies as the enzyme rotates, with the γ subunit acting like a camshaft and alternately distorting the β subunits, which can cause cycling of the three sites. Interactions between the *a* subunit and the *c* ring provide a ratchet that couples proton transfer with a ring rotation in a counterclockwise direction only.

Although many aspects of the mechanism of ATP synthesis have been determined, the stoichiometry of the number of protons required to be transferred for each ATP synthesized remains an open question. A full rotation of the enzyme should lead to the synthesis of three molecules of ATP. If we assume that each *c* subunit translocates one proton per cycle, then the H^+/ATP ratio is given by the number of *c* subunits divided by three. However, the number of protons is related to the rotation of the *c* ring but the number of *c* subunits in the F_0 domain is not the same for enzymes isolated from different organisms. For an enzyme with 12 *c* subunits, the ratio is $12H^+$/3ATP or $4H^+$/ATP. With a range of 10–14 for the number of *c* subunits, the calculated ratio is 4.67:3.33 H^+/ATP. Whereas an integer value for H^+/ATP would make life simple as the gearing is direct, the ratio may be a non-integer due to nonlinear responses which have not yet been identified.

REFERENCES

Abrahams, J.P., Leslie, A.G.W., Lutter, R., and Walker, J.E. (1994) Structure at 2.8 Å resolution of F_1-ATPase from bovine heart mitochondria. *Nature* **370**, 621–8.

Berry, E.A. and Huang, L. (2003) Observations concerning the quinol oxidation site of the cytochrome bc_1 complex. *FEBS Letters* **555**, 13–20.

Boyer, P.D. (2000) Catalytic site forms and controls in ATP synthase catalysis. *Biochimica Biophysica Acta* **1438**, 252–62.

Crofts, A.R. and Berry, E.A. (1998) Structure and function of the cytochrome bc_1 complex of mitochondria and photosynthetic bacteria. *Current Opinions in Structural Biology* **8**, 501–9.

Darrouzet, E., Moser, C.C., Dutton, P.L., and Daldal, F. (2001) Large scale domain movement in cytochrome bc_1: a new device for electron transfer in proteins. *Trends in Biochemical Sciences* **26**, 445–51.

Iverson, T.M., Luna-Chavez, C., Cecchini, G., and Rees, D.C. (1999) Structure of the *Escherichia coli* fumerate reductase respiratory complex. *Science* **284**, 1961–6.

Iwata, S., Lee, J.W., Okada, K. et al. (1998) Complete structure of the 11-subunit bovine mitrochondrial cytochrome bc_l complex. *Science* **281**, 64–71.

Jagendorf, A.T. (1998) Chance, luck, and photosynthesis research: an inside story. *Photosynthesis Research* **57**, 217–29.

Junge, W. and Nelson, N. (2005) Nature's rotary electromotors. *Science* **308**, 642–4.

Kinosita, K., Adachi, K., and Itoh, H. (2004) Rotation of F1-ATPase: how an atp-driven molecular machine may work. *Annual Review of Biophysics and Biomolecular Structure* **33**, 245–68.

Lancaster, C.R.D., Kröger, A., Auer, M., and Michel, H. (1999) Structure of fumerate reductase from *Wolinella succinogenes* at 2.2 Å resolution. *Nature* **402**, 377–85.

Lange, C. and Hunte, C. (2002) Crystal structure of the yeast cytochrome bc_1 complex with its bound substrate cytochrome *c*. *Proceedings of the National Academy of Sciences USA* **99**, 2800–5.

Meier, T., Polzer, P., Diederichs, K., Welte, W., and Dimroth, P. (2005) Structure of the rotot ring of F-type Na^+-ATPase from *Iyobacter tartaricus*. *Science* **308**, 659–62.

Murata, T., Yamato, I., Kakinuma, Y., Leslie, A.G.W., and Walker, J.E. (2005) Structure of the rotor of the V-type Na^+-ATPase from *Enterococcus hirae*. *Science* **308**, 654–9.

Sarasta, M. (1999) Oxidative phosphorylation at the *fin de siecle*. *Science* **283**, 1488–93.

Sazanov, L.A. and Hinchiffe, P. (2006) Structure of the hydrophilic domain of respiratory complex I from *Thermus thermophilus*. *Science* **311**, 1430–6.

Xia, D., Yu, C.A., Kim, H. et al. (1997) Crystal structure of the cytochrome bc_l complex from bovine heart mitochondria. *Science* **277**, 60–6.

Yankovskaya, V., Horsefield, R., Tornroth, S. et al. (2003) Architecture of succinate dehydrogenase and reactive oxygen species generation. *Science* **299**, 700–4.

Zhang, Z., Huang, L., Schumleister, V.M. et al. (1998) Electron transfer by domain movement in cytochrome bc_1. *Nature* **392**, 677–84.

PROBLEMS

6.1 In an electrochemical cell, where does an oxidation reaction occur?

6.2 Calculate the potential for the following reaction at 25°C using the following standard potentials:

$$Zn(solid) + CuSO_4(aqueous) \leftrightarrow Cu(solid) + ZnSO_4(aqueous)$$
$$Zn^{2+} + 2e^- \leftrightarrow Zn(solid) \quad E_m = -0.76V$$
$$Cu^{2+} + 2e^- \leftrightarrow Cu(solid) \quad E_m = +0.34V$$

6.3 How is the midpoint potential of a protein related to its ability to oxidize another protein?

6.4 Assuming 100% efficiency and standard conditions, calculate how many photons, at 700 nm or equivalently 2.8×10^{19} J, a chloroplast needs to absorb to make one ATP molecule from ADP and inorganic phosphate.

6.5 What is the ratio between oxidized and reduced species when the potential is equal to the midpoint potential?

6.6 If the typical energized thylakoid membrane has the estimated parameters of $\Delta V = 0$ mV, $[P_i] = 0.01$ M, and $\Delta pH = 2.5$ (out–in), and the proton/ATP stoichiometry is 4, what is the ATP/ADP ratio? Use $G^{\circ}_{ATP} = 8$ kcal mol^{-1}.

6.7 If a protein has a midpoint potential of 125 mV, calculate the potential when the reduced/oxidized ratio is (a) 10, (b) 1, and (c) 0.01.

6.8 Calculate the midpoint potential for a protein that was measured to have the following redox data. The parameter R is the ratio of the reduced form divided by the total amount of protein. Use the table to guess which protein this is.

Potential (mV)	R
100	1.0
150	0.92
200	0.83
250	0.49
300	0.12
350	0.02
400	0.00

6.9 The standard free energy of hydrolysis of ATP is -30.5 kJ mol^{-1}. However, the concentrations of ATP, ADP, and P_i are not at standard conditions in the cell. Calculate the Gibbs energy difference using concentrations of 2.25, 0.25, and 1.75 mM for ATP, ADP, and P_i.

6.10 Calculate the change in the Gibbs energy when condensation of P_i with ADP ($\Delta G = +30.5$ kJ mol^{-1}) is coupled with cleavage of P_i from phosphoenolpyruvate ($\Delta G = -61.9$ kJ mol^{-1}).

6.11 Calculate the change in the Gibbs energy when ADP is formed from ATP, using the ΔG values of -45.6 and -19.2 kJ mol^{-1} for the first and second bond cleavages.

$$ATP + 2H_2O \rightarrow AMP + 2P_i$$

6.12 The midpoint potential of a protein was found to have the following pH dependence. If this is interpreted as arising from a change in protonation of the cofactor, estimate the pK_A value.

pH	E_m (mV)
2	450
3	430
4	370
5	305
6	250
7	220
8	210
9	210

6.13 If reduced cytochrome is added to the oxidized bacterial reaction center, can an electron be transferred? Assume that the two proteins can dock properly.
6.14 If reduced cytochrome is added to the oxidized photosystem II, can an electron be transferred? Assume that the two proteins can dock properly.
6.15 Calculate the ionic strength of 0.5 M NaCl.
6.16 Calculate the ionic strength of 0.5 M $MgCl_2$.
6.17 Discuss why the proposed motion of the Rieske FeS protein plays a role in the function of the cytochrome bc_1 complex.
6.18 How would the results from fluorescently labeled actin attached to ATP synthase compare with the biochemical data?

7

Kinetics and enzymes

Thermodynamics can be used to determine whether a reaction is spontaneous and how much energy is involved in the reaction. However, answers are not determined to questions such as how long the process will take and which intermediate states are formed. Although the properties of molecules can be studied from understanding static properties, much of our knowledge of the reactions that proteins and biological molecules perform is obtained from measurement of the time dependence of each process. Kinetic studies of a process involve correlating the time evolution of each molecular species to a model of the mechanism of the reaction. The model will of necessity propose specific intermediate states and specific rates for each step. Such an analysis reveals the limitations to the rate and yield of the reaction as well as the energetic barriers that must be overcome for the reaction to proceed. For reactions that proceed through a series of steps, the rates for each step can be determined and the slowest step, termed the rate-limiting step, can be identified. By performing kinetic measurements using different concentrations of each molecule one can determine whether the rate-limiting step is dependent or independent of complex formation. Once established, a model of the mechanism provides a platform for probing the biochemical factors that control the functions of proteins. Examples of using these concepts in biological settings are the ability of enzymes to accelerate specific chemical processes and the components that allow proteins to serve as electron-transfer carriers in the cell.

THE RATE OF A CHEMICAL REACTION

Before proceeding too far, we consider how a reaction rate is determined experimentally. For simplicity, this question is addressed for the simple irreversible reaction of molecule A converting to molecule B:

$$A \rightarrow B \tag{7.1}$$

The rate of a reaction involving a molecule A, which at any time t has the concentration $A(t)$, is the change in the concentration or population of the molecule with time:

$$\text{Rate} = \frac{dA(t)}{dt} \tag{7.2}$$

For this simple reaction, the rate can be expressed in terms of either the change in [A] with respect to time or the change in [B] with time. Since the only source of molecule B is from those molecules of A that convert to molecule B, the rates at which molecules A and B change with time must be equal, but with opposite signs:

$$\frac{dA(t)}{dt} = -\frac{dB(t)}{dt} \tag{7.3}$$

Rates are typically concentration-, and therefore time-, dependent. In such cases there is no single rate for a reaction. To compare rates it would be necessary to specify all concentrations of each component. For example, consider the rate at which gumballs come out of a small hole in a container (Figure 7.1). If the container has many gumballs in it, then many gumballs are likely to hit the hole and pop out in any given period of time. However, if the number of gumballs is small, then the rate at which the gumballs will come out is slower. For simplicity, rather than try to use rates that are dependent upon the specific concentrations, parameters are determined that describe the reaction kinetics but which are concentration-independent.

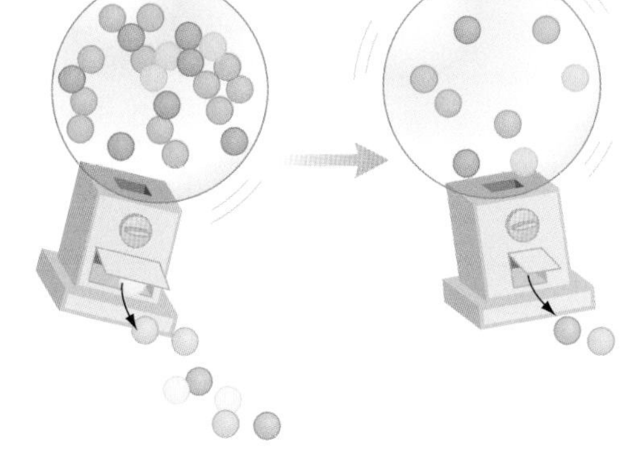

Figure 7.1 The number of gumballs leaving the dispenser in a period of time is dependent on the total number of gumballs.

For the case of the irreversible reaction (eqn 7.1), to relate $A(t)$ to a rate constant we need to make the fundamental assumption that all of the A molecules have the same probability of converting to B in a given period of time regardless of their history. If this is true, then the instantaneous probability that any particular molecule of A will change to B is independent of time. The instantaneous probability is a useful constant for characterizing irreversible first-order reactions. The *fractional change in the concentration of* $A(t)$, $dA(t)/A(t)$, is equal to the *instantaneous probability* of A converting to B at a rate k, multiplied by *the time interval* dt:

$$\frac{dA(t)}{A(t)} = -k\,dt \quad \text{or}$$

$$\frac{dA(t)}{dt} = -kA(t) \tag{7.4}$$

In this context, the instantaneous probability is equivalent to a first-order rate constant. The time dependence of the concentration of molecule A can be determined by integrating this expression:

$$\int \frac{dA(t)}{A(t)} = -\int k dt \quad (7.5)$$

$$\ln A(t) = -kt + c$$

$$A(t) = e^{-kt+c} = e^c e^{-kt}$$

$$\int \frac{dx}{x} = \ln x \qquad \int k dx = k \int dx = kx$$

$$e^{a+b} = e^a e^b$$

The time dependence of A is seen to be exponential with the rate multiplying the time in the exponent. To fully determine the dependence, it is necessary to identify the value of the *constant of integration*, c. The value of c can be found by realizing that at time zero the exponential term is 1 and so the constant of integration represents the amount of the molecule A at the initial time:

$$A(t = 0) = e^c e^{-k(0)} = e^c \quad (7.6)$$

With this substitution for the constant c, the time dependence can be rewritten as:

$$A(t) = A(t = 0)e^{-kt} \quad (7.7)$$

A plot of the time dependence of these two states shows an exponential decay of A and a corresponding increase of B (Figure 7.2). A classic example of a first-order process is radioactive decay in which the rate is often expressed in terms of the *half-life*, $t_{1/2}$, which represents the time required for molecule A to decay to half of its value. The time at which this happens can be written in terms of the rate constant by substituting a value of $A(t = 0)/2$ for $A(t)$ into eqn 7.7:

$$A(t = 1/2) = \frac{A(t = 0)}{2} = A(t = 0)e^{-kt_{1/2}} \quad (7.8)$$

$$\frac{1}{2} = e^{-kt_{1/2}} \text{ or } \ln(\frac{1}{2}) = -0.69 = -kt_{1/2}$$

$$t_{1/2} = \frac{0.69}{k}$$

The half-life and rate constant are inversely related to each other and correspondingly have inverse units. Whereas the half-life is commonly used to characterize radioactive processes, it is not usually used to describe chemical reactions. Rather, a process is described as having a lifetime that is simply the inverse of the rate constant, or equivalently the time at which $1/e$ of the initial population has changed.

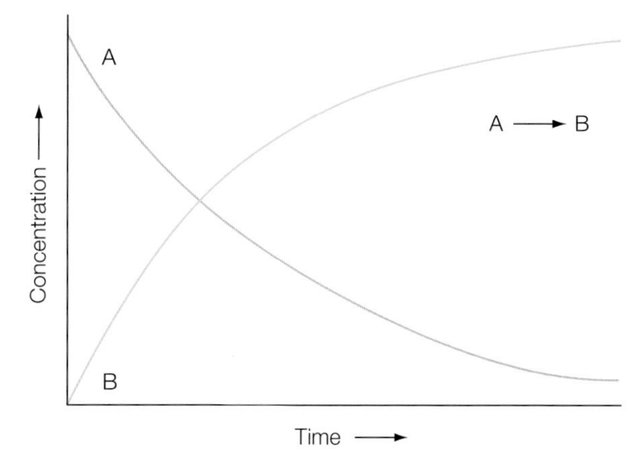

Figure 7.2 Kinetic curves for a simple first-order reaction.

PARALLEL FIRST-ORDER REACTIONS

In many biological processes, first-order rates are observed when a reaction involves a high-energy molecule that relaxes spontaneously to a lower-energy state without the involvement of another molecule. For example, such high-energy states are created in photosynthesis due to the absorption of light energy (see Chapter 20). However, the decay may be possible by more than one pathway and the kinetics will reflect the possible formation of several different products.

Consider the gumballs again, with the dispenser having two slots instead of one (Figure 7.3). Unbeknownst to the children awaiting the dispensation of the gumballs, the slots are different sizes and so the gumballs leave with two different rates, k_1 and k_2. The rate at which the initial state A decays and states B and C increase are identified by use of a scheme in which two independent paths proceed:

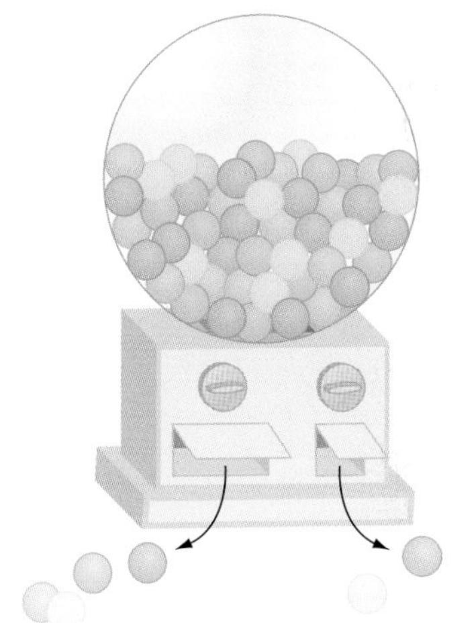

Figure 7.3 A multislot dispenser with gumballs being delivered through two different slots.

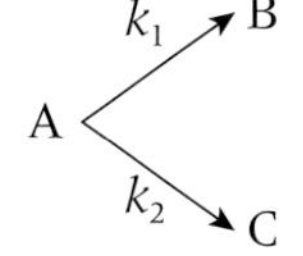

(7.9)

In this case, the initial concentration of gumballs in the dispenser, [A], can change due to loss to population B with a rate constant k_1 and simultaneous loss to population C with a rate constant k_2. The rate of change of [A] is then described by two terms that are the product of either rate constant k_1 or k_2 and the concentration of A (eqn 7.10). The increases in the amounts of B and C are then given by the product of [A] with the rate constants k_1 and k_2, respectively.

$$-\frac{d[A]}{dt} = k_1[A] + k_2[A] \quad \frac{d[B]}{dt} = +k_1[A] \quad \frac{d[C]}{dt} = +k_2[A] \tag{7.10}$$

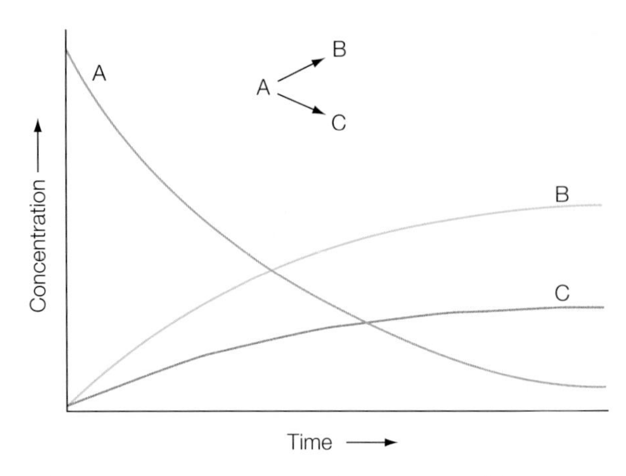

Figure 7.4 Kinetic curves for two parallel processes.

Since each of the individual rates is first order, the observed rate constant, k_{obs}, for the exponential loss of state A, will be given by the sum of the individual rates:

$$-\frac{d[A]}{dt} = (k_1 + k_2)[A] = k_{obs}[A] \tag{7.11}$$

$$A(t) = A(t = 0)e^{-k_{obs}t}$$

Thus, the state A decays exponentially with an observed rate that is the sum of the individual rates (Figure 7.4). The time dependencies of states B and C can be solved by substitution of eqn 7.11 into eqn 7.10:

$$\begin{aligned} \frac{d[B]}{dt} &= +k_1[A] = +k_1[A(t = 0)]e^{-k_{obs}t} \\ \frac{d[C]}{dt} &= +k_2[A] = +k_2[A(t = 0)]e^{-k_{obs}t} \end{aligned} \tag{7.12}$$

If the states B and C are assumed to not be present initially but only be generated by the decay of state A, then $[B(t = 0)] = [C(t = 0)] = 0$ and eqn 7.12 can be revised by separating variables and integrating to yield:

$$\begin{aligned} [B(t)] &= \frac{k_1[A(t = 0)]}{k_{obs}}(1 - e^{-k_{obs}t}) \\ [C(t)] &= \frac{k_2[A(t = 0)]}{k_{obs}}(1 - e^{-k_{obs}t}) \end{aligned} \tag{7.13}$$

Thus, both B and C start at zero concentration and increase exponentially with a rate k_{obs} (Figure 7.4). The ratio of these two states is always equal to the ratio of the two forward rates:

$$\frac{[B]}{[C]} = \frac{k_1}{k_2} \tag{7.14}$$

The concentration of A decreases exponentially while the concentrations of B and C increase exponentially. Assuming that k_1 is larger than k_2, the amount of B is always greater than that of C, as shown in Figure 7.4. Since A is being converted into both B and C, the final concentrations of B and C individually will always be less than the initial amount of A.

SEQUENTIAL FIRST-ORDER REACTIONS

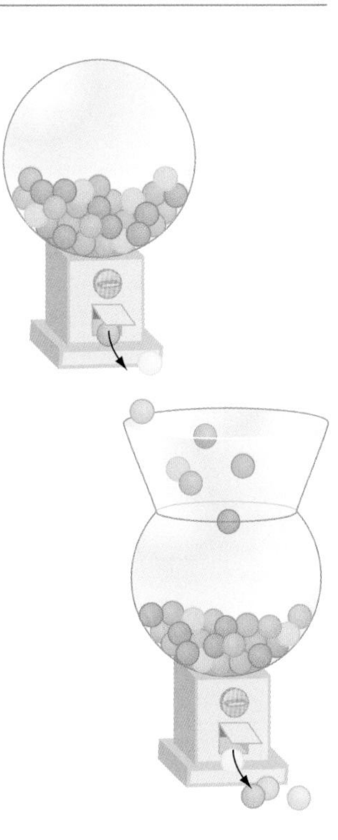

Figure 7.5 The sequential transfer of gumballs.

Another possible reaction is one that consists of sequential processes. In terms of the gumballs, there are two dispensers but an anxious child must wait until the gumballs pass from the first to the second and then out of the second before receiving the gumball (Figure 7.5). Such a reaction involving two sequential first-order steps can be written as beginning in state A, and progressing through state B, followed by state C:

$$A \underset{k_{b1}}{\overset{k_{f1}}{\leftrightarrow}} B \underset{k_{b2}}{\overset{k_{f2}}{\leftrightarrow}} C \qquad (7.15)$$

Since each of the two steps is reversible, there are two forward rate constants, identified as k_{f1} and k_{f2}, and two backward rate constants, identified as k_{b1} and k_{b2} for the first and second steps, respectively. Rate equations can be written for each species. The change in the concentration of A is caused by both loss to state B with the rate constant k_{f1} and gain due to the reverse reaction from B with the rate constant k_{b1}. The change in C arises from the step of B going to C with the rate constant k_{f2} or the decay of C back to B with the rate constant k_{b2}. The most intricate changes occur for state B as the amount can increase due to the forward step of A to B or the backward step of C to B, whereas the amount of B can decrease due to both the reverse step of B to A and the forward step of B to C:

$$\frac{dA}{dt} = -k_{f1}[A] + k_{b1}[B] \qquad (7.16)$$

$$\frac{dB}{dt} = k_{f1}[A] - k_{b1}[B] - k_{f2}[B] + k_{b2}[C]$$

$$\frac{dC}{dt} = k_{f2}[B] - k_{b2}[C]$$

These relationships can be solved to yield the time dependence of each species in terms of the four rate constants. Although this reaction can be solved explicitly, more complex reactions are best solved as numerical solutions using a computer program. An understanding of complex reactions can be achieved if certain assumptions are utilized. For example, if the reverse rates are all much smaller than the forward rates, the process is essentially irreversible. If one of the forward rates is much smaller than the other, than the overall rate will be determined by the slowest rate that is often termed the rate-limiting rate. Another example is the steady-state approximation in which the intermediate state is assumed to not change concentration due to a very fast forward rate for the second step with a backward rate so small that it can be neglected:

$$A \underset{k_{b1}}{\overset{k_{f1}}{\leftrightarrow}} B \overset{k_2}{\rightarrow} C \tag{7.17}$$

$$\frac{dB}{dt} = 0 = k_{f1}[A] - k_{b1}[B] - k_2[B] = k_{f1}[A] - (k_{b1} + k_2)[B]$$

$$\frac{[B]}{[A]} = \frac{k_{f1}}{k_{b1} + k_2}$$

$$\frac{dC}{dt} = k_2[B] = k_2\frac{k_{f1}}{k_{b1} + k_2}[A]$$

While the specifics of sequential reactions may be complex, the changes in states will follow a general trend. The concentration of A decreases in an exponential manner. Assuming that B is not present initially, it will start to increase from zero, reach a maximum, and then decline to zero as C begins to form. Assuming that C also is not present initially, the concentration of C will start at zero and slowly begin to rise. In contrast to the pattern of B reaching a maximum and then decreasing, the concentration of C will continue to increase with time. After a long time, the concentrations for each state will reach limiting values as the reactions approach equilibrium.

For reactions involving sequential steps, assignment of the specific reaction scheme is often difficult as concentration profiles of all states are often not all available. For example, if the only experimental observable is B, then it is often difficult to distinguish between the case when $k_{f1} \gg k_2$ and the opposite case when $k_{f1} \ll k_2$. In other cases when only A and C are observable, it may be difficult to verify experimentally the presence of the intermediate state. If the rate for the second step is much faster than the forward rate for the first step, then as state B is formed it is transformed rapidly into C and the concentration of B remains low (Figure 7.6). Only if the second rate is slow will the intermediate state build up. For these reasons, proper assignment of the mechanism for complex biological processes remains a challenging issue.

SECOND-ORDER REACTIONS

A simple second-order reaction is usually considered to involve two steps: the two components, A and B, must first form a complex AB, and then the reaction proceeds to form the state C:

$$A + B \leftrightarrow AB \rightarrow C \tag{7.18}$$

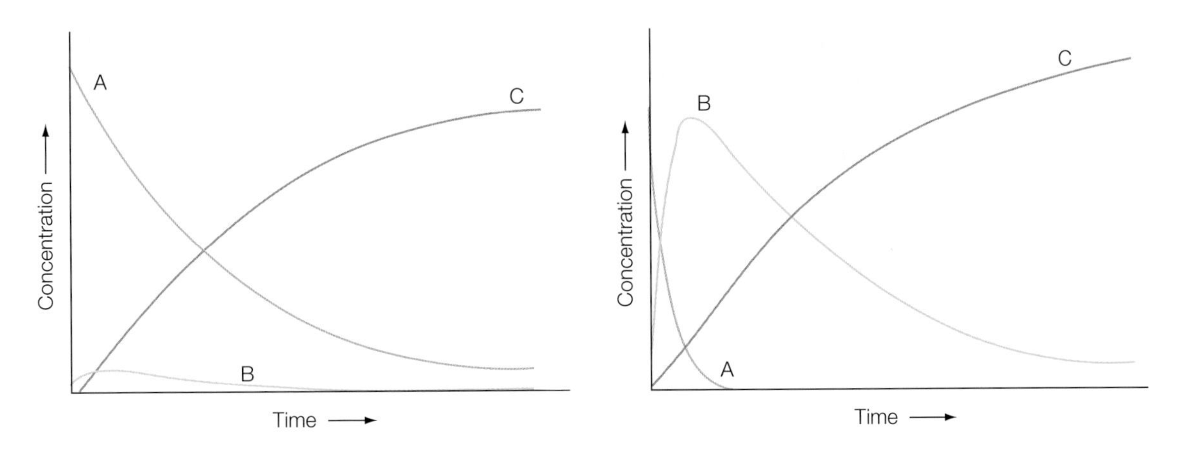

Figure 7.6 The kinetic curves for two sequential processes when (right) $k_{f1} \gg k_2$ and (left) $k_{f1} \ll k_2$.

The second step of the process is a first-order reaction with some instantaneous probability. However, to consider the overall process, this probability must be multiplied by the probability that the complex forms during an *interval of time*, dt. From the standpoint of molecule B, if any molecule A enters an interaction volume around B, then the complex may be formed. For simplicity, molecule A is assumed to enter the interaction volume and either form the complex or leave rapidly compared to the time required for the second process to occur. In this case, the distribution of A will always be random, and the number of A molecules within the interaction volume is equal to the product of the concentration of A and the *interaction volume around B*, V_B. The formation of the product, C, is given by the first-order rate expression:

$$\frac{\mathrm{dC}}{\mathrm{d}t} = k[\mathrm{AB}] \tag{7.19}$$

The concentration of the complex is given by the product of the concentration of B and the number of molecules of A within V_B, yielding:

$$\frac{\mathrm{dC}}{\mathrm{d}t} = k[\mathrm{AB}] = kV_B[\mathrm{A}][\mathrm{B}] \tag{7.20}$$

the term kV_B is usually referred to as the second-order rate constant for the reaction.

THE ORDER OF A REACTION

Although the reaction order may be considered from a molecular standpoint, in practice the order is an empirical quantity and may have a range of values. In general, the rate may depend upon the concentrations of the reactants according to:

$$\frac{dC}{dt} \propto k[A]^n[B]^m \qquad (7.21)$$

where the values of n and m will depend upon the specifics of the reaction. For example, if the complex formation involves two molecules of A and one of B, then $n = 2$ and $m = 1$:

$$2A + B \rightarrow C$$

$$\frac{d[A]}{dt} = -2k[A]^2[B] \qquad (7.22)$$

If the reaction is reversible then both the forward rate constant, k_f, and the backward rate constant, k_b, must be considered:

$$A + B \underset{k_b}{\overset{k_f}{\leftrightarrow}} C$$

$$\frac{d[A]}{dt} = -k_f[A] + k_b[B] \qquad (7.23)$$

In each case the units will match the order of the rate constant, with first-order rates having units of s^{-1}, and second-order rates have units of $M^{-1}\ s^{-1}$.

REACTIONS THAT APPROACH EQUILIBRIUM

From a thermodynamic viewpoint, a reaction reaches equilibrium when the ratio of the products and reactants is at the lowest Gibbs energy for the system (Chapter 6). Equilibrium can also be viewed from a kinetic viewpoint as occurring when the rate of the forward reaction is equal to the reverse reaction. For example, the reaction of A converting to B is at equilibrium when the rate of change of both components is zero:

$$A \underset{k_b}{\overset{k_f}{\leftrightarrow}} B \qquad (7.24)$$

$$\frac{dA}{dt} = \frac{dB}{dt} = 0$$

The equilibrium constant can be related to the rates by expressing the change in A in terms of the forward and backward reactions (eqn 7.16) and setting this term equal to zero:

$$\frac{d[A]}{dt} = -k_f[A] + k_b[B] = 0 \qquad (7.25)$$

$$k_f[A] = k_b[B]$$

$$K_{eq} = \frac{[B]}{[A]} = \frac{k_f}{k_b}$$

Thus, the *equilibrium constant* for a reaction, K_{eq}, is equal to the ratio of the forward and backward rates for a reaction.

ACTIVATION ENERGY

For some reactions, the change in the Gibbs energy is a large negative number and hence the overall reaction is thermodynamically favorable. However, the rate of product formation may still be slow. In these cases, the reaction usually requires the formation of an intermediate or transitional state that is energetically unfavorable. For enzymes, the intermediate state is not a real state but only a transitional one that lives for a short time. For a reaction with the reactants A and B and product C, the short-lived intermediate is denoted as $[AB]^{\ddagger}$:

$$A + B \leftrightarrow AB^{\ddagger} \rightarrow C \qquad (7.26)$$

The reaction can be shown schematically by plotting the energy of each step against what is termed the reaction coordinate, which represents changes in the nuclear conformation of each state. The intermediate state is assumed to be in rapid equilibrium with the product state due to the large free energy difference (Figure 7.7).

The overall rate is limited by the formation of the intermediate state because the increase in Gibbs energy for the intermediate represents an energy barrier. The rate to overcome the energy difference between the initial and intermediate state, termed the *activation energy*, E_A, is given by:

$$k = Ae^{-E_A/k_BT} \qquad (7.27)$$

where A is the rate that would be observed if $E_A = 0$. This rate dependence arises from a statistical determination of the probability that the system has an energy greater than E_A to overcome the barrier associated with the formation of an intermediate state. For an activated process, the activation energy is

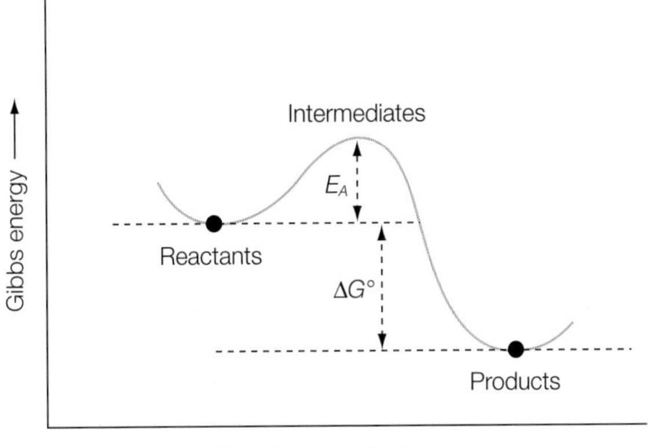

Figure 7.7 The energetics of a reaction showing the decrease in the Gibbs energy $\Delta G°$ and the activation energy, E_A, which can be determined by measuring the temperature dependence of the rate.

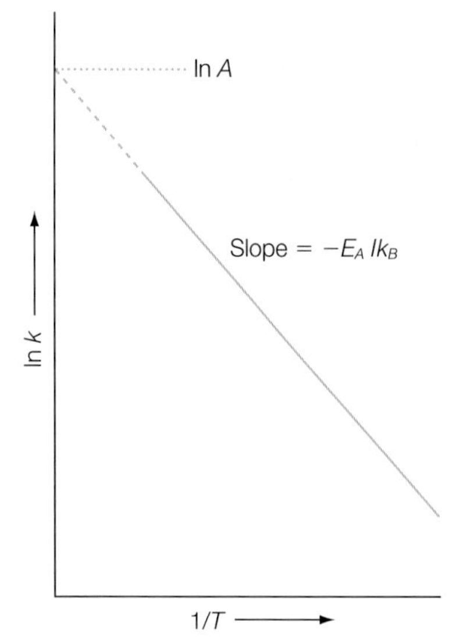

Figure 7.8 When a reaction proceeds via an activation energy, ln$k$ varies inversely with temperature, with the slope being proportional to the activation energy.

usually determined by measurement of the temperature dependence of the reaction. The temperature dependence, known as an Arrhenius equation, is most easily expressed as a linear equation by using the logarithm of the rate (Figure 7.8):

$$\ln k = \ln A - \left(\frac{E_A}{k_B}\right)\frac{1}{T} \tag{7.28}$$

$y = mx + b$; $y = \ln k$; $m = -(E_A/k_B)$, and $x = 1/T$

The activation energy represents an energy barrier that must be overcome if the reaction is to proceed. The dependence of the time evolution of the component concentrations is highly dependent upon the height of this barrier. Whereas the detailed changes are highly specific to the various rates, provided that the reaction can proceed to completion, some general comments can be made (Figure 7.8). The concentration of A will consistently decrease with time and the concentration of C will increase. The amount of the intermediate AB state will initially increase but will reach a peak at some point and then begin to decrease as the amount of C increases. The logarithm of the rate decreases linearly with a slope given by E_A/k_B, with a large slope corresponding to a large activation energy.

RESEARCH DIRECTION: ELECTRON TRANSFER I: ENERGETICS

Electron transfer plays a key role in many metabolic processes, including respiration (Chapter 9) and photosynthesis (Chapter 20). In some cases, electron transfer occurs as a second-order reaction when the electron donor is not normally part of the protein that contains the electron acceptor. Such cases are found when proteins are part of a metabolic pathway, as in respiration where cytochrome serves as an electron carrier between two complexes. For these electron-transfer reactions, the overall rate will be dictated by the diffusion of the carrier as the electron transfer does not occur until a complex is formed. In other cases, a protein has more than one cofactor as both the electron donor and acceptor are part of a large complex. The theory presented is applicable to either case: a protein with a bound donor and acceptor or a protein–protein complex that has formed transiently.

Electron transfer occurs between an *electron donor*, D, and *electron acceptor*, A, that can be separated by relatively large distances of up to 25 Å (Marcus & Sutin 1985; Murphy et al. 1993; Giese 2002; Page et al. 2003; Gray & Winkler 2005; Lin et al. 2005; Miyashita et al. 2005):

$$\text{DA} \rightarrow \text{D}^+\text{A}^- \quad (7.29)$$

In Marcus theory, the transfer is an activated process and the *electron transfer rate, k_{et}*, can be divided into two components, the *electronic coupling, V*, and the *Franck–Condon term*, which describes the energetics:

$$k_{et} = \frac{2\pi}{h} V^2(\text{Franck–Condon}) \quad (7.30)$$

Electron transfer is a quantum-mechanical process and the coupling, which has factors such as the distance between the donor and acceptor, will be considered in Chapter 10. The Franck–Condon term arises from a consideration of the states DA and D^+A^- in terms of harmonic oscillators (Figure 7.9). The two curves are displaced along the reaction coordinate corresponding to changes in the structure after transfer. These curves share the crossing point as the single common point where the two states have a common nuclear configuration. The crossing point represents the nuclear configuration at which the reactants are poised in the transitional state.

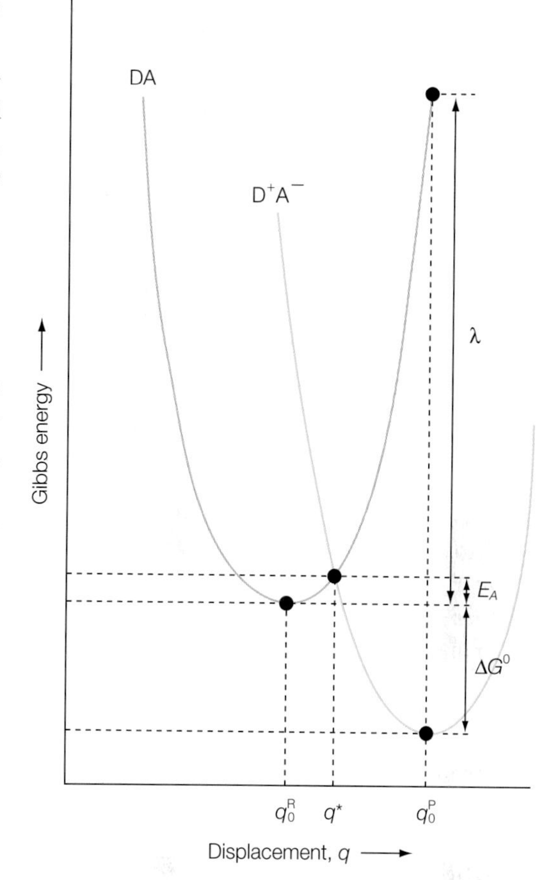

Figure 7.9 Gibbs energy surfaces for a DA pair and a D^+A^- pair that each are parabolas characteristic of harmonic oscillators (Chapter 13). Electron transfer usually makes use of the reorganization energy λ rather than the activation energy.

Electron transfer will depend upon the coupling and the probability that the product state will achieve the transition state. As discussed above, the probability of achieving the transition state can be expressed in terms of probability using the activation energy. For the specific case of Marcus theory, the activation energy can be written in terms of the difference in the Gibbs energy and the *reorganization energy, λ*:

$$E_A = \frac{(\Delta G^\circ + \lambda)^2}{4k_B} \quad (7.31)$$

The value of λ represents the energy that would need to be added to the system to force the state DA to have the same nuclear configuration as D^+A^-, and hence is a vertical line between the two curves. Substituting the activation energy into eqn 7.27 and accounting for the various factors yields:

$$\text{Franck–Condon} = \frac{1}{4\pi\lambda k_B T} e^{-(\Delta G^\circ + \lambda)^2/4k_B T} \quad (7.32)$$

One of the key aspects of Marcus theory is that it makes a basic prediction that the specific properties of the donor and acceptor control the coupling and that the dependence of the rate on the free-energy difference is parabolic (Figure 7.10):

$$k_{et} = \frac{2\pi}{h} V^2(\text{Franck–Condon}) \propto e^{-(\Delta G^\circ + \lambda)^2/4k_BT}$$
$$\ln k_{et} = \ln k_{max} - \frac{(\Delta G^\circ + \lambda)^2}{4k_BT} \tag{7.33}$$

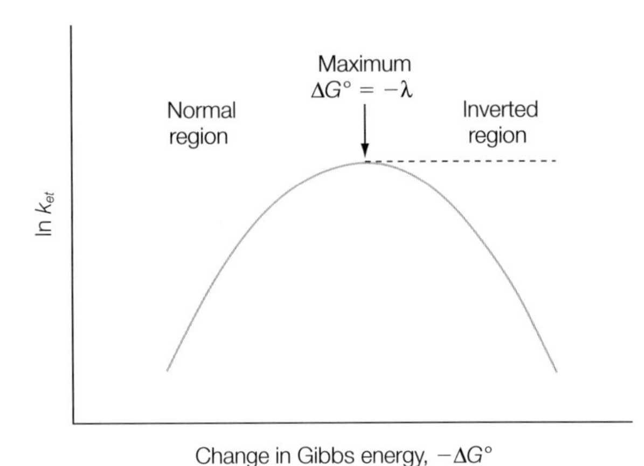

Figure 7.10 The dependence of rate on the free-energy difference in Marcus theory.

In writing this equation remember that ΔG° is a negative number for a favorable reaction. When ΔG° is smaller then λ, the reaction is slow due to the activation energy. In this region, the normal region, the rate increases as the free-energy difference increases. The optimal value for electron transfer occurs when the free-energy difference matches the reorganization energy and the activation energy has become zero. The theory predicts that when ΔG° is larger than λ, increasing the free-energy difference will result in slowing the reaction. This region has been termed the inverted region, and is not normally relevant to biological systems. While experimental tests have largely confirmed the validity of this dependence in the normal region, the predicted dependence in the inverted region is generally not observed due to the influence of additional factors involving the assumptions of harmonic oscillators. For his development of these ideas, Rudolph Marcus received the Nobel Prize in Chemistry in 1992.

Derivation box 7.1

Derivation of the Marcus relationship

The Marcus relationship can be derived by making use of the geometric nature of the parabolic relationships assigned to each state. Marcus assumed that the two curves were identical parabolas displaced relative to each other. Since the origin can be set without restriction, the *dependence for the initial state, y,* can be described with the assumption that the curve passes through the origin:

$$y = x^2 \tag{db7.1}$$

The *curve for the charge-separated state, y′*, can be written as a parabola that has been displaced horizontally by a *distance, d,* and displaced vertically by the *free-energy difference,* $\Delta G°$:

$$y' = (x - d)^2 + \Delta G° \tag{db7.2}$$

When x is equal to d, then the initial curve has a value of λ. Thus, the final curve must have a value of $\lambda + \Delta G°$ when x is equal to zero, yielding:

$$\begin{aligned} y' &= \lambda + \Delta G° = (0 - d)^2 + \Delta G° \\ \lambda &= d^2 \\ d &= \sqrt{\lambda} \end{aligned} \tag{db7.3}$$

At the crossing point, the two curves are equal so:

$$\begin{aligned} y &= y' \\ x^2 &= (x - \sqrt{\lambda})^2 + \Delta G° \\ x^2 &= x^2 - 2x\sqrt{\lambda} + \lambda + \Delta G° \\ 2x\sqrt{\lambda} &= \lambda + \Delta G° \\ x &= \frac{\lambda + \Delta G°}{2\sqrt{\lambda}} \end{aligned} \tag{db7.4}$$

Finally, we also know that at the crossing point the y value is equal to the activation energy E_A:

$$\begin{aligned} y = x^2 &= \left(\frac{\lambda + \Delta G°}{2\sqrt{\lambda}}\right)^2 = E_A \\ E_A &= \frac{(\lambda + \Delta G°)^2}{4\lambda} \end{aligned} \tag{db7.5}$$

Although these assumptions generally hold true in the normal region, they may not be applicable in the inverted region. Rather, the contribution of additional vibrational modes leads to the rate becoming close to independent of the free-energy difference.

ENZYMES

One of the fundamental conditions for life is that an organism must be able to catalyze chemical reactions efficiently and selectively. Such functions are performed in cells by highly specialized proteins called enzymes. Enzymes

Table 7.1

Classification of enzymes.

Enzyme class	Enzyme function
Oxidoreductase	Transfer of electrons
Transferase	Group-transfer reactions
Hydrolase	Hydrolysis reactions
Lyase	Addition of groups to double bonds or formation of double bonds by removal of groups
Isomerase	Transfer of groups within molecules to yield isomeric forms
Ligase	Formation of C–C, C–S, C–O, and C–N bonds coupled to ATP

not only have a remarkable degree of specificity for their substrates, but they also accelerate reactions tremendously under mild conditions of pH, temperature, and pressure. As early as the 1800s, scientists such as Louis Pasteur noted that biological reactions such as the fermentation of sugar into alcohol required biological catalysts to be present in organisms. In 1897, Eduard Buchner discovered that yeast extracts could perform fermentation, establishing that the catalysts were functional outside of the cell and hence that activity did not require that the catalyst be part of a living cell. In the early 1900s, the term enzyme was coined by Frederick Kuhne and these catalysts were established as being proteins. By the latter part of the century, thousands of enzymes were isolated and general mechanisms of enzymes were elucidated.

While some enzymes consist of only polypeptide chains, many enzymes contain *additional chemical components*, called *cofactors*. A cofactor may be a simple inorganic ion, such as Cu^{2+} or Mg^{2+}, or may be a cluster of inorganic ions as found in nitrogenase (Chapter 15). Alternatively, the cofactor may be a complex organic or metallo-organic molecule, such as the heme cofactor (ferriprotoporphyrin) found in catalase. A complete, catalytically active enzyme is referred to as a holoenzyme, and the protein part alone is termed the apoenzyme. Many enzymes are named according to their enzymatic function. Thus, the enzyme urease catalyzes the hydrolysis of urea while DNA polymerase catalyzes the DNA polymerization from nucleotides. In general, enzymes can be grouped into six major classes (Table 7.1).

Enzymes lower the activation energy

Enzymes accelerate reactions that have a substantial activation energy by modifying the reaction rates. The Gibbs energy difference between the initial and final states is not altered and the equilibrium is not changed. Rather, enzymes alter the transitional state of the reaction such that the activation energy is significantly decreased (Figure 7.11). Since the rate is exponentially dependent upon the activation energy, reductions of E_A lead to substantial increases in the rate. For example, catalase is an enzyme that catalyzes the decomposition of hydrogen peroxide:

$$H_2O_2\ (aq) \rightarrow H_2O\ (liquid) + \frac{1}{2}O_2\ (gas) \qquad (7.34)$$

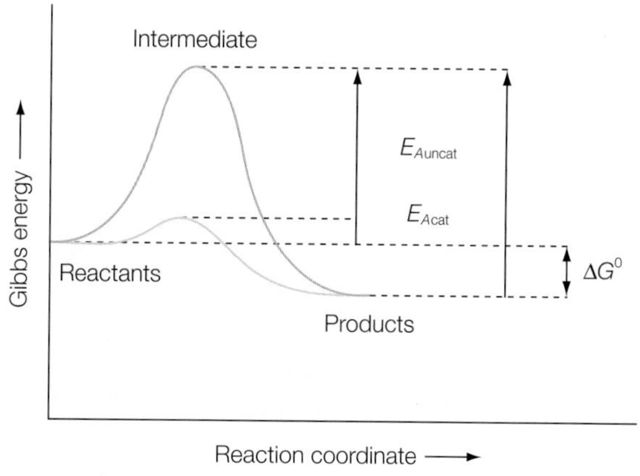

Figure 7.11 An energy diagram showing how enzymes increase reaction rates by lowering the activation energy from an uncatalyzed value E_{Auncat} to a catalyzed value E_{Acat}.

The reaction is exergonic with a Gibbs energy difference of −103.1 kJ mol^{-1}. However, the reaction essentially does not proceed due to a large activation energy of 71 kJ. The enzyme lowers E_A from 71 to 8 kJ, resulting in an increase in rate by a factor of more than 10^{15}.

While the degree of rate increase is unusually large for catalase, enzymes generally improve reaction rates by many orders of magnitude. Consider the overall reaction to be described by an equilibrium between the *initial* and *final states, A* and *B,* as described in eqns 7.35 and 7.36:

$$A \underset{k_b}{\overset{k_f}{\leftrightarrow}} B \qquad (7.35)$$

$$K_{eq} = \frac{[B]}{[A]} = \frac{k_f}{k_b} \qquad (7.36)$$

The equilibrium constant can be expressed in terms of the Gibbs energy change for the reaction (eqn 3.20):

$$K_{eq} = e^{-\Delta G/kT} \qquad (7.37)$$

where the product RT has a value of (8.315 J/(K mol))(298 K) or 2.47 kJ mol^{-1} at room temperature. By lowering the Gibbs energy difference, the equilibrium constant shifts towards the products (Table 7.2). The decrease in the equilibrium constant corresponds primarily to a decrease in the forward rate constant for the reaction. The rate enhancements achieved by enzymes are in the range of 5–17 orders of magnitude. Enzymes are able to achieve this significant rate enhancement while remaining very discriminating among substrates.

Table 7.2
Representative values of K_{eq} and $\Delta G°$ using eqn 3.20.

K_{eq}	$\Delta G°$ (kJ mol^{-1})
10^{-5}	28.5
10^{-3}	17.1
10^{-1}	5.7
1	0
10^{1}	−5.7
10^{3}	−17.1
10^{5}	−28.5

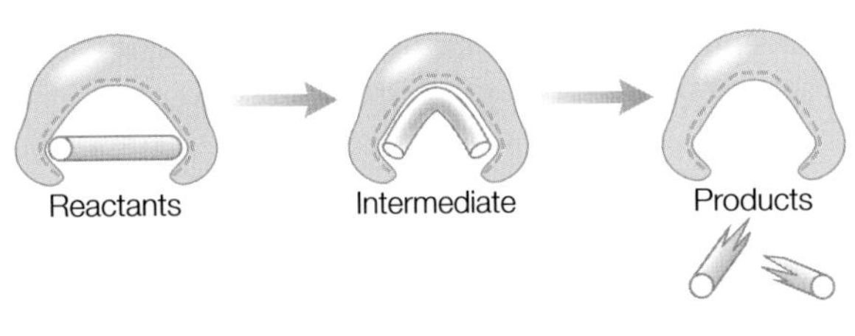

Figure 7.12 A schematic diagram of how enzymes facilitate the formation of an intermediate state of a reaction by favoring rearrangement of the reactants into the intermediate state.

Enzyme mechanisms

Although the specific mechanism by which enzymes stabilize the transitional state is unique for each protein, the general mechanism can be understood in terms of an enhancement of the stability of the transition state. In a simple picture (Figure 7.12), consider an enzyme designed to break a bond in a molecule. Normally the molecule is stable but binding of the molecule to the enzyme bends the molecule such that a specific bond is positioned for catalytic cleavage. The enzyme provides multiple weak interactions between the enzyme and substrate that are specifically positioned such that binding is optimized for the intermediate state. These interactions not only facilitate the reaction but also are designed for specificity, with binding constants several orders of magnitude larger for the substrate than for analogous molecules.

RESEARCH DIRECTION: DYNAMICS IN ENZYME MECHANISM

Proteins are not static molecules, they undergo dynamical rearrangements constantly (Frauenfelder et al. 1991). Proteins can undergo large-scale conformational changes on a millisecond timescale. For example, oxygen transport by hemoglobin is regulated by cooperative conformational changes. Even if proteins do not undergo large structural changes, they do undergo constant fluctuations with motions occurring on timescales as short as picoseconds (see Chapter 8). To some degree, such motions are inherent due to the nature of molecules in solution. However, enzymes may have evolved to make use of such motions in order to more efficiently perform catalysis.

Structural fluctuations may not only influence the binding of substrates to the active sites but they may also play a fundamental role in establishing the energetics of catalysis (Eisenmesser et al. 2005; Boehr et al. 2006; Vendruscolo & Dobson 2006). An enzyme with no substrate may be regarded as an inherently flexible molecule; however, the fluctuations may not be entirely random. Rather, the energies of the different dynamical conformations may map out a well-defined energy profile (Figure 7.13). The binding of the substrate results in a different energy landscape for the enzyme–substrate complex. A model for the role of dynamics in facilitating catalysis is for these two energy landscapes to have similar motions that have different relative populations. Thus, an enzyme with no substrate would have a dominant population, A, but only sample with small probability other, higher-energy states. Binding of the substrate would shift the dominant population from A, through an intermediate state B, towards the catalytic state C. According to a NMR study (Boehr et al. 2006), the enzyme dihydrofolate reductase has been found to undergo such motions. Dihydrofolate reductase is a well-characterized enzyme that catalyzes the reduction of 7,8-dihydrofolate to 5,6,7,8-tetrahydrofolate. The kinetic mechanism of this enzyme is coupled with the conversion of nicotinamide adenine dinucleotide phosphate, NADPH, to $NADP^+$. Substrate and cofactor exchange were found to occur through excited energy states of the enzyme with the modulation of the energy landscape funneling the enzyme through its kinetic path.

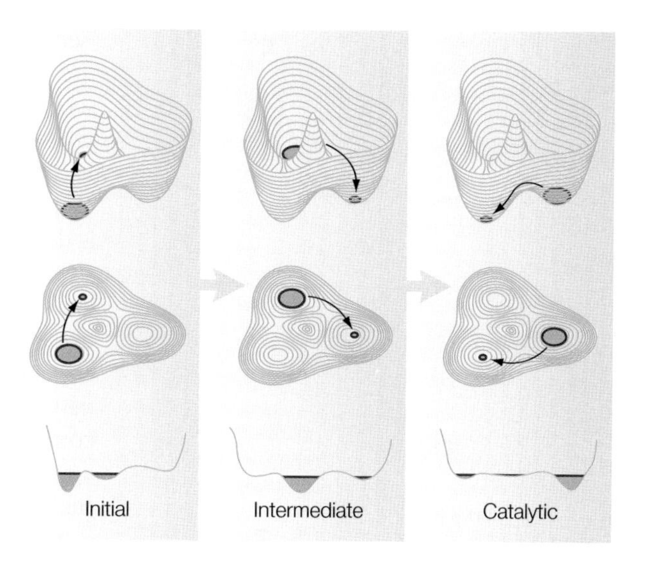

Figure 7.13
The energetics of enzymes may evolve as an enzyme undergoes dynamical changes that favor the transitional state (C) instead of the initial state (A) after binding of a substrate. Modified from Vendruscolo and Dobson (2006).

MICHAELIS–MENTEN MECHANISM

The basis for mechanisms involving formation of an enzyme–substrate complex was developed originally in the early 1900s, with a general theory of enzyme action proposed by Leonor Michaelis and Maud Menten in 1913. They postulated that the enzyme, E, and the substrate, S, form a complex, ES, in a fast reversible step (Figure 7.14) that yields the free enzyme and product, P, in a slower step:

$$E + S \underset{k_{b1}}{\overset{k_{f1}}{\leftrightarrow}} ES \underset{k_{b2}}{\overset{k_{f2}}{\leftrightarrow}} E + P \tag{7.38}$$

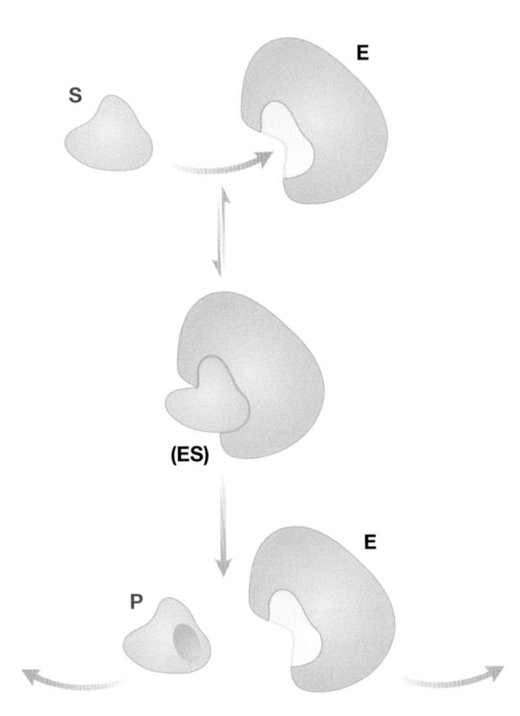

Figure 7.14 The basis of the Michaelis–Menten mechanism is the transient formation of a enzyme–substrate complex, ES.

Because the rate-limiting step is the second reaction, the rate of the overall reaction is determined by the second step and is proportional to the concentration of the complex.

Experimentally, enzyme reactions are often probed by measuring the *initial rate*, or *initial velocity*, which is denoted by V_0, when the concentration of the substrate is much greater than the concentration of the enzyme. The use of the initial velocity allows the use of the assumption that changes of the substrate concentration are negligible. The initial velocity usually has a linear dependence on the substrate concentration at low substrate concentrations and approaches an asymptotic value denoted by V_{max} at high concentrations (Figure 7.15).

This dependence of the initial velocity on the substrate concentration can be qualitatively understood in terms of the Michaelis–Menten model. At any given point in time, the enzyme is present both as a free enzyme and a complex with the substrate. At low substrate concentrations, most of the enzyme is present in the free form and the rate is proportional to the substrate concentration because the complex formation is favored as the substrate concentration is increased. The maximum velocity is approached at high substrate concentrations when essentially all of the enzyme is present as the complex. Under these conditions the enzyme is said to be saturated with its substrate, so the changes in the substrate concentration have very little effect.

Figure 7.15 The effect of substrate concentration on the initial velocity for an enzyme-catalyzed reaction.

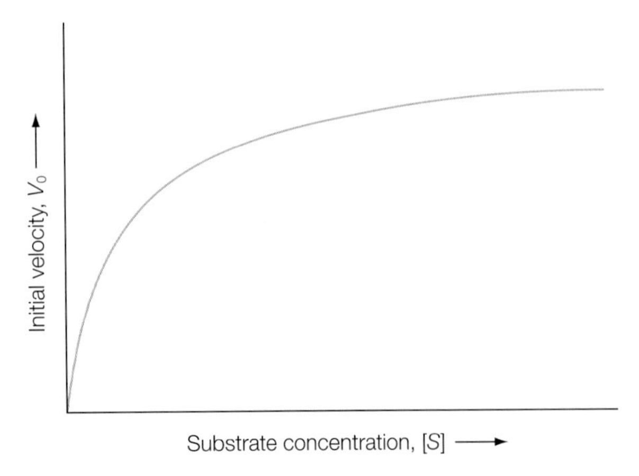

The dependence can also be quantified using the Michaelis–Menten model. For simplicity, assume that once the product is released from the enzyme, rebinding is unlikely and so the back reaction for the second step, k_{b2}, is negligible. The initial velocity is then determined by the product of the forward rate constant for the second step, k_{f2}, and the concentration of the complex, [ES]:

$$V_0 = k_{f2}[\mathrm{ES}] \tag{7.39}$$

Since the concentration of the complex is usually not determined readily, an expression for this concentration in terms of the experimental observables must be determined. The rate of complex formation is given by the product of the first forward rate constant, k_{f1}, the concentration of the free enzyme, [E], and the concentration of the substrate, [S]:

$$\frac{d}{dt}[ES] = k_{f1}[E][S] \qquad (7.40)$$

Whereas the first step results in an increase in the amount of the complex in the forward direction, the complex concentration will decrease due to the back reaction of the first step and the product formation in the second reaction. Thus, the rate of loss of the concentration of the complex is given by the sum of these two terms:

$$-\frac{d}{dt}[ES] = k_{b1}[ES] + k_{f2}[ES] \qquad (7.41)$$

In order to make use of these relationships, a critical assumption is invoked, termed the *steady-state assumption*. The initial rate of the reaction is assumed to occur with a constant concentration of the complex; that is, the rates of formation and loss of the complex are equal. With this assumption the two rates can be equated:

$$\frac{d}{dt}[ES] = -\frac{d}{dt}[ES] \qquad (7.42)$$

$$k_{f1}[E][S] = k_{b1}[ES] + k_{f2}[ES]$$

The concentration of the free enzyme can be written as the total concentration of the enzyme, $[E_{total}]$, minus the amount of the complex, and this term can be substituted into eqn 7.42 and the relationship can be rewritten to provide the concentration of the complex in terms of the experimental observables, the total enzyme concentration, and the amount of the substrate:

$$[E] = [E_{total}] - [ES] \qquad (7.43)$$

$$k_{f1}([E_{total}] - [ES])[S] = k_{b1}[ES] + k_{f2}[ES]$$

$$k_{f1}[E_{total}][S] - k_{f1}[ES][S] = (k_{b1} + k_{f2})[ES]$$

$$k_{f1}[E_{total}][S] = (k_{b1} + k_{f2} + k_{f1}[S])[ES]$$

Solving this equation for the concentration of the complex yields:

$$[ES] = \frac{k_{f1}[E_{total}][S]}{(k_{b1} + k_{f2} + k_{f1}[S])} = \frac{[E_{total}][S]}{(k_{b1} + k_{f2})/k_{f1} + [S]} = \frac{[E_{total}][S]}{K_M + [S]} \quad K_M = \frac{k_{b1} + k_{f2}}{k_{f1}}$$

(7.44)

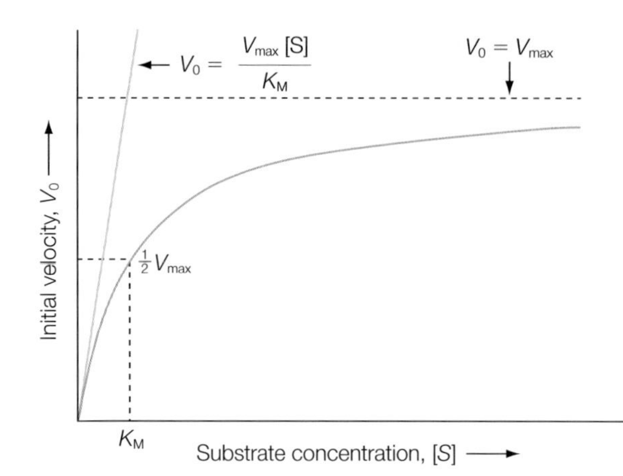

Figure 7.16 Michaelis–Menten dependence of the initial velocity showing the values of the maximum velocity, V_{max}, and the velocity at half maximum.

In the last expression, the term $(k_{b1} + k_{f2})/k_{f1}$ has been replaced with the *Michaelis constant*, K_M. Since the initial velocity is proportional to the concentration of the complex (eqn 7.39), the initial velocity can now be written in terms of the total enzyme and substrate concentrations. Because the *maximum velocity*, V_{max}, occurs when the enzyme is saturated – [ES] = $[E_{total}]$ – the maximum velocity defined in terms of the total enzyme concentration can substituted into the expression for the initial velocity:

$$V_{max} = k_{f2}[ES_{saturation}] = k_{f2}[E_{total}] \tag{7.45}$$

$$V_0 = k_{f2}[ES] = \frac{k_{f2}[E_{total}][S]}{K_M + [S]} = \frac{V_{max}[S]}{K_M + [S]}$$

This final expression for the initial velocity is termed the *Michaelis–Menten equation*. The interpretation of this relationship in terms of the observables (Figure 7.16) can be established using two different cases. First, at very high concentrations of the substrate, the substrate concentration is much larger than K_M, and so the initial velocity is seen to approach the maximum velocity as expected:

$$V_0 = \frac{V_{max}[S]}{K_M + [S]} \approx \frac{V_{max}[S]}{[S]} = V_{max} \quad \text{when} \quad K_M << [S] \tag{7.46}$$

The second special situation is when the initial velocity is exactly half the maximum velocity, as at this point the substrate concentration exactly equals K_M. First, the initial velocity is set equal to half of the maximum velocity:

$$V_0 = \frac{V_{max}}{2} = \frac{V_{max}[S]}{K_M + [S]} \tag{7.47}$$

then both sides of the equation are divided by V_{max}:

$$\frac{1}{2} = \frac{[S]}{K_M + [S]} \rightarrow K_M = [S] \tag{7.48}$$

$$\frac{1}{2} = \frac{[S]}{K_M + [S]} \rightarrow K_M + [S] = 2[S]$$

Thus, the value of the maximum velocity can be found by an extrapolation of the curve, and the value of K_M is equal to the substrate concentration at half-maximum velocity.

LINEWEAVER–BURK EQUATION

The Michaelis–Menten equation can be transformed into a linear relationship by making use of parameters other than the initial velocity and substrate concentration for the graph. One common relationship is derived by taking the reciprocal of the Michaelis–Menten equation (eqn 7.45):

$$(V_0)^{-1} = \left(\frac{V_{max}[S]}{K_M + [S]}\right)^{-1} \qquad (7.49)$$

$$\frac{1}{V_0} = \left(\frac{K_M + [S]}{V_{max}[S]}\right) = \frac{K_M}{V_{max}[S]} + \frac{[S]}{V_{max}[S]} = \frac{K_M}{V_{max}[S]} + \frac{1}{V_{max}}$$

$y = mx + b$ where $y = 1/V_0$, $m = K_M/V_{max}$, and $b = 1/V_{max}$

This form of the relationship is known as the *Lineweaver–Burk equation.* For enzymes following the Michaelis–Menten mechanism, a plot of the reciprocal of the initial velocity against the reciprocal of the substrate concentration, a so-called *double-reciprocal plot,* produces a straight line (Figure 7.17). The slope of the plot provides the value of K_M and the y intercept provides the value of the maximum velocity. The advantage of this type of plot is that it allows the data to be interpreted in terms of a simple linear relationship and provides an accurate estimate of the maximum velocity. Other transformations of the Michaelis–Menten equation have been derived, each of which is useful in analyzing certain types of enzyme data.

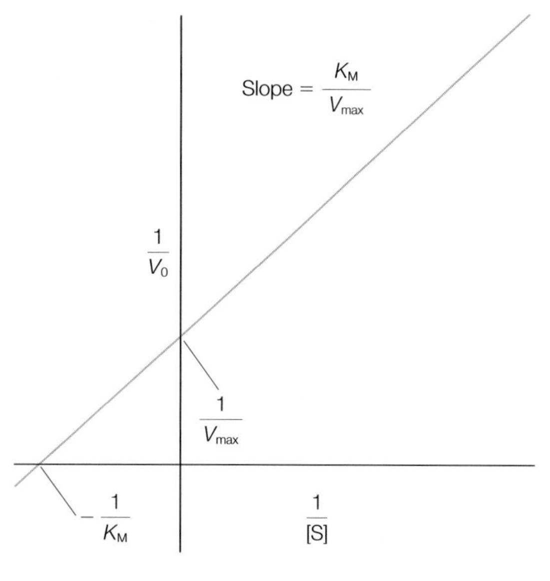

Figure 7.17 A double-reciprocal plot for enzymes with the slope equal to K_M/V_{max} and a y intercept of $1/V_{max}$.

ENZYME ACTIVITY

The mechanism of any given enzyme may be much more complex than the simple two-step model. However, enzymes usually do follow the relationship described by the Michaelis–Menten equation, although the interpretation of the resulting parameters depends upon the specific mechanism. For enzyme reactions that proceed in two steps, the parameter K_M is given by the ratio of the rates (eqn 7.44).

$$K_M = \frac{k_{f2} + k_{b1}}{k_{f1}} \tag{7.50}$$

When the enzyme operates under the condition that $k_{f2} \ll k_{1b}$, then the equation reduces to:

$$K_M = \frac{k_{f2}}{k_{f1}} \tag{7.51}$$

In this case, K_M is a measure of the affinity of the enzyme for its substrate. However, for many enzymes this limit of the rates does not hold and the interpretation of K_M is not straightforward, especially when the enzyme undergoes a multistep process.

To describe the enzyme activity, a more general rate constant, k_{cat}, is used to describe the limiting rate under saturating conditions. When the mechanism is complex, this parameter represents a function of several rate constants. However, for a simple two-step reaction, the new parameter can be equated to the forward rate of the second step ($k_{cat} = k_2$). In this case the Michaelis–Menten equation (eqn 7.45) can be written in terms of k_{cat}:

$$V_0 = \frac{k_{f2}[E_{total}][S]}{K_M + [S]} = \frac{k_{cat}[E_{total}][S]}{K_M + [S]} \tag{7.52}$$

The parameter k_{cat} is a *first-order rate constant* and is called the *turnover number* as it is equivalent to the number of substrate molecules that are converted to product within a certain time when the enzyme is saturated with substrate. Together with the K_M value, the k_{cat} provides a measure of the kinetic efficiency of the enzyme (Table 7.3). When the substrate concentration is small compared to K_M, this relationship can be simplified:

$$V_0 = \frac{k_{cat}[E_{total}][S]}{K_M + [S]} \approx \frac{k_{cat}[E_{total}][S]}{K_M} = \left(\frac{k_{cat}}{K_M}\right)[E_{total}][S] \tag{7.53}$$

The initial velocity is now dependent upon the concentrations of the two reactants: the total enzyme and substrate. This relationship is a second-order rate equation with the ratio k_{cat}/K_M being a second-order rate constant. One way to compare the efficiencies of different enzymes or the turnover of different substrates by one enzyme is to make use of the ratio k_{cat}/K_M, termed the specificity constant. The upper limit of this ratio is the rate at which the enzyme and substrate can diffuse together in an

Table 7.3
Experimental parameters for some enzymes.

Enzyme	Substrate	k_{cat} (s^{-1})	K_M (M)	k_{cat}/K_M (M^{-1} s^{-1})
Carbonic anhydrase	CO_2	1×10^6	1.2×10^{-2}	8.3×10^7
Catalase	H_2O_2	4×10^7	1.1	4×10^7
Fumerase	Fumarate	8×10^2	5×10^{-6}	1.6×10^8
Fumerase	Malate	9×10^2	2.5×10^{-5}	3.6×10^7

From Fersht, A. (1999) *Structure and Mechanism in Protein Science*. W.H. Freeman, New York, p. 166.

aqueous solution. The maximum limit of 10^8–10^9 M^{-1} s^{-1} has been achieved by some enzymes.

Enzymes can often catalyze reactions with different substrates, and molecules that resemble the substrates are sometimes able to occupy the catalytic site and act as competitive inhibitors. Competitive inhibitors are able to bind reversibly to the active site and to prevent catalysis of the substrate. Even if the inhibitor, I, binds only temporarily, the enzyme efficiency will decrease and effectively change the value of K_M, yielding an apparent K_M:

$$V_0 = \frac{k_{cat}[E_{total}][S]}{\alpha_{comp}K_M + [S]} \quad \alpha_{comp} = 1 + \frac{1}{K_I} \quad K_I = \frac{[E][I]}{[EI]} \tag{7.54}$$

Alternatively, an inhibitor may bind at a site distinct from the substrate active site and only bind to the enzyme–substrate complex. In this case, the binding of the inhibitor effectively changes the value of the substrate concentration:

$$V_0 = \frac{k_{cat}[E_{total}][S]}{K_M + \alpha_{uncomp}[S]} \quad \alpha_{uncomp} = 1 + \frac{1}{K_I'} \quad K_I' = \frac{[ES][I]}{[ESI]} \tag{7.55}$$

The type of inhibitor can be identified by use of the double-reciprocal plots (Figure 7.18). For a competitive inhibitor, the competition between the substrate and inhibitor for the binding site can be biased to favor the substrate by adding more substrate. When the substrate concentration exceeds the inhibitor concentration, the probability of an inhibitor binding instead of the substrate is minimized. The enzyme shows the same value of the maximum velocity but the substrate concentration required to produce the half-maximum velocity has increased, resulting in the apparent increase in K_M. On the double-reciprocol plot, this effect is evident as the lines have the same y intercept, which is proportional to $1/K_M$, but the

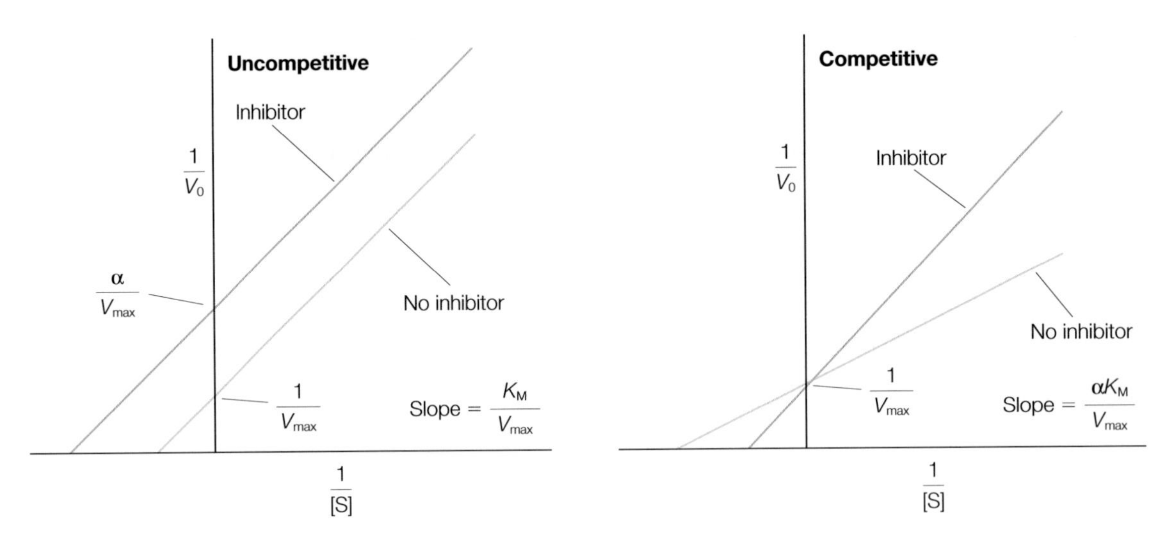

Figure 7.18 A competitive inhibitor results in a shift in the slope but not the y intercept (right) whereas an uncompetitive inhibitor shifts the y intercept but not the slope (left).

slopes have changed according to the apparent change in K_M. For the uncompetitive inhibitor, the apparent maximum velocity is decreased while the value of K_M is unchanged. On the double-reciprocal plot, the apparent change in the maximum velocity will alter the y intercept of the line but the slope of the line will remain unchanged. In practice, enzymes may have more than one substrate-binding position and so the presence of an inhibitor may have mixed contributions.

Inhibitors may also be irreversible and thus destroy an enzyme function. A special class of irreversible inhibitors are suicide inactivators. These compounds are unreactive until they bind to the active site of a particular enzyme. By design, the compound is not transformed into the normal product but into a reactive compound that combines irreversibly to the enzyme. Suicide inactivators play a critical role in the design of drugs as such inhibitors are specific and are less likely to have side effects than drugs with nonspecific action.

RESEARCH DIRECTION: THE RNA WORLD

The prebiotic world of the primitive Earth was significantly different than the Earth of today. Among the many unanswered questions concerning the earliest times of the Earth is how did life begin? The prebiotic environment may have contained basic building blocks, such as amino acids, sugars, purines, and pyrimidines (Figure 7.19). Although these may have combined to form short-lived polymers, the Earth at that time would not have had the chemically complex molecules found today, such as proteins or chromosomes. Of particular interest concerning early times

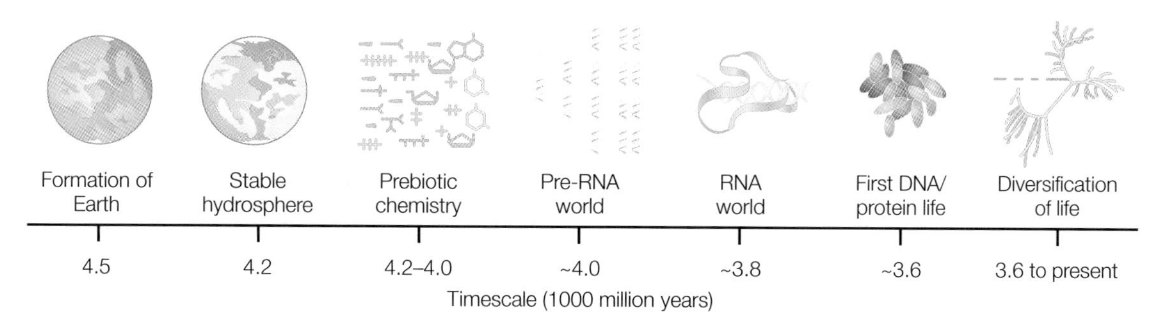

Figure 7.19 A timeline of events pertaining to the early history of the Earth. Modified from Joyce (2002).

is how these early molecules gained the capacity for self-replication. In organisms today, nucleic acids encode genetic information, which specifies the synthesis of proteins, and proteins perform cellular reactions, including the replication of nucleic acids. This interrelationship raises the question of which came first, nucleic acids or proteins?

A possible answer came about with the discovery that *some RNAs*, termed *ribozymes*, can have a *catalytic role in the transcriptional processing of RNA*; this was recognized with the Nobel Prize for Chemistry to Thomas Cech and Sidney Altman in 1989 (Guerrier-Takada & Altman 1984; Lantham & Cech 1989). Most of the roles identified for ribozymes are based upon splicing of RNA (Doudna & Cech 2002). In eukaroytic DNA, genes are often divided into segments, termed exons, with intervening noncoding regions termed introns. One prominent role of ribozymes is to facilitate the splicing of the RNA to remove these noncoding regions. To achieve this role, ribozymes have a wide range of sizes and structures. The best-characterized ribozyme is RNase P, which promotes the self splicing of a small RNA segment as part of replication. These RNA segments are termed the hammerhead ribozyme as their structures are in the shape of the head of a hammer (Figure 7.20). Other ribozymes are much larger, with some containing over 400 nucleotides that form complex structures. The ability of RNA molecules to act as catalysts in their own formation suggests that RNA may have been both the first source of genetic information and catalytic activity. According to this scenario, the primitive Earth was an RNA world. One of the earliest stages of evolution was the formation of RNA. As is true for protein enzymes, the three-dimensional structure of ribozymes is critical in their ability to perform catalytic functions. Activity is lost when ribozymes are heated to high denaturing temperatures or when essential nucleotides are changed. The enhancement rate constant for ribozymes can be substantial; for example, the self-cleaving hepatitis delta virus has a constant of about $10^3\ s^{-1}$. As found for their protein counterparts, ribozymes can make use of cofactors, such as imidazole, and can be regulated by small-molecule allosteric effects.

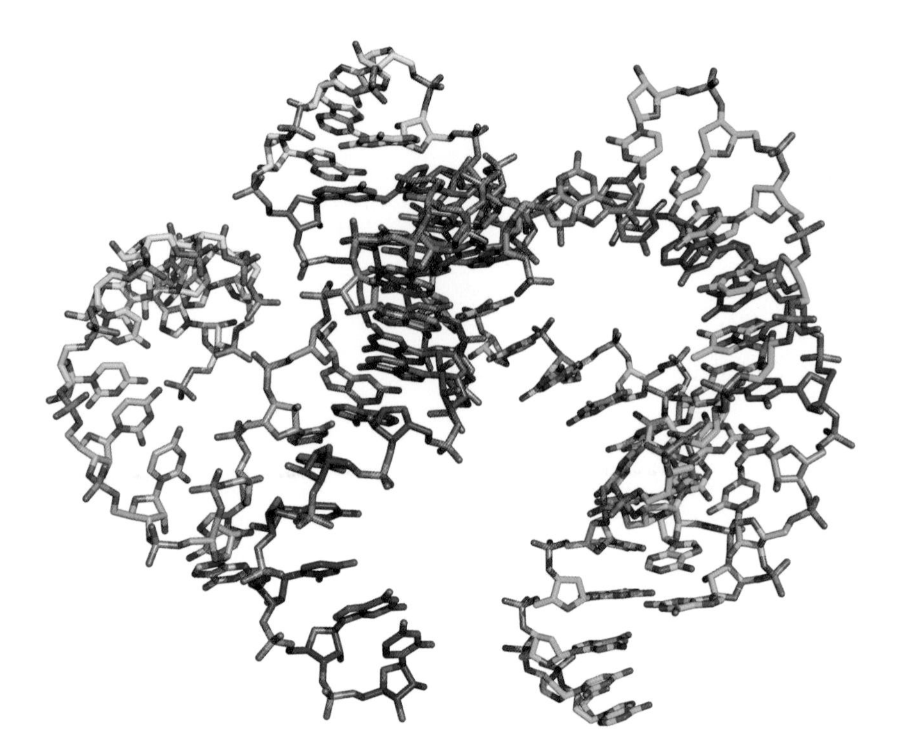

Figure 7.20 Certain virus-like elements called virusoids have small segments called hammerhead ribozymes due to the shape of their secondary structure. These RNA segments fold into well-defined three-dimensional structures.

Shortly after Cech and Altman discovered the catalytic properties of RNA, the ideal of an RNA world in which early life consisted on self-replicating RNA molecules was first proposed by Wallace Gilbert (Gilbert 1986). One of the earliest stages of evolution was the formation of catalytic RNA in the primordial soup of the primitive Earth. The amount of catalytic RNA would have increased exponentially with variants providing the opportunity for the selection of ribozymes with enhanced catalytic ability. In the RNA world hypothesis, the division of function into genetic information storage with DNA and catalysis by proteins would have arisen at a later stage of evolution. Proteins would have arisen after variants of self-replicating RNA molecules developed the ability to catalyze the condensation of amino acids into peptides. Some time after the development of a primitive protein-synthesis system, DNA molecules would have developed with sequences that were complementary to those of the self-replicating RNA. Eventually, DNA took over the role of conserving the genetic information and RNA became an intermediate in the process of converting information stored in DNA into the production of proteins. Despite significant issues concerning the RNA world, such as the stability of RNA and the lack of detailed mechanisms, the idea of the RNA world continues to be embraced by scientists and serves as a platform for research (Orgel 1998; Huttenhofer & Schattner 2006).

REFERENCES

Boehr, D.D., McElheny, D., Dyson, H.J., and Wright, P.E. (2006) The dynamic energy landscape of dihydrofolate reductase catalysis. *Science* **313**, 1638–42.

Doudna, J.A. and Cech, T.R. (2002) The chemical repertoire of natural ribozymes. *Nature* **418**, 222–8.

Eisenmesser, E.Z., Millet, O., Labeikovsky, W. et al. (2005) Intrinsic dynamics of an enzyme underlies catalysis. *Nature* **438**, 117–21.

Frauenfelder, H., Sligar, S., and Wolynes, P.G. (1991) The energy landscapes and motions of proteins. *Science* **254**, 1598–1603.

Giese, B. (2002) Electron transfer in DNA. *Current Opinion in Chemical Biology* **6**, 612–18.

Gilbert, W. (1986) The RNA world. *Nature* **319**, 618.

Gray, H.B. and Winkler, J.R. (2005) Long-range electron transfer. *Proceedings of the National Academy of Sciences USA* **102**, 3534–9.

Guerrier-Takada, C. and Altman, S. (1984) Catalytic activity of an RNA molecule prepared by transcription in vitro. *Science* **223**, 285–6.

Hüttenhofer, A. and Schattner, P. (2006) The principles of guiding by RNA: chimeric RNA-protein enzymes. *Nature Reviews Genetics* **7**, 475–82.

Joyce, G.F. (2002) The antiquity of RNA-based evolution. *Nature* **418**, 214–21.

Lantham, J.A. and Cech, T.R. (1989) Defining the inside and outside of a catalytic RNA molecule. *Science* **245**, 276–82.

Lin, J., Balabin, I.A., and Beratan, D.N. (2005) The nature of aqueous tunneling pathways between electron-transfer proteins. *Science* **310**, 1311–13.

Marcus, R.A. and Sutin, N. (1985) Electron transfers in chemistry and biology. *Biochimica Biophysica Acta* **811**, 265–322.

Miyashita, O., Okamura, M.Y., and Onuchic, J.N. (2005) Interprotein elctron transfer from cytochrome c_2 to photosynthetic reaction center: tunneling across an aqueous interface. *Proceedings of the National Academy of Sciences USA* **102**, 3558–63.

Murphy, C.J., Arkin, M.R., Jenkins, Y. et al. (1993) Long-range photoinduced electron transfer through a DNA helix. *Science* **262**, 1025–9.

Orgel, L.E. (1998) The origin of life-a review of facts and speculations. *Trends in Biochemical Sciences* **23**, 491–5.

Page, C.C., Moser, C.C., and Dutton, P.L. (2003) Mechanism for electron transfer within and between proteins. *Current Opinion in Chemical Biology* **7**, 551–6.

Vendruscolo, M. and Dobson, C.M. (2006) Dynamic visions of enzymatic reactions. *Science* **313**, 1586–7.

PROBLEMS

7.1 What are the units for the rate of a chemical reaction?

7.2 For a reaction $A \leftrightarrow B$ that has an equilibrium constant of 10, what can be said about the rates?

7.3 For the irreversible reaction $A \rightarrow B$, what effect will doubling the concentration of B have on the rate of change of A?

7.4 The time dependence of the conversion of a substrate by an enzymatic reaction is characterized by a half-life of 150 s. How long is required for the concentration of the substrate to decrease from 16 to 1 nM?

7.5 The time dependence of the conversion of a substrate by an enzymatic reaction is characterized by a rate of 0.0046 s^{-1}. How long is required for the concentration of the substrate to decrease from 16 to 1 nM?

7.6 Initially a system has species A at a concentration of 0.1 M. If a process proceeds from A to both B and C in parallel first-order reactions, and the two forward rates are identical, what are the final amounts of species B and C?

7.7 Initially a system has only species A at a concentration of 0.1 M. If a process proceeds from A to both B and C in parallel first-order reactions, and the forward rate to produce B is ten times larger than the rate to produce C, what are the final amounts of species B and C?

7.8 Consider the reaction $A + B \xrightarrow{k_{f1}} C \underset{k_{b2}}{\overset{k_{f2}}{\leftrightarrow}} D$, with rate constants of 2.0 M^{-1} s^{-1}, 1.0 s^{-1}, and 5.0 s^{-1} for k_{f1}, k_{f2}, and k_{b2} respectively. The initial concentration of all four components is 0.1 M. (a) Write the differential rate equations for each component. (b) Evaluate the initial rates for each component assuming that the initial concentrations are unchanged. (c) Determine the final equilibrium concentration of each component.

7.9 Consider the simple first-order reaction A → B. If you start with 1 M A and 0 M B and it takes 10 s to go to a concentration of 0.5 M for A and 0.5 M for B, what is the rate constant for this reaction?

7.10 Using 0.1 M for all initial conditions, for the irreversible reaction $A \xrightarrow{k_f} B$ with $k_f = 1\ s^{-1}$, write the (a) rate equations, (b) initial rates, and (c) final concentrations.

7.11 Using 0.1 M for all initial conditions, for the reversible reaction $A \underset{k_b}{\overset{k_f}{\leftrightarrow}} B$ with rate constants both equal to 1 s^{-1}, write the (a) rate equations, (b) initial rates, and (c) final concentrations.

7.12 Using 0.1 M for all initial conditions, for the sequential reaction $A \xrightarrow{k_{f1}} B \xrightarrow{k_{f2}} C$ with the two rates equal to 1 s^{-1}, write the (a) rate equations; (b) the initial rates; (c) the final concentrations.

7.13 If the rate for a reaction increases from 10 to 20 s^{-1} as the temperature increases from 298 to 330 K, what is the activation energy?

7.14 If the rate for a reaction increases from 7.0×10^{-6} to 3.0×10^{-5} s^{-1} as the temperature increases from 20 to 30°C, what is the activation energy?

7.15 If a process has a rate of 10^3 s^{-1} at 295 K and an activation energy of 5 kJ mol^{-1}, what is the rate when the system is cooled to 277 K?

7.16 The presence of an enzyme is observed to increase a reaction rate at 298 K from 1 to 10^3 s^{-1}. What can be said about the difference in the activation energies?

7.17 If the initial velocity of an enzyme changes in the presence of an inhibitor but the maximum velocity does not change, what can be said about the nature of the inhibitor?

7.18 If the initial velocity of an enzyme changes in the presence of an inhibitor and the maximum velocity changes, what can be said about the nature of the inhibitor?

7.19 From the following data measured for an enzyme, estimate the values of both V_{max} and K_m.

Substrate concentration (M)	**V_0 (μM min^{-1})**
2.5×10^{-6}	56
1×10^{-5}	140
4×10^{-5}	225
3×10^{-4}	265
1×10^{-3}	278
1×10^{-2}	280

7.20 How is an electron-transfer rate predicted to change with decreasing temperature in the Marcus theory?

7.21 Why are rates expected to reach a maximum in Marcus theory? What is the activation energy at this point?

8

The Boltzmann distribution and statistical thermodynamics

The concepts of thermodynamics have been presented in a classical way in terms of the laws of thermodynamics and fundamental concepts such as Gibbs energy and entropy. These same central concepts can also be explored from the more mathematical foundation of statistical thermodynamics. The statistical approach is capable of reproducing the thermodynamic properties of matter from a microscopic viewpoint. Systems are described in terms of the probabilities of different states that are described by the Boltzmann distribution and partition functions. A particular usefulness of this approach is its ability to provide a formal definition for entropy. The description of objects in terms of distributions is useful for understanding biological systems that involve multiple states, such as proteins as they fold, and that can exist in more than one conformation, as found for prions.

PROBABILITY

Probability theory was developed in the late 1600s as a formalism to describe games of chance. In probability theory, variables are quantities that can change in value throughout a series of events. Discrete variables can assume only a limited number of specific values. For example, the outcome of a coin toss is one of two possibilities, heads or tails. Once the variables are established, the goal is to determine the probability that a variable will have a certain value. Probabilites are defined such that for any given event with i possible outcomes, each of which has a probability P_i, the sum of all probabilities is one:

$$\sum_i P_i = 1 \tag{8.1}$$

Figure 8.1 All possible outcomes of four spins; this is analogous to tossing a coin four times.

Given all of the probabilities for any given event, what are the probabilities for some outcome after a series of events? Consider the question for the probabilities of different outcomes after four tosses. The outcomes can be determined by counting all possible combinations. After tossing a coin four times, there are a total of 16 possible outcomes (Figure 8.1). The probability of obtaining all heads is only one in 16, whereas the probability of obtaining at least two heads is 11 in 16. In general, if there are N total possible outcomes and the number of these outcomes corresponding to the desired result is N_i, then the probability of that result P_i is:

$$P_i = \frac{N_i}{N} \tag{8.2}$$

Although the counting of all possible outcomes can be done for simple cases, determining all possible outcomes for complex situations is not practical. For example, consider a deck of 52 cards. How many different arrangements of five cards are possible? There are 52 possibilities for the first card, 51 for the second, 50 for the third, 49 for the fourth, and 48 for the fifth. Each arrangement of the cards is called a permutation. The total number of permutations is given by the product of the possible outcomes for each card:

$$(52)(51)(50)(49)(48) = 311{,}875{,}200 \tag{8.3}$$

The product of numbers that decrease sequentially is called a factorial and identified by an exclamation point !, assuming that the numbers decrease to 1:

$$n! = n(n-1)(n-2)(n-3)\ldots(3)(2)(1) \tag{8.4}$$

The total number of possible configurations of a deck of 52 cards is then given by 52!. But what about the previous example, of the possible configurations of five cards from the deck of 52? In that case the factorials can be used if the contributions of the remaining 47 cards are removed:

$$(52)(51)(50)(49)(48) = \frac{52!}{47!} \tag{8.5}$$

In general, consider a set of n objects. The *number of permutations possible* for a *subset of j objects, $P(n,j)$* is given by:

$$P(n,j) = (n)(n-1)(n-2)\ldots(n-j+1) \tag{8.6}$$

But this expression can be rewritten using factorials by dividing out the remaining contributions:

$$P(n,j) = \frac{n!}{(n-j)!} \tag{8.7}$$

BOLTZMANN DISTRIBUTION

Instead of tossing coins, consider the different possible outcomes of four electron spins that can be either up or down. Assume that the energy of the up state is higher than the energy of the down state. There are still 16 possible outcomes, as shown in Figure 8.1, but in this case the total energy of the four spins will differ depending upon the number of up and down spins, with four up spins having the highest energy and four down spins the lowest. The different configurations can then be ranked in terms of the energy (Figure 8.2). The total energy of each of the configurations reflects the number of up and down spins, with the most probable value being the energy for two up and two down spins.

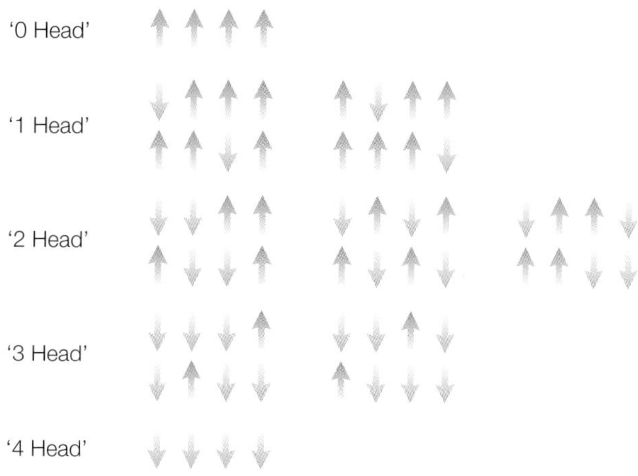

'0 Head'
'1 Head'
'2 Head'
'3 Head'
'4 Head'

Figure 8.2 The possible configurations of electron spins ranked by counting the number of up and down spins.

For real systems, the 16 different outcomes are not equally probable, as the lower-energy states are always favored. The probability of outcomes must then be weighted by the energy of each configuration, with lower-energy states receiving a larger weight. The weighting is performed by use of the Boltzmann distribution, named after Ludwig Boltzmann. In this case, the *probability of occupancy*, P_i, at any given energy E_i is given by:

$$P_i = \frac{e^{-E_i/k_BT}}{\sum_i e^{-E_i/k_BT}} \tag{8.8}$$

If the four spins are now weighted by the Boltzmann distribution, the number of configurations at the lowest energy will be much higher than at the other energies (Figure 8.3). The Boltzmann factor can be used to determine relative numbers of molecules in different energy states. For example, consider a system of N molecules that has two energies, E_1 and E_2. The ratio of the number of molecules in the two energy states is given by:

$$\frac{N_2}{N_1} = \frac{Ne^{-E_2/k_BT}/(e^{-E_1/k_BT} + e^{-E_2/k_BT})}{Ne^{-E_2/k_BT}/(e^{-E_1/k_BT} + e^{-E_2/k_BT})} = \frac{e^{-E_2/k_BT}}{e^{-E_1/k_BT}} = e^{-(E_2-E_1)/k_BT} \tag{8.9}$$

$$\frac{e^a}{e^b} = e^a e^{-b} = e^{a-b}$$

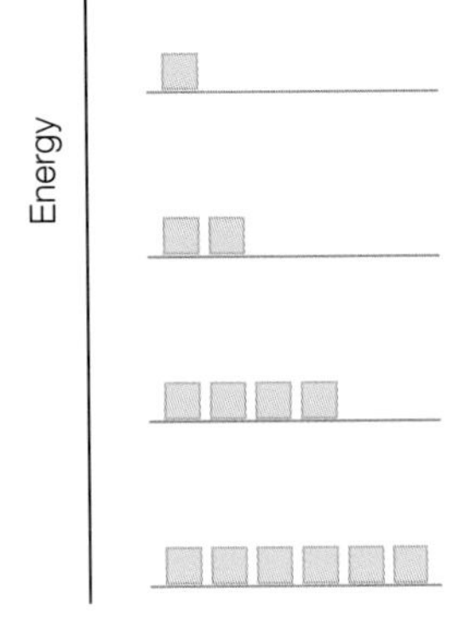

Figure 8.3
The number of configurations at each energy with weighting by the Boltzmann factor.

Thus, the relative population of the higher-energy state will exponentially decrease according to the difference in energies. Consider the probability of an electron occupying different electron orbitals (which will be discussed in detail in Chapter 12). The difference in energy for the two lowest orbitals of the hydrogen atom, the 1s and 2s orbitals, is 1.64×10^{-18} J, so the population ratio is:

$$e^{-(E_2-E_1)/k_BT} = e^{-(1.64\times10^{-18}\text{ J})/[(1.38\times10^{-23}\text{ J/K})(298\text{ K})]} = e^{-397} \approx 0 \qquad (8.10)$$

Essentially all of the electrons will occupy the lowest-energy states of the hydrogen atom model. When an unpaired electron is in the presence of a magnetic field, the energy levels will split for the two spin states. In this case the difference in energy is much smaller, at about 6.7×10^{-25} J, yielding a probability of just less than one:

$$e^{-(E_2-E_1)/k_BT} = e^{-(6.7\times10^{-25}\text{ J})/[(1.38\times10^{-23}\text{ J/K})(298\text{ K})]} = e^{-0.0016} = 0.9984 \qquad (8.11)$$

Note that the lower-energy state will be only slightly more populated. For this reason, spectroscopic measurements of the electron spins using magnetic fields (Chapter 16) often make use of low-temperature measurements that have an increase in the population difference compared to room temperature. For example, lowering the temperature from 298 K to 4 K results in the population ratio becoming measurably smaller:

$$e^{-(E_2-E_1)/k_BT} = e^{-(6.7\times10^{-25}\text{ J})/[(1.38\times10^{-23}\text{ J/K})(4\text{ K})]} = e^{-0.12} = 0.887 \qquad (8.12)$$

PARTITION FUNCTION

One convention in problems with states possessing different energies is that only the relative energies are considered, with the zero energy usually defined to be the ground state. The partition factor, q, is the sum of all terms that describe the probability associated with the variable of interest. For the Boltzmann factor, the *partition function*, q, is given by summation of the exponential terms at the energies E_i:

$$q = \sum_i e^{-E_i/k_BT} \qquad (8.13)$$

The *probability* P_i of occupying the *energy level* E_i can be written in terms of the partition function by inserting eqn 8.13 into eqn 8.8:

$$P_i = \frac{e^{-E_i/k_BT}}{\sum_i e^{-E_i/k_BT}} = \frac{e^{-E_i/k_BT}}{q} \tag{8.14}$$

The partition function then provides a measure of the occupancy of a certain energy level relative to the ground state. As an example, consider the simple harmonic oscillator. It will be shown in Chapter 11 that oscillators have energy levels that are equally separated with a value of $h\nu$, where h is Planck's constant. Defining the ground state to have zero energy yields the following for the partition function using eqn 8.13:

$$q = 1 + e^{-h\nu/k_BT} + e^{-2h\nu/k_BT} + e^{-3h\nu/k_BT} + \ldots \tag{8.15}$$

This equation can be rewritten using the series approximation:

$$\frac{1}{1+x} = 1 + x + x^2 + x^3 + x^4 + \ldots \tag{8.16}$$

Substituting the exponential term of eqn 8.15 for x in eqn 8.16 yields:

$$q = \frac{1}{1 - e^{-h\nu/k_BT}} \tag{8.17}$$

This partition function was used by Max Planck in developing the quantum theory, as discussed in Chapter 9. The partition term is dependent upon energy dependence with motions, such as rotation, having a different associated partition function.

STATISTICAL THERMODYNAMICS

Partition functions serve as the platform for the calculation of the thermodynamic properties of molecules. For example, the total energy of a system, E, is given by the sum of the products of the *number of molecules, N_i, at each energy, E_i*:

$$E = \sum_i N_i E_i \tag{8.18}$$

This expression can be rewritten since the number of molecules at an energy E_i is given by the partition function using eqn 8.14:

$$E = \sum_i \frac{Ne^{-E_i/k_BT}}{q} E_i = \frac{N}{q} \sum_i E_i e^{-E_i/k_BT} \tag{8.19}$$

The entropy can also be written using statistical arguments without the use of terms such as randomness. As originally shown by Ludwig Boltzmann in the late 1800s, the *entropy* S of a system can be written in terms of the *parameter* W:

$$S = k_B \ln W \tag{8.20}$$

The parameter W represents the *number of different configurations of molecules that result in the same energy*. If there is only one configuration for a molecule the entropy is zero since $\ln 1 = 0$. As the number of possible configurations increases, the entropy will increase. With these expressions, it is possible to derive relationships among different parameters that are identical to those found based upon the laws of thermodynamics.

RESEARCH DIRECTION: PROTEIN FOLDING AND PRIONS

Statistical approaches toward understanding biological systems are becoming increasingly important in biology and biochemistry. In genomics, the expression levels of thousands of genes are now routinely monitored using so-called DNA chips. As the number of organisms with their entire genome sequenced increases, the next step of using the availability of every gene to understand the properties of organisms remains a challenge. The genes can be individually placed on microarrays, providing the opportunity for massive parallel monitoring of the degree of gene expression in response to outside stimuli, such as temperature changes. For a number of organisms, commercial products are available that allow the monitoring of most or all genes of certain organisms ranging from *Escherichia coli* to humans. Meaningful interpretation of the output of these arrays requires careful consideration of the best statistical analysis of the data.

Aside from data analysis, the question of how states with conformations of different energy find the lowest-energy state is critical for protein folding. In cells, proteins are constantly produced and fold into unique conformations, despite the need for transport across cell membranes and assembly into large complexes with cofactors, which is often assisted by proteins called chaperones. If proteins were free to fold without direction the number of possible conformations would be too large for a unique shape to always be achieved, in what is termed the *Levinthal paradox* after Cyrus Levinthal who originally posed the paradox in 1968. For example, if a protein has 50 amino acid residues that each can have one of 10 different conformations, then the entire protein has 10^{50} different possible conformations. If each of these conformations had equal weight then the protein would never be able to randomly sample a sufficient number to determine the best conformation. Even a very rapid sampling time of 1 ps per conformation would require 10^{38} s, and the cell containing the protein would be long dead before even a modest set of possibilities were sampled.

The prediction of the folding pathway of a protein of a specified sequence remains a challenge (Onuchic & Wolynes 2004; Das et al. 2005; Hubner et al. 2005; Liwo et al. 2005; Cho et al. 2006). The folding process is complex and different models have been proposed. The folding can be hierarchical, with local structural elements, such as α helices, folding first followed by longer-range interactions driving the folding of helices together and eventually large domains folding into the full structure. Alternatively, the initial folding may be sequence-specific, with hydrophobic amino acid residues collapsing into a glassy state mediated by hydrophobic interactions with hydrophilic residues on the exterior. Thermodynamically, each of these conformations has a certain energy and the distribution can be plotted (Figure 8.4). The unfolded states are at high energy and have a large degree of conformational entropy. As folding proceeds the energy decreases and the distribution of states is funneled into an ensemble of folding intermediates until the single lowest-energy state is achieved. In the course of the folding, the protein must have enough energy available to overcome barriers and to continuously sample states, even if a local minimum is reached.

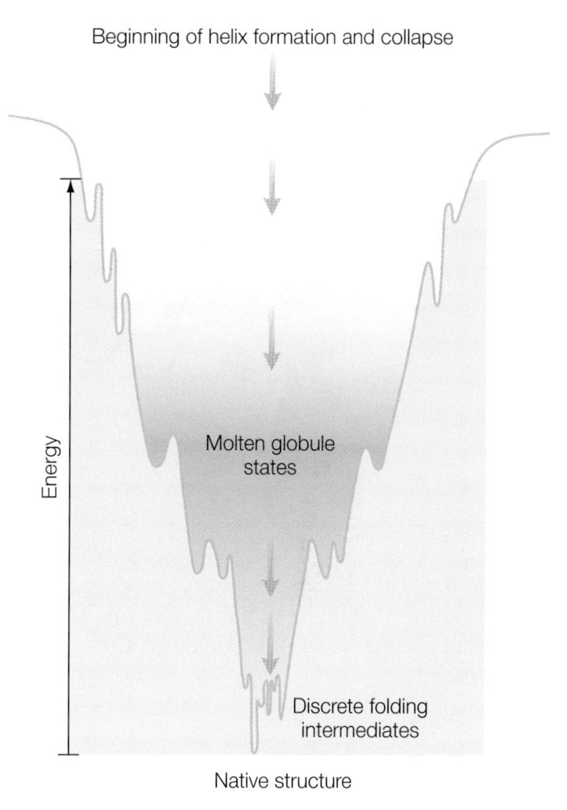

Figure 8.4 The energy landscape for protein folding showing the energies of a protein in different configurations with the fully folded state having the lowest energy.

In principle, for a given sequence, if all of the interactions involving amino acid residues were properly modeled, it should be possible to predict the folding process and hence the three-dimensional structure. Effectively, the energy landscape shown in Figure 8.4 could be mapped and the lowest-energy state predicted. The availability of very fast computers has led to the development of different programs that are increasingly becoming more effective in their predictions. The plethora of sequences provides a testing ground for predicted structures and the possibility of refining critical interactions for folding based upon sequence comparisons. Furthermore, the experimental feasibility to generate peptides, whether chemically or through expression systems, provides the opportunity to test the effects of sequence changes on protein folding.

PRIONS

One of the intriguing questions in biology is how the conversion of proteins with intricate folds, appearing to be at a stable energy minima, can alter their shape and form long aggregates. These transformations are not only of interest in understanding the energy landscapes of proteins but are

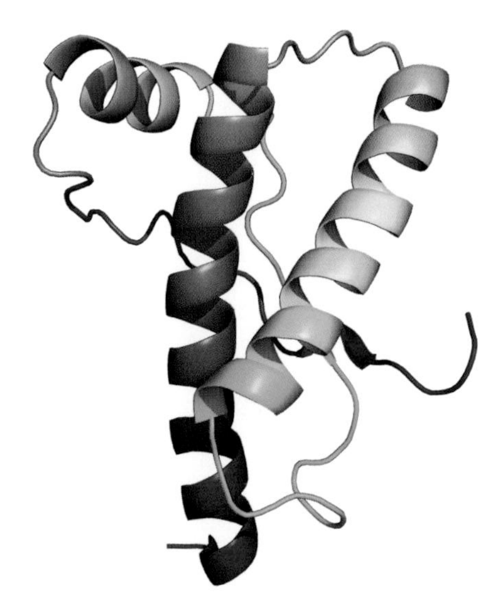

Figure 8.5 The NMR structure of a prion showing a well-defined globular domain. The open, extended region was not resolved in the NMR data.

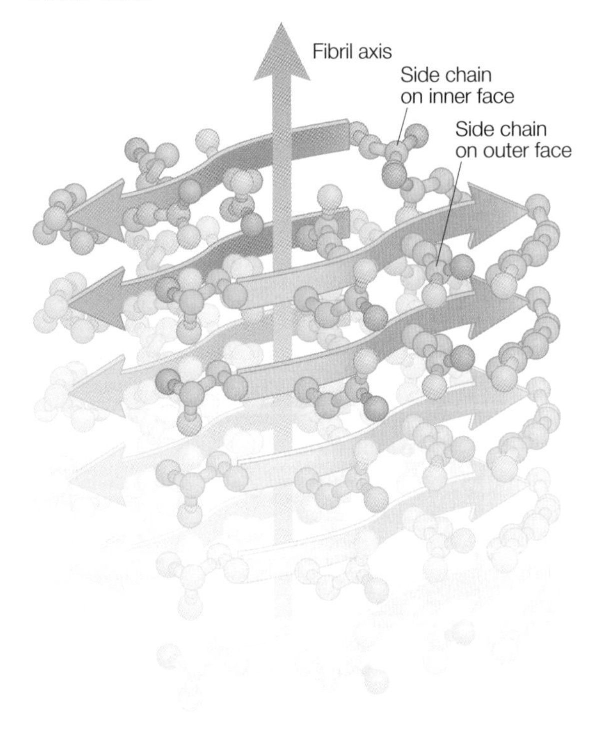

Figure 8.6 The crystal structure of a seven-residue peptide from the yeast prion Sup35, showing the β-sheet arrangement of the peptide. From Nelson et al. (2005).

of very real importance as such aggregates are linked to many diseases that cannot be treated, including Alzheimer's disease, Creutzfeldt–Jakob disease, and bovine spongiform encephalopathy (so-called mad-cow disease). The involvement of misfolded proteins giving rise to these diseases was initially met with skepticism, but the efforts of Stanley Pruisner in the 1980s (Prusiner 1987) have led to a general acceptance as recognized by a Nobel Prize in Medicine in 1997.

Whereas these proteins, which are termed prions, for proteinaceous infectious particles, usually have a globular shape, they can also adopt a structure that leads to formation of amyloid-like fibrils (Zahn et al. 2000; Dobson 2003; Masison 2004; May et al. 2004; Dyson & Wright 2005; Krishnan & Lindquist 2005; Nelson et al. 2005). Proteins in amyloid fibrils are folded to form continuous arrays of β sheets. A large portion of a prion is folded in a compact globular arrangement with α helices and β sheets, with a sizable portion of the protein missing due to disorder, including most of the first 100 amino acid residues (Zahn et al. 2000; Figure 8.5).

The determination of the arrangement of prions in amyloid fibrils has been hampered by the limited order of fibrils isolated from diseased tissues. A seven-residue fragment has been shown by X-ray diffraction to form β sheets in the crystal structure (Figure 8.6; Nelson et al. 2005). One of the mysteries of prions is why one misfolded protein can drive a conformational chain reaction resulting in other folded proteins becoming misfolded. Although the mechanism of self-assembly remains unknown, the tendency of peptide fragments to form β sheets suggests the involvement of specific parts of the protein. The process should involve the N-terminal domain of the prion forming an intermediate state similar to those proposed above. However, this intermediate state is driven away from a globular form to a structure that results in formation of amyloid fibroids.

REFERENCES

Cho, S.S., Levy, Y., and Wolynes, P.G. (2006) P versus Q: structural reaction coordinates capture protein folding on smooth landscapes. *Proceedings of the National Academy of Sciences USA* **103**, 586–91.

Das, P., Matysiak, S., and Clementi, C. (2005) Balancing energy and entropy: a minimalist model for the characterization of protein folding landscapes. *Proceedings of the National Academy of Sciences USA* **102**, 10141–6.

Dobson, C.M. (2003) Protein folding and misfolding. *Nature* **426**, 884–90.

Dyson, H.J. and Wright, P.E. (2005) Intrinsically unstructured proteins and their functions. *Nature Reviews Molecular Cell Biology* **6**, 197–208.

Hubner, I.A., Deeds, E.J., and Shakhnovich, E.I. (2005) High-resolution protein folding with a transferable potential. *Proceedings of the National Academy of Sciences USA* **102**, 18914–19.

Krishnan, R. and Lindquist, S.L. (2005) Structural insights into a yeast prion illuminate nucleation and strain diversity. *Nature* **435**, 765–72.

Liwo, A., Khalili, M., and Scheraga, H.A. (2005) *Ab initio* simulations of protein-folding pathways by molecular dynamics with the united-residue model of polypeptide chains. *Proceedings of the National Academy of Sciences USA* **102**, 2362–7.

Masison, D.C. (2004) Designer prions. *Nature* **429**, 37–8.

May, B.C.H., Govaerts, C., Prusiner, S.B., and Cohen, F.E. (2004) Prions: so many fibers, so little infectivity. *Trends in Biochemical Sciences* **29**, 162–5.

Nelson, R., Sawaya, M.R., Babirnie, M. et al. (2005) Structure of the cross-β spine in amyloid-like fibrils. *Nature* **435**, 773–8.

Onuchic, J.N. and Wolynes, P.G. (2004) Theory of protein folding. *Current Opinion in Structural Biology* **14**, 70–5.

Prusiner, S.B. (1987) Prions and neurodegenerative diseases. *New England Journal of Medicine* **317**, 1571–81.

Zahn, B., Liu, A., Lührs, T. et al. (2000) NMR solution structure of the human prion protein. *Proceedings of the National Academy of Sciences USA* **97**, 145–50.

PROBLEMS

8.1 What is the probability of drawing a spade from a standard deck of 52 cards?

8.2 What is the probability of drawing a 4 from a standard deck of 52 cards?

8.3 A coin is tossed three times. What is the probability of (a) three heads and (b) two heads and one tail?

8.4 How many different arrangements of three cards from a standard deck of 52 cards are possible?

8.5 Four bases (A, C, G, and T) appear in DNA. Assume that the appearance of each base is random (which is not actually true). What is the probability of observing the following sequences? (a) AAG; (b) GGG; (c) GGGGAAG; (d) CATCATCATCAT.

8.6 Consider three spins with energies that are ranked with the energy of the up states being higher than the energy of the down states. For the three spins: (a) how many outcomes are possible; (b) how many energy states are possible; (c) how does each outcome fit into the energy ranking?

8.7 Consider two closely lying electronic states that are separated in energy by 9×10^{-24} J. What is the ratio of the number of molecules in the two energy states according to the Boltzmann distribution at (a) 298K; (b) 4K.

8.8 Consider two closely lying electronic states that are separated in energy by 5×10^{-27} J. What is the ratio of the number of molecules in the two energy states according to the Boltzmann distribution at (a) 298 K and (b) 4 K?

8.9 For a molecule with a vibrational mode with a CO stretch frequency of 500 cm^{-1} (corresponding to 1.5×10^{12} s^{-1}), what is the partition function at 298 K?

8.10 Why did Ludwig Boltzmann have $S = k_B \ln W$ written on his gravestone?

8.11 Use $S = k_B \ln W$ to calculate the entropy for a system that can exist in (a) one configuration, (b) two configurations, and (c) 10 configurations.

8.12 Why does the Levinthal paradox not prevent proteins from folding in the cell?

8.13 What is an energy landscape?

8.14 How does the time required for proteins to fold into their secondary structural elements compare with time required for overall protein folding?

8.15 Suppose a protein can exist in two conformations that have an energy difference of 2.0 kJ mol^{-1}. What is the estimated ratio of the conformations?

8.16 Why was the involvement of prions in disease so controversial?

8.17 How is the possible formation of β sheets related to the properties of prions?

8.18 Why are prion-based diseases difficult to treat?

Part 2

Quantum mechanics and spectroscopy

9

Quantum theory: introduction and principles

To understand the properties of cell membranes, proteins, and nucleic acids, it is necessary to have knowledge of how the molecules forming these biological components respond to the different interactions in their environment. At the turn of the nineteenth century much of what is now called classical physics had been developed, allowing scientists to understand the world around them. For example, the laws of motion introduced by Isaac Newton in the seventeenth century explained the motion of everyday objects as well as planets. However, several experimental results that were reproducible could not be explained based upon the classical science. This led some to re-examine the basic scientific principles and to ask how these principles could be modified to explain the experimental results. In a tremendous burst of intellectual effort, modification of the existing scientific theories lead to quantum mechanics, which is discussed in these next few chapters, as well as to relativity and gravitational theory. The new theories not only provided answers to the previously inscrutable experimental results but also led to many new directions for scientific inquiry.

In this chapter, the way in which classical physics characterized the physical world is summarized followed by a description of how certain experiments were in conflict with the behavior predicted using the classical theory. Presented next is the introduction of several nonclassical concepts that were used to understand these experimental results and gave rise to the development of quantum mechanics, which is applied in the following chapters to describe the properties of atoms and molecules.

CLASSICAL CONCEPTS

A classical particle is described by a number of parameters: mass, m, position, r, velocity, v, and charge, q. For simplicity, these are regarded as simple variables rather than as vectors that are needed to describe three-dimensional motion. The kinetic energy of the particle, KE, and the total energy, E, are related to these parameters and the potential energy, V, by:

$$KE = \frac{1}{2}mv^2$$
$$E = KE + V \tag{9.1}$$

According to classical mechanics, the energy of the system is always conserved. Also conserved is the linear momentum, p, which is equal to:

$$p = mv \tag{9.2}$$

Since the linear momentum is proportional to the velocity, the kinetic energy and total energy can be rewritten as:

$$KE = \frac{1}{2}mv^2 = \frac{m^2v^2}{2m} = \frac{p^2}{2m}$$
$$E = \frac{p^2}{2m} + V \tag{9.3}$$

Various possible interactions between particles, such as gravitation or electrostatic interactions, are described by well-defined relationships involving the parameters that describe the particles. For example, the electrostatic force, F, between two charged particles, q_1 and q_2, separated by the distance, r_{12}, is given by:

$$F = \frac{1}{4\pi\varepsilon_o}\frac{q_1q_2}{r_{12}^2} \tag{9.4}$$

where ε_o is a constant, the vacuum permittivity. According to classical mechanics, once all of the interactions and the initial conditions have been established, then the time evolution of the system can be predicted for all times using the principle of the conservation of energy, which states that the total energy of the system always remains the same, and using the laws of mechanics, such as the relationship between force, F, mass, m, and position, x:

$$F = ma = m\frac{\mathrm{d}^2x}{\mathrm{d}t^2} \tag{9.5}$$

where a is the acceleration of the particle and t is the time.

In classical physics, in addition to particles there are also objects called waves, which are physical objects extended throughout space. Because waves are not confined at a specific location they cannot be described by the same parameters as used for particles. Waves can be described by the parameters wavelength, λ, frequency, ν, and amplitude, A (Figure 9.1).

These parameters describe stationary waves that do not change with time, but waves can also travel with a velocity v. In this case, the wavelength is related to the frequency and speed according to:

$$\lambda\nu = v \qquad (9.6)$$

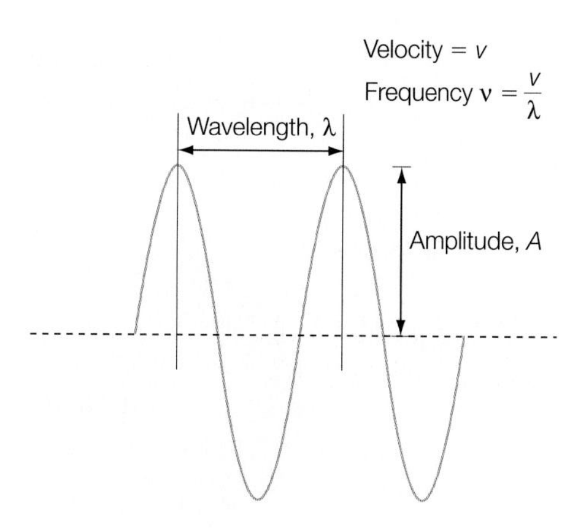

Figure 9.1
A classical wave with the wavelength, λ, and amplitude, A, marked. The frequency, ν, is inversely proportional to λ.

For the specific case of light, the velocity in a vacuum is denoted by the constant c and the equation becomes:

$$\lambda\nu = c \qquad (9.7)$$

The energy of a wave is given by its intensity, which is proportional to the square of the amplitude, and is independent of the frequency or wavelength:

$$E \propto A^2 \qquad (9.8)$$

According to classical theory, once all of these parameters are known for a particle or wave, the theory should be able to predict all properties of the object. Classical theory was indeed very successful in such predictions; however, there were some dramatic failures. We will present some of the failures and discuss how these lead to the development of quantum mechanics.

EXPERIMENTAL FAILURES OF CLASSICAL PHYSICS

Blackbody radiation

When objects get hot they emit radiation. The type of radiation emitted depends upon the temperature. For example, when you heat something the color can be a red or blue or even white. In most cases, the color emitted at different parts of the object will vary due to differences in the temperature. For an ideal emitter called a blackbody, the emitted radiation is in thermal equilibrium with the object, resulting in a uniform emission of radiation. A blackbody can be modeled as a sphere in which the light emitted by the interior walls is trapped – that is, absorbed and re-emitted – except for a small portion that can escape through a pinhole (Figure 9.2).

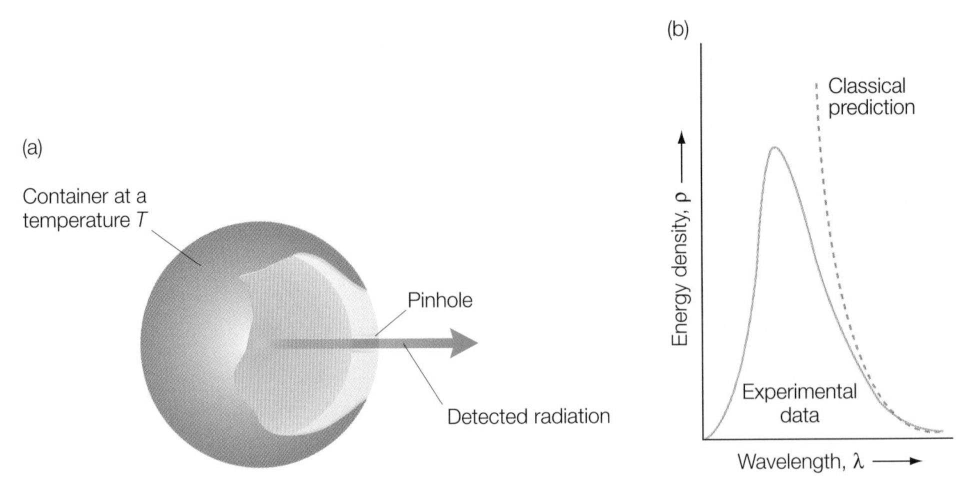

Figure 9.2 Blackbody radiation. (a) In the sphere light undergoes a series of reflections causing the light radiated through the pinhole to be uniform, with a spectrum characteristic of the temperature of the sphere. (b) The wavelength dependence for the energy density at both a high temperature and a low temperature. Also shown is the classical prediction for the dependence (dashed line).

A blackbody can be constructed, allowing the radiation output to be measured experimentally. The amount of light emitted at any given temperature is characterized by the energy density, ρ, which is a measure of how much energy is emitted within a specific wavelength region λ to $\lambda + \delta\lambda$, where $\delta\lambda$ is a small number. The energy density of a blackbody has a characteristic dependence upon both wavelength and temperature (Figure 9.2). At small values of wavelength, the amount of energy emitted is low. As the wavelength is increased, the amount of energy emitted increases until a peak value is reached. Increasing the wavelength further results in smaller amounts of energy being emitted.

The dependence of ρ on the wavelength was determined for many materials and was found to always follow the same general dependence. The dependence of the energy distribution can be modeled using statistical arguments for classical thermodynamics. Assuming that the radiated energy follows the classical dependence on the amplitude and is independent of the wavelength (eqn 9.8), the long-wavelength part of the dependence is predicted but fails for short wavelengths. Instead of predicting that the density will drop to zero, the theory predicts the so-called ultraviolet catastrophe: that the energy density will become infinite:

$$\rho = \frac{8\pi kT}{\lambda^4} \tag{9.9}$$

In 1900, Max Planck realized that he could derive the observed distribution if he made an unusual change to classical theory. Light is emitted from the material forming the blackbody as that material vibrates due to its thermal energy. Classically, it was assumed that the vibrators could have any frequency and any value of energy. Instead of using this assumption, Planck proposed that the energy was proportional to the frequency, according to:

$$E = nh\nu \quad \text{where} \quad n = 0,1,2,\ldots \quad \text{and} \quad h = 6.626 \times 10^{-34}\ \text{Js} \tag{9.10}$$

where h is now known as Planck's constant. Unlike classical mechanics, which allows objects to have any energy, this equation predicts that light can have energy only at certain discrete values, or, in other words, it is quantized according to the frequency.

Using statistical arguments (eqn 8.17) but with the different dependence for the energy, Planck derived a new dependence for the energy density that agreed with the experimental data (Figure 9.2):

$$\rho = \frac{8\pi hc}{\lambda^5}\left(\frac{1}{e^{hc/\lambda kT} - 1}\right) \tag{9.11}$$

At long wavelengths, this equation agrees with the classical prediction (eqn 9.8). At long wavelengths the exponential term is very small and the exponential can be written approximately as:

$$e^x - 1 \approx x \quad \text{when} \quad x \ll 1 \tag{9.12}$$

Using this approximation, the energy density can be written as:

$$\rho \approx \frac{8\pi hc}{\lambda^5}\left(\frac{hc}{\lambda kT}\right)^{-1} = \frac{8\pi kT}{\lambda^4} \tag{9.13}$$

At long wavelengths, the new equation gives the classical prediction but at small wavelengths only the quantum model correctly predicts that the energy density will decrease with decreasing wavelength. Why does this new assumption predict that the distribution will decrease at small wavelengths? Blackbody radiation is associated with a temperature and arises from motion of the walls of the blackbody. According to classical mechanics, the atoms can vibrate at all wavelengths and so energy should be emitted at all wavelengths. According to quantum mechanics the wavelength of the vibration is coupled to the energy of the system. At a given temperature there is only a limited amount of thermal energy available to vibrate the atoms and not enough energy for high-energy vibrations. The effect of the quantum theory is to remove the high-energy vibrations from consideration and hence to decrease the predicted amount of energy emitted at large frequencies or correspondingly small wavelengths. For his theoretical work, Planck received the Nobel Prize in Physics in 1918.

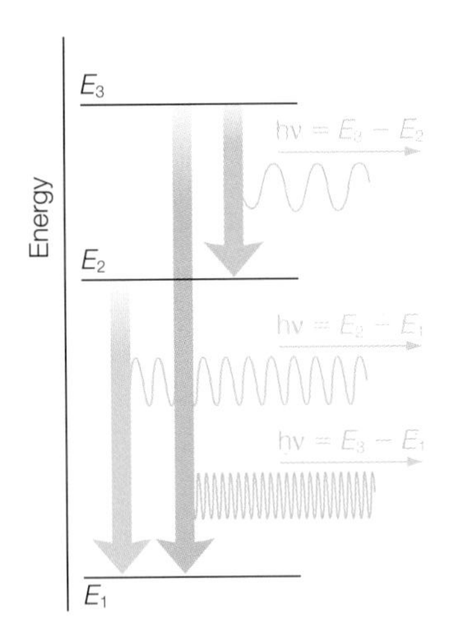

Figure 9.3 The presence of the discrete lines in the emission spectra of atoms suggests that the electrons are making transitions between orbitals with discrete energy levels.

Photoelectric effect

When light of certain wavelengths strikes a metal surface, an electron is ejected from the metal in the photoelectric effect. Classically, the kinetic energy of the ejected electron should be related to the light intensity, or amplitude squared, and is independent of the frequency used. Experimentally it is possible to measure this property for different metals and different light conditions (Figure 9.3). It is found that no electrons are ejected regardless of the intensity if the light frequency is below a certain value that is characteristic of the metal. For light above this critical frequency, the kinetic energy is linearly dependent upon the frequency, and the intensity only changes the number of electrons ejected (Figure 9.3).

Thus, the kinetic-energy dependence of the ejected electron does not follow the classical prediction. Albert Einstein resolved the discrepancy between the classical prediction and the experimental observation in 1905. Using the same ideas as put forward by Planck, he proposed that the light energy was not related to the amplitude but rather was proportional to the frequency according to eqn 9.8. A minimal energy, ϕ, called the work function, is required to eject the electron from the metal. According to the conservation of energy, the kinetic energy of the ejected electron, $\frac{1}{2}m_e v^2$, is equal to the energy of the light, $h\nu$, minus the energy required to remove the electron:

$$\frac{1}{2}m_e v^2 = h\nu - \phi \tag{9.14}$$

With this model, no electron can be ejected until the light has enough energy to expel the electron from the metal. The critical frequency for the photoelectric effect is when the energy of the light just matches the work function. Above that frequency, the ejected electron will have energy that is in excess of the energy required to leave the metal. For different metals, the work function is different due to the different affinities of each metal for its electrons. The slope of the increase in energy with respect to frequency is just Planck's constant h and is therefore the same for all metals. For these insights, Einstein was awarded the Nobel Prize in Physics in 1921.

Atomic spectra

When objects are heated they emit light that is characteristic for each element. Experimentally, the emitted light is observed to be not continuous but discrete (Figure 9.4). This feature suggests that the energy associated with the atoms is discrete or quantized. In classical physics, there is no reason for this property of quantized light emission. However,

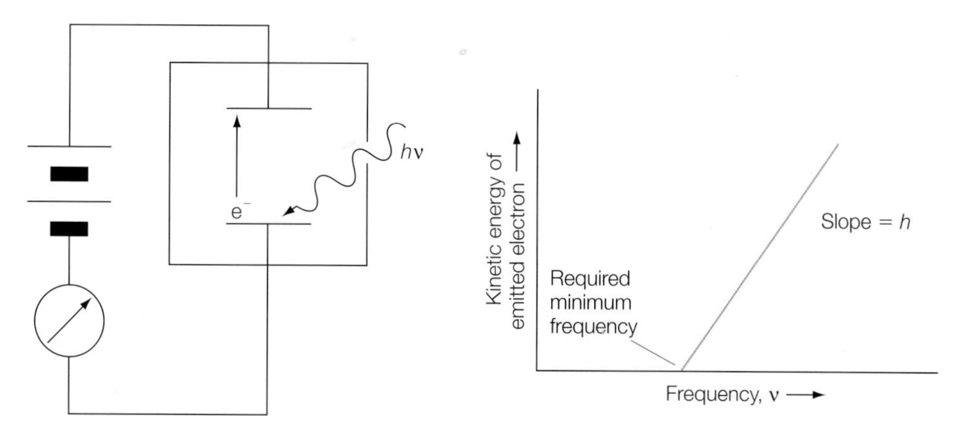

Figure 9.4 A schematic diagram of the experimental arrangement for measuring the photoelectric effect. Light strikes the metal, which is serving as the cathode, causing an electron to be emitted. The electron is attracted by the positive electrode and the flow of electrons is measured by the meter. By varying the strength of the electric field it is possible to determine the kinetic energy of the emitted electrons.

it was realized in 1885 that the frequencies associated with the lines of the spectrum of hydrogen could be mathematically related to each other by the relationship:

$$\tilde{\nu} = \frac{1}{2^2} - \frac{1}{n^2} \tag{9.15}$$

where the frequencies are written as wavenumbers (see Chapter 11). These lines are now known as the Balmer lines. After the lines in the ultraviolet region were discovered, yielding the Lyman and Paschen series, Johannes Rydberg noted in 1890 that all of the lines could be described using:

$$\tilde{\nu} = R_H\left(\frac{1}{n_1^2} - \frac{1}{n_2^2}\right) \tag{9.16}$$

where $n_1 = 1$ for the Lyman series and $n_1 = 2$ for the Balmer series. The constant R_H is now known as the Rydberg constant and has a value of 109,677 cm^{-1}.

Despite knowing that the frequencies were related to each other, an understanding of the physical principle did not follow from the simple pictures of the atomic structure as groupings of particles. Neils Bohr proposed that discrete emission naturally follows from the concept of electrons revolving around the nucleus in well-defined orbitals. An electron normally does not emit light, but when it makes a transition between two orbitals with different energies, light is emitted to conserve energy. The lines in the emission spectra then have energies corresponding to the energy differences between different electronic orbitals. Since there are only a limited number of orbitals, the light is emitted only at a limited number of frequencies.

For reasons discussed in Chapter 10, this model was qualitatively useful but could not provide an accurate prediction for complex multi-electron atoms. However, the Bohr model represented a fundamentally new paradigm for understanding the atom and provided a compelling reason for the development of detailed theories of quantum mechanics.

PRINCIPLES OF QUANTUM THEORY

The inability of classical mechanics to provide explanations for experiments such as the ones described above led to the introduction of several postulates concerning the behavior of matter. These various postulates were developed into a coherent theory in the 1920s. There were two parts to this development. First, a physical picture of the theory was developed by Louis de Broglie, called the wave–particle duality. Second, classical theory was modified by Erwin Schrödinger to provide a mathematical formalism describing the quantum effects leading to the Nobel Prize in Physics in 1933. This new formalism introduced the ideal that objects are not fully described according to the classical picture but the wave and particle nature of objects must be combined in a new description called the wavefunction. These developments had some unexpected predictions whose merits were debated for many years, with one example known as Schrödinger's cat, discussed at the end of this chapter.

Wave–particle duality

Classically, particles and waves have distinct parameters:

- Particles: mass m, position r, velocity v, charge q
- Waves: wavelength λ, frequency ν, velocity v, amplitude A

However, the experiments described above could only be interpreted if light was assumed to have particle-like properties; that is, if light was composed of discrete objects with quantized energies. This led to the idea that particles and waves could not be treated as distinct objects. Experimentally this could be tested directly as it implies that particles would have wave properties. Diffraction was a property that had been thought to be unique for waves. In 1925, Clinton Davisson and Lester Germer tested the possibility of a wave property for particles using diffraction experiments and found that electrons would diffract in a grating (Figure 9.5). Subsequently, it was shown that other types of particle could also diffract.

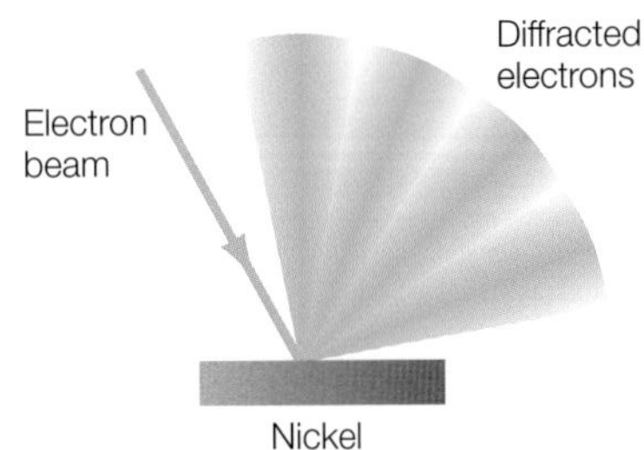

Figure 9.5 A beam of electrons strikes a nickel crystal and the outward beam shows the variation of intensity characteristic of a diffraction pattern.

The theoretical basis for relating the classical and wave parameters was provided by Louis de Broglie (Nobel Prize winner in Physics in 1923), who realized that the wavelength could be related to the momentum according to:

$$\lambda = \frac{h}{p} \tag{9.17}$$

As a particle moves faster the momentum increases and the wavelength decreases. Because the proportionality constant is a very small constant, *h*, the wavelength properties are normally not observed. For example, a tennis ball with a mass of 50 g moving at 90 miles/h (40 m s^{-1}) has the wavelength:

$$\lambda = \frac{h}{mv} = \frac{6.6 \times 10^{-34}\ \text{Js}}{(0.050\ \text{kg}) \times (40\ \text{m s}^{-1})} = 3.3 \times 10^{-34}\ \text{m} \tag{9.18}$$

$$90\,\frac{\text{miles}}{\text{h}} = 90\,\frac{\text{miles}}{\text{h}}\,\frac{5280\ \text{ft}}{\text{mile}}\,\frac{0.305\ \text{m}}{\text{ft}}\,\frac{\text{h}}{3600\ \text{s}} = 40\,\frac{\text{m}}{\text{s}}$$

For small particles the wavelength is very different. For an electron orbiting a nucleus with a velocity of 10^6 m s^{-1}, the wavelength is:

$$\lambda = \frac{h}{mv} = \frac{6.6 \times 10^{-34}\ \text{Js}}{(9.109 \times 10^{-31}\ \text{kg}) \times (10^{6}\ \text{m s}^{-1})} = 7.2 \times 10^{-10}\ \text{m} \tag{9.19}$$

This wavelength is comparable to the size of the orbital, and hence the wave properties play an important role in describing the properties of electrons in atoms.

The wavelength of an electron that is free, rather than orbiting in an atom, is determined by the velocity. Since the electron is charged, the velocity can be manipulated using electric fields. For an electron moving through a potential difference, *V*, of 40 kV, as found in an electron microscope, the potential energy gained by the electron is the product of the potential difference and the charge of the electron, *eV*. This energy can be equated to the kinetic energy, yielding:

$$eV = \frac{1}{2} m_e v^2 \tag{9.20}$$

Solving this equation for the velocity yields:

$$v = \sqrt{\frac{2eV}{m_e}} \tag{9.21}$$

Using this expression for velocity from eqn 9.17 gives us the wavelength as:

$$\lambda = \frac{h}{p} = \frac{h}{mv} = \frac{h}{\sqrt{2m_e eV}}$$

$$= \frac{6.626 \times 10^{-34}\ \text{Js}}{\sqrt{2(9.109 \times 10^{-31}\ \text{kg})(1.609 \times 10^{-19}\ \text{C})(4 \times 10^{4}\ \text{V})}}$$

$$= 6.1 \times 10^{-12}\ \text{m} \qquad (9.22)$$

The ability to manipulate the wavelength of an electron by use of an electric field has led to the development of electron microscopes that are commonly used for biological samples, as discussed in Chapter 18.

Schrödinger's equation

Independently, Werner Heisenberg and Erwin Schrödinger both developed mathematical formulations that, despite the differences in their details, were found to yield the same basic results. In the Schrödinger formulation, the key concepts are wavefunctions and operators. Each object is characterized by a wavefunction that describes the distribution of the object over space, reflecting its wave-like nature. The determination of the wavefunction for an object in a particular environment is done by use of a second-order differential equation called Schrödinger's equation, as described below. Once the wavefunction has been determined, then the various properties of the object, such as position and momentum, can be described.

Classically, the energy of a particle is given by the sum of the kinetic and potential energies:

$$E = \frac{1}{2}mv^2 + V(x) = \frac{p^2}{2m} + V(x) \qquad (9.23)$$

In quantum mechanics this expression is modified with the introduction of operators, as provided in Table 9.1. Some of these operators involve $\hbar$ and i, which are defined as:

$$\hbar = \frac{h}{2\pi} \quad \text{and} \quad i = \sqrt{-1} \qquad (9.24)$$

In some cases, the operator is simply a variable, for example the position operator is the position variable x. In other cases, the operator is a differential term that operates on the wavefunction. For example, the momentum is substituted by the derivative with respect to position, $\frac{\partial}{\partial x}$,

Table 9.1
Physical variables and the corresponding quantum operators.

Variable	Operator
X	x
V	V
p_x	$-i\hbar\frac{\partial}{\partial x}$
t	t
E	$i\hbar\frac{\partial}{\partial t}$

with a factor of $i\hbar$. Since the wavefunctions usually involve more than one dimension, for example, x, y, and z for a three-dimensional problem, the derivatives are written as partial derivatives rather than full derivatives. To show a simple derivation of Schrödinger's equation, these operators are substituted into the classical expression for energy, with each side of the equation multiplied by the wavefunction $\psi(x,t)$:

$$E = \frac{p^2}{2m} + V(x) \tag{9.25}$$

$$\left(i\hbar\frac{\partial}{\partial t}\right)\psi(x,t) = \left[\frac{1}{2m}\left(-i\hbar\frac{\partial}{\partial x}\right)^2 + V(x)\right]\psi(x,t) \tag{9.26}$$

$$i\hbar\frac{\partial}{\partial t}\psi(x,t) = -\frac{\hbar^2}{2m}\frac{\partial^2}{\partial x^2}\psi(x,t) + V(x)\psi(x,t) \tag{9.27}$$

In dealing with the hydrogen atom, we will need to expand the equation from a one-dimensional problem to a three-dimensional problem. By writing the equation in terms of three positional coordinates, the most general form of Schrödinger's equation is obtained:

$$i\hbar\frac{\partial}{\partial t}\psi(r,t) = -\frac{\hbar^2}{2m}\nabla^2\psi(r,t) + V(r)\,\psi(r,t) \tag{9.28}$$

where ∇^2 represents the second derivative with respect to all of the spatial variables. The three-dimensional positional vector $\vec{r}$ can be expressed in Cartesian coordinates, (x,y,z), or spherical coordinates, (r,θ,ϕ) (Figure 9.6), that are related according to:

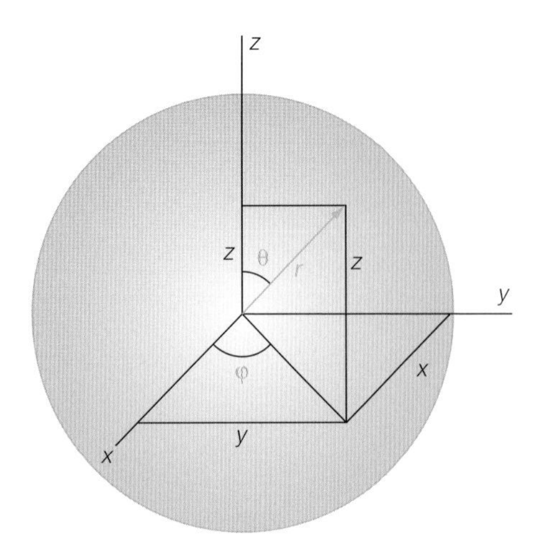

Figure 9.6
A point in space can be described by Cartesian coordinates, (x,y,z), or by spherical coordinates, (r,θ,ϕ).

$$\begin{aligned} x &= r \sin\theta \cos\phi \\ y &= r \sin\theta \sin\phi \\ z &= r \cos\theta \end{aligned} \tag{9.29}$$

To make use of these two coordinate systems in Schrödinger's equation it is necessary to write out the full expression for ∇^2. When using Cartesian coordinates it can be written as the sum of the second partial derivatives:

$$\nabla^2 = \frac{\partial^2}{\partial x^2} + \frac{\partial^2}{\partial y^2} + \frac{\partial^2}{\partial z^2} \tag{9.30}$$

For spherical coordinates the expression for ∇^2 can be shown to be:

$$\nabla^2 = \frac{\partial^2}{\partial r^2} + \frac{2}{r}\frac{\partial}{\partial r} + \frac{1}{r^2}\left[\frac{1}{\sin^2\theta}\frac{\partial^2}{\partial \phi^2} + \frac{1}{\sin\theta}\frac{\partial}{\partial \theta}\sin\theta\frac{\partial}{\partial \theta}\right] \tag{9.31}$$

Although the use of Cartesian coordinates is much easier and generally preferred, in some circumstances spherical coordinates are used. When we examine Schrödinger's equation for the hydrogen atom, the potential term for the electrostatic energy is given by the product of the charges on the proton and electron divided by the separation distance. In this case, the separation distance can be written in Cartesian coordinates as the square root of the sum of the squares of x, y, and z or in spherical coordinates simply as r. As we will see, the advantage of using spherical coordinates to set up the problem outweighs the disadvantage of using the more complicated mathematical form of the derivatives.

For the applications discussed in the following chapters, there will be no time dependence involved. The time dependence arises when we consider systems that change with time, for example an electron that travels near the speed of light until it hits a target at a synchrotron (see Chapter 15). All of the problems considered here are stationary problems that do not change with time. For these cases, the equation simplifies because we can write a general form for the wavefunction.

First, Planck's relationship (eqn 9.10) is used with the definition for angular frequency;

$$\omega = 2\pi\nu \tag{9.32}$$

yielding

$$E = h\nu = \frac{h}{2\pi}(2\pi\nu) = \hbar\omega \tag{9.33}$$

$$\psi(r,t) = \psi(r)e^{-i\omega t} = \psi(r)e^{-iEt/\hbar} \quad \text{as} \quad \omega = E/\hbar \tag{9.34}$$

With this substitution we can write:

$$i\hbar\frac{\partial}{\partial t}[\psi(r)e^{-iEt/\hbar}] = -\frac{\hbar^2}{2m}\nabla^2[\psi(r)e^{-iEt/\hbar}] + V(r)[\psi(r)e^{-iEt/\hbar}] \tag{9.35}$$

Using the relationship that:

$$\frac{\mathrm{d}}{\mathrm{d}x}e^{ax} = ae^{ax} \tag{9.36}$$

gives:

$$\psi(r)i\hbar\frac{\partial}{\partial t}(e^{-iEt/\hbar}) = e^{-iEt/\hbar}\left[-\frac{\hbar^2}{2m}\nabla^2\psi(r) + V(r)\psi(r)\right] \tag{9.37}$$

$$e^{-iEt/\hbar}\left[i\hbar\frac{(-iE)}{\hbar}\psi(r)\right] = e^{-iEt/\hbar}\left[-\frac{\hbar^2}{2m}\nabla^2\psi(r) + V(r)\psi(r)\right] \tag{9.38}$$

$$E\psi(r) = -\frac{\hbar^2}{2m}\nabla^2\psi(r) + V(r)\psi(r) \tag{9.39}$$

This last equation is the time-independent form of Schrödinger's equation and will be the form that we use to solve various applications including the hydrogen atom (Chapter 11).

The two relationships in quantum mechanics, the de Broglie and Schrödinger equations, are consistent with each other, as can be seen by substituting into Schrödinger's equation a typical wave. For simplicity, one-dimensional expressions can be used. The oscillation behavior of a wave can be expressed by a number of equivalent formulations, for example:

$$\psi(x) = A\sin(kx) \quad \text{or} \quad A\cos(kx) \quad \text{where} \quad k = \frac{2\pi}{\lambda} \tag{9.40}$$

The difference for these two expressions is the phase; that is, whether the wave has a peak at 0 or 90°. Eqn 9.40 can be rewritten in terms of exponentials since:

$$e^{ikx} = \cos(kx) + i\sin(kx)$$

and

$$e^{-ikx} = \cos(kx) - i\sin(kx) \tag{9.41}$$

so equivalently:

$$\psi(x) = Ae^{ikx} \tag{9.42}$$

Substitution into eqn 9.39 gives:

$$E(Ae^{ikx}) = -\frac{\hbar^2}{2m}\frac{\partial^2}{\partial x^2}(Ae^{ikx}) + V(r)(Ae^{ikx}) \tag{9.43}$$

$$(Ae^{ikx})E = -\frac{\hbar^2}{2m}(-k^2)(Ae^{ikx}) + V(r)(Ae^{ikx}) \tag{9.44}$$

$$E - V = \frac{\hbar^2 k^2}{2m} = \text{kinetic energy} = \frac{p^2}{2m} \tag{9.45}$$

$$\rightarrow p = \hbar k = \frac{h}{2\pi}\frac{2\pi}{\lambda} = \frac{h}{\lambda} \tag{9.46}$$

Born interpretation

The physical interpretation of quantum mechanics, in particular the interpretation of the wavefunction, was developed by many scientists, most notably Max Born (Nobel Prize winner in Physics in 1954). Since all particles are also waves, particles are always distributed in space. The wavefunction in Schrödinger's equation has no direct physical meaning and can be a complex function rather than a real function. In the Born interpretation, the probability of finding any particle at a particular location is not given by the wavefunction itself but rather by the square of the wavefunction. These ideas led to several fundamental postulates of quantum mechanics, as follows.

1 A particle is never at a specific location but only has a probability of being there. The probability of finding a particle at a specific position is given by the square of the wavefunction times the volume $d\tau$ as:

$$\psi^*(r)\psi(r)\mathrm{d}\tau$$
$$\text{where} \quad \mathrm{d}\tau = \mathrm{d}x\,\mathrm{d}y\,\mathrm{d}z = r^2 \sin\theta\,\mathrm{d}r\,\mathrm{d}\vartheta\,\mathrm{d}\phi \tag{9.47}$$

In this equation ψ^* is the complex conjugate. Since any wavefunction ψ can be written in terms of two real functions, A and B, the conjugate can be defined as:

$$\psi = A + iB \quad \text{and} \quad \psi^* = A - iB \tag{9.48}$$

The probability of finding a particle within a volume V is then:

$$\int_V \psi^*(r)\psi(r)\,\mathrm{d}\tau \tag{9.49}$$

where $\psi^*(x)$ is the complex conjugate of the wavefunction.

2 The wavefunction may be a complex function but the probability is always real, since:

$$\psi(x) = A(x) + iB(x)$$
$$\psi^*(x) = A(x) - iB(x) \tag{9.50}$$
$$\text{so} \quad \psi^*(x)\psi(x) = [A(x) + iB(x)][A(x) - iB(x)] = A^2(x) + B^2(x) \geq 0$$

3 The probability is always a positive and real number (Figure 9.7). The total probability of finding an object anywhere in space must be equal to one. The sum of all of the probabilities is mathematically written as the integral of the probability. Therefore, the integral of the probability over all space must be equal to one:

$$1 = \int_0^\infty \psi^*(r)\psi(r)\,\mathrm{d}\tau \tag{9.51}$$

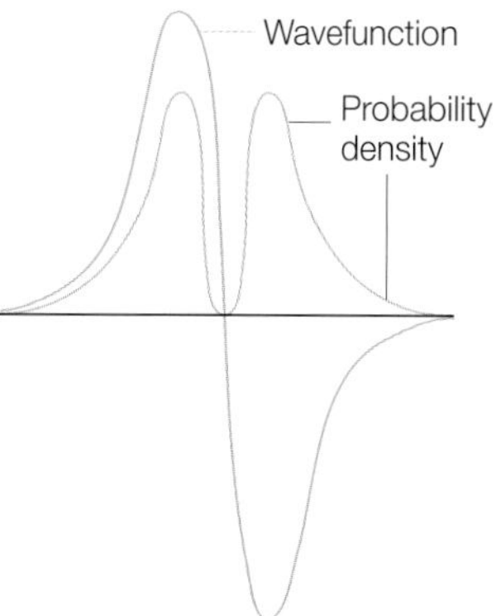

Figure 9.7 The sign of a wavefunction may be either positive or negative, but the probability is always zero or positive.

4 Particles do not have a specific position or momentum but rather there is a distribution of values that reflect the distribution of the particle. The physically relevant quantity is the average, or expectation, value. Every physical observable p has an associated operator (eqn 9.17) and the average, or expectation, value of the observable is given by:

$$\langle p \rangle = \int \psi^*(r)\hat{p}\psi(r)\,\mathrm{d}\tau \tag{9.52}$$

where $\hat{p}$ is the operator.

For example, the average values of the position and momentum of a particle can be calculated by substituting the operators for position r and x component of the momentum:

$$\langle r \rangle = \int \psi^*(r) r \psi(r)\,\mathrm{d}\tau$$
$$\langle x^2 \rangle = \int \psi^*(x) x^2 \psi(x)\,\mathrm{d}x \tag{9.53}$$

$$\langle p_x \rangle = \int \psi^*(r)\left(-i\hbar\frac{\partial}{\partial x}\right)\psi(r)\,\mathrm{d}\tau \tag{9.54}$$

5 To be physically reasonable, some restrictions can be placed on the allowed values of the wavefunction. Since the probability must be less than one everywhere, the wavefunction must be finite. The probability must have a single, unique solution at every position (Figure 9.8). Solutions of Schrödinger's equation that yield more than one solution at a certain position are not allowed. When comparing the solutions of the wavefunction at two close locations, there cannot be any discontinuities

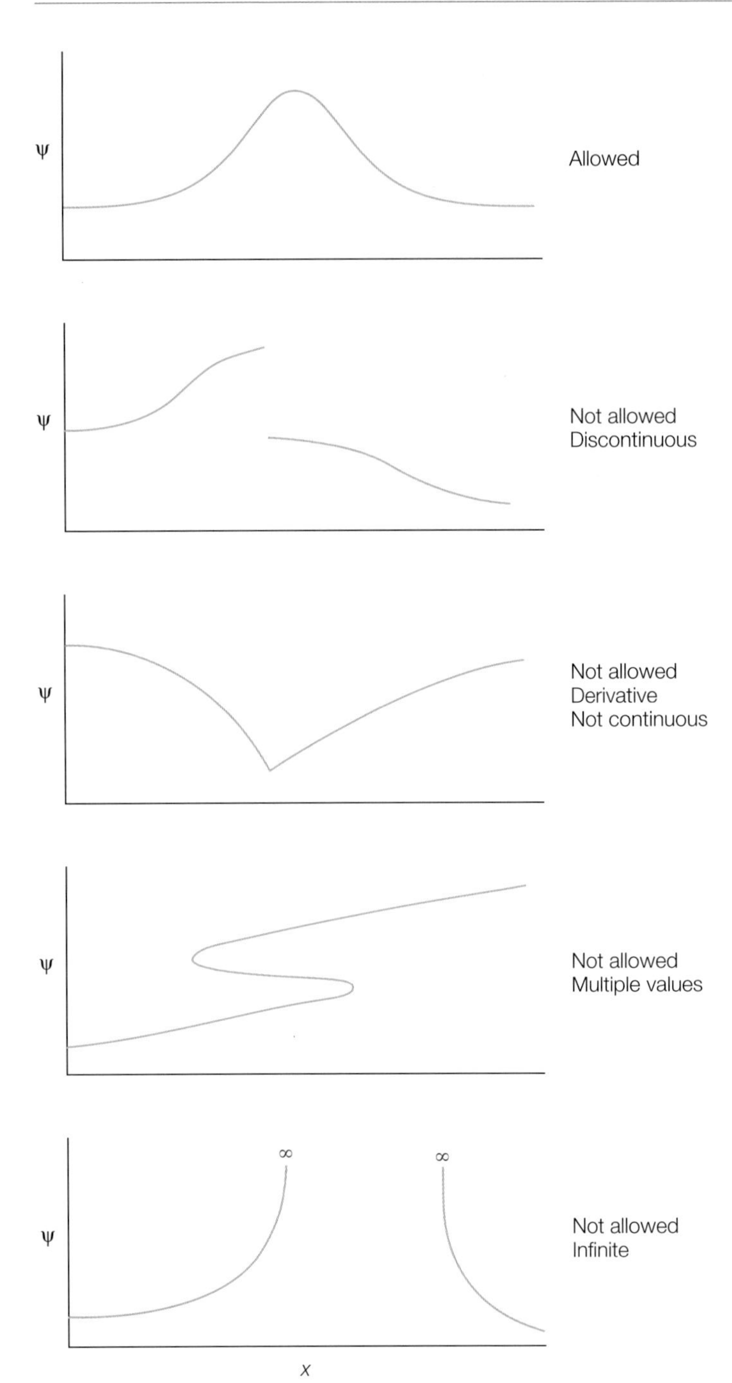

Figure 9.8 The wavefunction $\psi(x)$ must be a smooth, continuous, and single-valued function. Solutions are unacceptable if they have a discontinuity or more than one value for a given value of x.

and no sharp turns are allowed; that is, the wavefunction must be smooth and continuous.

6 The solution to Schrödinger's equation will always be a series of related solutions because the equation is an *eigenfunction*. The solutions to all eigenfunctions have common properties. Each of the solutions is *orthogonal*, meaning that for each of the wavefunctions the solution is unique and, when the wavefunctions for the hydrogen atom are found, they will each represent a different possible orbital for the electrons.

7 Quantum mechanics does not replace classical mechanics but complements it when considering small energies and particles. According to the Correspondence Principle, the results from quantum mechanics always must agree with classical mechanics. Thus, the results from quantum mechanics must agree with the classical mechanics in the appropriate limits. For example, quantum mechanics must still conserve energy of the observed states.

GENERAL APPROACH FOR SOLVING SCHRÖDINGER'S EQUATION

There is a general strategy that will be used to solve Schrödinger's equation for the different applications. The first step is to determine the dependence of the potential energy upon the distance. We start with the classical expression of the potential and then substitute the appropriate operators, which for distance are simply the variables for distance. We also will need to determine the number of degrees of freedom, whether the object is constrained to move only in one dimension or in three dimensions. For example, for the first problem, we will consider an object trapped in a one-dimensional box with a potential of zero inside of the box. As a result, only one variable, x, is used in Schrödinger's equation and the potential term is always zero. This is followed by investigating how the problem is changed when the object is free to move in two dimensions. For the subsequent applications, a different potential term is used and, consequently, different differential equations are solved. For the harmonic oscillator, the problem is again one-dimensional but the potential term is that for a spring. For the hydrogen atom, we will need to use three dimensions with an electrostatic potential.

Once the potential is substituted and the number of dimensions is chosen, we will determine the wavefunction by solving Schrödinger's' equation. In each case, a solution to the equation will be presented to have a better understanding of what gives rise to the specific forms of the wavefunctions. As we will see, each of these equations is a well-known second-order differential equation that was studied and solved by mathematicians in the 1800s. Indeed, at that point in time, many of the differential equations with solutions relevant to chemistry and physics were solved, although their application was not known until many years later. For the initial applications, the solutions will be functions that have no meaning in themselves. However, for the hydrogen atom, the wavefunctions will be shown to be related to the well-known orbitals, identified as s, p, d, etc.

Once the wavefunction has been determined, then it is possible to predict all of the properties of the system, within the constraints of quantum mechanics. For example, the wavefunction can be used to determine the probability of finding the object at a certain position in space, the average momentum of the object, or the average position. Thus in principle, all of the functional properties of the object can be predicted, at least as an average. For the hydrogen atom, this provides a means to predict what happens in spectroscopy when light strikes an atom.

INTERPRETATION OF QUANTUM MECHANICS

Quantum mechanics has been accepted because, in each case, the calculated values have been found to be correct when compared with experimental data, while classical mechanics has failed in many cases. This led to the idea that classical physics was correct only for those situations involving the physical world, whereas for small particles, namely electrons, protons, and neutrons, quantum mechanics applied. Unlike classical mechanics, the object in quantum mechanics is never at a specific location but rather only has a certain probability of being at a certain location. Likewise, the object has only average values rather than absolute values of position, momentum, and energy. Thus, the classical ideal of precise determination of the properties of objects gave way to an interpretation in which there is only a probability of knowing the properties. The correctness of this interpretation was heavily debated but was accepted eventually.

Heisenberg Uncertainty Principle

Due to the wave nature of all objects, we cannot determine both the position and momentum of a particle precisely. For example, if we try to observe where a particle is, we must probe the particle and the probe will move the particle in an unknown manner. Likewise, if we try to observe what velocity the particle has, then when we probe it the location will change. This led to a fundamental concept by Werner Heisenberg in 1927 (Nobel Prize winner in Physics in 1932) that relates the uncertainty of the parameters according to:

$$(\Delta p_x)(\Delta x) \geq \hbar/2$$
$$(\Delta E)(\Delta t) \geq \hbar/2 \tag{9.55}$$

where Δp_x, Δx, ΔE, and Δt are the uncertainties in momentum, position, energy, and time, respectively.

How critical is this? For a 1 g bullet moving with a small uncertainty at a velocity of 10^{-6} m s^{-1}, the estimated uncertainty in position is 10^{-26} m, and so is undetectable. However, for very small objects, such as electrons, the uncertainties become important. For example, an electron moving with an uncertainty at a speed of 5×10^5 m s^{-1} has an uncertainty in position of 1.1×10^{-10} m, which is twice the Bohr radius and so is comparable to the size of an atomic orbital. Experimentally this uncertainty becomes evident when processes become very fast. A laser that emits light as a fast time pulse loses its monochromatic precision as the precision in time causes a corresponding uncertainty in energy:

$$\Delta E = h\Delta \nu \geq \frac{\hbar}{2\Delta t} \quad \text{or} \quad \Delta \nu \geq \frac{1}{4\pi\Delta t} \tag{9.56}$$

Using optical spectroscopy, the function of proteins that contain pigments can be followed in time by using lasers (Chapter 17). For a laboratory experiment, extremely fast reactions are studied using lasers that can operate with a time of 0.1 fs, or 10^{-16} s, yielding an uncertainty in frequency of approximately 8×10^{14} s^{-1} or equivalently an uncertainty in wavelength of 380 nm that is comparable to the wavelength itself. By comparison, the same laser operating in a steady-state mode would have an accuracy in wavelength that is a small fraction of a per cent. Thus, the decrease in the time interval results in the laser light changing from being a monochromatic color such as red or green to having many colors, or effectively becoming white.

A quantum-mechanical world

Under common circumstances, people do not encounter effects such as the wavelength nature of matter or the Uncertainty Principle, due to the size and speed of everyday objects. To better understand the issues raised by these concepts, consider a quantum-mechanical world where everyone is so small that the wave nature of matter must be dealt with in common activities. The wave nature of matter becomes important when h/p yields a wavelength that is the size of the objects. To achieve this limit, let's have a population of people whose height is about the size of an electron orbital, or 0.1 nm, and who have a mass of m_e. If the people can run at a speed of 7×10^6 m s^{-1}, then the wavelength is the same as the height and the effects are observed daily. For example, when a person walks through a door that has an opening of 0.1 nm, then the person will not be able to walk through the door in a straight line but will undergo diffraction.

$$\lambda = \frac{h}{mv} = \frac{6.6 \times 10^{-34} \text{ Js}}{(9.109 \times 10^{-31} \text{ kg})(7 \times 10^6 \text{ m s}^{-1})} \tag{9.57}$$
$$= 10^{-10} \text{ m} = 0.1 \text{ nm}$$

To illustrate the complexities, consider people in this quantum-mechanical world playing the sport baseball. A ball is hit deep and the outfielder chases after the ball. He gets to the spot where the trajectory predicts that the ball should land but can he catch the ball? To catch the ball he must be able to predict the landing position of the ball within the accuracy determined by the size of his glove, which has a size of 0.01 nm. Although the overall trajectory can be predicted, the path becomes fuzzy due to the wave nature of the ball, and the predicted landing spot is limited by the Heisenberg Uncertainty Principle. If the ball has a mass of 0.1 m_e and the speed has an uncertainty of 5×10^5 m s^{-1}, then the predicted landing spot will have an uncertainty of 0.1 nm. Since the uncertainty in the landing spot is 10 times larger than the size of the glove, then the outfielder will have only a 10% chance of catching the ball. This example, of course, does not address the other difficulties, including hitting the ball in the first place (Figure 9.9).

Figure 9.9 Outfielders have a difficult time catching a baseball in a quantum-mechanical world.

$$\Delta x = \frac{\hbar}{2(\Delta p_x)} = \frac{1.1 \times 10^{-34} \text{ Js}}{2(9.109 \times 10^{-31} \text{ kg})(5 \times 10^5 \text{ m s}^{-1})}$$
$$= 10^{-10} \text{ m} = 0.1 \text{ nm} \tag{9.58}$$

RESEARCH DIRECTION: SCHRÖDINGER'S CAT

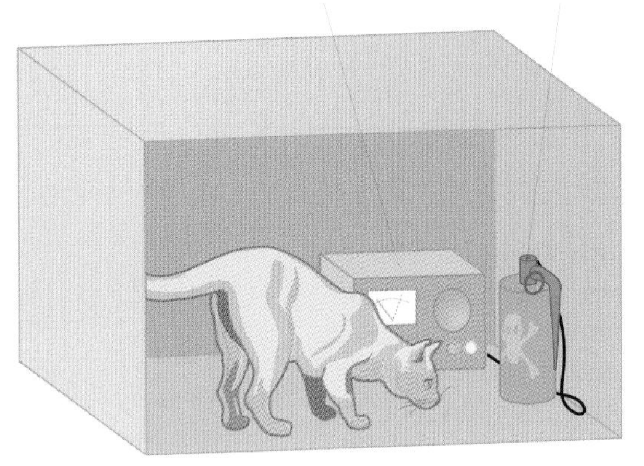

Figure 9.10 Schrödinger's cat. This is a thought experiment in which a cat is trapped inside a sealed box. The box contains a poisonous gas that is released after a radioactive decay. Does the cat live or die?

The unusual features of quantum theory resulted in heated debate about the correctness of the theory, and scientists discussed the relative merits of classical and quantum theory throughout the 1920s and 1930s. One approach that they used to question the usefulness of each concept was the use of *gedenken experiments*; that is, thought experiments that are not performed in the laboratory. One of the most famous examples of these thought experiments was originally poised by Schrödinger in 1935 and is known as Schrödinger's cat (Schrödinger 1935). In the original version of his experiment, a cat is placed into a sealed chamber (Figure 9.10). In the box is a tiny amount of radioactive material so small that perhaps one of the atoms will decay within an hour. If the atom does decay, a Geiger counter measures the event, which triggers the release of a hammer that then shatters a flask containing a deadly poison. Therefore, if the atom decays then the vial will break releasing the deadly gas and killing the poor cat but if the decay does not occur the cat lives. The question is: does the cat live or die? How can the cat live without allowing for a trick, such as an unseen hole in the box? The answer is that in quantum mechanics, after the box is sealed, then we no longer know what happened to the cat. The atom may either decay or not decay, which means that the atom has the properties of both states at the same time. Thus, the cat can be considered to be in a superposition of two states, alive and dead. For the real world the cat cannot be in both states. However, in the world of quantum mechanics such possibilities are not only allowed but must be considered in order to understand the possible outcomes. For example, when we consider chemical bonds, electrons are thought to be present in orbitals involving two atoms rather than belonging to either one individually.

Historically, Schrödinger used this gedenken to test the unusual features of quantum mechanics as the outcome of the superposition of the macroscopically distinguishable states did not seem reasonable. Recently, the thought experiment involving Schrödinger's cat has been tested experimentally. In these experiments the production of Schrödinger-cat-like states is hindered by the phenomenon that the superposition of states is destroyed quickly under most circumstances. The larger the object the better it must be isolated from outside influences to remain unchanged. In one experiment (Blatter 2000; Friedman et al. 2000) the system was not a cat but rather a special instrument called a SQUID, or superconducting quantum-interference device. In superconducting devices, electrons can circulate without any decay; that is, with no resistance.

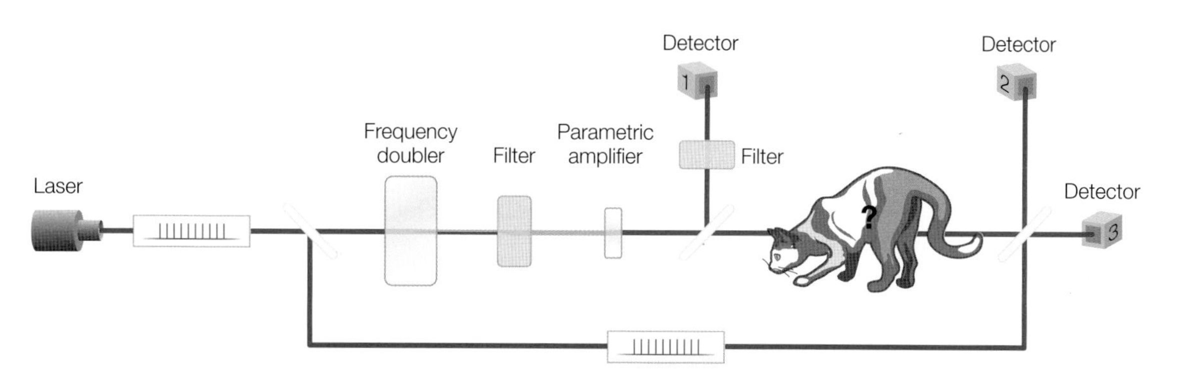

Figure 9.11 Femtosecond laser pulses are generated that induce the formation of photons corresponding to small Schrödinger kittens. Modified from Gisin (2006).

Superconductors play a critical role in many instruments, including magnets used for NMR experiments (Chapter 14). In a SQUID, the electrons circulate in either a clockwise or counterclockwise motion without any resistance that would alter the motion. The experiment showed that the electrons could be considered to be a superposition of the states involving rotation in either direction. Although this system is not quite as large as a cat, it does show that billions of electrons collectively can behave with a strictly quantum-mechanical property. In another experiment (Gisin 2006; Ourjoumtsev et al. 2006), a light pulse, representing the cat, is generated as a superposition of two coherent states (Figure 9.11). A single photon, representing a kitten, is removed from the pulse, which still contains information concerning the pulse (as what is termed a Wigner function). While these experiments may appear to be fanciful, they represent the development of quantum states of propagating light beams. Such beams could be used for quantum computing and communication, providing greater speeds and new means for encryption of data (Bennett & DiVincenzo 2000).

REFERENCES

Bennett, C.H. and DiVincenzo, D.P. (2000) Quantum information and computation. *Nature* **404**, 247–55.

Blatter, G. (2000) Schrödinger's cat is now fat. *Nature* **406**, 25–6.

Friedman, J.R., Patel, V., Chen, W., Tolpygo, S.K., and Lukens, J.E. (2000) Quantum superposition of distinct macroscopic states. *Nature* **406**, 43–6.

Gisin, N. (2006) New additions to the Schrödinger cat family. *Science* **312**, 63–4.

Ourjoumtsev, A., Tualle-Brouri, R., Lauret, J., and Grangler, P. (2006) Generating optical Schrödinger kittens for quantum information processing. *Science* **312**, 83–6.

Schrödinger, E. (1935) Die genenwartige situation in der quantenmechanik. *Naturwissenschaftern* **23**, 807–12; 823–8; 844–9.

PROBLEMS

9.1 Calculate the velocity of an electron if it has a wavelength of (a) 2.0 Å, (b) 1.0 Å, and (c) 10.0 Å.

9.2 Calculate the velocity of a proton if it has a wavelength of (a) 2.0 Å, (b) 1.0 Å, and (c) 10.0 Å.

9.3 Calculate the wavelength of an electron moving at a velocity of (a) 2×10^5 m s^{-1}, (b) 1.0×10^7 m s^{-1}, and (c) 1.5×10^8 m s^{-1}.

9.4 Calculate the wavelength of a proton moving at a velocity of (a) 2×10^5 m s^{-1}, (b) 1.0×10^7 m s^{-1}, and (c) 1.5×10^8 m s^{-1}.

9.5 Calculate the wavelength of a 5-g snail moving at a speed of (a) 1 mm h^{-1}, (b) 1 μm h^{-1}, and (c) 100 mm h^{-1}.

9.6 Calculate the wavelength of an object that is moving at a speed of 100 km h^{-1} and has a mass of (a) 5 μg, (b) 1 g, and (c) 5 kg.

9.7 Why is the observed blackbody radiation in conflict with classical physics?

9.8 Qualitatively, why does quantum theory give the correct expression for blackbody radiation?

9.9 Why does the presence of discrete lines in the emission spectra of atoms support the ideas of quantum theory?

9.10 For each of the following experimental results, write whether classical theory can explain the result or whether quantum theory is required: (a) light can diffract; (b) electrons can diffract; (c) for blackbody radiation, the energy density is small at small wavelengths; (d) for blackbody radiation, the energy density is small at long wavelengths; (e) light has wavelength; (f) electrons are in atomic orbitals; (g) electrons have mass; and (h) electrons have a wavelength.

9.11 For the photoelectric effect, what does the classical theory predict for the dependence of the kinetic energy of the emitted electron on the frequency of the light? Does this agree with the observed dependence?

9.12 For the photoelectric effect, what does the quantum theory predict for the dependence of the kinetic energy of the emitted electron on the frequency of the light? Does this agree with the observed dependence?

9.13 For the photoelectric effect with a metal that has a work function of 2.0 eV, what is the kinetic energy of the ejected electron if the light has a wavelength of (a) 200 nm, (b) 250 nm, and (c) 350 nm?

9.14 For the photoelectric effect with a metal that has a work function of 2.3 eV, what is the kinetic energy of the ejected electron if the light has a wavelength of (a) 200 nm, (b) 250 nm, and (c) 350 nm?

9.15 For the photoelectric effect, what is the maximal wavelength of light that will still result in ejection of an electron from a metal that has a work function with the value of (a) 2.25 eV, (b) 2.0 eV, and (c) 2.5 eV?

9.16 Why is there a minimal frequency needed to cause the photoelectric effect?

9.17 What is the probability of finding a particle within a volume V_0?

9.18 What frequency dependence does classical theory predict for the energy, E, of light? What dependence is predicted using quantum theory?

9.19 If the classical term for kinetic energy was $4\,m^3v$ instead of $mv^2/2$ what would Schrödinger's equation be? What if the classical term for kinetic energy was mv^4?

9.20 What is the uncertainty in position of a proton that has an uncertainty in velocity of (a) 10 Å s^{-1}, (b) 1 Å s^{-1}, and (c) 1000 Å s^{-1}?

9.21 What is the uncertainty in position of an electron that has an uncertainty in velocity of (a) 10 Å s^{-1}, (b) 1 Å s^{-1}, and (c) 1000 Å s^{-1}?

9.22 What is the uncertainty in position of a 5-g snail that has an uncertainty in velocity of (a) 10 cm s^{-1}, (b) 1 cm s^{-1}, and (c) 1 μm s^{-1}?

9.23 Write the quantum operator for (a) momentum, (b) position, and (c) momentum squared.

9.24 Write Schrödinger's equation for the potential $V(x)$ equal to (a) Ax, (b) Ax^2, and (c) $Ax + Bx^2$.

9.25 Write Schrödinger's equation for the potential $V(x,y)$ equal to (a) $Ax^3 + By$, (b) $Ax + By^3$, and (c) Axy.

9.26 Write Schrödinger's equation for the potential $V(x,y,z)$ equal to (a) $Ax + By + Cz$, (b) $Ax^3 + By + Cz^2$, and (c) $Axyz$.

9.27 Explain why the observation of a different diffraction for two slits compared to one raises questions about the nature of a particle.

9.28 Why can both the position and velocity of an object not be determined to high accuracy?

9.29 Consider a one-dimensional problem that yields a wavefunction of zero everywhere except between $x = 0$ and $x = a$, where it has a constant value of A. Determine the value of A using the normalization condition.

9.30 Consider a one-dimensional problem that yields a wavefunction of zero everywhere except between $x = 0$ and $x = a$, where it has a value of Ax. Determine the value of A using the normalization condition.

9.31 In a quantum-mechanical world, a baseball game is being played. A batter hits the ball and an outfielder runs to catch the ball. He gets to the spot where the ball will land and opens his glove. What is the probability that he will catch the ball? Assume the ball is moving at a speed of 200 ± 50 nm s^{-1}, that it weighs 1.054×10^{-22} kg, and that the glove has a diameter of 5 μm.

9.32 In this quantum-mechanical world, a person who has a mass of 6.6×10^{-18} kg is running through a tiny door that is only 100 nm wide. How fast does the person need to run such that quantum effects become an important influence in determining where he ends up after running through the door?

9.33 Explain what unusual effects would occur if an electron was present in a three-dimensional box that had very small sides of less than 0.1 Å.

10

Particle in a box and tunneling

An understanding of the properties of biological molecules requires knowledge of quantum mechanics. This chapter presents two examples of how Schrödinger's equation can be used to solve problems. First, we consider the one-dimensional particle in a box, in which a particle is trapped inside a box with infinite walls. The second problem is the simple harmonic oscillator. For this problem, we consider how quantum mechanics describes vibrational motion. Using these problems as typical examples, we show how to use Schrödinger's equation and discuss how various properties of objects can be described. In both cases we see that quantum theory makes surprising predictions, such as tunneling. The usefulness of these problems is shown by their application to understanding the properties of conjugated molecules commonly found in biological systems, such as carotenoids, as well as the vibrational properties of molecules.

ONE-DIMENSIONAL PARTICLE IN A BOX

A particle of mass m is confined between two walls at $x = 0$ and $x = L$. The potential energy is zero inside the box but rises immediately to infinity at the walls (Figure 10.1). Because the potential is infinite outside the box, the particle cannot be outside the box and the wavefunction is zero there. To determine the properties of the particle inside the box, Schrödinger's equation is solved. In this case, the potential energy $V(x)$ is zero inside of the box, so the equation can be simply written as:

$$-\frac{\hbar^2}{2m}\frac{d^2}{dx^2}\psi(x) = E\psi(x) \tag{10.1}$$

The solutions to this equation can be written in terms of exponential, sine, or cosine functions. The choice of

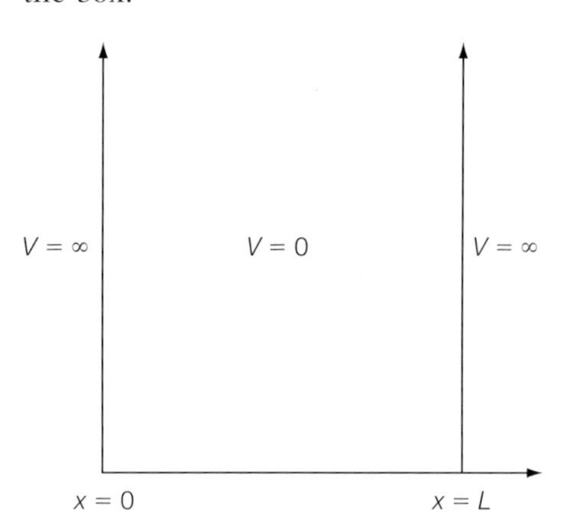

Figure 10.1
A particle is confined to be within a box with a potential of zero inside the box and an infinite potential outside the box.

the solution to use is based upon experience in solving differential equations. For this situation, consider the general solution:

$$\psi(x) = A \sin kx + D \cos kx \tag{10.2}$$

where A and B are the amplitudes of the sine and cosine functions, respectively, and k is related to the wavelength according to $k = 2\pi/\lambda$ where λ is the wavelength. These parameters will be determined by substitution of eqn 10.2 into Schrödinger's equation (eqn 10.1). Remember that the derivatives of the trigonometric terms are given by:

$$\frac{\mathrm{d}}{\mathrm{d}x}\sin kx = k\cos kx \quad \text{and} \quad \frac{\mathrm{d}}{\mathrm{d}x}\cos kx = -k\sin kx \tag{10.3}$$

Taking the derivative of eqn 10.2 yields:

$$\frac{\mathrm{d}}{\mathrm{d}x}(A\sin kx + B\cos kx) = k(A\cos kx - B\sin kx) \tag{10.4}$$

and

$$\begin{aligned} \frac{\mathrm{d}^2}{\mathrm{d}x^2}(A\sin kx + B\cos kx) &= \frac{\mathrm{d}}{\mathrm{d}x}(Ak\cos kx - Bk\sin kx) \\ \frac{\mathrm{d}^2}{\mathrm{d}x^2}(A\sin kx + B\cos kx) &= (-Ak^2\sin kx - Bk^2\cos kx) = -k^2\psi(x) \end{aligned} \tag{10.5}$$

Notice that the second derivative was written in terms of the wavefunction; this is something that we will do for all of the problems. Substitution of the second derivative into Schrödinger's equation gives:

$$-\frac{\hbar^2}{2m}\frac{\mathrm{d}^2}{\mathrm{d}x^2}\psi(x) = +\frac{\hbar^2}{2m}k^2\psi(x) = E\psi(x) \tag{10.6}$$

$$\rightarrow \quad E = \frac{\hbar^2 k^2}{2m}$$

We now have a solution to the problem and have found that the parameter k is related to the energy. To determine the values of the other parameters we make use of the boundary conditions. As discussed in Chapter 9, the wavefunctions are required to be everywhere continuous. Due to the infinite potential outside of the box, the wavefunction outside of the box is everywhere zero. To be continuous, we must set the wavefunctions to also be zero at the walls of the box. There we can write:

$$\psi\,(x = 0) = 0 = A\sin 0 + B\cos 0 \rightarrow B = 0 \tag{10.7}$$

$$\psi\,(x = L) = 0 = A\sin kL \rightarrow kL = n\pi \quad \text{where} \quad n = 1, 2, 3\ldots \tag{10.8}$$

We now have determined the parameters B and k, and have the following solutions:

$$\psi(x) = A \sin\left(\frac{n\pi x}{L}\right) \quad n = 1, 2, 3 \ldots \tag{10.9}$$

$$E = \frac{\hbar^2 k^2}{2m} = \frac{\hbar^2}{2m}\left(\frac{n\pi}{L}\right)^2 = \frac{n^2 h^2}{8mL^2} \tag{10.10}$$

To determine the last parameter, A, we make use of the normalization condition that requires the total probability to be equal to one (Chapter 9). Since there is only one particle in the box, the integral of the probability over the length of the box must be equal to one, allowing us to write:

$$1 = \int_0^L \psi^*(x)\psi(x)\,\mathrm{d}x = \int_0^L A^2 \sin^2\left(\frac{n\pi x}{L}\right)\mathrm{d}x \tag{10.11}$$

$$1 = A^2 \int_0^L \frac{1}{2}\left[1 - \cos\left(\frac{2n\pi x}{L}\right)\right]\mathrm{d}x \tag{10.12}$$

$$1 = \frac{A^2}{2}\left[x - \left(\sin\frac{2n\pi x}{L}\right)\left(\frac{L}{2\pi n}\right)\right]_0^L \tag{10.13}$$

$$1 = \frac{A^2}{2}\left[(L-0)-(0-0)\right] = \frac{A^2 L}{2} \rightarrow A = \sqrt{\frac{2}{L}} \tag{10.14}$$

This gives a final solution of:

$$\psi(x) = \sqrt{\frac{2}{L}} \sin\left(\frac{n\pi x}{L}\right) \quad n = 1, 2, 3 \ldots \tag{10.15}$$

$$E = \frac{\hbar^2}{2m}\left(\frac{n\pi}{L}\right)^2 = \frac{n^2 h^2}{8mL^2} \tag{10.16}$$

PROPERTIES OF THE SOLUTIONS

Energy and wavefunction

The solutions of Schrödinger's equation for the particle in a box are a series of wavefunctions and energies related by a quantum number, n, that must be a positive integer. The allowed energy levels are quantized and increase as n^2 with their separation increasing as the quantum number increases

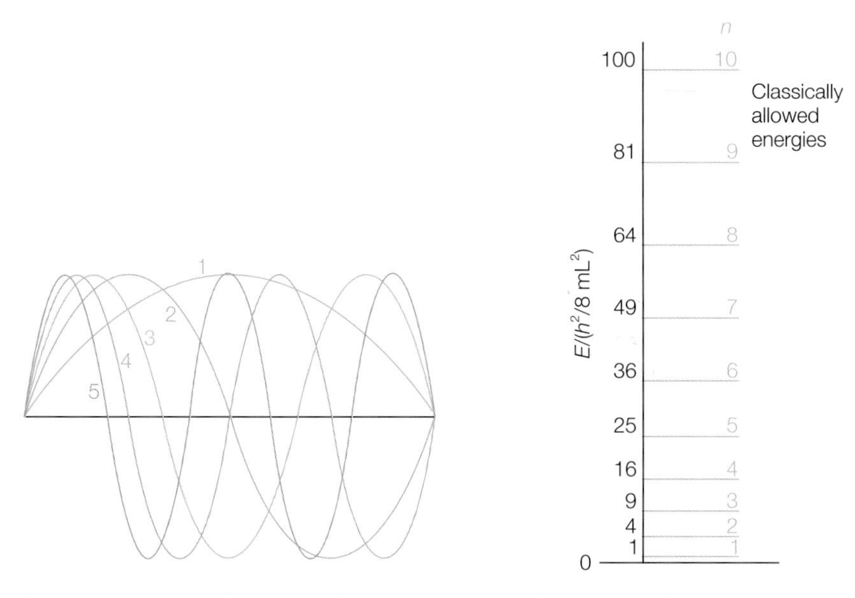

Figure 10.2 Wavefunctions, for n = 1 to 5, and energies, for n = 1 to 10, for the particle in a box.

(Figure 10.2). The wavefunctions are related to each other by n, with the different solutions corresponding to harmonics of the lowest-energy solution. Obtaining a series of solutions that are related to each other rather than a single solution is expected since Schrödinger's equation is an eigenfunction (Chapter 9).

The quantum number n must be a positive integer. The value of $n = 0$ is not allowed as for this value of n the wavefunction is zero everywhere and so this corresponds to the case of no particle. Because n cannot be zero, the lowest energy that the particle may possess is not zero as is allowed by classical physics. Instead, there is a *zero-point energy* that is the minimal value that the particle can have. We will find that all of the solutions discussed will have a zero-point energy due to the basic principle that the particle can never be both stopped and at a single location, but rather must be moving according to the Heisenberg Uncertainty Principle. For the hydrogen atom, this concept will be important as it will mean that the electrons in atoms will not always have a certain minimal energy at all temperatures.

Symmetry

The potential energy is symmetrical around the center of the box (Figure 10.3). The resulting wavefunctions must also reflect this symmetry. The ground-state wavefunction is symmetrical about the center with the value of the wavefunction being the same at both x and $x + L/2$. This corresponds to the wavefunction having positive parity. The first excited

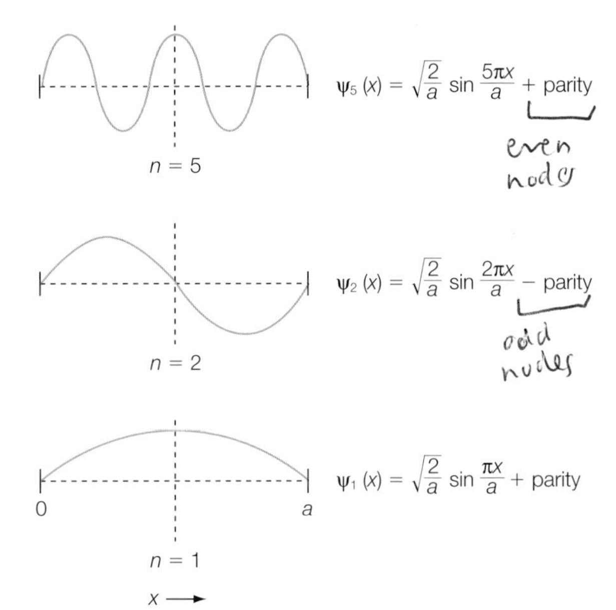

Figure 10.3 Symmetry of the wavefunctions for the particle in a box for n = 1, 2, and 5.

state is also symmetrical but with the wavefunction having a change in sign, corresponding to negative parity. The change in sign for the wavefunction as it passes through $x = L/2$ means that the wavefunction must be exactly zero at $x = L/2$. The higher-order wavefunctions all have parity, alternating between positive and negative.

Wavelength

Each solution to Schrödinger's equation possesses a wavelength given by:

$$L = n\frac{\lambda}{2} \quad \text{or} \quad \lambda = \frac{2L}{n} \tag{10.17}$$

According to the de Broglie relation, the wavelength is related to the momentum:

$$p = \frac{h}{\lambda} = \frac{nh}{2L} \tag{10.18}$$

Since the momentum is also related to kinetic energy, the de Broglie relation predicts the following values of energy:

$$E = \frac{p^2}{2m} = \frac{1}{2m}\left(\frac{nh}{L}\right)^2 = \frac{n^2h^2}{8mL^2} \tag{10.19}$$

As expected, these energies agree exactly with those derived using Schrödinger's equation.

Probability

The probability of finding a particle at any given position in the box varies depending upon the position and the quantum number of the wavefunction. For the ground-state wavefunction, the probability is zero at $x = 0$ and increases until it reaches a maximum at $x = L/2$. Due to the symmetry, the probability is equal for finding a particle at equal distances from the center. The total probability of finding the particle is set to one so the probability of finding the particle in a smaller region must be less than one. We can calculate the probability for any region: for example, the probability between $x = 0$ and $x = l$ is given by:

$$\int_0^l \psi^*(x)\psi(x)\,\mathrm{d}x = \frac{2}{L}\int_0^l \sin^2\left(\frac{n\pi x}{L}\right)\mathrm{d}x \tag{10.20}$$

$$= \frac{2}{L}\frac{1}{2}\left[x - \sin\left(\frac{2n\pi x}{L}\right)\left(\frac{L}{2\pi n}\right)\right]_0^l \tag{10.21}$$

$$= \frac{1}{L}\left[l - \frac{L}{2\pi n}\sin\left(\frac{2\pi n l}{L}\right)\right]$$

This term equals 1 when l = L, 0.5 when l = $L/2$, and 0.05 when l = $0.2L$.

Orthogonality

One of the special properties of eigenfunctions is that the solutions are orthogonal. This means that although the solutions are related to each other their overlap is zero; that is

$$\int_{\substack{\text{all}\\\text{space}}} \psi_n^* \psi_{n'} \mathrm{d}\tau = 0 \tag{10.22}$$

Average or expectation value

In quantum mechanics, one can only determine the expectation, or average, value of a parameter. For example, for this problem let's calculate first the average momentum $\bar{p}$:

$$\bar{p} = \int_0^L \psi^*(x)\left(\frac{\hbar}{i}\frac{\partial}{\partial x}\right)\psi(x)\,\mathrm{d}x \tag{10.23}$$

$$\bar{p} = \frac{2}{L}\int_0^L \sin\left(\frac{n\pi x}{L}\right)\left(\frac{\hbar}{i}\frac{\partial}{\partial x}\right)\sin\left(\frac{n\pi x}{L}\right)\mathrm{d}x \tag{10.24}$$

$$\bar{p} = \frac{2\hbar}{iL}\frac{n\pi}{L}\int_0^L \sin\frac{n\pi x}{L}\cos\frac{n\pi x}{L}\,\mathrm{d}x = \frac{i\hbar}{L}\left[\sin^2\left(\frac{n\pi x}{L}\right)\right]_0^L = 0 \tag{10.25}$$

The average momentum is zero because the particle will travel both to the right and the left with equal probability, resulting in no net momentum. The mean square of the momentum $\bar{p}^2$ does not average to zero and is calculated as:

$$\bar{p}^2 = \int_0^L \psi^*(x)\left(\frac{\hbar}{i}\frac{\partial}{\partial x}\right)^2 \psi(x)\,\mathrm{d}x \tag{10.26}$$

$$\bar{p}^2 = \frac{2\hbar^2}{L}\int_0^L \sin\left(\frac{n\pi x}{L}\right)\left(\frac{n\pi}{L}\right)^2 \sin\left(\frac{n\pi x}{L}\right)\mathrm{d}x \tag{10.27}$$

$$\bar{p}^2 = \frac{h^2 n^2}{4L^2} = \frac{h^2}{\lambda^2} \tag{10.28}$$

So the expectation value agrees with the de Broglie relationship as expected. We can also calculate the expectation value of position $\bar{x}$:

$$\bar{x} = \int_0^L \psi^*(x) x \psi(x)\,\mathrm{d}x = \frac{2}{L}\int_0^L \sin\left(\frac{n\pi x}{L}\right) x \sin\left(\frac{n\pi x}{L}\right)\mathrm{d}x \tag{10.29}$$

$$\bar{x} = \frac{2}{L}\int_0^L x \sin^2\left(\frac{n\pi x}{L}\right)\mathrm{d}x = \frac{L}{2} \tag{10.30}$$

The particle has an average position of $L/2$; that is, the particle is on average in the middle of the box.

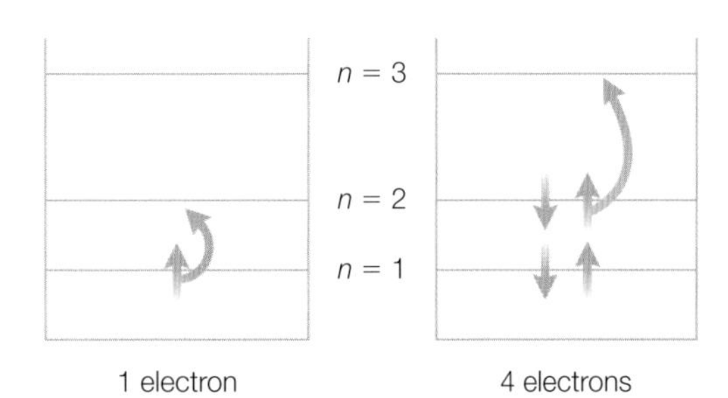

Figure 10.4 The longest-wavelength transitions for one and four electrons in a particle in a box.

Transitions

A transition between states is associated with absorption or emission of energy. The particle is normally present in the lowest-energy state. If energy is introduced into the box then the particle can make a transition to a higher-energy state. If a particle changes from the $n = 2$ to the $n = 1$ state (Figure 10.4), the change in energy is:

$$\Delta E = E_2 - E_1 = \frac{h^2}{8mL^2}(2^2 - 1^2) = \frac{3h^2}{8mL^2} \tag{10.31}$$

For an electron $m = 9.11 \times 10^{-31}$ kg. Assuming that $L = 1$ nm then

$$\Delta E = \frac{3(6.63 \times 10^{-34}\ \mathrm{Js})^2}{8(9.11 \times 10^{-31}\ \mathrm{kg})(10^{-9}\ \mathrm{m})^2} = 1.8 \times 10^{-19}\ \mathrm{J} \tag{10.32}$$

Using conservation of energy we can set the energy of the light emitted equal to the energy change between states, giving:

$$\Delta E = h\nu = \frac{hc}{\lambda} \tag{10.33}$$

$$\lambda = \frac{hc}{\Delta E} = \frac{(6.63 \times 10^{-34}\ \mathrm{Js})(3 \times 10^8\ \mathrm{m\ s^{-1}})}{1.8 \times 10^{-19}\ \mathrm{J}} = 1.1 \times 10^{-6}\ \mathrm{m} = 1100\ \mathrm{nm} \tag{10.34}$$

If there is more than one electron in the box, then the transition is not the same, but rather proceeds from the highest occupied level to the lowest unoccupied level, with each level containing two electrons. For example, if there are four electrons in the box, then the first two levels are filled and the transition is from the second to the third level (Figure 10.4) and the difference in the energy is:

$$\Delta E = E_3 - E_2 = \frac{h^2}{8mL^2}(3^2 - 2^2) = \frac{5h^2}{8mL^2} \tag{10.35}$$

RESEARCH DIRECTION: CAROTENOIDS

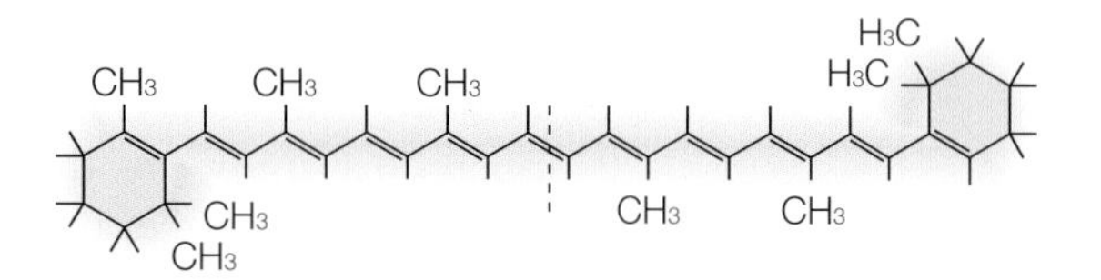

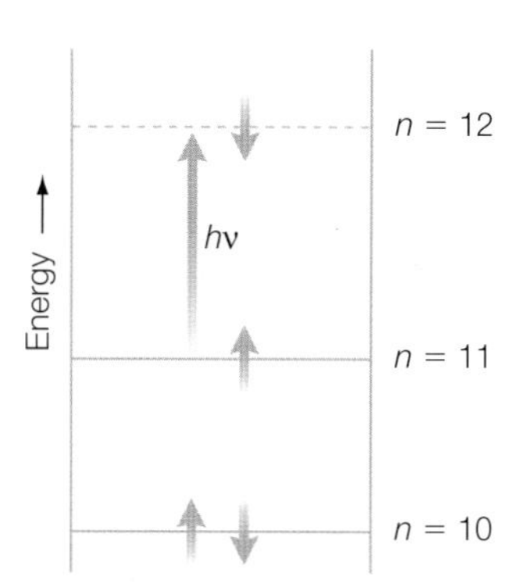

Figure 10.5 The chemical structure of a carotenoid, β-carotene, and the transition for the particle-in-a-box model of the conjugated system.

For larger systems the calculation is the same but there are more electrons. Carotenoids are conjugated molecules that are found widely in biological systems. In some systems, their function is to serve as a pigment giving color to birds, fish, or lobsters (the red color of lobsters when cooked arises from changes in the carotenoid). Carotenoids are often good choices for coloration as they absorb light very strongly. In humans, carotenoids are necessary precursors for the formation of vitamin A, or retinal, which is an important molecule in vision (Chapter 17). In photosynthetic organisms, carotenoids play critical roles in the conversion of light energy into chemical energy. The ability of carotenoids to strongly absorb light also serves them well in performing critical roles in energy transfer; that is, absorbing light energy and transferring it to other molecules which can convert the energy into chemical energy. There are many chemically distinct carotenoids; however, all carotenoids have extended conjugated systems with delocalized π electrons.

Let us consider the specific example of one of the common carotenoids, β-carotene, which is a linear polyene with 11 double bonds alternating with 10 single bonds (Figure 10.5). The electrons can move within the total length of the conjugation so this will represent the length L of the box. Typical lengths for single and double bonds between carbon atoms are 1.46 and 1.35 Å respectively. By adding up the 11 double bonds and 10 single bonds we estimate the length to be 29 Å. There are 22 π electrons, so the 11 lowest-energy levels are filled with electrons. The first transition between a filled level and an unfilled level is between the eleventh and twelfth levels, yielding:

$$\Delta E = E_{12} - E_{11} = \frac{h^2}{8mL^2}(12^2 - 11^2) = \frac{23h^2}{8mL^2} = \frac{hc}{\lambda} \tag{10.36}$$

$$\lambda = \frac{8m_e cL^2}{23h} = 1207 \text{ nm} \tag{10.37}$$

In practice, this transition is not observed because it is forbidden as we have not yet considered factors that influence the strength of the transition, such as symmetry (Chapter 14).

Carotenoids have several essential functions in photosynthetic organisms. This calculation is useful for understanding how the length of a carotenoid would influence its absorption spectrum. However, to understand the role of carotenoids in the conversion of light into chemical energy requires a detailed analysis of the electronic structure. For example, during the afternoon on a sunny day, the amount of sunlight available is in significant excess and plants dispose of the excess light by a process called nonphotochemical quenching (see below). One role of carotenoids in plants is to help avoid over-excitation of the photosynthetic system by dissipating energy when light levels are very high (Deming-Adams 1990; Kulheim et al. 2002; Li et al. 2002). This is called the xanthophyll cycle because the carotenoid composition in the plant changes as the light level changes (Figure 10.6).

Under low light levels the light energy causes excitation of the chlorophyll followed by electron transfer, forming a charge-separated state involving oxidized chlorophyll and a reduced quinone. These reactions are part of the overall process known as the Z scheme involving many different proteins.

$$\text{Low light: Chl Q} \xrightarrow{h\nu} \text{Chl* Q} \rightarrow \text{Chl}^+ \text{ Q}^- \tag{10.38}$$

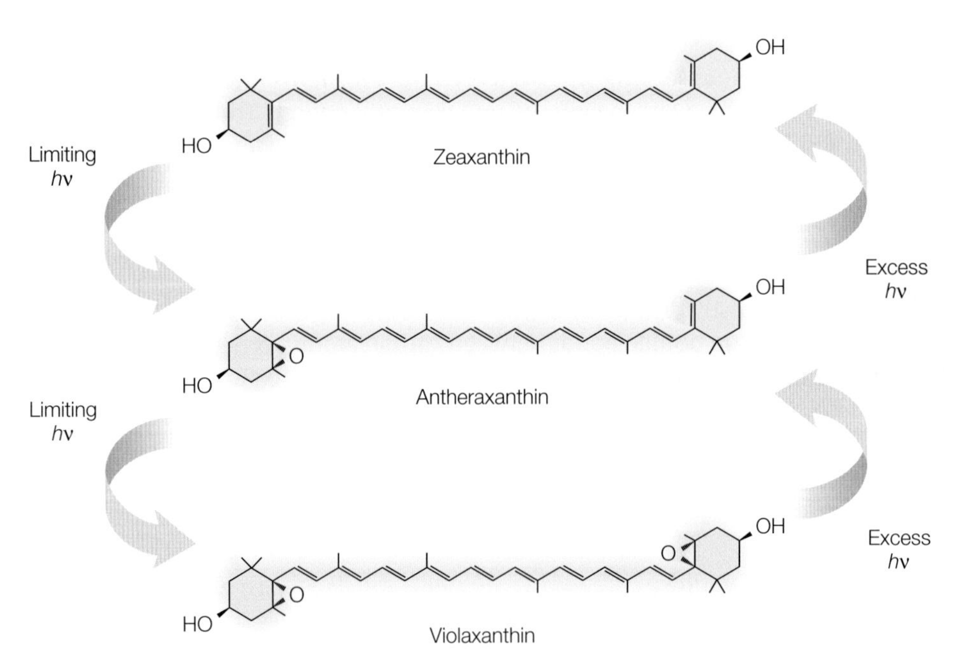

Figure 10.6 The carotenoid composition in plants changes in response to light intensity in the xanthophyll cycle.

However, under high light conditions, the excess light cannot be used to form the charge separated state but could result in photodamage. To avoid such harmful reactions, plants use nonphotochemical quenching, which involves the xanthophyll cycle to quench the excess energy (Figure 10.7). The process is termed nonphotochemical since it is distinct from the normal light-induced reactions. The mechanism is debated as the quenching can arise either from an energy transfer from excited chlorophyll to another molecule that could safely dissipate the energy or by electron transfer followed by recovery to the ground state.

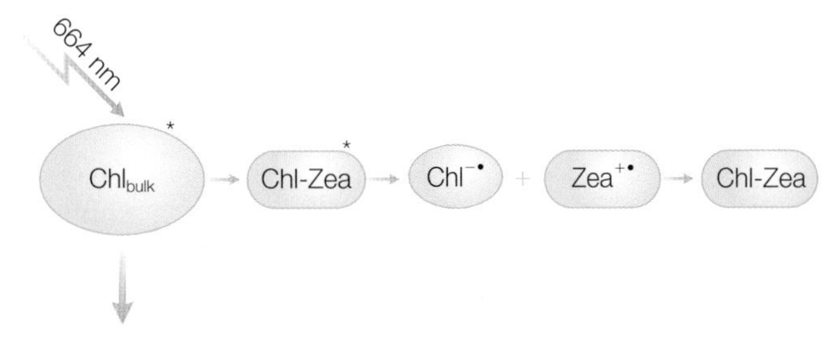

Figure 10.7 A possible mechanism of energy quenching in the xanthophyll cycle. Chl, chlorophyll; Zea, zeaxanthin. Modified from Holt et al. (2005).

How are energy and electron transfer related to the change in carotenoids? In the particle-in-a-box model, the energy of the electron, E, is inversely proportional to the square of the length, L, of the box:

$$E \propto \frac{1}{L^2} \tag{10.39}$$

When the epoxide groups of the carotenoid are removed to form zeaxanthin, the conjugated π system is extended from nine to 10 and then 11 double bonds. This has the effect of lowering the energy levels of the carotenoid, so allowing the new reactions to occur.

To investigate these different possible reactions, scientists have performed many different spectroscopic measurements. Recently, scientists have made picosecond measurements of these processes on mutants with different quenching capabilities (Holt et al., 2005). The resulting spectra were consistent with the formation of an oxidized carotenoid under quenched conditions, and hence the mechanism may involve electron transfer.

TWO-DIMENSIONAL PARTICLE IN A BOX

Although the problem of a one-dimensional particle in a box may be suitable for simple conjugated polyenes, most molecules are more complex. Many biological complexes are composed of amino acid residues with conjugated rings, such as tyrosine, or even large cofactors such as chlorophylls. Theoretical treatment of these molecules requires the additional dimensions to be considered. The properties of particles confined to two- or three-dimensional boxes can be solved following the approach used for the one-dimensional case.

Consider a particle confined to a rectangular area where the potential is zero, with the potential being infinite outside the box (Figure 10.8). Schrödinger's equation becomes:

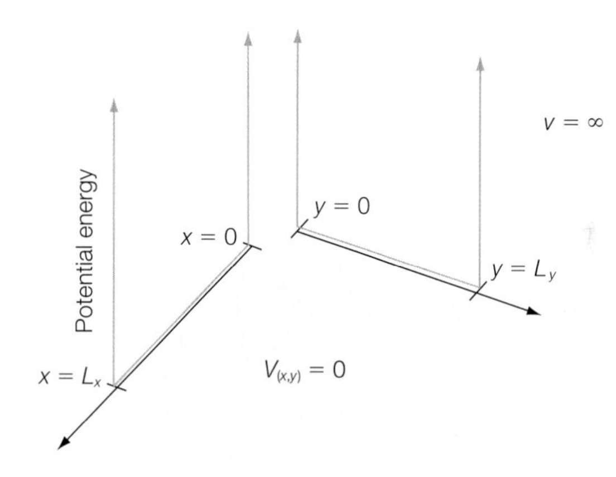

Figure 10.8
A particle in a two-dimensional box with a potential of zero inside the box and an infinite potential outside the box.

$$-\frac{\hbar^2}{2m}\left(\frac{\partial^2}{\partial x^2}\psi(x,y) + \frac{\partial^2}{\partial y^2}\psi(x,y)\right) = E\psi(x,y) \qquad (10.40)$$

Two- and three-dimensional problems can be solved by using the approach of separation of variables. With this approach, the equation is rewritten to separate the contributions from each variable so that the partial differential equation becomes two full second-order equations. Assume that the wavefunction can be written as the product of two wavefunctions, each of which is dependent only upon one variable:

$$\psi(x,y) = X(x)Y(y) \qquad (10.41)$$

We do not know *a priori* whether this can be used successfully; we will simply try substituting eqn 10.41 into Schrödinger's equations (eqn 10.40) yielding:

$$-\frac{\hbar^2}{2m}\left[Y(y)\frac{d^2}{dx^2}X(x) + X(x)\frac{d^2}{dy^2}Y(y)\right] = EX(x)Y(y) \qquad (10.42)$$

To simplify this equation, we group the constants with the energy and remove the wavefunction from the right-hand side by dividing by the wavefunction. So dividing this expression by:

$$-\frac{2m}{\hbar^2}\frac{1}{X(x)Y(y)} \qquad (10.43)$$

gives:

$$\left(\frac{1}{X(x)}\frac{d^2X(x)}{dx^2}\right) + \left(\frac{1}{Y(y)}\frac{d^2Y(y)}{dy^2}\right) = -\frac{2mE}{\hbar^2} \qquad (10.44)$$

The right-hand term is a simple constant, independent of x or y. Since the two left-hand terms are dependent upon different variables that are always equal to a constant, then each must be equal to a constant. We can then set:

$$\frac{1}{X(x)}\frac{d^2X(x)}{dx^2} = -\frac{2mE_x}{\hbar^2} \quad \text{or} \quad -\frac{\hbar^2}{2m}\frac{d^2X(x)}{dx^2} = E_xX(x) \qquad (10.45)$$

$$\frac{1}{Y(y)}\frac{d^2Y(y)}{dy^2} = -\frac{2mE_y}{\hbar^2} \quad \text{or} \quad -\frac{\hbar^2}{2m}\frac{d^2Y(y)}{dy^2} = E_yY(y) \qquad (10.46)$$

where $E_x + E_y = E$.

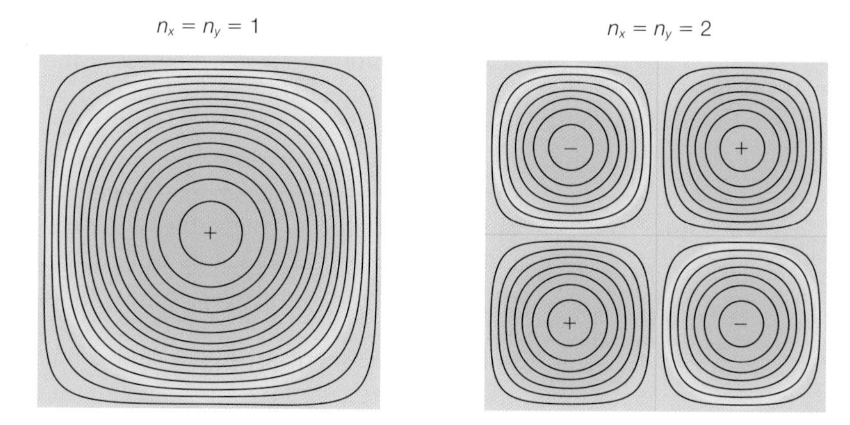

Figure 10.9 The solutions for the two-dimensional particle in a box.

These equations are the same as for the one-dimensional case, so we can write the solutions (Figure 10.9) directly as:

$$\psi(x,y) = X(x)Y(y) = \left(\sqrt{\frac{2}{L_x}}\sin\left(\frac{n_x\pi x}{L_x}\right)\right)\left(\sqrt{\frac{2}{L_y}}\sin\left(\frac{n_y\pi y}{L_y}\right)\right) \quad (10.47)$$

$$E = E_x + E_y = \frac{h^2}{8m}\left[\left(\frac{n_x}{L_x}\right)^2 + \left(\frac{n_y}{L_y}\right)^2\right] \quad (10.48)$$

Unlike the one-dimensional problem, two states now can have equal energies – that is, they are degenerate – if L_x and L_y are equal. For example, the $n_x = 2$, $n_y = 1$ and $n_x = 1$, $n_y = 2$ states are degenerate when $L_x = L_y$. This degeneracy in energy is reflective of the symmetry.

TUNNELING

For this problem, there was zero probability of finding the particle outside the box because we had said that the walls were infinitely high. What happens when we relax this condition and merely say that the walls are high but not infinitely high? From a classical viewpoint, a ball has a certain amount of energy. If the ball is in a valley between two hills it will simply roll back and forth between the hills. The ball will have its energy all in the form of kinetic energy at the bottom of the hills and be moving at its maximum speed. As it moves up the hill the ball will slow down until it stops and begins back down the hill. Because the ball does not have enough energy to go over the hill it will always be found between the two hills.

In quantum mechanics, this problem corresponds to having two wells that are separated by a wall that is finite (Figure 10.10). This is formally

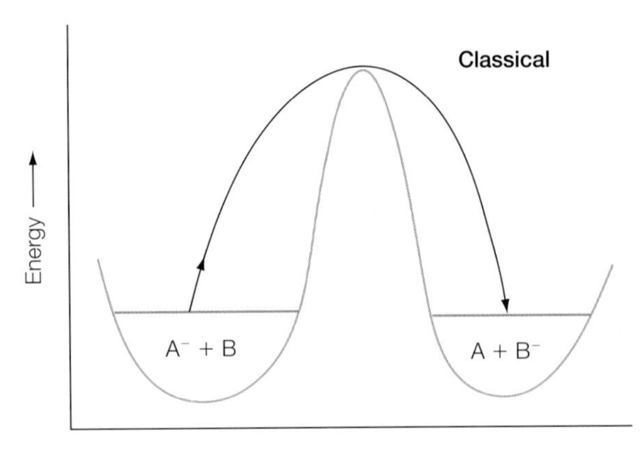

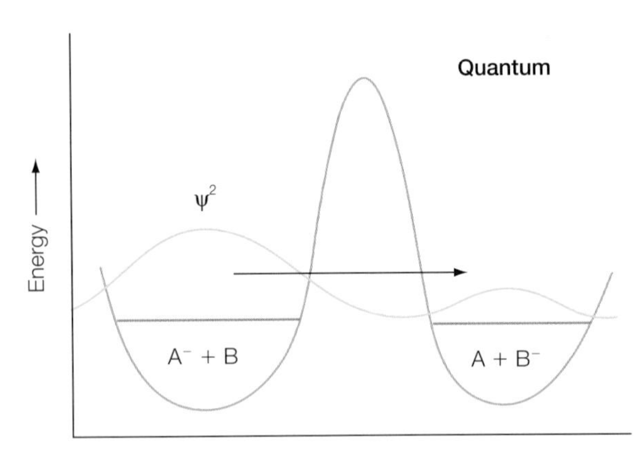

Figure 10.10 Tunneling across an energy barrier permits an electron to have a probability to be located in both wells.

expressed by saying that the potential energy at $x = 0$ is finite with a specified value that is greater than the energy of the ball. Since the walls at $x = -L$ and $x = +L$ are still infinite, we know that the wavefunction must have a value of zero at these positions. However, we can no longer specify the value of the wavefunction at $x = 0$. Rather, this value must be solved based upon the value of the potential at this position. We know that the probability must be very small at this position, but in quantum mechanics the wavefunction has a non-zero value. The ball cannot be found at $x = 0$ for any period of time since that would violate classical mechanics and the correspondence principle. However, because it has a non-zero value, it has a finite probability of being at $x = 0$ for a fleeting moment that allows the ball to move into the second well. Thus, in quantum mechanics, it is possible for a particle to enter a "classically forbidden region" provided the visit is only temporary.

To determine the conditions that permit tunneling of an object, consider a particle with an energy E in a large box (Figure 10.11). Within most of the box the potential is zero except for a small region that has a barrier with a small width a and a potential V_0. A wavefunction outside the barrier is a traveling wave since the potential is zero in this region. A wavefunction incident to the barrier can then be described as a simple sine function. In the barrier region, the energy E of the particle is less than the potential V_0, and hence the particle is classically forbidden to pass through the barrier. To determine the wavefunction, Schrödinger's equation is written for that region:

$$(E - V_0)\psi(x) = -\frac{\hbar^2}{2m}\frac{d^2}{dx^2}\psi(x) \quad 0 \leq x \leq a \tag{10.49}$$

Since the energy E and potential V_0 are constants, the solution of this equation can be written as an exponential function:

$$\psi(x) \propto e^{-\kappa x} \quad 0 \leq x \leq a \tag{10.50}$$

$$\kappa = \sqrt{\frac{2m(V_0 - E)}{\hbar^2}}$$

Since $V_0 > E$, the wavefunction inside the barrier is an exponential decaying according to the constant κ, termed the decay length. The extent of the decay inside the barrier depends upon the parameters a and V_0. As the barrier width increases, the extent of the exponential decay increases. Tunneling decreases the value of the decay length, and is therefore more likely for particles with larger energies.

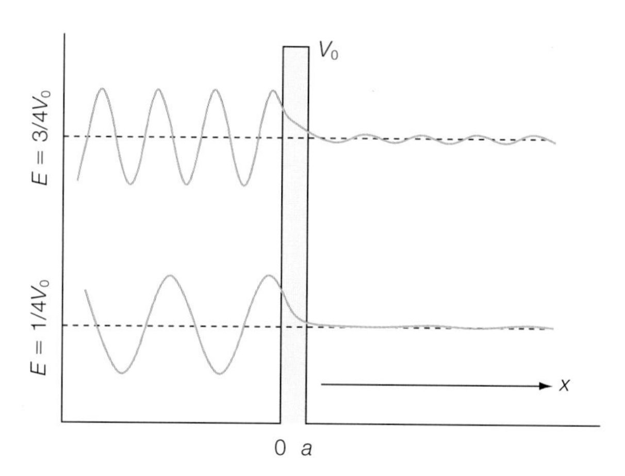

Figure 10.11 Two traveling wavefunctions incident to a barrier. The wavefunctions have a finite probability of passing through the barrier despite having an energy E, which is less than the barrier height V_0. The extent of the decay is dependent upon the barrier width a and the difference $E - V_0$.

RESEARCH DIRECTION: PROBING BIOLOGICAL MEMBRANES

In the early 1980s, the phenomenon of tunneling of electrons was adopted as a new spectroscopic technique to probe the surfaces of materials. For the development of this spectroscopic tool, Gerd Binnig and Heinrich Rohrer received a Nobel Prize for Physics in 1986 (Binning et al. 1982, 1986). The technique of scanning tunneling microscopy, or STM, allows the imaging of a surface at atomic resolution with a relatively simple conceptual approach. An essential feature of an STM is the probe tip, which must approach a surface very closely without touching (Figure 10.12). If the tip is close enough, the space, which may be a vacuum, air, or water, between the probe and the surface will represent a barrier that can be crossed by electrons by tunneling. Following the discussion above, the approach must be a few decay lengths which, from a practical viewpoint, represents a distance within 1 nm. Ideally, the tip is very sharp and the tunneling will occur from the closest point of the tip to the surface. The tip is mounted on an instrument called a piezoelectric device that can translate the tip across the surface in increments of less than 1 nm. If the tip scans strictly horizontally over the surface then a rise in the surface will decrease the spacing and the current will increase. Alternatively, the probe can be positioned such that the current is fixed by use of a feedback circuit that can move the probe vertically as needed.

Since the mid-1990s, a related technique, atomic force microscopy (AFM), has increasingly been applied to biological systems. The goal of this technique is also to characterize the surface of a material, and the equipment is similar to that used for STM, but AFM operates by measuring the force between the tip and sample. In a sense, AFM resembles the operation of a stylus passing over a vinyl record to produce sounds in a record player. In AFM the sensitivity is achieved by the use of a cantilever whose

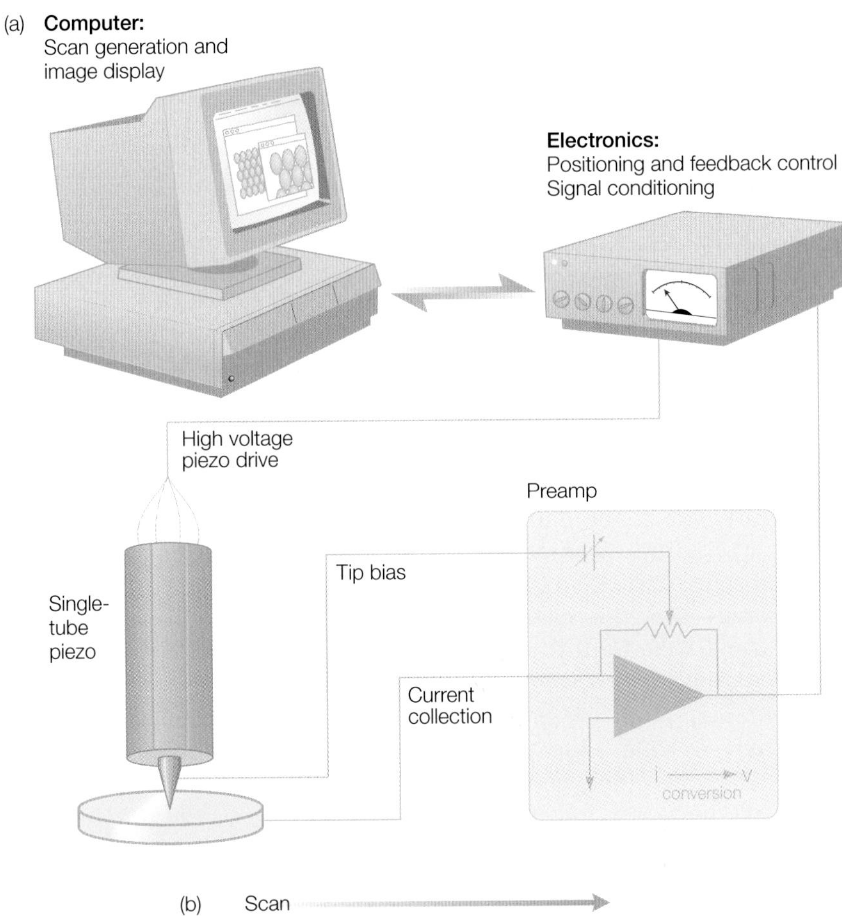

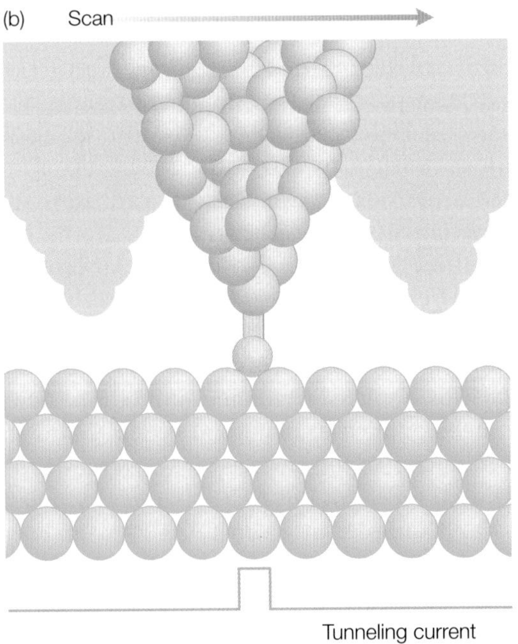

Figure 10.12
A schematic representation of (a) a scanning tunneling microscope with (b) an expanded view of the probe.

position is determined sensitively by measuring the deflection of a laser beam off a cantilever that contains the tip (Figure 10.13). The deflection is magnified over 1000-fold, allowing a near-atomic resolution to be achieved. The extent of the deflection provides a measure of the force. Unlike the stylus system of a record player, the probe is controlled by the use of a feedback loop that not only allows the measurement of the force but also regulates it to allow the acquisition of accurate measurements at very low forces.

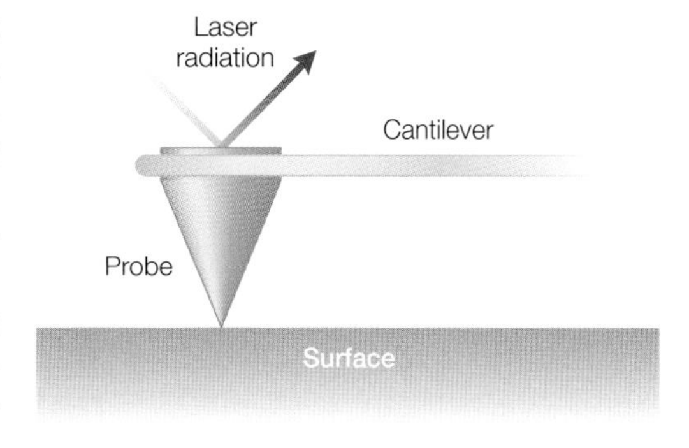

Figure 10.13 AFM is similar to STM in being able to measure at the atomic level except that the probe works with a cantilever to probe the force between the tip and surface, which may be noncoducting as no current is involved.

In addition to probing the surface of a material, the technique can also be used to measure the conductivity of an electron through a single molecule. A monolayer of molecules is assembled on a gold surface linked through thiol groups at the end of each molecule (Figure 10.14). An approximate vertical orientation arises from the choice of molecules that prefer not to bind together. Electrical contact is made between the probe tip and the monolayer, and the current is measured as a function of the applied voltage. In a set of experiments on 1,8-octanedithiol, the measurements revealed only a limited number of distinct curves that were modeled as arising from quantization of the current, allowing the resistance of a single molecule to be determined at 900 MΩ (Cui et al. 2001).

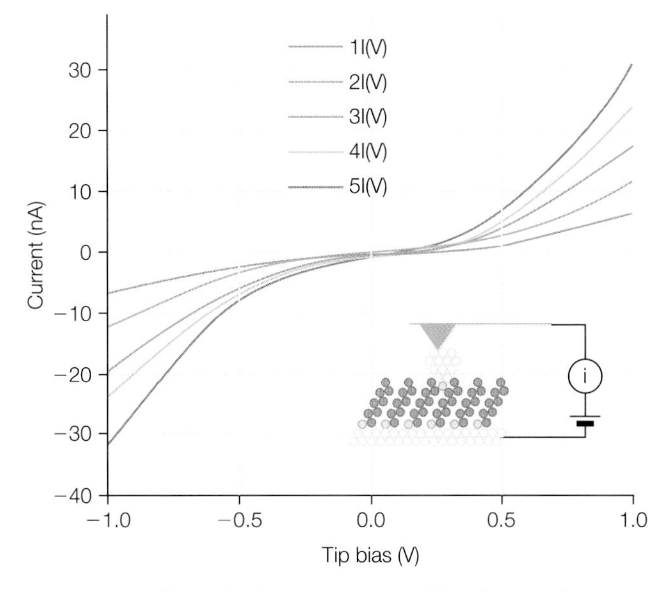

Figure 10.14 A tip is brought into contact with a layer of 1,8-octanedithiol [$HS(CH_2)_8SH$] molecules that are attached to the gold surface though the thiol groups. The electrical current is measured as a function of the applied voltage, which yields the conductivity for a single molecule. Modified from Cui et al. (2001).

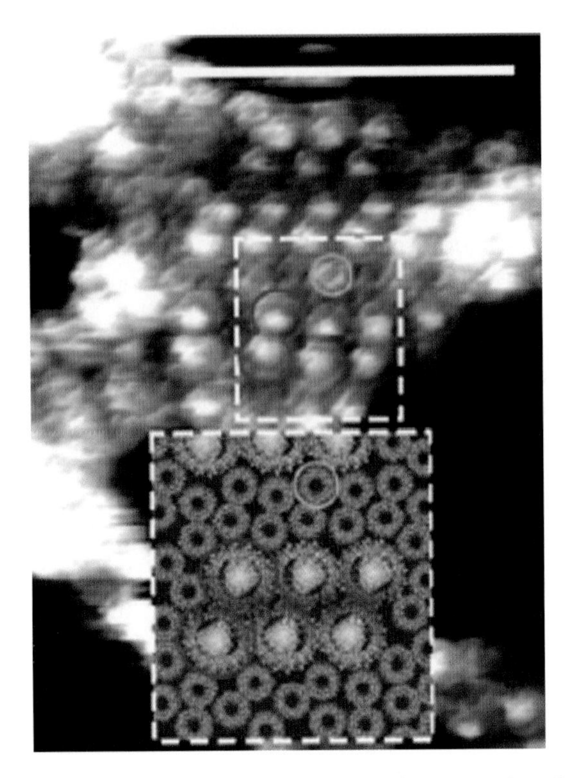

Figure 10.15 AFM images of membranes showing that they have large structures in circular configurations. From Bahatyrova et al. (2004).

Resistance = (voltage/current)

The ability of AFM to provide detailed images of the organization of biological cells has been demonstrated by its use to obtain images of membranes from photosynthetic bacteria (Figure 10.15; Bahatyrova et al. 2004; Scheuring & Sturgis 2005; Muller et al. 2006). In photosynthesis, the harvesting of light energy is performed by large arrays of chlorophyll–protein complexes (Chapter 20). In purple bacteria, two types of light-harvesting complex are present, identified as light-harvesting complexes I and II. Both of these complexes contain bacteriochlorophylls and carotenoids that absorb the light and transfer the energy to a central protein complex, the reaction center, that performs the initial photochemistry. All three of these protein complexes are embedded within the cell membrane. The light-harvesting complexes I and II are oligomeric with eight to nine and 16 pairs of subunits arranged in a ring, with a single reaction center found within each light-harvesting I ring. The structures of the individual, purified complexes have been determined using protein crystallography but the cellular arrangements have been largely unknown.

Electron-microscopic studies have shown the presence of a highly invaginated surface. Biochemical studies have shown that the invaginations are packed with the light-harvesting complexes. To optimize the efficiency of light capture and energy conversion, the cells contain many light-harvesting II complexes that surround the light-harvesting I reaction center. AFM has revealed in fine detail the packing of groups of the light-harvesting II complex surrounding the light-harvesting complex I and reaction center. The arrangement of the rings observed in the AFM images fits extremely well with the known structural models. Such studies demonstrated the utility of AFM and how it can be used to delineate the organization of protein complexes in cell membranes.

RESEARCH DIRECTION: ELECTRON TRANSFER II: DISTANCE DEPENDENCE

Electron transfer plays a key role in many metabolic processes including respiration (Chapter 9) and photosynthesis (Chapter 20). In some cases, electron transfer occurs as a second-order reaction, for example, when the electron donor belongs to one protein that must dock to a second protein that has the electron acceptor. Such cases are found when proteins are part of a metabolic pathway, such as respiration, where cytochrome serves as an electron carrier between two complexes. For these electron-transfer reactions, the overall rate will be dictated by the diffusion of the carrier as the electron transfer does not occur until a complex is formed. In other cases, a protein has more than one cofactor as both the electron donor and acceptor are part of a large complex. The theory presented is applicable to either case: a protein with a bound donor and acceptor or a protein–protein complex that has formed transiently.

Electron transfer occurs between an electron donor, D, and electron acceptor, A, that can be separated by relatively large distances of up to 25 Å:

$$\mathrm{DA} \rightarrow \mathrm{D^+A^-} \tag{10.51}$$

In Marcus theory (Marcus & Sutin 1985), the transfer is an activated process and the rate k_{et} can be divided into two components, the electronic coupling V and the Franck–Condon term that described the energetics:

$$k_{\mathrm{et}} = \frac{2\pi}{h} V^2(\text{Franck–Condon}) \tag{10.52}$$

In Chapter 7 the energetics of electron transfer were discussed. In this section the factors that influence the coupling will be discussed. In a simple representation of electron transfer, an electron on a donor must make a transition to the acceptor. Both the donor and acceptor can be considered to be local regions with stable potentials (Figure 7.9). In order to make the transition, the electron must travel through regions that have varying potentials, but are generally less favorable, and therefore less stable, than those associated with the donor and acceptor. The problem then can be represented as calculating the probability of an electron making a transition through a series of potential barriers. The presence of the barriers is modeled in terms of the coupling constant V. For each barrier, there will be a decrease in the transmission of the electron, and this is modeled as giving rise to a decrease in the coupling strength. The effect of the multiple barriers is to give rise to a product of terms, each of which, identified as ε_I, contributes to the overall strength of the coupling with a proportionality constant A:

$$|V|^2 = A^2\left(\prod_i \varepsilon_I\right)^2 \tag{10.53}$$

The value of each contribution ε_I can be estimated using two different possible approaches. In one approach, each of the steps is considered to be of equal strength, $\varepsilon_I = \varepsilon_{av}$ for all steps i. In this case, the coupling can be written in terms of the average barrier for each step and expressed in terms of the physical separation between the donor and acceptor, r_{DA}:

$$|V|^2 = A^2\varepsilon_{av}^{2N} = A^2 e^{-\beta R_{DA}} \quad \beta = -\frac{2N}{R_{DA}}\ln\varepsilon_{av} \approx 1\ \text{Å}^{-1} \tag{10.54}$$

Thus, the prediction is that the coupling, and hence the electron-transfer rate, will decay exponentially with increasing separation between the donor and acceptor. The alternative approach is that each step has a distinctive effect on the coupling. In this scheme, the individual coupling is determined by the nature of the space that the electron is traversing. In general, the steps can be classified into three groups, with the best coupling arising for travel through covalent bonds, weaker coupling for travel through hydrogen bonds, and the weakest coupling being for travel through space. The net coupling is calculated as the product of these three contributions:

$$|V|^2 = A^2\left[\prod_i \varepsilon_I(\text{covalent bonds})\right]\left[\prod_j \varepsilon_j(\text{hydrogen bonds})\right]\left[\prod_k \varepsilon_k(\text{through space})\right] \tag{10.55}$$

To calculate the coupling, a specific pathway must be chosen and then all of the individual contributions are evaluated and the coupling is determined. In practice, computer algorithms sample many different possible pathways to determine the optimal path.

Different sets of experimental data were considered in evaluating these different models. In the laboratory of Harry Gray and coworkers (Gray & Winkler 2005), ruthenium compounds, serving as electron donors, were attached at different locations on a cytochrome and the electron transfer to the heme was measured. The variation of the measured rate was well described in terms of the coupling for the optimal path coupling in each case. Alternatively, Leslie Dutton and coworkers (Page et al. 2003) considered the dependence of many different electron-transfer rates and found that the rate could be described using only the separation distance R_{DA}. In both cases, electron transfer is expected to decrease exponentially with distance, but the distance is either the closest distance or the pathway distance. The appropriateness of each model to describe the transfer of electrons in proteins is still under discussion, although the distinction between the models is lessened if electrons are considered to sample several different comparable pathways (Page et al. 2003; Gray & Winkler 2005; Lin et al. 2005; Miyashita et al. 2005). While the nature of electron transfer in proteins was being investigated, experiments of electron transfer were being performed under more artificial circumstances, namely in DNA. The initial data indicated that electrons could travel large distances at rates that indicated much larger coupling values than in proteins, but later experiments showed that the electron transfer did not involve long-range transfer but, rather, a series of short steps that yielded parameters in keeping with those observed for proteins (Murphy et al. 1993; Giese 2002).

All of these models are for electron transfer within a protein complex, but in cells electrons are often transferred between two different proteins, representing a second-order process rather than a first-order process (Chapter 7). In such cases, the observed rate of electron transfer is usually limited by the rate of diffusion and formation of the protein–protein complex prior to the electron transfer. The formation of the protein–protein complex is often driven by electrostatic interactions involving a primarily negatively charged surface on one protein and a largely positively charged surface on the second. Once the proteins are spatially close, other interactions, such as hydrophobic interactions and steric considerations, stabilize the position of one protein relative to the other. Investigations continue into understanding the contributions of different factors that influence formation of the complex, such as reorganization of the solvent surrounding the surface of the proteins (Lin et al. 2005; Miyashita et al. 2005) and protein dynamics (Wang et al. 2007).

REFERENCES

Bahatyrova, S., Frese, R.N., Siebert, C.A. et al. (2004) The native architecture of a photosynthetic membrane. *Nature* **430**, 1058–62.

Binning, G., Quate, C.F., and Gerber, Ch. (1986) Atomic force microscope. *Physical Review Letters* **56**, 930–3.

Binning, G., Rohrer, H., Gerber, Ch., and Weibel (1982) Surface studies by scanning tunneling microscopy. *Physical Review Letters* **49**, 57–61.

Cui, X.D., Primak, A., Zarate, X. et al. (2001) Reproducible measurement of single-molecule conductivity. *Science* **294**, 571–4.

Deming-Adams, B. (1990) Carotenoids and photoprotection in plants: a role for the xanthophyll zeaxanthin. *Biochimica Biophysica Acta* **1020**, 1–24.

Giese, B. (2002) Electron transfer in DNA. *Current Opinion in Chemical Biology* **6**, 612–18.

Gray, H.B. and Winkler, J.R. (2005) Long-range electron transfer. *Proceedings of the National Academy of Sciences USA* **102**, 3534–9.

Holt, N.E., Zigmantas, D., Valkunas, L. et al. (2005) Carotenoid cation formation and the regulation of photosynthetic light harvesting. *Science* **307**, 433–6.

Kulheim, C., Agren, J., and Jansson, J. (2002) Rapid regulation of light harvesting and plant fitness in the field. *Science* **297**, 91–3.

Li, X.P., Muller-Moule, P., Gilmore, A.M., and Niyogi, K.K. (2002) PsbS dependent enhancement of feedback de-excitation protects photosystem II from photoinhibition. *Proceedings of the National Academy of Sciences USA* **99**, 15222–7.

Lin, J., Balabin, I.A., and Beratan, D.N. (2005) The nature of aqueous tunneling pathways between electron-transfer proteins. *Science* **310**, 1311–13.

Marcus, R.A. and Sutin, N. (1985) Electron transfers in chemistry and biology. *Biochimica Biophysica Acta* **811**, 265–322.

Miyashita, O., Okamura, M.Y., and Onuchic, J.N. (2005) Interprotein electron transfer from cytochrome c_2 to photosynthetic reaction center: tunneling across an aqueous interface. *Proceedings of the National Academy of Sciences USA* **102**, 3558–63.

Muller, D.J., Sapra, K.T., Scheuring, S. et al. (2006) Single-molecule studies of membrane proteins. *Current Opinion in Structural Biology* **16**, 489–95.

Murphy, C.J., Arkin, M.R., Jenkins, Y. et al. (1993) Long-range photoinduced electron transfer through a DNA helix. *Science* **262**, 1025–9.

Page, C.C., Moser, C.C., and Dutton, P.L. (2003) Mechanism for electron transfer within and between proteins. *Current Opinion in Chemical Biology* **7**, 551–6.

Scheuring, S. and Sturgis, J.N. (2005) Chromatic adaption of photosynthetic membranes. *Science* **309**, 484–7.

Wang, H., Lin, S., Allen, J.P. et al. (2007) Protein dynamics control the kinetics of initial electron transfer in photosynthesis. *Science* **316**, 747–50.

PROBLEMS

The first two questions are for the one-dimensional particle in a box.

10.1 What is the wavelength of the first excited state state in terms of L?

10.2 If the length L is 8 Å, what is the wavelength of the (a) ground state and (b) second excited state of an electron?

10.3 Write (but do not solve) an integral expression for the average value, or expectation value, of (a) the momentum and (b) the momentum squared for the ground state.

10.4 Write (but do not solve) an integral expression for the probability of finding the particle between $x = 0$ and $x = L/4$.

10.5 Write (but do not solve) an integral expression for the average value, or expectation value, of x^2 for the ground state.

10.6 Substitute the ground state into Schrödinger's equation and determine the energy of this state.

10.7 If the length L is 1.0 Å, determine the probability of finding the particle in the ground state between $x = L/4$ and $x = 3L/4$.

10.8 If the length L is 1.0 Å, calculate the longest wavelength transition for an electron that is initially in the lowest-energy state.

10.9 If the length L is 1.0 Å, calculate the wavelength of a photon that is absorbed after an electron makes a transition from the ground state to the fourth excited state.

10.10 Assume there are six electrons in a box with a length of 10 nm: (a) What is the highest occupied state? (b) What is the wavelength of the highest occupied state? (c) What is the lowest unoccupied state? (d) What is the wavelength of the lowest unoccupied state? (e) What is wavelength of the longest wavelength transition?

10.11 For the following questions assume that the potential is given by:

$$V\ (x \leq -a) = \infty$$
$$V\ (x \geq +3a) = \infty$$
$$V\ (-a \leq x \leq +3a) = 0$$

(a) Write the boundary conditions for the problem.
(b) Write Schrödinger's equation for this problem.
(c) Write the general solution to this problem.
(d) In terms of a, what is the wavelength of the ground state?

10.12 For the following questions, assume that a particle is trapped in a box such that:

$$V\ (x < b) = \infty$$
$$V\ (x \geq b \text{ and } x \leq 3b) = V_0$$
$$V\ (x > 3b) = \infty$$

where V_0 is a constant greater than zero.

(a) Write Schrödinger's equation for this potential.
(b) What boundary conditions can be placed on the wavefunctions?
(c) What are the wavefunctions for this problem?
(d) What is the wavelength of the ground state?
(e) What is the energy of the ground state?

10.13 Assume that a two-dimensional box is rectangular with a length of 10 Å along x and 20 Å along y: (a) write Schrödinger's equation for this problem; (b) write the ground-state wavefunction; (c) calculate the ground-state energy for a single electron; and (d) calculate the longest wavelength transition when there is one electron in the box.

10.14 For the following two-dimensional problem assume that the potential is given by:

$$V\ (x \leq -2a) = \infty$$
$$V\ (x \geq +3a) = \infty$$
$$V\ (-a \leq x \leq +3a) = 0$$
$$V\ (y \leq -b) = \infty$$
$$V\ (y \geq +2b) = \infty$$
$$V\ (-b \leq y \leq +2b) = 0$$

(a) Write the boundary conditions for the problem.
(b) Write Schrödinger's equation for the problem.
(c) Write the general solution to the problem.

10.15 Write Schrödinger's equation for a three-dimensional cubic box with a length l.

10.16 Explain the basic experimental observation of nonphotochemical quenching. How is this process coupled to a change in carotenoids?

10.17 What is the structural difference between zeaxanthin and violaxanthin? How can this structural difference contribute to nonphotochemical quenching?

10.18 What is tunneling? How does tunneling depend upon experimental parameters?

10.19 What is STM?

10.20 What is AFM?

10.21 Give an example of how STM or AFM can be used for a biological sample.

10.22 A protein has one electron donor but two identical electron acceptors. The distance between each acceptor and the donor is the same but significantly different electron-transfer rates are measured between the donor and acceptor in each case. Explain this rate difference using Marcus theory.

10.23 For a protein, the electron transfer rate k was measured for mutants with different free energy differences as listed below. Estimate the value of the reorganization energy.

k (s^{-1})	**1.0×10^5**	**7.0×10^5**	**3.0×10^6**	**8.0×10^6**	**1.2×10^7**
$-\Delta G°$ (meV)	100	200	300	400	500

11

Vibrational motion and infrared spectroscopy

In this chapter the properties of vibrational motion are described. First, the classical treatment of the simple harmonic oscillator is reviewed. The classical theory is followed by the equivalent analysis of vibrational motion using Schrödinger's equation. This chapter has a detailed derivation of the equation and you are encouraged to work through the analysis to get a better understanding of how to solve problems in quantum mechanics. After the derivation, the various properties of the simple harmonic oscillator are discussed, such as the calculation of expectation values. The usefulness of these concepts is demonstrated by examining how the structure and function of proteins can be probed by measuring their vibrational properties using infrared spectroscopy.

SIMPLE HARMONIC OSCILLATOR: CLASSICAL THEORY

Classically, when a mass, m, is attached to an immovable object by a spring, the mass undergoes harmonic motion; that is, the mass will vibrate back and forth as the spring alternatively pulls and pushes on the mass (Figure 11.1). This motion is expressed using the classical laws by describing the properties of the spring with the spring constant, k. The stiffer the spring the larger the spring constant will be and the faster the mass will vibrate. The motion is described using Newton's laws. As the mass stretches the spring, the spring exerts a force, F, in the opposite direction to the motion. Since the force opposes the motion the mass slows down until it reaches the maximal extension, or amplitude, A. The mass does not stay at that position, owing to the force that pulls the mass back to the center. When the

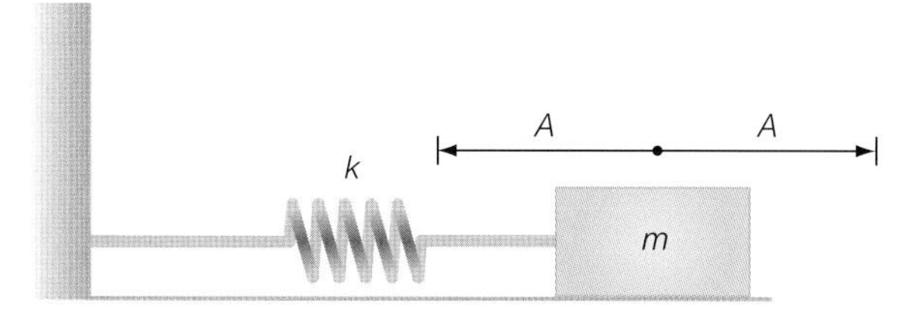

Figure 11.1 Classical model for the simple harmonic oscillator showing the amplitude of motion, A, the mass, m, and the spring constant, k.

mass reaches the center, its speed causes it to continue moving and the spring begins to compress. The compression of the spring causes a force to be exerted towards the center and opposite the motion. As before, the opposing force leads to the mass slowing down, stopping, and then returning back to the center where the process begins again.

To formally describe the motion, the force exerted on the spring is related to the displacement of the mass, x, from the equilibrium position:

$$F = -kx \tag{11.1}$$

The negative sign is included so that the force is always directed towards the center. The value of $x = 0$ is defined as the equilibrium position where the spring is neither stretched nor compressed and there is no force exerted on the mass. By Newton's laws, force is equal to the product of the mass and acceleration, a, of a particle:

$$F = ma \tag{11.2}$$

The acceleration of a particle is equal to the change of velocity, v, of the particle with respect to time. Since the velocity is equal to the change in position with respect to time, we can write:

$$F = ma = m\frac{\mathrm{d}v}{\mathrm{d}t} = m\frac{\mathrm{d}^2x}{\mathrm{d}t^2} \tag{11.3}$$

We can equate the two expressions for the force, eqns 11.1 and 11.2, and derive the following second-order differential equation that describes the motion of the mass:

$$m\frac{\mathrm{d}^2x}{\mathrm{d}t^2} = -kx \tag{11.4}$$

As we saw for the other problems that we have already studied, the solutions to this equation are of the form sin t, cos t, and e^t. For this case, the most useful expression is to use the sine function, including the amplitude, A, and the frequency ω:

$$x = A \sin \omega t \tag{11.5}$$

To verify that this is a correct solution, we will put this solution into eqn 11.3. To do so, we need to first calculate the second-order derivative:

$$\frac{\mathrm{d}^2x}{\mathrm{d}t^2} = \frac{\mathrm{d}}{\mathrm{d}t}\left(\frac{\mathrm{d}}{\mathrm{d}t}A \sin \omega t\right) = \frac{\mathrm{d}}{\mathrm{d}t}(+A\omega \cos \omega t) = -A\omega^2 \sin \omega t = -\omega^2 x \tag{11.6}$$

Note that the second derivate of x yields x multiplied by a constant. Substituting this into eqn 11.3 yields:

$$m(-\omega^2 x) = -kx \quad \text{or} \quad \omega = \sqrt{\frac{k}{m}} \tag{11.7}$$

So the mass oscillates with a frequency ω that is equal to the square root of the spring constant divided by the mass. The frequency will be slower for stiffer springs or for larger masses.

Potential energy for the simple harmonic oscillator

To solve Schrödinger's equation for this problem, we must first determine the classical expression for the potential energy, V. Classically, for a given force, the potential energy is given by the integral of the force over the displacement:

$$V = -\int F\,\mathrm{d}x = -\int (-kx)\,\mathrm{d}x = k\int x\,\mathrm{d}x = \frac{k}{2}x^2 \tag{11.8}$$

Thus, the potential energy of the simple harmonic oscillator is parabolic. When there is no displacement x has a value of zero and the potential energy is zero (Figure 11.2). As the spring is either expanded or compressed, the potential energy increases and the mass begins to slow down. When the mass reaches the point where the kinetic energy is zero and all of the energy is potential energy, the mass stops and turns back toward the center of the motion.

SIMPLE HARMONIC OSCILLATOR: QUANTUM THEORY

Using the potential energy for the harmonic oscillator (eqn 11.8), Schrödinger's equation can written as:

$$-\frac{\hbar^2}{2m}\frac{\mathrm{d}^2}{\mathrm{d}x^2}\psi(x) + \frac{k}{2}x^2\psi(x) = E\psi(x) \tag{11.9}$$

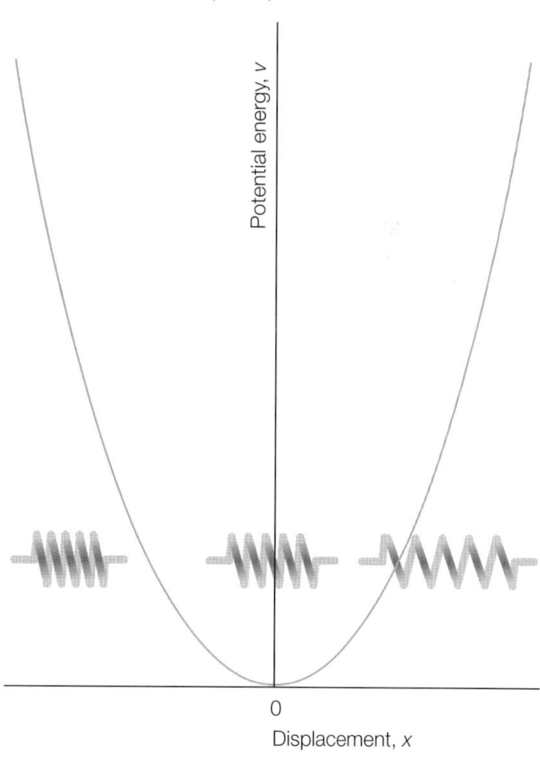

Figure 11.2 The potential energy of a simple harmonic oscillator has a parabolic dependence on the displacement of the mass.

The solutions of this equation are a series of functions called Hermite polynomials, with energies that are quantized. The mathematical analysis for solving the equation is provided in Derivation box 11.1. The reader may skip to the main text following the box, where detailed description of the solutions is given.

Derivation box 11.1

Solving Schrödinger's equation for the simple harmonic oscillator

To solve this equation, we first rewrite it by letting:

$$y = \frac{x}{\alpha} \quad \frac{\mathrm{d}}{\mathrm{d}x} = \frac{\mathrm{d}}{\mathrm{d}y}\frac{\mathrm{d}y}{\mathrm{d}x} = \frac{1}{\alpha}\frac{\mathrm{d}}{\mathrm{d}y} \quad \text{and} \quad \frac{\mathrm{d}^2}{\mathrm{d}x^2} = \frac{\mathrm{d}}{\mathrm{d}x}\frac{\mathrm{d}}{\mathrm{d}x} = \frac{1}{\alpha^2}\frac{\mathrm{d}^2}{\mathrm{d}y^2} \tag{db11.1}$$

Substitution of y for x yields:

$$-\frac{\hbar^2}{2m}\frac{1}{\alpha^2}\frac{\mathrm{d}^2}{\mathrm{d}y^2}\psi(y) + \frac{k}{2}(\alpha y)^2\psi(y) = E\psi(y) \tag{db11.2}$$

Multiplication of this equation by $-\frac{2m}{\hbar^2}\alpha^2$ yields:

$$\frac{\mathrm{d}^2}{\mathrm{d}y^2}\psi(y) - \frac{2m\alpha^2}{\hbar^2}\frac{k}{2}(\alpha y)^2\psi(y) = -\frac{2m\alpha^2}{\hbar^2}E\psi(y) \tag{db11.3}$$

or

$$\frac{\mathrm{d}^2}{\mathrm{d}y^2}\psi(y) + \frac{2m\alpha^2}{\hbar^2}E\psi(y) - \frac{mk\alpha^4}{\hbar^2}y^2\psi(y) = 0 \tag{db11.4}$$

Now, let:

$$1 = \frac{mk\alpha^4}{\hbar^2} \quad \text{then} \quad \alpha^4 = \frac{\hbar^2}{mk} \quad \text{or} \quad \alpha = \left(\frac{\hbar^2}{mk}\right)^{1/4} \tag{db11.5}$$

$$\varepsilon = \frac{2m\alpha^2}{\hbar^2}E = \frac{2mE}{\hbar^2}\left(\frac{\hbar^2}{mk}\right)^{1/2} = \frac{2E}{\hbar}\left(\frac{m}{k}\right)^{1/2} = \frac{2E}{\hbar\omega} \tag{db11.6}$$

The expression can then be written in the simpler form of:

$$\frac{\mathrm{d}^2}{\mathrm{d}y^2}\psi(y) + (\varepsilon - y^2)\,\psi(y) = 0 \tag{db11.7}$$

This is called the Hermite equation. The general solution of this equation would require special approaches involving the use of summations of polynomials. However, the critical part can be determined by realizing that a solution can be written by looking at the limit of $y \rightarrow \infty$ then $\varepsilon \ll y^2$ and $\varepsilon - y^2 \approx -y^2$. For this limiting case, the last equation reduces to:

$$\frac{d^2}{dy^2}\psi(y) - y^2\psi(y) = 0 \qquad \text{(db11.8)}$$

The appearance of this equation is similar to what we have seen before with the second derivative of the wavefunction, yielding the wavefunction times another term, so we expect that the solution case be expressed in terms of an exponential. In this case, the multiplying term is not a constant but rather a variable, so the exponential term is somewhat more involved than being a constant times the variable. The solutions for this are given by:

$$\psi(y) = e^{-y^2/2} \qquad \text{(db11.9)}$$

so

$$\frac{d}{dy}\psi(y) = e^{-y^2/2}(-y) \qquad \text{(db11.10)}$$

$$\frac{d^2}{dy^2}\psi(y) = \frac{d}{dy}[e^{-y^2/2}(-y)] = e^{-y^2/2} + ye^{-y^2/2}(-y) \approx y^2e^{-y^2/2} = y^2\psi(y) \qquad \text{(db11.11)}$$

The correctness of this solution is shown by substitution of the second derivative into eqn db11.8:

$$[y^2\psi(y)] - y^2\psi(y) = 0 \qquad \text{(db11.12)}$$

This shows us that the solution to this problem will always have an exponential term. The form used above actually represents the ground-state wavefunction. The higher-level wavefunctions are given by the exponential times a polynomial term as described in the main text.

PROPERTIES OF THE SOLUTIONS

The quantum-mechanical problem was solved by substituting the potential for the simple harmonic oscillator:

$$V(x) = \frac{1}{2}kx^2 \qquad \text{(11.10)}$$

into Schrödinger's equation, yielding

$$-\frac{\hbar^2}{2m}\frac{d^2}{dx^2}\psi(x) + \frac{k}{2}x^2\psi(x) = E\psi(x) \tag{11.11}$$

As shown above, this second-order differential equation is called the Hermite equation and has solutions of the form:

$$\psi_v(y) = N_v H_v e^{-y^2/2} \tag{11.12}$$

where

$$y = \frac{x}{\alpha}, \quad \alpha = \left(\frac{\hbar^2}{mk}\right)^{1/4}, \quad v = 0, 1, 2, \ldots \tag{11.13}$$

N_v is the normalization constant and is different for each term. It is equal to:

$$N_v = \frac{1}{\sqrt{\alpha\pi^{1/2}2^v v!}} \tag{11.14}$$

where $v!$ is the factorial term:

$$v! = v(v-1)(v-2)(v-3)\ldots(1) \tag{11.15}$$

The *Hermite functions*, $H_v(y)$, are polynomials. The first four are listed in Table 11.1. The ground-state wavefunction and its probability distribution are shown in Figure 11.3. Both terms have a maximal value at the origin. Since the potential is symmetrical about $x = 0$, the solutions are also symmetrical. Note that because of the exponential dropoff both functions quickly approach a zero value but remain positive and non-zero for all values of x.

Table 11.1
The first four solutions for the Hermite equation.

v	$H_v(y)$
0	1
1	$2y$
2	$4y^2 - 2$
3	$8y^3 - 12y$

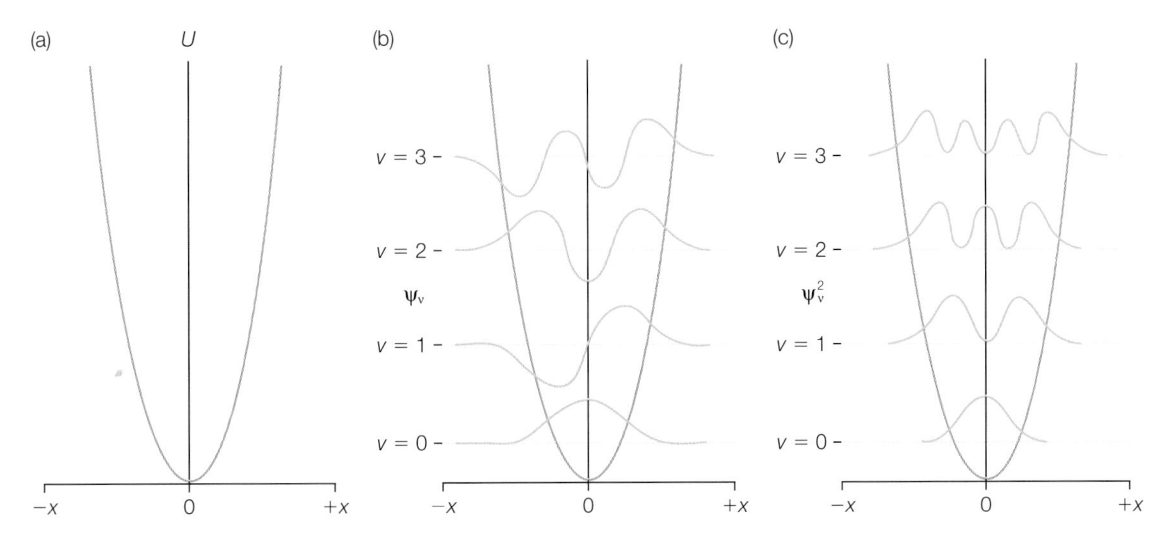

Figure 11.3 Wavefunctions and their energies of the simple harmonic oscillator.

We can check the solutions for all values of x by substituting the different solutions into Schrödinger's equation. For example, let us put the ground-state solution into the equation and show that it holds. First, we calculate the second derivative of the wavefunction. In taking these derivatives note that the two terms are obtained and can be rewritten in terms of the wavefunction.

$$\psi_0(x) = N_0 e^{-x^2/2\alpha^2} \psi_0(x) = N_0 e^{-x^2/2\alpha^2} \tag{11.16}$$

$$\frac{\mathrm{d}}{\mathrm{d}x}\psi_0(x) = N_0 \frac{\mathrm{d}}{\mathrm{d}x} e^{-x^2/2\alpha^2} = N_0 e^{-x^2/2\alpha^2}\left(-\frac{x}{\alpha^2}\right) \tag{11.17}$$

$$\begin{aligned}\frac{\mathrm{d}^2}{\mathrm{d}x^2}\psi_0(x) &= N_0 \frac{\mathrm{d}}{\mathrm{d}x}\left(-\frac{x}{\alpha^2} e^{-x^2/2\alpha^2}\right) = N_0 e^{-x^2/2\alpha^2}\left(\frac{x^2}{\alpha^4} - \frac{1}{\alpha^2}\right) \\ &= \psi_0(x)\left(\frac{x^2}{\alpha^4} - \frac{1}{\alpha^2}\right)\end{aligned} \tag{11.18}$$

Substitution of the second derivative into Schrödinger's equation yields:

$$-\frac{\hbar^2}{2m}\left[\psi_0(x)\left(\frac{x^2}{\alpha^4} - \frac{1}{\alpha^2}\right)\right] + \frac{kx^2}{2} x^2\psi_0(x) = E_0\psi_0(x) \tag{11.19}$$

$$\psi_0(x)\left[x^2\left(-\frac{\hbar^2}{2m\alpha^4} + \frac{k}{2}\right) + \left(\frac{\hbar^2}{2m\alpha^2} - E_0\right)\right] = 0 \tag{11.20}$$

There are a total of four terms in the equation. Two are multiplying x^2 (the terms in the left-hand parentheses) and can be rewritten by substituting the value of α from eqn 11.9:

$$-\frac{\hbar^2}{2m\alpha^4} + \frac{k}{2} = -\frac{\hbar^2}{2m}\left(\frac{mk}{\hbar^2}\right) + \frac{k}{2} = 0 \tag{11.21}$$

So these two terms cancel and we are left with the terms in the right-hand parentheses:

$$\psi_0(x)\left(\frac{\hbar^2}{2m\alpha^2} - E_0\right) = 0 \tag{11.22}$$

The product of the wavefunction and the terms in the parentheses must always zero for all values of the wavefunction, including all non-zero values. This can only be true if the term in the parentheses is always zero. Thus, we can write:

$$E_0 = \frac{\hbar^2}{2m\alpha^2} = \frac{\hbar^2}{2m}\left(\frac{mk}{\hbar^2}\right)^{1/2} = \frac{\hbar}{2}\left(\frac{k}{m}\right)^{1/2} = \frac{\hbar\omega}{2} \tag{11.23}$$

Thus, substitution of the wavefunction $\psi_0(x)$ yields a specific energy of $\frac{\hbar\omega}{2}$. This is the ground-state energy. Substitution of the vth wavefunction will yield the energy:

$$E_v = \left(v + \frac{1}{2}\right)\hbar\omega \tag{11.24}$$

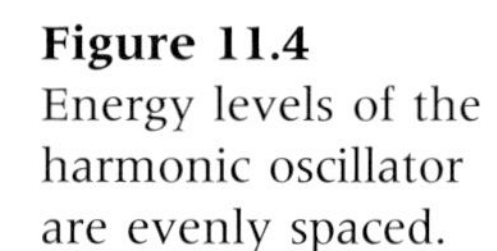

Figure 11.4
Energy levels of the harmonic oscillator are evenly spaced.

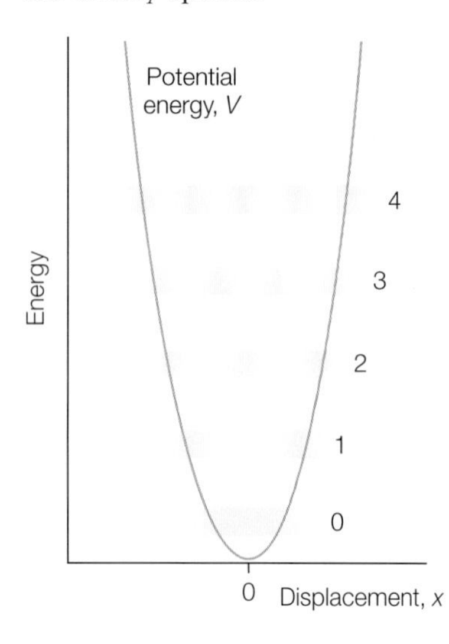

In summary, for the simple harmonic oscillator, the energies of the wavefunctions are proportional to the quantum number and separated by a constant factor of $\hbar\omega$ (Figure 11.4).

Forbidden region

Classically, the mass attached to the spring vibrates back and forth and is restricted to a narrow region. The maximum displacement of the mass from the equilibrium position, x_{TP}, is where the total energy is all potential energy, so:

$$E = \frac{kA^2}{2} \quad \text{so} \quad A = \sqrt{\frac{2E}{k}} \tag{11.25}$$

The probability, P, of finding the mass outside the classically allowed region is then:

$$P = \int_{A}^{\infty} \psi^*(x)\psi(x)\,dx = \int_{A}^{\infty} \psi^*(y)\psi(y)\alpha\,dy \tag{11.26}$$

The value of y_A is given by:

$$y_A = \frac{A}{\alpha} = \sqrt{\frac{2E}{k}}\left(\frac{mk}{\hbar^2}\right)^{1/4} = \sqrt{\frac{2(v+1/2)\hbar\omega}{k}}\left(\frac{mk}{\hbar^2}\right)^{1/4} \tag{11.27}$$

Substitution of ω yields:

$$y_A = \sqrt{2v+1}\left(\frac{\hbar}{k}\right)^{1/2}\left(\frac{k}{m}\right)^{1/4}\left(\frac{mk}{\hbar^2}\right)^{1/4} = \sqrt{2v+1} \tag{11.28}$$

For the ground state, $v = 0$ and $y_A = 1$. Substitution of the wavefunction gives:

$$\psi_0(y) = \sqrt{\frac{1}{\alpha\pi^{1/2}}}\,e^{-y^2/2} \tag{11.29}$$

$$P = \int_{1}^{\infty} \frac{1}{\alpha\pi^{1/2}}\,e^{-y^2}\alpha\,dy = \frac{1}{\pi^{1/2}}\int_{1}^{\infty} e^{-y^2}\,dy \tag{11.30}$$

This integral is related to the error function, *erf z*:

$$erf\,z = 1 - \frac{2}{\pi^{1/2}}\int_{z}^{\infty} e^{-y^2}\,dy \tag{11.31}$$

For the case presented above, $z = 1$ and $P = 0.079$. Thus, there is a 7.9% probability of finding the mass past the classic turning point on each side or a total probability of 15.8% of the mass being in the forbidden region. If chemical bonds are pictured as springs holding atoms, then there is a considerable probability of the bonds having large bond distances.

Transitions

Unlike the particle in a box, the energy levels for the harmonic oscillator are evenly spaced (Figure 11.4). The difference between adjacent levels is proportional to the frequency and independent of the quantum number:

$$\Delta E = (v + 1 + 1/2)\hbar\omega - (v + 1/2)\hbar\omega = \hbar\omega \tag{11.32}$$

Let us determine what energies and wavelengths are relevant for transitions between adjacent energy states. Consider a proton held to a spring with $k = 500$ N m^{-1}:

$$\omega = \sqrt{\frac{500 \text{ N m}^{-1}}{1.67 \times 10^{-27} \text{ kg}}} = 5.5 \times 10^{14} \text{ s}^{-1} \tag{11.33}$$

$$E = \hbar\omega = (1.05 \times 10^{-34} \text{ Js})(5.5 \times 10^{14} \text{ s}^{-1}) = 5.8 \times 10^{-20} \text{ J} \tag{11.34}$$

$$\lambda = \frac{hc}{v} = 3.4 \ \mu\text{m} \tag{11.35}$$

In general, transitions of vibrational states are always associated with infrared light and involve energies that are much smaller than the energies associated with the electronic levels.

Figure 11.5 Some vibrational modes for carbon dioxide: (a) two stretching modes, (b) a symmetrical and antisymmetrical mode, and (c) two perpendicular modes.

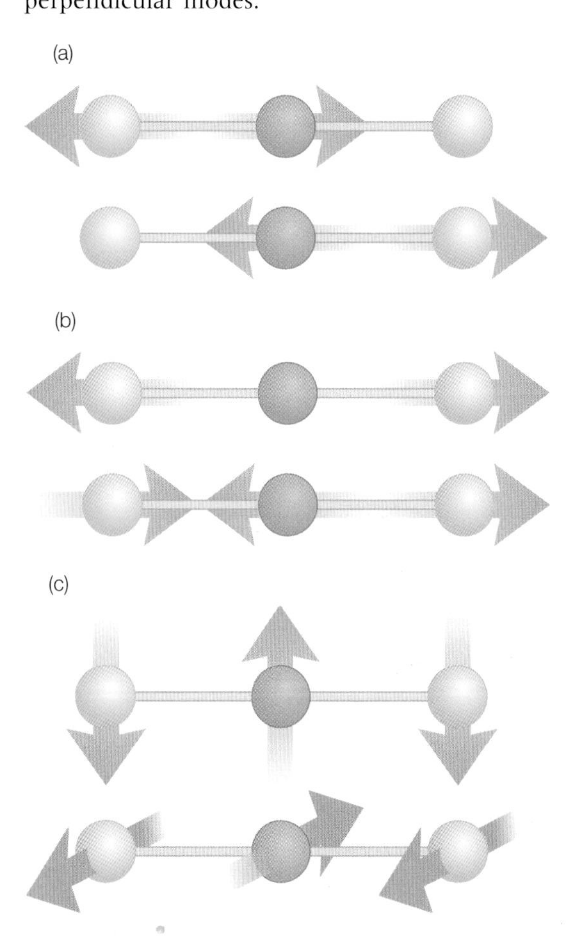

VIBRATIONAL SPECTRA

In small molecules, the vibrational modes of the molecules are well defined and predictable. For a molecule of N atoms, there are $3N$–6 independent modes if the molecule is nonlinear and $3N$–5 if it is linear. For example, CO_2 has four vibrational modes, as shown in Figure 11.5.

Once the molecules become larger predicting the vibrational modes becomes more complex because of anharmonicities, the effects of molecular rotation, and collisions. However, because different groups of the molecule have certain spectral features at characteristic frequencies, the infrared spectrum can still often be used for identification. Since proteins and DNA are built from repeating units, their vibrational spectra will reflect these repeating units, allowing assignments of the individual vibrational bands. Typical values for these modes are: N–H stretch at 3200–3500 cm^{-1}; N—H deformation at 1500–1600 cm^{-1}; and C=O stretch at 1600–1800 cm^{-1}. The C=O stretch is usually called the amide I band and the N—H deformation is called the amide II band.

The presence of a hydrogen bond will cause shift of a vibrational mode to a lower frequency compared to the same group that does not have the hydrogen bond. Because the secondary structure has well-defined hydrogen-bonding patterns, this results in highly characteristic vibrational spectra. For example, the amide I mode has typical frequencies at: 1650–1660 cm^{-1} for an α helix, 1630–1640 cm^{-1} for a β sheet, and 1680–1700 cm^{-1} for no hydrogen bond.

The effect of hydrogen bonding can be understood by reviewing the properties of vibrational states. For a mass, m, attached to a wall by a spring with a characteristic constant, k, the mass will vibrate with a vibrational frequency ω (eqn 11.7):

$$\omega = \sqrt{\frac{k}{m}} \tag{11.36}$$

We can represent a simple bond between two atoms as a spring, with a spring constant k attached to two masses, m_1 and m_2. In this case, we need to include the contribution of the mass of the second atom to the vibrational frequency ω. This is done through the reduced mass μ:

$$\mu = \frac{m_1 + m_2}{m_1 m_2} \quad \text{or} \quad \frac{1}{\mu} = \frac{1}{m_1} + \frac{1}{m_2} \tag{11.37}$$

then

$$\omega = \sqrt{\frac{k}{\mu}} \tag{11.38}$$

The effect of hydrogen reducing the vibrational frequency can be modeled as decreasing the bond strength or increasing the effective mass. The sensitivity of the vibrational frequency to such effects allows identification of the contributions that individual atoms make to the vibrational spectra through isotopic substitution.

Consider a CO bond in a protein. Normally it is composed of $^{12}C^{16}O$ and a typical bond strength would be 1900 N m^{-1}. This leads to a vibrational frequency of:

$$\mu = \frac{m_C m_O}{m_C + m_O} = \frac{12 \times 16}{12 + 16} m_p = 6.06 \times (1.67 \times 10^{-27}\ \text{kg}) = 1.14 \times 10^{-26}\ \text{kg} \tag{11.39}$$

$$\bar{\nu} = \frac{\nu}{c} = \frac{1}{c}\frac{\omega}{2\pi} = \frac{1}{2\pi c}\sqrt{\frac{k}{\mu}} \tag{11.40}$$

$$\bar{\nu} = \frac{1}{2\pi(3 \times 10^{10}\ \text{cm s}^{-1})}\sqrt{\frac{1900\ \text{N m}^{-1}}{1.14 \times 10^{-26}\ \text{kg}}} = 2167\ \text{cm}^{-1} \tag{11.41}$$

Now consider the molecule with the isotopic substitution $^{13}C^{16}O$:

$$\mu = \frac{m_C m_O}{m_C + m_O} = \frac{13 \times 16}{13 + 16} m_p = 7.17 \times (1.67 \times 10^{-27}\ \text{kg}) = 1.19 \times 10^{-26}\ \text{kg} \tag{11.42}$$

$$\bar{\nu} = \frac{1}{2\pi(3 \times 10^{10}\ \text{cm s}^{-1})} \sqrt{\frac{1900\ \text{N m}^{-1}}{1.19 \times 10^{-26}\ \text{kg}}} = 2122\ \text{cm}^{-1} \tag{11.43}$$

The change in frequency due to the isotope substitution is 45 cm^{-1}. This is easily measured as infrared spectrometers have a resolution of 1 cm^{-1}. Thus, by selectively substituting specific atoms of a molecule, the entire assignment of the vibrational spectrum can be made unambiguously.

RESEARCH DIRECTIONS: HYDROGENASE

Various organisms contain the enzyme hydrogenase, which catalyzes the simple reaction of two protons and two electrons, forming hydrogen gas:

$$2H^+ + 2e^- \leftrightarrow H_2 \tag{11.44}$$

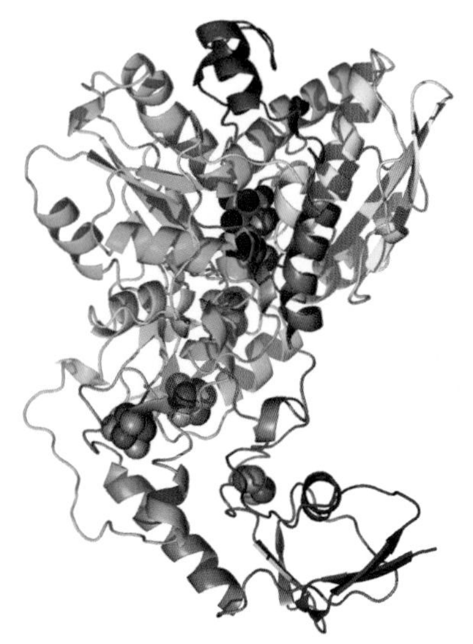

Figure 11.6 The general structural features of the iron hydrogenase showing the positions of the iron–sulfur and H cofactors.

Representatives of most prokaryotic genera as well as some eukaryotes contain hydrogenase, which was identified originally in the 1930s. Despite the simplicity of this reaction and the many years of research, scientists still are probing its molecular mechanism to both better understand the biological processes involved and for application in the hydrogen economy (Chapter 12). Hydrogenases are proteins that can be divided into two classes: those that contain only iron and those that have both nickel and iron. The iron-only enzymes are highly evolved and can produce nearly 10,000 molecules of hydrogen per second. Despite the simplicity of the reaction, the molecular machinery is quite sophisticated. Hydrogenases typically contain one or two protein subunits surrounding the metal cofactors. Some of the metal cofactors are organized into standard iron–sulfur cofactors that are aligned to direct electrons from outside the protein (Figure 11.6). The protein also has a potential pathway involving protonatable amino acid residues for the transport of protons from the solvent into the protein.

The most remarkable feature of hydrogenases is the active site of the enzyme, known as the H or Ni–Fe cluster; which is the point where the electrons and protons are delivered. The composition of this cofactor differs for the two types of enzyme, containing either nickel and iron or two irons (Figure 11.7). The arrangement of the atoms of the H cluster in the iron-only enzyme revealed by X-ray diffraction experiments (Nicolet et al. 2000) was unexpected, showing a thiocubane iron–sulfur cluster bridging through a sulfur of a cysteine to a surprising dinuclear iron subcluster. The Ni–Fe cluster was found to have a remarkably similar arrangement with a nickel atom at the site corresponding to one of the iron positions in the H cluster.

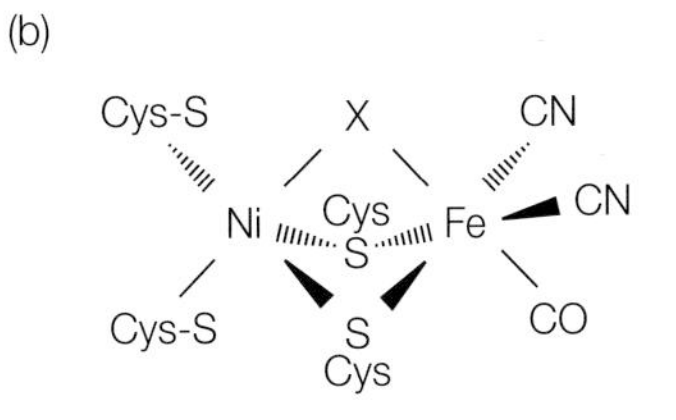

Figure 11.7 The arrangement of the Fe–Fe and Ni–Fe clusters of hydrogenase with the nonprotein ligands CO and CN, and the unassigned nonprotein ligands identified as Y and L.

Both clusters have coordinating cysteine residues but the nature of the other ligands was found to be very unusual. When the coordination was first being elucidated, the scientists were surprised to see that the ligands did not appear to be contributed by the surrounding protein. One of the key spectroscopic techniques used to identify the ligands for the iron-only enzyme was Fourier transform infrared spectroscopy, or FTIR (Chen et al. 2002). The enzyme was prepared in different states: both a reduced state and the oxidized state, prepared using thionin. The possible contribution of CO was probed by examining the spectra in the presence of both added ^{12}CO and ^{13}CO. The spectra of the normal oxidized state shows infrared bands at 2086, 2072, 1971, 1948, and 1802 cm^{-1} (Figure 11.8). Upon incubation with ^{12}CO (actually using naturally abundant isotopes), the infrared bands were observed at 2095, 2077, 1974, 1971, and 1810 cm^{-1}. When the enzyme was incubated with ^{13}CO, the infrared bands at 2017, 1974, and 1971 cm^{-1} were replaced with bands at 2000, 1971, and 1947 cm^{-1}, whereas the bands at 2095, 2077, and 1810 cm^{-1} were unaffected.

The largest shifts were observed for the bands in the 1947–2017 cm^{-1} region of the spectra. In the presence of ^{13}C, these bands shifted consistently to lower wavenumbers, as would be expected, due to the presence of the heavier isotope. The shifting arises from contributions of two different CO ligands that are sensitive to the introduction of CO as well as another CO that does not exchange in the presence of added CO. Based upon additional studies, including the temperature dependence of the spectral features, it was possible to assign the bands in the 1947–2017 cm^{-1} region to two different CO ligands. The 2017 and 1974 cm^{-1} bands detected in the ^{12}CO samples shifted to 2000 and 1947 cm^{-1} in the presence of ^{13}CO but the 1971 cm^{-1} band is unaffected. Due to the presence of the isotope, the other infrared bands of the spectrum that were not associated with the exchangeable CO ligand essentially did not change. The bands between 2072 and 2095 cm^{-1} were identified as arising from a stretching mode involving another nonprotein ligand (CN). The band at 1802–1810 cm^{-1} was identified with a bridging CO.

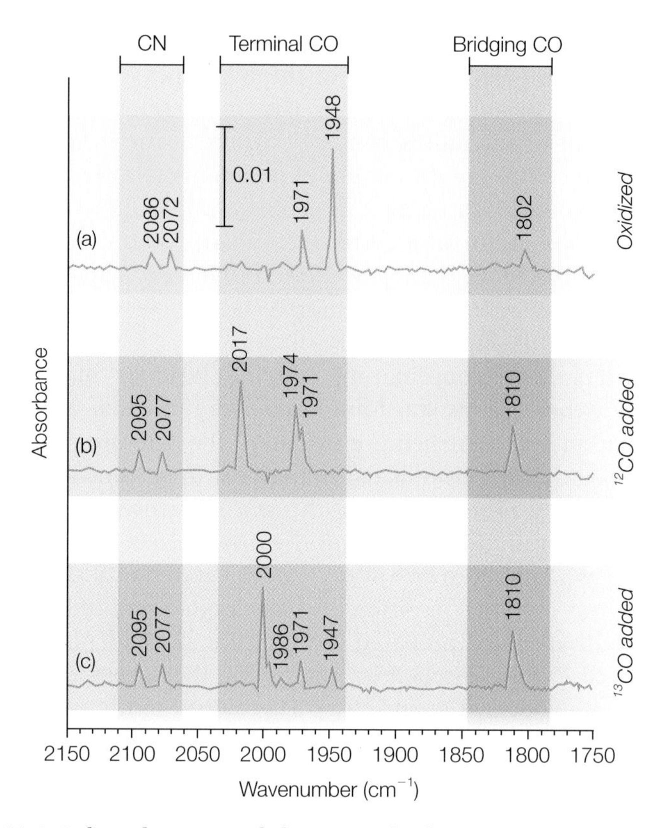

Figure 11.8 Infrared spectra of the Fe–Fe hydrogenase in the oxidized state, the ^{12}CO-added oxidized state, and the ^{13}CO-added oxidized state. Modified from Chen et al. (2002).

The molecular structure of the H cluster represented a new type of bioinorganic cluster. The involvement of two toxic compounds, carbon monoxide and cyanide, in the reduction reaction to form the H_2 bond was without precedent. Research to unambiguously identify all of the ligands and to understand the enzymatic mechanism is ongoing (Armstrong 2004). Efforts to mimic this unusual cluster have been successful as a synthetic compound was synthesized in a configuration similar to that of the H cluster (Figure 11.9). By activating an initial Fe–Fe unit with a thioacetyl group and protecting the 4Fe–4S cluster with a large ligand, the cluster could be assembled despite the sensitivity of these inorganic materials, as well as of the natural enzyme, to oxygen (Tanti et al. 2005). Although the efficiency was limited, the synthetic cluster was able to catalyze the reduction of H^+ to H_2. Comparison of the properties, including the infrared spectra, of the synthesized compound with that of the natural system should greatly aid in our understanding of hydrogenases.

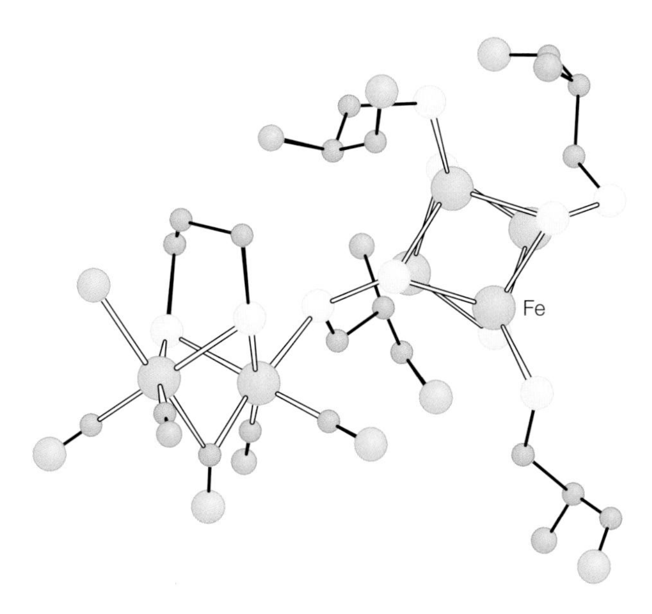

Figure 11.9 The structure of the synthesized H cluster. Modified from Armstrong (2004).

REFERENCES

Armstrong, F.A. (2004) Hydrogenases: active site puzzles and progress. *Current Opinion in Chemical Biology* **8**, 133–40.

Cammack, R. (1999) Hydrogenase sophistication. *Nature* **397**, 214–15.

Chen, Z., Lemon, B.J., Huang, S. et al. (2002) Infrared studies of the CO-inhibited form of the Fe-only hydrogenase from *Chlostridium pasteurianum*. *Biochemistry* **41**, 2036–43.

Nicolet, Y., Lemon, B.J., Fontecilla-Camps, J.C., and Peters, J.W. (2000) A novel FeS cluster in Fe-only hydrogenases. *Trends in Biochemical Sciences* **25**, 138–43.

Tard, C., Liu, X., Ibrahim, S.K. et al. (2005) Synthesis of the H-cluster framework of iron-only hydrogenase. *Nature* **433**, 610–13.

PROBLEMS

11.1 Write the smallest energy that the oscillator can have in terms of k and m.

11.2 Write the zero-point energy for an electron in terms of k and m.

11.3 Write the energy of the second excited state for an electron in terms of k and m.

11.4 Write the wavelength of light absorbed for a transition of an electron between adjacent levels in terms of k and m_e.

11.5 Calculate the reduced mass of a $^{12}C^{16}O$ bond.

11.6 Calculate the reduced mass of a $^{13}C^{16}O$ bond.

11.7 Calculate the vibrational frequency for a proton if $k = 600 \text{ N m}^{-1}$.

11.8 For a $^{12}C^{16}O$ bond, if $k = 500 \text{ N m}^{-1}$ calculate (a) the reduced mass and (b) the vibrational frequency.

11.9 For a $^{13}C^{16}O$ bond, if $k = 500 \text{ N m}^{-1}$ calculate (a) the reduced mass and (b) the vibrational frequency.

11.10 If the mass is doubled, what is the change in vibrational frequency?

11.11 Write (but do not solve) the probability of the ground state being in the classically forbidden region.

11.12 Water (H_2O) absorbs infrared radiation at 3 μm (3400 cm^{-1}) due to the O–H stretch. What wavelength would you predict is absorbed by D_2O?

11.13 Write (but do not solve) the average squared position of the ground state.

11.14 Write (but do not solve) the average squared position of the $v = 1$ state.

11.15 Calculate the energy of the ground state by putting the ground-state wavefunction into Schrödinger's equation.

11.16 For a simple $^{16}O_2$ molecule, calculate (a) the reduced mass and (b) the frequency change when the isotope ^{15}O is present.

11.17 Write Schrödinger's equation for the potential $V(x) = Ax^2 + Be^{-x}$. If $B \ll A$ what will the wavefunctions be approximately?

11.18 Suppose that the mass is attached to two springs with spring constants k_1 and k_2, one on each side of the mass. Write Schrödinger's equation for this system.

11.19 For a simple harmonic oscillator show (but do not solve) in integral form the probability of the ground state being in the classically forbidden region.

11.20 Calculate the energy of the $v = 1$ state of the simple harmonic oscillator by substituting the wavefunction into Schrödinger's equation.

11.21 For the following questions, assume that a particle of mass m is in a valley between two hills. You model the potential as:

$$V(x) = a(x^2 + bx^3) \quad |x| \leq L/2$$

$$V(x) = cL^2 \quad |x| \geq L/2$$

where cL^2 is larger than the total energy of the particle.

(a) Write Schrödinger's equation for these potentials.

(b) How are a and b related to c?

(c) If L is very small, with a value of 0.1 Å and m is very small, with a value of 1.054×10^{-34} kg, what can be said about the uncertainity in velocity?

(d) Estimate the probability of finding the particle in the $|x| \leq L/2$ and $|x| \geq L/2$ regions. Be sure to explain your reason for the estimates.

(e) If b is very small and $c = 0$, what can be said about the wavefunctions in the region $|x| \leq L/2$?

11.22 For the following, assume that the molecules are linear and that all bonds are equal and can be described as a spring with a constant k.

(a) Consider an ethene molecule, $H_2C{=}CH_2$. Calculate the reduced mass, assuming that each CH_2 behaves as a rigid object.

(b) Write Schrödinger's equation for ethene using the harmonic oscillator approximation.

(c) Write (but do not solve) the probability of finding the C=C bond to be longer than 2 Å.

(d) Now consider a carotenoid with 40 carbon atoms in a simple linear conjugated chain. Write the force that the nth carbon experiences.

(e) Write Schrödinger's equation for the nth carbon atom using the harmonic oscillator approximation.

12

Atomic structure: hydrogen atom and multi-electron atoms

The primary motivation in studying quantum mechanics is that the key constituents of biochemical reactions, namely electrons, have properties that are described by quantum mechanics rather than classical mechanics. Quantum mechanics will serve as the foundation for understanding the properties of electrons and how different spectroscopies can be used to probe proteins and other biological samples. In this chapter, Schrödinger's equation is solved for the hydrogen atom. The concepts of atomic orbitals are interpreted in terms of the resulting wavefunctions of the equation. As was done for the particle in a box and simple harmonic oscillator, the wavefunction allows the calculation of all of the properties of the hydrogen atom, such as the average distance or energy of any given orbital. These results for the hydrogen atom with one electron are extended to understanding the properties of multi-electron atoms in terms of both empirical constants and as detailed calculations. Finally, the solutions are used to understand the organization of elements in the periodic table. The biological question addressed in this chapter is how the use of hydrogen by biological organisms can contribute to energy policies.

SCHRÖDINGER'S EQUATION FOR THE HYDROGEN ATOM

The hydrogen atom is solved using Schrödinger's equation, as has been used for the other problems – the particle in a box and the harmonic oscillator. First, we must determine the potential to be used in the problem. For the hydrogen atom, the potential is assumed to be due to the electrostatic interaction between the negatively charged electron and the positively charged nucleus, giving a potential, $V(r)$, of:

$$V(r) = \frac{-e^2}{4\pi\varepsilon_0 r} \tag{12.1}$$

where r is the distance between the electron and nucleus, ε_0 is the vacuum permittivity, $-e$ is the electron charge, and $+e$ is the proton charge. Substituting this potential into Schrödinger's equation gives:

$$-\frac{\hbar^2}{2m}\nabla^2(r,\theta,\varphi)\psi(r,\theta,\varphi) + \frac{-e^2}{4\pi\varepsilon_0 r}\psi(r,\theta,\varphi) = E\psi(r,\theta,\varphi) \tag{12.2}$$

The mathematical analysis of this equation is presented in Derivation box 12.1: the reader may skip to the main text following the box, which discusses the properties of the general solution.

Derivation box 12.1

Solving Schrödinger's equation for the hydrogen atom

Since the potential energy is determined by the radial distance r, we must substitute the radial form of the gradient:

$$\nabla^2 = \frac{\delta^2}{\delta r^2} + \frac{2}{r}\frac{\delta}{\delta r} + \frac{1}{r^2}\left[\frac{1}{\sin^2\theta}\frac{\delta^2}{\delta\varphi^2} + \frac{1}{\sin\theta}\frac{\delta}{\delta\theta}\sin\theta\frac{\delta}{\delta\theta}\right] \tag{db12.1}$$

To solve the equation we will use the separation-of-variables approach to produce three separate equations, each with only one of the three variables. This will lead to three eigenequations and three quantum numbers, which are called n, l, and m_l. Since the potential has only a radial dependence, we first will consider the radial part separately and define the angular component as:

$$\Lambda^2(\theta,\phi) = \frac{1}{\sin^2\theta}\frac{\delta^2}{\delta\phi^2} + \frac{1}{\sin\theta}\frac{\delta}{\delta\theta}\left(\sin\theta\frac{\delta}{\delta\theta}\right) \tag{db12.2}$$

Using this definition we can write the gradient squared as:

$$\nabla^2(r,\theta,\phi)\psi(r,\theta,\phi) = \frac{1}{r}\frac{\delta^2}{\delta r^2}(r\psi(r,\theta,\phi)) + \frac{1}{r^2}\Lambda^2(\theta,\phi)\psi(r,\theta,\phi) \tag{db12.3}$$

This gives, for the Schrödinger equation (eqn 12.2), the following:

$$-\frac{\hbar^2}{2m}\left\{\frac{1}{r}\frac{\delta^2[r\psi(r,\theta,\varphi)]}{\delta r^2} + \frac{1}{r^2}\Lambda^2(\theta,\phi)\psi(r,\theta,\varphi)\right\} + \frac{-e^2}{4\pi\varepsilon_0 r}\psi(r,\theta,\varphi) = E\psi(r,\theta,\varphi) \tag{db12.4}$$

Separation of variables

Now we shall use the separation-of-variables approach and define:

$$\psi(r,\theta,\phi) = R(r)Y(\theta,\phi) \tag{db12.5}$$

Substitution of this into the Schrödinger equation gives:

$$-\frac{\hbar^2}{2m}\left\{Y(\theta,\phi)\frac{1}{r}\frac{\delta^2[rR(r)]}{\delta r^2}+R(r)\frac{1}{r^2}\Lambda^2(\theta,\phi)Y(\theta,\phi)\right\}+\frac{-e^2}{4\pi\varepsilon_0 r}R(r)Y(\theta,\phi)=ER(r)Y(\theta,\phi) \quad \text{(db12.6)}$$

Now, multiply by:

$$r^2/[R(r)Y(\theta,\phi)] \quad \text{(db12.7)}$$

The result is:

$$-\frac{\hbar^2}{2m}\left\{\frac{r}{R(r)}\frac{\delta^2[rR(r)]}{\delta r^2}+\frac{1}{Y(\theta,\phi)}\Lambda^2(\theta,\phi)Y(\theta,\phi)\right\}+\frac{-e^2r}{4\pi\varepsilon_0}=Er^2 \quad \text{(db12.8)}$$

The terms with the same variables are grouped, giving:

$$\left(-\frac{\hbar^2}{2m}\frac{r}{R(r)}\frac{\delta^2[rR(r)]}{\delta r^2}+\frac{-e^2r}{4\pi\varepsilon_0}-Er^2\right)-\left(\frac{\hbar^2}{2m}\frac{1}{Y(\theta,\phi)}\Lambda^2(\theta,\phi)\,Y(\theta,\phi)\right)=0 \quad \text{(db12.9)}$$

The term in the first bracket depends only upon the variable r, whereas the second depends only upon θ and ϕ, so they must both be constants for the sum to always be a constant. Let us define the separation constant as such that:

$$\frac{\Lambda^2(\theta,\phi)\,Y(\theta,\phi)}{Y(\theta,\phi)}=-l(l+1) \quad \text{(db12.10)}$$

Then the radial equation becomes:

$$\frac{-\hbar^2}{2m}\frac{r}{R(r)}\frac{d^2[rR(r)]}{dr^2}+\frac{-e^2\,r}{4\pi\varepsilon_0}-r^2E+\frac{\hbar^2}{2m}l(l+1)=0 \quad \text{(db12.11)}$$

Angular solution

The angular part of the equation is:

$$\Lambda^2(\theta,\phi)\,Y(\theta,\phi)=-l(l+1)\,Y(\theta,\phi) \quad \text{(db12.12)}$$

We need to substitute the definition for Λ:

$$\frac{1}{\sin^2\theta}\frac{\delta^2}{\delta\phi^2}Y(\theta,\phi)+\frac{1}{\sin\theta}\frac{\delta}{\delta\theta}\left(\sin\theta\frac{\delta}{\delta\theta}\right)Y(\theta,\phi)=-l(l+1)\,Y(\theta,\phi) \quad \text{(db12.13)}$$

Now let us use the separation of variables again. Define:

$$Y(\theta,\phi) = \Theta(\theta)\Phi(\phi) \tag{db12.14}$$

After substituting this into the equation db 12.13 multiply by:

$$\frac{\sin^2\theta}{\Theta(\theta)\,\Phi(\phi)} \tag{db12.15}$$

This yields:

$$\left(\frac{1}{\Phi(\phi)}\frac{\delta^2}{\delta\phi^2}\Phi(\phi)\right) + \left[\frac{\sin\theta}{\Theta(\theta)}\frac{\delta}{\delta\theta}\left(\sin\theta\frac{\delta}{\delta\theta}\Theta(\theta)\right) + l(l+1)\sin^2\theta\right] = 0 \tag{db12.16}$$

These two equations can now be separated into:

$$\frac{1}{\Phi(\phi)}\frac{d^2}{d\phi^2}\Phi(\phi) = -m_l^2 \quad \text{or} \quad \frac{d^2}{d\phi^2}\Phi(\phi) = -m_l^2\Phi(\phi) \tag{db12.17}$$

and

$$\sin\theta\frac{d}{d\theta}\left[\sin\theta\frac{d\Theta(\theta)}{d\theta}\right] + \left[l(l+1)\sin^2\theta - m_l^2\right]\Theta(\theta) = 0 \tag{db12.18}$$

The solutions to the equation for ϕ should be familiar to you already, as the equation is the same as for the particle in a box. For this case, let us use exponential terms:

$$\Phi(\phi) = Ae^{im_l\phi} \tag{db12.19}$$

You can check that this is correct by substitution. Note that we needed to include i because of the negative sign in the equation. The normalization constant is determined using the condition that integration over every value of ϕ should yield a value of 1:

$$1 = \int_0^{2\pi} \Phi^*(\phi)\,\Phi(\phi)\,d\phi = A^2\int_0^{2\pi} e^{-im_l\phi}e^{im_l\phi}\,d\phi = A^2\int_0^{2\pi} d\phi = A^2 2\pi \tag{db12.20}$$

Thus, the constant is:

$$A = \frac{1}{\sqrt{2\pi}} \tag{db12.21}$$

The values of the constant m_l are fixed by the physical constraint that, if the object is fully rotated by 2π, then the solution must be the same. This gives:

$$\Phi(\phi + 2\pi) = \Phi(\phi) \quad \text{(db12.22)}$$

$$Ae^{im_l\phi} = Ae^{im_l(\phi+2\pi)} = Ae^{im_l\phi}e^{im_l2\pi} \quad \text{(db12.23)}$$

For this to be true then:

$$1 = e^{im_l2\pi} = \cos(2\pi m_l) + i\sin(2\pi m_l) \quad \text{(db12.24)}$$

with $m_l = 0, \pm1, \pm2, \pm3, \ldots$.

The Θ term is called the Legendre equation and it has a series of solutions of the form commonly denoted as $P_l(\cos\Theta)$ for which there are restrictions on the separation constants:

$$\sin\theta\frac{\mathrm{d}}{\mathrm{d}\theta}\left[\sin\theta\frac{\mathrm{d}\Theta(\theta)}{\mathrm{d}\theta}\right] + \left[l(l+1)\sin^2\theta - m_l^2\right]\Theta(\theta) = 0 \quad \text{(db12.25)}$$

Here, $l = 0, 1, 2, 3, 4, \ldots$ and $m_l = -l, -l+1, -l+2, \ldots, l-1, l$. The combined angular terms are called spherical harmonics, with some solutions given in Table 12.1.

Some of the solutions can be easily seen. If we substitute:

$$\Theta(\theta) = A \quad \text{(db12.26)}$$

where A is a constant, then the equation is simply:

$$0 + [l(l+1)\sin^2\theta - m_l^2]A = 0 \quad \text{(db12.27)}$$

Table 12.1
Some solutions to the angular part of Schrodinger's equation.

l	m_l	$Y(\Theta,\Phi)$
0	0	$\frac{1}{\sqrt{4\pi}}$
1	1	$-\sqrt{\frac{3}{8\pi}}\sin\theta e^{i\phi}$
1	0	$-\sqrt{\frac{3}{4\pi}}\cos\theta$
2	0	$\sqrt{\frac{15}{16\pi}}(3\cos^2\theta - 1)$

This will be true if $l = 0$ and $m_l = 0$. Another possible solution is:

$$\Theta(\theta) = B\cos\theta \qquad \text{(db12.28)}$$

Substitution gives:

$$\sin\theta(-2B\sin\theta\cos\theta) + [l(l+1)\sin^2\theta - m_l^2]\Theta(\theta) = 0 \qquad \text{(db12.29)}$$

This is true if $l = 1$ and $m_l = 0$. In general the solutions are polynomials of trigonometric functions.

Radial solution

The radial equation was found previously (eqn db 12.11) to be:

$$\frac{-\hbar^2}{2m}\frac{r}{R(r)}\frac{d^2[rR(r)]}{dr^2} + \frac{-e^2 r}{4\pi\varepsilon_0} - r^2E + \frac{\hbar^2}{2m}l(l+1) = 0 \qquad \text{(db12.30)}$$

Multiplying this by $R(r)/r$ yields:

$$\frac{-\hbar^2}{2m}\frac{d^2[rR(r)]}{dr^2} + \frac{-e^2}{4\pi\varepsilon_0 r}[rR(r)] + \frac{\hbar^2}{2m}l(l+1)\frac{1}{r^2}[rR(r)] = [rR(r)]E \qquad \text{(db12.31)}$$

These are called Laguerre functions and are eigenfunctions with a series of solutions. To determine the general functional form of the solutions, let $\Pi(r) = rR(r)$. Then:

$$-\frac{\hbar^2}{2m}\frac{d^2}{dr^2}\Pi(r) + \left[\frac{\hbar^2 l(l+1)}{2mr^2} - \frac{e^2}{4\pi\varepsilon_0 r}\right]\Pi(r) = E\Pi(r) \qquad \text{(db12.32)}$$

Consider the case where $l = 0$ then, after multiplying by $-2m/\hbar$, the equation reduces to:

$$\frac{d^2\Pi(r)}{dr^2} + \frac{2m}{\hbar^2}\left(\frac{e^2}{4\pi\varepsilon_0 r} + E\right)\Pi(r) = 0 \qquad \text{(db12.33)}$$

To solve this equation, try a solution with exponentials with a constant, α:

$$\Pi(r) = re^{-\alpha r} \qquad \text{(db12.34)}$$

$$\frac{d}{dr}\Pi(r) = r(-\alpha)e^{-\alpha r} + e^{-\alpha r} \qquad \text{(db12.35)}$$

$$\frac{d^2}{dr^2}\Pi(r) = -\alpha(-\alpha re^{-\alpha r} + e^{-\alpha r}) - \alpha e^{-\alpha r} = \alpha^2 re^{-\alpha r} - 2\alpha e^{-\alpha r} \qquad \text{(db12.36)}$$

Then

$$\alpha^2 re^{-\alpha r} - 2\alpha e^{-\alpha r} + \frac{2m}{\hbar^2}\left(\frac{e^2}{4\pi\varepsilon_0 r} + E\right) re^{-\alpha r} = 0 \quad \text{(db12.37)}$$

$$re^{-\alpha r}\left(\alpha^2 + \frac{2mE}{\hbar^2}\right) + e^{-\alpha r}\left(-2\alpha + \frac{e^2}{4\pi\varepsilon_0}\frac{2m}{\hbar^2}\right) = 0 \quad \text{(db12.38)}$$

Each of these two terms must be zero. The second term yields:

$$\alpha = \frac{m}{\hbar^2}\frac{e^2}{4\pi\varepsilon_0} \quad \text{(db12.39)}$$

so

$$R(r) = e^{-\alpha r} = \exp\left(-\frac{m}{\hbar^2}\frac{e^2}{4\pi\varepsilon_0}\right) \quad \text{(db12.40)}$$

The first term yields:

$$E = -\frac{me^4}{32\pi^2\varepsilon_0\hbar^2} \quad \text{(db12.41)}$$

This is the solution for a specific set of quantum numbers ($n = 1$, $l = 0$, $m_l = 0$). In general, the solutions for the Laguerre equation, $L_s^k(x)$, can be expressed as in the following form:

$$L_s^k(x) = \frac{e^x x^{-k}}{s!}\frac{\mathrm{d}}{\mathrm{d}x^s}(e^{-x}x^{s+k}) \quad \text{(db12.42)}$$

where s is the index with integral values starting with 0, and k is a second index greater than -1.

PROPERTIES OF THE GENERAL SOLUTION

The solutions to the Schrödinger equation for the hydrogen atom should in principle allow us to calculate all of the properties of molecules. We will now use the solutions to address three related questions: what are the physical properties of the orbitals, how do they change when there is more than one electron, and how do they change when there is more

Table 12.2
Wavefunctions for some of the lower-energy states of the hydrogen atom.

n	l	m_l	$\psi(r,\theta,\pi)$
1	0	0	$\frac{1}{\sqrt{\pi}}\left(\frac{1}{a_0}\right)^{3/2} e^{-r/a_0}$
2	0	0	$\frac{1}{4\sqrt{2\pi}}\left(\frac{1}{a_0}\right)^{3/2}\left(2-\frac{r}{a_0}\right)e^{-r/2a_0}$
2	1	0	$\frac{1}{4\sqrt{2\pi}}\left(\frac{1}{a_0}\right)^{3/2}\left(\frac{r}{a_0}\right)e^{-r/2a_0}(\cos\theta)$
2	1	±1	$\frac{1}{8\sqrt{\pi}}\left(\frac{1}{a_0}\right)^{3/2}\left(\frac{r}{a_0}\right)e^{-r/2a_0}(\sin\theta)e^{\pm i\phi}$
3	0	0	$\frac{1}{81\sqrt{3\pi}}\left(\frac{1}{a_0}\right)^{3/2}\left(27-18\frac{r}{a_0}+2\frac{r^2}{a_0^2}\right)e^{-r/3a_0}$
3	1	0	$\frac{1}{81}\sqrt{\frac{2}{\pi}}\left(\frac{1}{a_0}\right)^{3/2}\left(6\frac{r}{a_0}-\frac{r^2}{a_0^2}\right)e^{-r/3a_0}(\cos\theta)$
3	1	±1	$\frac{1}{81\sqrt{\pi}}\left(\frac{1}{a_0}\right)^{3/2}\left(6\frac{r}{a_0}-\frac{r^2}{a_0^2}\right)e^{-r/3a_0}(\sin\theta)e^{\pm i\phi}$
3	2	0	$\frac{1}{81\sqrt{6\pi}}\left(\frac{1}{a_0}\right)^{3/2}\left(\frac{r^2}{a_0^2}\right)e^{-r/3a_0}(3\cos^2\theta-1)$
3	2	±1	$\frac{1}{81\sqrt{\pi}}\left(\frac{1}{a_0}\right)^{3/2}\left(\frac{r^2}{a_0^2}\right)e^{-r/3a_0}(\sin\theta\cos\theta)e^{\pm i\phi}$
3	2	±2	$\frac{1}{162\sqrt{\pi}}\left(\frac{1}{a_0}\right)^{3/2}\left(\frac{r^2}{a_0^2}\right)e^{-r/3a_0}(\sin^2\theta)e^{\pm 2i\phi}$

than one nucleus? For the hydrogen atom, the solutions are given by the product of the three terms, yielding the solutions listed in Table 12.2 with the form:

$$\psi_{n,l,m_l}(r,\theta,\phi) = R_{n,l}(r)Y_l^{m_l}(\theta,\phi) \quad \text{with} \quad E_n = -\frac{me^4}{32\pi^2\varepsilon_0^2\hbar^2}\frac{1}{n^2} = -\frac{hcR_H}{n^2} \qquad (12.3)$$

where R_H is the Rydberg constant.

The solutions can be written in terms of the individual solutions that we have found and the three quantum numbers:

$$\psi_{n,l,m_l}(r,\theta,\phi) = A_{n,l,m_l} P_l^{m_l}(\cos\theta) e^{im_l\phi} R_{n,l}(r) e^{-zr/(na_0)} \quad (12.4)$$

where $a_0 = \dfrac{h^2\varepsilon_0}{\pi me^2}$, and Z is the number of protons ($Z = 1$ for hydrogen atoms). The quantum numbers are as follows: the principal quantum number is $n = 1, 2, 3, \ldots$, the angular-momentum quantum number is $l = 0, 1, 2, \ldots, n - 1$, and the magnetic quantum number is $m_l = l, l - 1, l - 2, \ldots, -l$. The states, or orbitals, are dependent only upon the quantum number n and so the orbitals are degenerate in energy. There are two electrons per orbital, one spin up and one spin down. Note that the fourth quantum number, spin, does not arise from these equations but will appear when relativistic effects are considered. The functional forms for several of the lower-energy wavefunctions are provided in Table 12.2.

Angular momentum

Classically the angular momentum of a particle is:

$$\vec{L} = \vec{r} \times \vec{p} \text{ where } \vec{r} = (x,y,z),\ \vec{p} = (p_x,p_y,p_z) \quad (12.5)$$

or, considering the z component only,

$$L_z = xp_y - yp_x \quad (12.6)$$

For quantum mechanics the operators are substituted, giving:

$$l_z = x\left(\frac{\hbar}{i}\frac{\partial}{\partial y}\right) - y\left(\frac{\hbar}{i}\frac{\partial}{\partial x}\right) \quad (12.7)$$

By substituting for x, y, and z the variables r, θ, and Φ it is possible to show that:

$$L_z = \frac{\hbar}{i}\frac{\partial}{\partial\phi} \quad \text{and} \quad L^2 = L_x^2 + L_y^2 + L_z^2 = \hbar^2\Lambda^2 \quad (12.8)$$

Since the ϕ dependence of the solutions are given by the exponential part, we can write in general:

$$L_z\psi(r,\theta,\phi) = A_{n,l,m_l} P(\theta) R(\rho) e^{-\rho/2}\left(\frac{\hbar}{i}\frac{\partial}{\partial\phi}\right) e^{im_l\phi} = m_l\hbar\psi(r,\theta,\phi) \quad (12.9)$$

$$L^2\psi(r,\theta,\phi) = \Lambda^2\hbar^2[A_{n,l,m_l} Y_l^{m_l}(\theta,\phi) R(r)] = \hbar^2 l(l + 1)\psi(r,\theta,\phi) \quad (12.10)$$

So, the total angular momentum is quantized by l and the projection of the angular momentum along the z direction is quantized by m_l.

Orbitals

The solutions to Schrödinger's equation consist of the product of a radial term and an angular term:

$$\psi_{n,l,m_l}(r,\theta,\phi) = R_{n,l}(r)\,Y_l^{m_l}(\theta,\phi) \qquad (12.11)$$

where each solution is uniquely defined by the three indices, n, l, m_l, that are restricted to be $n = 1, 2, 3, \ldots$, $l = 0, 1, 2, \ldots, n - 1$, and $m_l = l, l - 1, l - 2, \ldots, -l$, where n is the principal quantum number arising from the quantization of energy as:

$$E_n = -\frac{hcR_H}{n^2} \quad \text{where} \quad R_H = \frac{m_e e^4}{8\varepsilon_0^2 h^3 c} \qquad (12.12)$$

l is the angular-momentum quantum number arising from quantization of the angular momentum with magnitude $\sqrt{l\,(l+1)}\,\hbar$, and m_l arises from the quantization of the z component of angular momentum $m_l\hbar$. These solutions are degenerate in energy; that is, they are dependent only upon n and not l or m_l, as seen on the energy diagram (Figure 12.1).

All of the orbitals are directly related to these wavefunctions. Let us look first at the properties of the $l = 0$ and $l = 1$ wavefunctions that correspond to the s and p orbitals (see Figure 12.2).

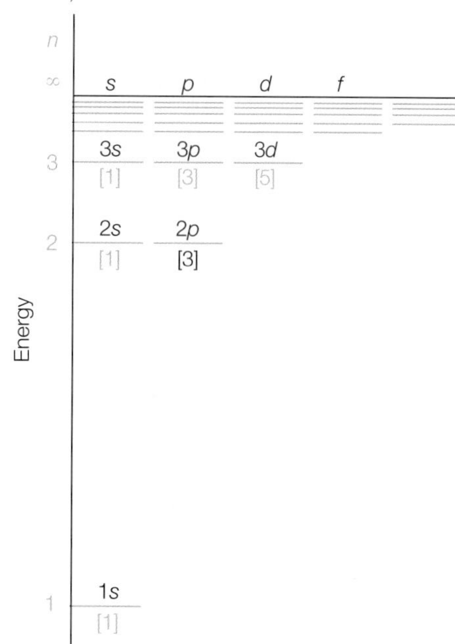

Figure 12.1 An energy diagram for the solutions of the hydrogen atom.

s Orbitals

An s orbital is one with $l = 0$. As a result the angular component is a constant and the dependence is strictly radial. For the 1s orbital, the wavefunction is:

$$\psi_{100} = \sqrt{\frac{1}{\pi a_0^3}}\, e^{-r/a_0} \qquad (12.13)$$

The higher orbitals all have the same general appearance as there is an exponential dependence multiplied by a polynomial term. For example, the 2s orbital is:

$$\psi_{200} = \frac{1}{2\sqrt{2}}\sqrt{\frac{1}{4\pi a_0^3}}\left(2 - \frac{r}{a_0}\right) e^{-r/2a_0} \qquad (12.14)$$

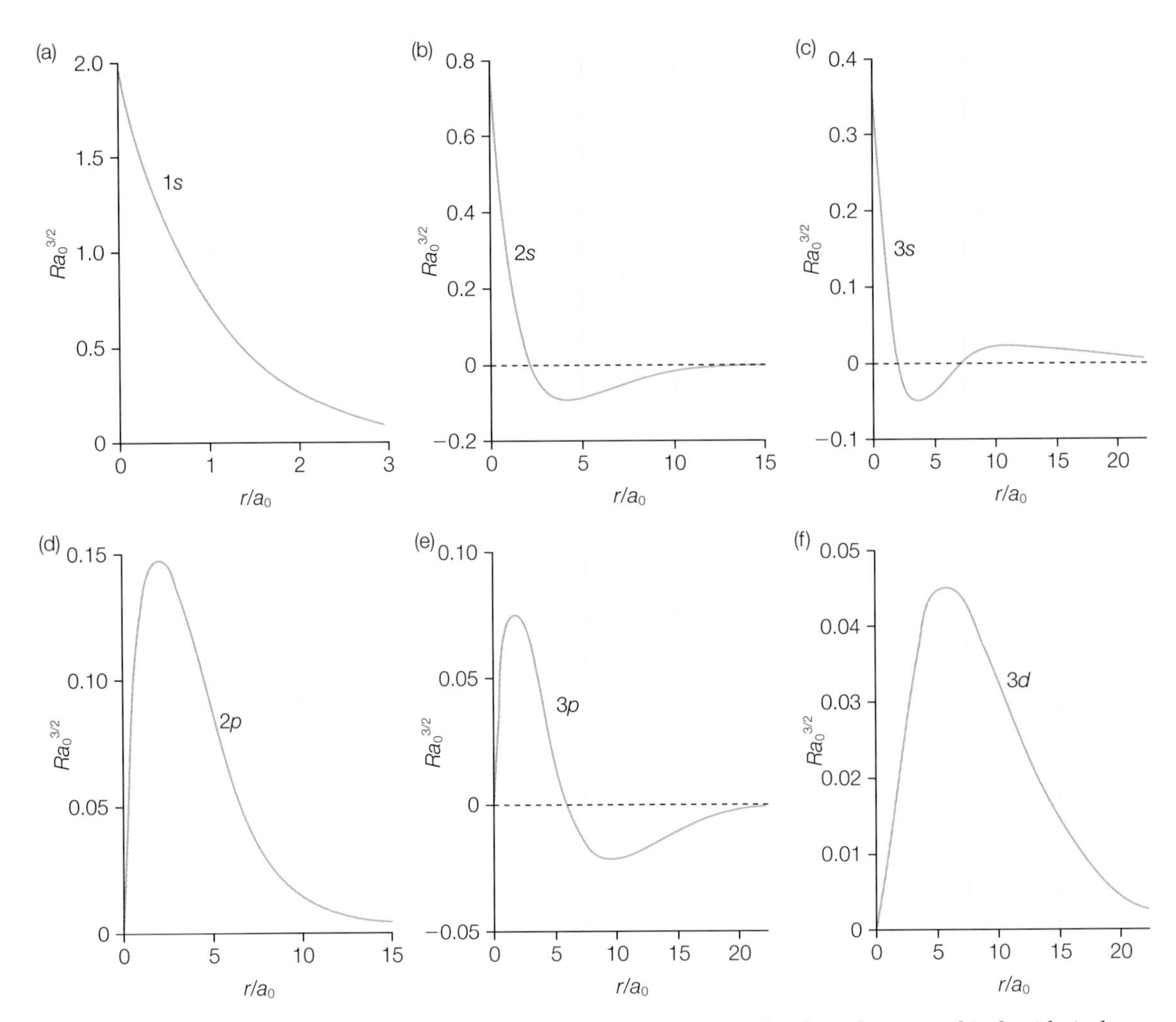

Figure 12.2 The radial wavefunctions for some solutions: (a–c) the first three s orbitals, (d,e) the first two p orbitals, and (f) the first d orbital.

We can plot the radial dependence of these wavefunctions and see that the higher orbitals have a more complex distribution with nodes appearing. For example, the 2s orbital has nodes at:

$$2 - \frac{r}{a_0} = 0 \quad \text{so} \quad r = 2a_0 \tag{12.15}$$

Since the solutions are non-zero for all values of r (excluding the nodes), the electron has a finite probability of being located anywhere in space. However, the probability is clearly highest near the nucleus. Let's calculate these probabilities using the wavefunctions just as we did for the particle in a box and harmonic oscillator. The probability of finding an electron at a certain point in space in $\psi^*\psi\,\mathrm{d}\tau$ that exponentially decreases as the radius increases (Figure 12.3).

We normally want to know the probability of finding the electron at a certain radius, not at a specific point, so we can average over the volume of a thin radial shell, giving the probability:

$$\psi^*\psi(4\pi r^2)\,\mathrm{d}r \tag{12.16}$$

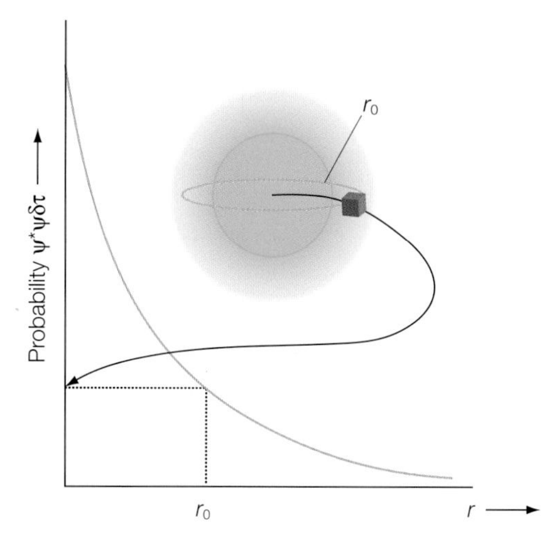

Figure 12.3 The dependence of the probability of the 1s state at a point located at a radius r.

The probability at a given radius has a very different dependence on the radius (Figure 12.4). For small values of the radius, the probability increases approximately as r^2 until the exponential term becomes dominant. As a result the function reaches a maximum value at an intermediate radius.

The most probable radial position of the electron is simply the peak position of this term. We can find the peak by finding the value for which the derivative is zero:

$$0 = \frac{\mathrm{d}}{\mathrm{d}r}(4\pi r^2\psi^*\psi) \tag{12.17}$$

For the ground state this can be calculated to give:

$$0 = \frac{\mathrm{d}}{\mathrm{d}r}\left(4\pi r^2\frac{1}{\pi a_0^3}e^{-2r/a_0}\right) = \frac{4}{a_0^3}\frac{\mathrm{d}}{\mathrm{d}r}(r^2e^{-2r/a_0}) \tag{12.18}$$

$$0 = \frac{8}{a_0^3}e^{-2r/a_0}r\left(1-\frac{r}{a_0}\right) \quad \text{or} \quad r = a_0 \tag{12.19}$$

What is the probability of finding the electron within this radius? We can calculate it.

$$\int_0^{a_0}\psi^*\psi\,\mathrm{d}\tau = \int_0^{a_0}\frac{1}{\pi a_0^3}e^{-2r/a_0}(4\pi r^2\,\mathrm{d}r) = \frac{4}{a_0^3}\int_0^{a_0}r^2e^{-2r/a_0}\,\mathrm{d}r \tag{12.20}$$

To solve this let $x = r/a_0$.

$$\frac{4}{a_0^3}\int_0^1(a_0x)^2e^{-2x}(a_0\,\mathrm{d}x) = 4\int_0^1 x^2e^{-2x}\,\mathrm{d}x \tag{12.21}$$

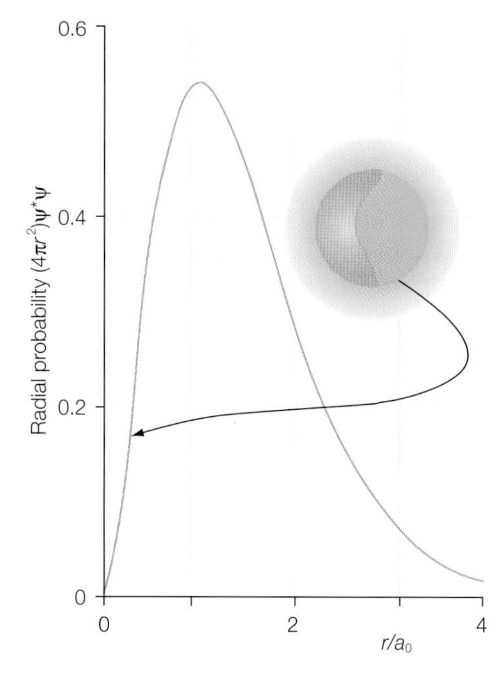

Figure 12.4 The radial distribution of the probability of finding 1s state within a shell of radius r.

This can be solved by using the product rule:

$$\int u\,\mathrm{d}v = uv - \int v\,\mathrm{d}u \tag{12.22}$$

letting

$$u = x^2 \quad \mathrm{d}u = 2x\,\mathrm{d}x \quad \mathrm{d}v = e^{-2x}\,\mathrm{d}x \quad v = -e^{-2x}/2 \tag{12.23}$$

$$4\int_0^1 x^2e^{-2x}\,\mathrm{d}x = 4\left[-x^2\frac{e^{-2x}}{2} - \int_0^1 \frac{e^{-2x}}{2}\,2x\,\mathrm{d}x\right] = -2xe^{-2x} + 4\int_0^1 e^{-2x}\,\mathrm{d}x$$

and another substitution

$$u = x \quad \mathrm{d}u = \mathrm{d}x \quad \mathrm{d}v = e^{-2x}\,\mathrm{d}x \quad v = -e^{-2x}/2 \tag{12.24}$$

$$\int_0^1 e^{-2x}x\,\mathrm{d}x = -\frac{x}{2}e^{-2x} + \int_0^1 \frac{1}{2}e^{-2x}\,\mathrm{d}x = -\frac{x}{2}e^{-2x} + \frac{1}{2}e^{-2x}\left(-\frac{1}{2}\right) \tag{12.25}$$

$$4\int_0^1 x^2e^{-2x}\,\mathrm{d}x = -e^{-2x}(2x^2 + 2x + 1)_0^1 = 0.323 \tag{12.26}$$

So there is only a 32% probability of finding the electron with the radius a_0.

Perhaps more useful is the average (or expectation) value of the radius. We can calculate it as:

$$\int_0^\infty \psi^* r\psi\,\mathrm{d}\tau = \int_0^\infty r\left(\frac{1}{\pi a_0}\right)e^{-2r/a_0}\,4\pi r^2\,\mathrm{d}r = \frac{4}{a_0}\int_0^\infty r^3e^{-2r/a_0}\,\mathrm{d}r \tag{12.27}$$

Let $x = r/a_0$ and then use the same approach as above:

$$\frac{4}{a_0}\int_0^\infty r^3e^{-2r/a_0}\,\mathrm{d}r = 4a_0\int_0^\infty x^3e^{-2x}\,\mathrm{d}x$$

$$= 4a_0e^{-2x}\left(-\frac{x^3}{2} - \frac{3x^2}{4} - \frac{3x}{4} - \frac{3}{8}\right)_0^\infty = \frac{3a_0}{2} \tag{12.28}$$

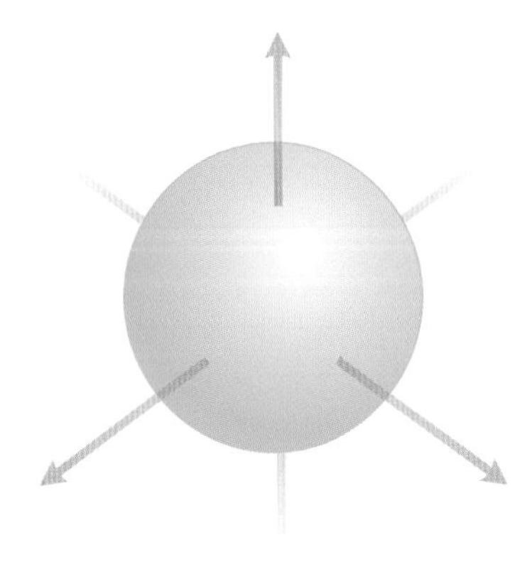

Figure 12.5 A boundary-surface representation of an s orbital.

Since the orbital extends for all values, there is a convention that orbitals should be represented by the 90% boundary; that is, the radius at which there is a 90% probability of finding the electron. You can substitute the value $3a_0$ into the integral above and see that, at this radius, you have a probability also of $3a_0$ (actually you are slightly over). It is this representation that is usually shown for orbitals (Figure 12.5).

All of these distances, the most probable radius (a_0), the average value ($3/2a_0$), and the radius at which a sphere encloses the electron with a 90% probability ($3a_0$) are measures of the distribution of the wavefunction (Figure 12.6). The usefulness of these values will depend upon what property of the wavefunction needs to be calculated.

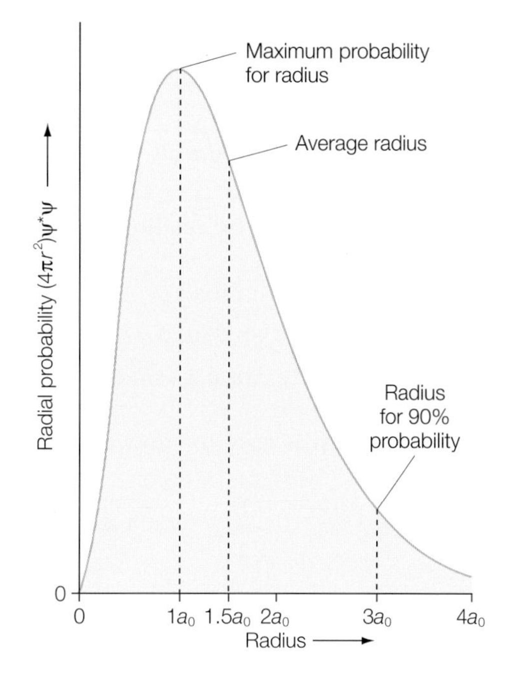

Figure 12.6 The radial distribution of the 1s orbital, showing the most probable radius, average value, and 90% probability value.

p Orbitals

For p orbitals, $l = 1$ and the electron has a non-zero angular momentum $\sqrt{2}\hbar$. This results in the distribution being no longer radially symmetrical as we saw for the s orbitals. Let us look at the $n = 2$ levels, of which there are three orbitals with $m_l = +1$, 0, and -1. The $m_l = 0$ orbital is denoted as p_0 and the tables give the wavefunction as being:

$$p_0 = \frac{1}{4\sqrt{2\pi}}\left(\frac{1}{a_0}\right)^{3/2}\left(\frac{r}{a_0}\right)e^{-r/2a_0}(\cos\theta) \tag{12.29}$$

But in spherical coordinates $z = r\cos\theta$, so we can write:

$$p_0 = z\left[\frac{1}{4}\sqrt{\frac{1}{2\pi}}\left(\frac{1}{a_0}\right)^{3/2}e^{-r/2a_0}\right] = zf(r) \tag{12.30}$$

Since the wavefunction can be written as the product of z times a radial exponentially decreasing function, it will have the appearance of a node at the origin, initially increasing for small values (where the exponential is approximately a constant), and eventually decreasing exponentially. The solutions are also seen to be symmetrical about the z axis.

We can in a similar manner write the two other orbitals using the tables. Since these orbitals are degenerate we are free to use linear combinations of the orbitals. The convention is to use the orbitals that give the form very similar to the term above. The two p orbitals are given by:

$$p_{\pm 1} = R_{2,1}Y_{1,\pm 1} = \left[\frac{1}{4\sqrt{6}}\left(\frac{1}{a_0}\right)^{3/2}\left(\frac{r}{a_0}\right)e^{-r/2a_0}\right]\left[\mp\left(\frac{3}{8\pi}\right)^{1/2}\sin\theta e^{\pm i\varphi}\right] \tag{12.31}$$

This reduces to:

$$p_{\pm 1} = \mp \frac{1}{8\sqrt{\pi}} \left(\frac{1}{a_0}\right)^{5/2} r \sin\theta e^{\pm i\phi} \tag{12.32}$$

Using the relationships

$$\begin{aligned} x &= r\sin\theta\cos\phi \\ y &= r\sin\theta\sin\phi \\ e^{\pm i\phi} &= \cos\phi \pm i\sin\phi \end{aligned} \tag{12.33}$$

it is easy to show that the p orbitals can be rewritten as

$$p_x = \frac{-1}{\sqrt{2}}(p_{+1} - p_{-1}) \propto xr \tag{12.34}$$

$$p_y = \frac{i}{\sqrt{2}}(p_{+1} + p_{-1}) \propto yr \tag{12.35}$$

These combinations have the z component of angular momentum canceling and have the same basic form as for p_0. These versions of the orbitals yield the conventional representation shown in Figure 12.7.

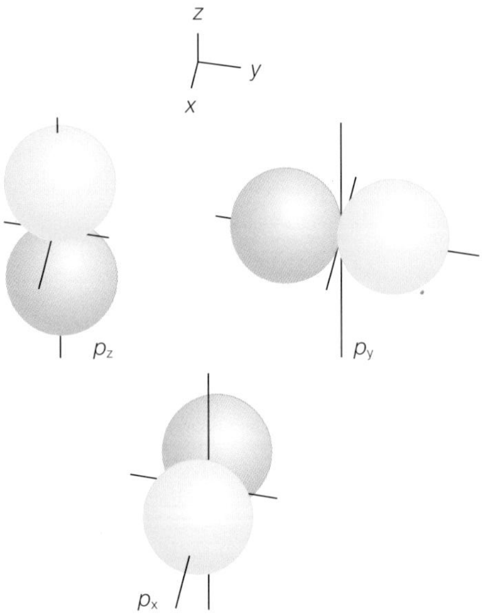

Figure 12.7 A representation of p orbitals of the hydrogen atom. A nodal plane separates the two lobes of each orbital.

d Orbitals

The d orbitals represent the $l = 2$ orbitals and arise when n is at least 3; the $n = 3$ shell contains one 3s orbital, three 3p orbitals, and five 3d orbitals. The electrons in the 3d orbitals have an angular momentum $\sqrt{6}\hbar$ with m_l equal to −2, −1, 0, +1, and +2. As was found for the p orbitals, the wavefunctions with opposite values of m_l can be combined in pairs to give rise to conventional standing orbitals (Figure 12.8), expressed as:

$$\begin{aligned} d_{xy} &= xyf(r) \\ d_{yz} &= yzf(r) \\ d_{zx} &= zxf(r) \\ d_{x^2-y^2} &= \frac{1}{2}(x^2 - y^2)f(r) \\ d_{z^2} &= \frac{1}{2\sqrt{3}}(3z^2 - r^2)f(r) \end{aligned} \tag{12.36}$$

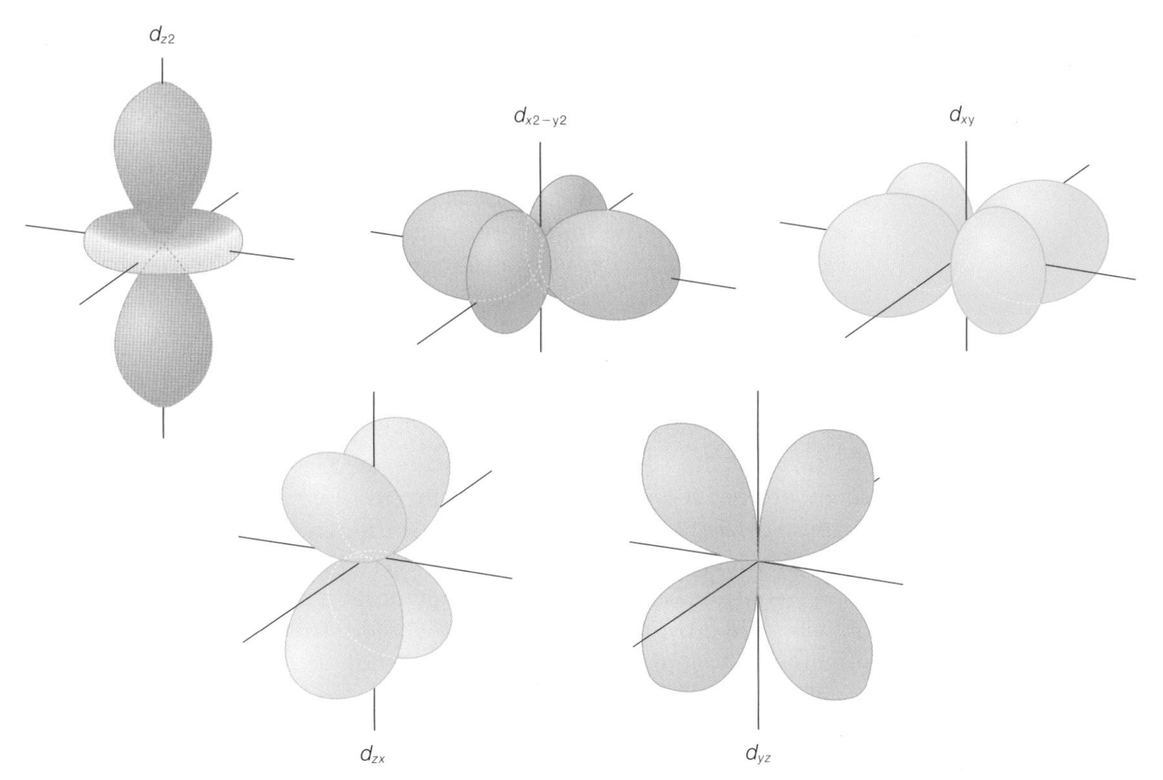

Figure 12.8 A representation of d orbitals of the hydrogen atom. Two nodal planes separate the lobes of each orbital.

TRANSITIONS

For the hydrogen atom, the electron will reside in the 1s orbital. As was true for the other applications, the electron can make a transition to a higher-energy orbital after it absorbs a photon whose energy matches the energy difference of the orbitals. Transitions between the orbitals are responsible for the presence of discrete lines in the atomic spectrum of hydrogen as well as other atoms (Figure 12.9). In addition to satisfying the requirement of energy conservation, for the transition to be allowed it also must obey the conservation of momentum. Photons have an intrinsic spin of 1 (see below). According to the conservation of momentum, the change in the angular momentum of the electron, due to the transition, must exactly compensate for the loss of the spin due to the photon absorption. Thus an electron in the 1s orbital, with an angular momentum of $l = 0$, can only make a transition to a p orbital with $l = 1$ but not to another s orbital or d orbital.

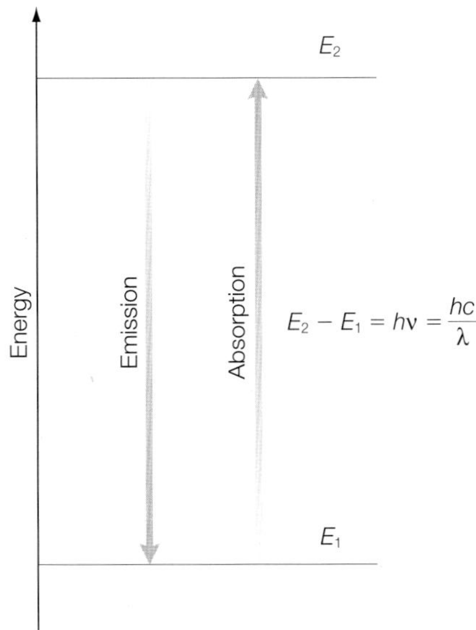

Figure 12.9 The spectral lines of atomic hydrogen arise from transitions between wavefunctions.

In general, for absorption, an electron in an orbital with the angular momentum l must make a transition to an orbital with an angular momentum of $l + 1$. Likewise, an electron that makes a transition from a higher-energy orbital with angular momentum l must end up in an orbital with the angular momentum $l - 1$ due to the emission of a photon. It follows that some transitions are allowed whereas others are forbidden. The restrictions on the transitions that are allowed are summarized as a selection rule. For electrons, transitions are allowed when the change in angular momentum l is 1 and correspondingly the change in m_l is 0 or 1.

RESEARCH DIRECTION: HYDROGEN ECONOMY

Hydrogen is the simplest of atoms with one electron in orbit around a single proton. It is the lightest and third most abundant element on the Earth's surface. Why has there been such a significant focus on this element by scientists and politicians? The reason is relatively simple. Global energy consumption is increasing dramatically, driven by rising standards of living and growing populations worldwide. This leads to the challenge of potentially doubling the global production of energy within 50 years coupled with increasing demands for clean energy sources that do not add more carbon dioxide and other pollutants to the environment, to minimize effects such as global warming. Today, the major source of fuel for transportation, though, is petroleum, with the accompanying release of unwanted exhaust fumes. By allowing hydrogen and oxygen to combine, hydrogen-based fuel systems are touted as generating enough energy to power a car without the production of exhaust fumes, only water (Hoffert et al. 2002; US Department of Energy 2003). Whereas hybrid cars will improve the fuel efficiency of a car, hydrogen-based cars use fuel cells that will eliminate the need for petroleum. In the USA, funds have been allocated to both academic institutes and industrial companies to produce hydrogen-based cars by the year 2015, replacing both gasoline-based and hybrid cars as the means of transportation (Figure 12.10).

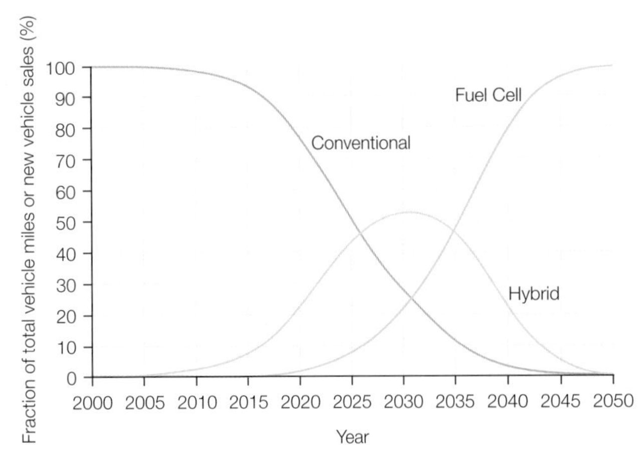

Figure 12.10 A graph showing relative projected production of cars with different fuel systems.

The attractiveness of hydrogen as an energy-storage material arises from the fact that hydrogen can be oxidized to water with the release of energy and then recovered from water with the input of energy while generating no environmentally detrimental side products. To be economically competitive with the present fossil-fuel economy, a number of technical aspects must be improved: for instance, the cost of the fuel cells must be significantly lowered, means must be established for the storage and delivery of hydrogen, and approaches for the industrial-scale production of hydrogen must be established. Although hydrogen is abundant, it is almost never found by itself, but is present in chemical compounds such as hydrocarbons or water.

Both the formation of hydrogen during electrolysis and its consumption in a fuel cell must be run as spontaneous reactions and therefore will involve some free-energy loss. However, in current practice, the amount of energy lost during these processes is considerably larger than simple thermodynamics would require. This is due to the substantial activation energy inherent in the redox half reactions performed at the surface of bare metal electrodes. One of the greatest losses occurs during the four-electron oxidation of water to generate molecular oxygen and protons during electrolysis (the half reaction at the oxygen-producing electrode). The overall reaction for electrolysis is:

$$2H_2O \rightarrow 2H_2 + O_2 \qquad (12.37)$$

For the simplest case, this overall chemical reaction at pH 7 can be expressed as two half reactions:

$$\begin{aligned} &2H_2O \rightarrow 4H^+ + O_2 + 4e^- \quad && E_0 = -0.82\ \text{V} \\ &4H^+ + 4e^- \rightarrow 2H_2 && E_0 = -0.42\ \text{V} \end{aligned} \qquad (12.38)$$

The half reaction midpoint potentials listed are given under standard conditions at pH 7. Considering only the free-energy difference between the reactants and products, the magnitude of the thermodynamic potential required for the overall reaction is about 1.24 V under standard conditions. However, in practice, electrolysis only occurs at a useful rate between two metal electrodes when much larger voltages, typically 2 V or more, are applied (US Department of Energy 2003). Because the power required to perform this process is simply given by the product of the voltage and the current (the current determines the rate of hydrogen production), the higher the voltage required to obtain a particular rate of hydrogen production, the more power is dissipated per quantity of hydrogen formed. The voltage that must be applied beyond that demanded by the free-energy difference between reactant and products is the so-called overpotential and the greater the overpotential used, the greater the loss of energy as heat.

The necessity for using a substantial overpotential during electrolysis largely results from the activation energy associated with the splitting of water on a metal surface, generating molecular oxygen and protons (eqn 12.37). The overpotential that is required for useful levels of hydrogen production from protons can be quite low using a platinum electrode (in the order of 0.1 V), but for oxygen evolution the numbers are considerably higher, typically 0.6 V or more on an unmodified surface (both overpotentials can be a serious consideration on lower-cost electrodes such as nickel, but the overpotential required for effective oxygen evolution via water splitting remains nearly a half volt higher than for hydrogen evolution via proton reduction; US Department of Energy 2003).

Nature has evolved catalysts that perform the reactions involved in electrolysis, with a very low overpotential and in relatively low-ionic-strength water near neutral pH. The oxygen-evolving complex of photosystem II in photosynthesis (Chapter 20) catalyzes the sequential removal of four electrons from two water molecules bound to the manganese cluster. In the photosystem II reaction center, optical excitation results in the formation of an oxidized chlorophyll complex with a midpoint potential of roughly 1 V. This is only slightly higher than what is required thermodynamically to split water, forming oxygen and hydrogen, especially considering the high H^+ concentration in the chloroplast lumen compared to the standard state. The ability of photosystem II to perform this reaction is discussed in detail in Chapter 20.

Hydrogen gas is naturally produced by a wide variety of natural organisms, including methanogenic archaea, sulfate- and nitrate-reducing bacteria, and some hyperthermophilic bacteria (Vignas et al. 2001; Armstrong 2004). In these cases, fermentation of an organic substrate, such as proteins, sugars, or lipids, yields hydrogen gas. Reduction of H^+ to hydrogen gas is used by fermentating bacteria to dispose of excess reducing equivalents and to maintain a suitable oxidation/reduction potential in the cell. The hydrogen gas is subsequently used as a low-potential reductant by bacteria living in the same environment as part of a biogeochemical cycle. A key component of such processes is the involvement of hydrogenases that catalyze the reversible conversion of molecular hydrogen (H_2) to protons and electrons and thus play a central role in microbial hydrogen metabolism. On the basis of their metal content, these enzymes can be grouped into three structural forms, the vast majority of which contain either iron and nickel ([Ni–Fe]-hydrogenases) or Fe only ([Fe–Fe]-hydrogenases) in their H_2-activating sites (Chapter 10). Although hydrogenases catalyze a very simple reaction, they do so in many different metabolic contexts and for a diversity of functions. In many microorganisms, the [Ni–Fe]-containing hydrogenases often catalyze the reaction in which H_2, generated by other metabolic sources, is consumed while the [Fe–Fe]-containing hydrogenases catalyze the reduction of protons (as terminal electron acceptors) to produce hydrogen. However, so-called bidirectional hydrogenases, having the capacity to both take up and produce hydrogen, are also known from a diversity of microbes. Although hydrogen production from natural systems is actively being studied, the sensitivity of these systems to the presence of oxygen coupled with difficulties in large-scale production makes the approach conceptually feasible but technically challenging.

SPIN

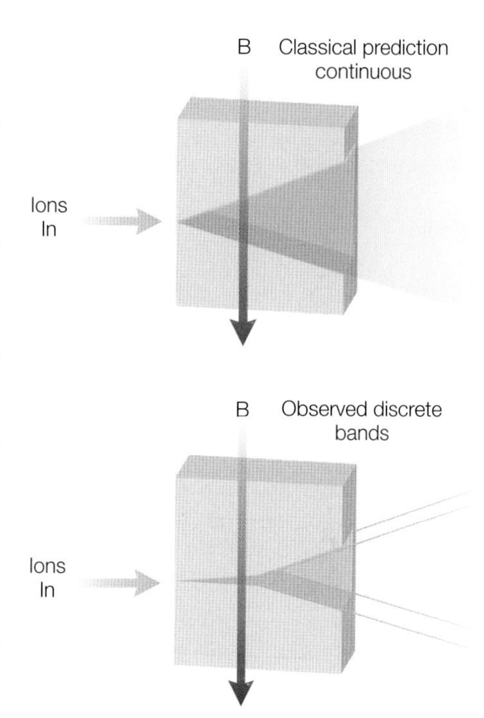

Figure 12.11 A schematic representation of the experimental arrangement for the Stern–Gerlach experiment showing the observed experimental outcome of two bands.

From Schrödinger's equation for the hydrogen atom we obtained three indices, n, l, and m_l, that represent the quantization of energy, angular momentum, and the z projection of angular momentum respectively. Based upon these three indices, there are, at most, two electrons per orbital. At the time that the hydrogen atom was originally being formulated, it was realized that there were, at most, two electrons per orbital; however, there was no reason for the pairing of electrons in each orbital. As you already know, there is a fourth quantum number, spin. Electrons are paired in orbitals with one electron as a spin up and the other as a spin down. Thus, each electron is defined uniquely by the four quantum numbers, n, l, and m_l, and the spin projection m_s.

The term spin derives from its contribution to the total angular momentum. Spin is observed experimentally in the Stern–Gerlach experiment in which a beam of silver atoms was shot through a heterogeneous magnetic field (Figure 12.11). If the atoms had a continuous range of angular momentum, as would be allowed classically, a broad, continuous band would be observed. However, only two bands are observed and this quantization must arise from two states of angular momentum. This must arise from another contribution to angular momentum that is quantized. For these experiments, silver atoms were used with a total of 47 electrons: 46 would be paired together, leaving one unpaired electron. The quantization must be from the unpaired electron. This was assigned as arising from a specific property of the electron, termed spin. Spin is a fundamental property of the electron, having a fixed value, like charge, that allows the electron to interact with a magnetic field.

The expressions that define the spin operators are written in a fashion similar to those for orbital angular momentum. In this case, both the total squared momentum and the z projection are quantized by the quantum numbers l and m_l:

$$L^2\psi = l(l+1)\hbar^2\psi \qquad (12.39)$$

$$L_z\psi = m_l\hbar\psi$$

For the spin, the corresponding quantum numbers are the *spin quantum number*, s, and the *z projection of the spin*, m_s. These two quantum numbers

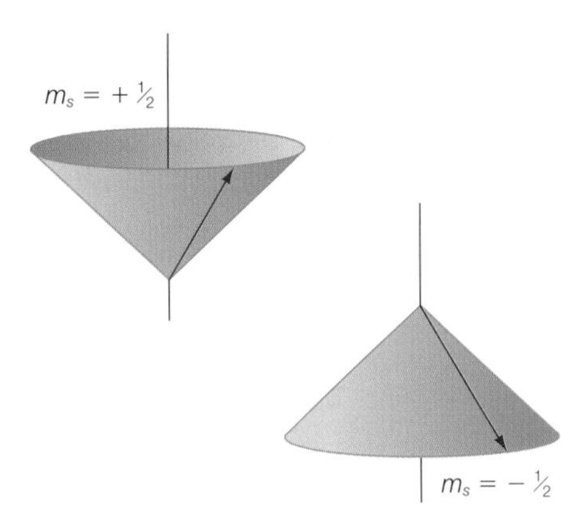

Figure 12.12 Particles with a value of $s = 1/2$ can have two spin projections along a direction, $m_s = +1/2$ or up, and $m_s = -1/2$ or down.

determine the total spin and z projection of spin according to the relationships:

$$S^2\psi = s(s+1)\,\hbar^2\psi = \frac{3\hbar^2}{2}\psi \tag{12.40}$$

$$S_z\psi = m_s\hbar\psi = \pm\frac{\hbar}{2}\psi$$

Electrons, protons, and neutrons all have a fixed value s of 1/2, yielding a total spin of $\sqrt{\frac{3\hbar^2}{2}}$. The allowed values of the z projection are restricted to m_s equal to +1/2 and −1/2, corresponding to spin values of $\pm\hbar/2$. Despite its name, this property does not represent a spinning motion of the electrons and, despite the differences in the masses of electrons and protons, their spin values are equal. This quantization of these parameters is often represented by use of diagrams, as shown in Figure 12.12.

Derivation box 12.2

Relativistic equations

Because its value is fixed, spin must represent an intrinsic property of the electron rather than a property that can change (such as the orbital angular momentum). Schrödinger's equation does not deal with this property because there is no true analogy with the properties of particles in classical physics. Rather, it comes from relativistic corrections to classical physics. Schrödinger's equation is derived from the classical law of mechanics defining the total energy E in terms of the kinetic and potential energy (Chapter 9):

$$\frac{p^2}{2m} + V = E \tag{db12.43}$$

The operators for momentum, p, potential, V, and energy, E (Table 9.1) are:

$$p \rightarrow \frac{\hbar}{i}\nabla \quad V \rightarrow V \quad \text{and} \quad E \rightarrow i\hbar\frac{\partial}{\partial t} \tag{db12.44}$$

Substitution of these operators into eqn db12.43 leads to the time-dependent form of Schrödinger's equation (eqn 9.28):

$$\frac{1}{2m}\left(\frac{\hbar}{i}\nabla\right)^2\psi + V\psi = i\hbar\frac{\partial}{\partial t}\psi \tag{db12.45}$$

which can be written for the time-independent case (eqn 9.39) as:

$$-\frac{\hbar^2}{2m}\nabla^2 \psi + V\psi = E\psi \tag{db12.46}$$

The classical laws are not valid for very fast motion, where relativity becomes important. As with quantum mechanics, the relativistic equations supplement the classical laws and agree with the classical laws in certain limits. The corresponding relativistic equation is:

$$c^2p^2 + m^2c^4 = E^2 \tag{db12.47}$$

This equation reduces to the familiar $E = mc^2$ for a particle at rest (and $p = 0$). Substitution of the operators for momentum and energy (eqn db12.44) into the relativistic expression for energy (eqn db12.47) leads to a relativistic formulation of quantum mechanics as expressed by:

$$c^2\left(\frac{\hbar}{i}\nabla\right)^2 \psi + m^2c^4\psi = \left(i\hbar\frac{\partial}{\partial t}\right)^2 \psi \tag{db12.48}$$

$$-\hbar^2c^2\nabla^2\psi + m^2c^4\psi = -\hbar^2\frac{\partial^2}{\partial t^2}\psi$$

This equation is known as the Klein–Gordon equation and unfortunately is not useful for electrons. It does, however, hold for a class of particles known as bosons that have integer spin. An example of a boson is the photon.

The equation can be revised for use with the other class of particles, fermions, that have half-integer spin, including the electron. The correct Fermi–Dirac equation looks like a linearized version of the Klein–Gordon equation:

$$\alpha c(i\hbar\nabla)\psi + \beta mc^2\psi = i\hbar\frac{\partial}{\partial t}\psi \tag{db12.49}$$

In this equation α and β are 4×4 matrices and the wavefunctions must then be vectors with four components. If the components are paired in up and down states, electrons of necessity have two new characteristics. Two of these components arise from the spin, which can be up or down. The second set arises from two states of energy, the normal-energy state and the negative-energy state, or, as we would say today, particles and antiparticles. Thus, there are four states: spin up and spin down for electrons and spin up and spin down for positrons.

Positrons are the antiparticle version of electrons and are the subject of many science fiction novels. They are not present normally but can exist in synchrotron experiments where energy is converted into an electron and positron pair for very short periods of time. In such cases, both the particle and antiparticle are created in equal balance to maintain the overall charge neutrality. The minimal input energy needed to create the pair is equal to the energy of each particle corresponding to their masses (technically the rest mass as the mass is dependent upon the motion in relativity). Electrons and positrons have the same mass, m_e, corresponding to energies of m_ec^2. Thus, a total energy of at least $2m_ec^2$ is needed to create an electron–positron pair at a synchrotron. Once created, the positron will rapidly recombine with an electron, releasing an energy of $2m_ec^2$, which is equal to 1022 keV, and it is this release of energy that is often discussed in literature. To conserve momentum, the energy will be released as two photons emitted at a relative angle of 180° (assuming that the particles are at rest). For example, in the book *Angels and Demons* by Dan Brown, scientists make use of the Conseil Europeen pour la Recherche Nucleaire, denoted by the acronym CERN, a center for particle physics that straddles the French–Swiss border. In that story, scientists at CERN generated 0.25 g of positrons that are trapped in a special container (a plot device that has never been accomplished). This mass corresponds to 0.25 g m_e^{-1}, or 2.7×10^{26} positrons, and hence each recombination of every positron with an equivalent amount of electrons will release an energy of 1.4×10^{29} keV or equivalently 5.3 kt, an amount comparable to that of the first atomic bombs.

$$i^2 = -1 \quad \frac{1}{i^2} = -1$$

$$m_e = 9.1 \times 10^{-31}\ \text{kg}$$
$$m_ec^2 = (9.1 \times 10^{-31}\ \text{kg})(3 \times 10^8\ \text{m s}^{-1})^2 = 511\ \text{keV}$$
$$1\ \text{eV} = 3.8 \times 10^{-32}\ \text{kt (kilotonnes)}$$

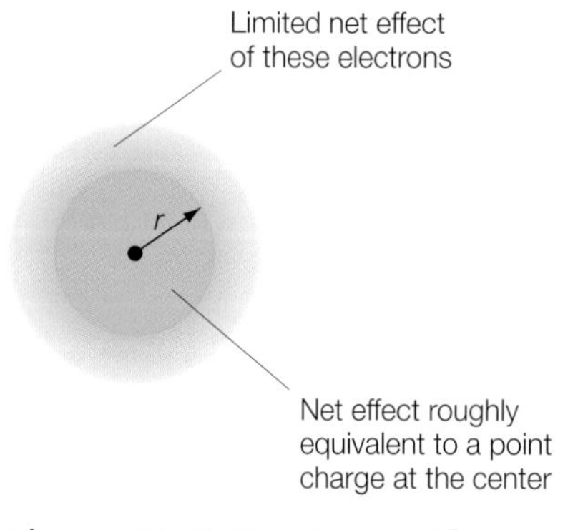

Figure 12.13 For atoms with more than one electron, the nuclear charge, $Z^{atom}e$, that any given electron experiences is effectively reduced by the presence of the other electrons, thus reducing the charge to the effective nuclear charge Z_{eff}^{atom}.

MULTI-ELECTRON ATOMS

The solution of Schrödinger's equation when more than one electron is present is usually solved by use of approximations to incorporate the interactions between the electrons. There are two basic approaches that can be used. Both rely on the idea that the wavefunctions derived for the hydrogen atom are basically correct and simply need minor corrections in order to be used for many electron atoms.

Empirical constants

One approach is to modify the constants used in the expressions to include the new interactions. This can easily be done by modifying the nuclear-charge parameter, Z. The presence of the additional negatively charged electrons effectively decreases the positive charge of the nucleus (Figure 12.13).

Using the hydrogen atom solution (eqn 12.3) with a nuclear charge Z the energy of an electron is:

$$E_n = -\frac{Z^2 m_e e^4}{32\pi^2\varepsilon_0^2\hbar^2}\frac{1}{n^2} \tag{12.41}$$

The value of the energy can be measured since the atomic spectrum has a series of lines and the energy required to completely remove the electron, called the ionization energy, is simply the limiting value of this series observed in atomic spectra (Chapter 9). Measurement of the spectral lines shows that the energy is lower than expected and, since the energy is a function of the square of Z, we can write:

$$Z_{eff}^{atom} = \sqrt{\frac{E^{atom}}{E^{H}}} \tag{12.42}$$

For helium, substitution of the measured value of 2372 kJ mol^{-1} compared to the value of 1312 kJ mol^{-1} for hydrogen yields a Z_{eff} value equal to 1.34. This makes sense since we would expect the effective value to be between the maximum value of 2 (no effect, due to the second electron) and 1 (complete screening, due to the second electron).

We need to be careful as we must compare equivalent orbitals. For lithium, we measure an ionization energy of 513 kJ mol^{-1}, and comparison with the 1312 kJ mol^{-1} value for hydrogen yields Z_{eff} = 0.62. This is much lower than expected as two electrons should not produce an effective screen of 2.37 charges. The problem is that the ionization energy for lithium is for the 2s orbital and not the 1s orbital. For lithium, the ionization energy for the 1s orbital is 7298 kJ mol^{-1}, yielding a Z_{eff} value of 2.4, which makes more sense. Use of the 2s value for hydrogen of 1.27 yields a Z_{eff} value of 1.27. Thus, correct comparisons yield effective charges in the expected ranges.

Self-consistent field theory (Hartree–Fock)

Another approach is to calculate how the wavefunctions should be changed in response to the interactions between electrons. The basic idea is that the potential is now:

$$V = -\sum_i \frac{Ze^2}{4\pi\varepsilon_0 r_i} + \frac{1}{2}\sum_{ij}\frac{e^2}{4\pi\varepsilon_0 r_{ij}} = V_0 + V_1 \tag{12.43}$$

where there are i electrons present. The first term V_0 is the potential between the nucleus and each electron, and the second term V_1 is the interactions between electrons (the 1/2 prevents double counting). The basic idea of this approach is to assume that the second term is much smaller than the first term. This means that we can write for the wavefunctions:

$$\psi = \psi_0 + \psi_1 \qquad (12.44)$$
$$E = E_0 + E_1$$

Again, the second term is much smaller than the first. Substitution of these terms into Schrödinger's equation yields:

$$-\frac{\hbar^2}{2m}\nabla^2(\psi_0 + \psi_1) + (V_0 + V_1)(\psi_0 + \psi_1) = (E_0 + E_1)(\psi_0 + \psi_1) \qquad (12.45)$$

This equation simplifies because ψ_0 is the solution to the 0 terms and we can write that:

$$\psi_1 = \sum a\psi_0 \qquad (12.46)$$

The actual solution to this problem would take some time to solve, but for our purposes the important aspect is that there is a well-defined procedure to perform the calculation. For biological systems, however, the calculation is difficult because they are very complex; simple molecules need to be compared in isolated situations (such as a vacuum).

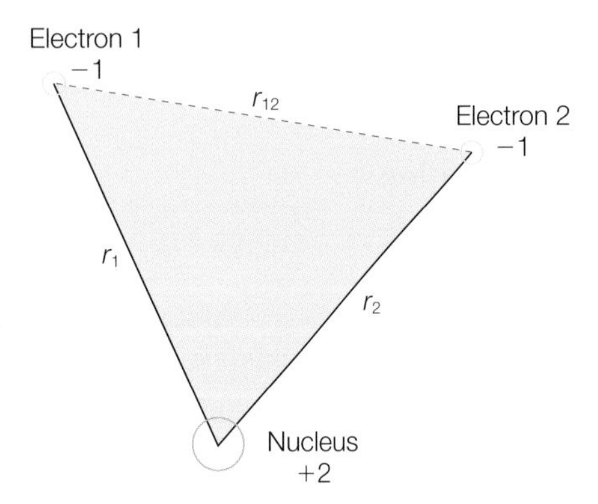

Figure 12.14 The potential for the helium atom is modeled as consisting of electrostatic interactions that are dependent on the relative distances between the charges.

HELIUM ATOM

Let us apply this approach to the simplest atom that has more than one electron, helium. In this case there are two electrons and one nucleus. As was true for the hydrogen atom, the potential is established by the electrostatic interactions among the charges (Figure 12.14). For helium, the atomic number must be included to account for the nuclear charge. There are a total of three interactions: two are between the nucleus and each electron, and one is between the two electrons. Since there are now two electrons, there are two sets of coordinates that must be accounted for in the equation. With this in mind, the classical expression for energy conservation and Schrödinger's equation is written as:

$$\frac{p_1^2}{2m} + \frac{p_2^2}{2m} - \frac{e^2}{4\pi\varepsilon_0}\left[\frac{Z}{r_1} + \frac{Z}{r_2} - \frac{1}{r_{12}}\right] = E \qquad (12.47)$$

$$-\frac{\hbar^2}{2m}(\nabla_1^2 + \nabla_2^2)\psi(r_1r_2) - \frac{e^2}{4\pi\varepsilon_0}\left[\frac{Z}{r_1} + \frac{Z}{r_2} - \frac{1}{r_{12}}\right]\psi(r_1r_2) = E\psi(r_1r_2)$$

$$p \rightarrow \frac{\hbar}{i}\nabla \quad r \rightarrow r$$

This is first solved at zeroth order; that is, neglecting the interactions between electrons. These solutions are denoted by the superscript zero:

$$-\frac{\hbar^2}{2m}(\nabla_1^2 + \nabla_2^2)\,\psi^0(r_1r_2) - \frac{e^2}{4\pi\varepsilon_0}\left[\frac{1}{r_1} + \frac{1}{r_2}\right]\psi^0(r_1r_2) = E^0\psi^0(r_1r_2) \tag{12.48}$$

Assume that the wavefunction can be separated as the product of two contributions:

$$\psi(r_1r_2) = \psi(r_1)\,\psi(r_2) \tag{12.49}$$

This yields two separate equations:

$$-\frac{\hbar^2}{2m}\nabla_1^2\psi^0(r_1) - \frac{e^2}{4\pi\varepsilon_0}\frac{Z}{r_1}\psi^0(r_1) = E_1^0\psi^0(r_1) \tag{12.50}$$

$$-\frac{\hbar^2}{2m}\nabla_2^2\psi^0(r_2) - \frac{e^2}{4\pi\varepsilon_0}\frac{Z}{r_2}\psi^0(r_2) = E_2^0\psi^0(r_2) \tag{12.51}$$

where

$$E_1^0 + E_2^0 = E^0 \tag{12.52}$$

The first-order correction for energy is determined by modifying Schrödinger's equation to include the interaction. First, rewrite Schrödinger's equation using:

$$H^0 = -\frac{\hbar^2}{2m}(\nabla_1^2 + \nabla_2^2) - \frac{e^2}{4\pi\varepsilon_0}\left[\frac{Z}{r_1} + \frac{Z}{r_2}\right]$$

$$H^1 = +\frac{e^2}{4\pi\varepsilon_0}\left[\frac{1}{r_{12}}\right] \tag{12.53}$$

$$(H^0 + H^1)(\psi^0 + \psi^1) = (E^0 + E^1)(\psi^0 + \psi^1) \tag{12.54}$$

Ignoring the two second-order terms yields:

$$H^1\psi^0 + H^0\psi^1 = E^1\psi^0 + E^0\psi^1 \tag{12.55}$$

$$H^0\psi^1 - E^0\psi^1 = -H^1\psi^0 + E^1\psi^0 \tag{12.56}$$

Multiply by ψ^{0*} and integrate:

$$\int \psi^{0*} H^0 \psi^1 \, d\tau - \int \psi^{0*} E^0 \psi^1 \, d\tau = -\int \psi^{0*} H^1 \psi^0 \, d\tau + \int \psi^{0*} E^1 \psi^0 \, d\tau \tag{12.57}$$

The operators used in Schrödinger's equation are called Hermetian operators and have the property:

$$\int \psi^{0*} H^0 \psi^1 \, d\tau = \int \psi^1 (H^0 \psi^0)^* \, d\tau = \int \psi^1 (E^0 \psi^0)^* \, d\tau = \int \psi^{0*} E^0 \psi^1 \, d\tau \tag{12.58}$$

and

$$\int \psi^{0*} E^1 \psi^0 \, d\tau = E^1 \int \psi^{0*} \psi^0 \, d\tau = E^1 \tag{12.59}$$

Resulting in the expression for the first-order energy correction:

$$E^1 = \int \psi^{0*} H^1 \psi^0 \, d\tau \tag{12.60}$$

For the $n_1 = 1$, $n_2 = 1$ state this is:

$$E^1 = \left(\frac{1}{\pi^2} \frac{Z^6}{a_0^6} \right) \iint_{1,2} e^{-2Zr_1/a_0} e^{-2Zr_2/a_0} \left(\frac{e^2}{4\pi\varepsilon_0} \frac{1}{r_{12}} \right) d\tau_1 \, d\tau_2 \tag{12.61}$$

$$d\tau_1 = r_1^2 \sin\theta_1 \, dr_1 \, d\theta_1 \, d\varphi_1 \tag{12.62}$$
$$d\tau_2 = r_2^2 \sin\theta_2 \, dr_2 \, d\theta_2 \, d\varphi_2$$

The term r_{12} must be rewritten in spherical coordinates, where the > or < signs denote the larger or smaller values of r_1 or r_2:

$$\frac{1}{r_{12}} = \sum_l \sum_{m_l} \frac{4\pi}{2l+1} \frac{r_<^l}{r_>^{l+1}} Y_l^{m_l}(\theta_1\varphi_1)^* \, Y_l^{m_l}(\theta_2\varphi_2) \tag{12.63}$$

Experimentally, a value of −79.01 eV is determined for the $n_1 = 1$ and $n_2 = 1$ state of helium. Substitution of these values into E^0 gives −108.8 eV. The first-order term has a value of +34.0 eV, giving −74.8 eV, and using up to the third order gives −78.9 eV.

SPIN–ORBITAL COUPLING

The electron spin influences the energies of the electronic states because associated with the spin is a magnetic dipole moment. Likewise, the orbital angular momentum, for the states with l greater than 0, will possess a dipole moment due to the angular momentum. The interaction of the spin magnetic moment with the magnetic field arising from orbital motion is termed *spin–orbit coupling*. One expression of the coupling makes use of the total angular momentum of the system, $\vec{J}$, which is the vector sum of the contributions from the spin, $\vec{S}$, and the orbital angular momentum, $\vec{L}$:

$$\text{Total angular momentum } \vec{J} = \vec{L} + \vec{S} \tag{12.64}$$

The total angular momentum can be described using the quantum numbers j and m_j, where $j = l + 1/2$ or $j = l - 1/2$ (the spin angular momentum is either aligned or opposite to the orbital angular momentum). For the s orbitals, $l = 0$ and the total angular momentum is simply the spin angular momentum. For the p orbitals, $l = 1$ and the total angular momentum is either 3/2 or 1/2. In this case, the energies of the two states are different as the $j = 1/2$ state, with the two moments in opposite directions, will have a lower energy than the $j = 3/2$ state, with the two moments aligned. This splitting gives rise to what is termed fine structure in atomic emission spectra. The splitting of the $2p^1$ state allows two distinct transitions to the 2s state and a corresponding two lines in spectrum, which are very close. For example, sodium has lines at 589.16 and 589.76 nm.

PERIODIC TABLE

We are now in a position to understand how quantum mechanics provides an opportunity to understand the arrangement of the periodic table (Figure 12.15). Electrons are always assumed to be present in the lowest-energy states available. Each electron will have a unique set of

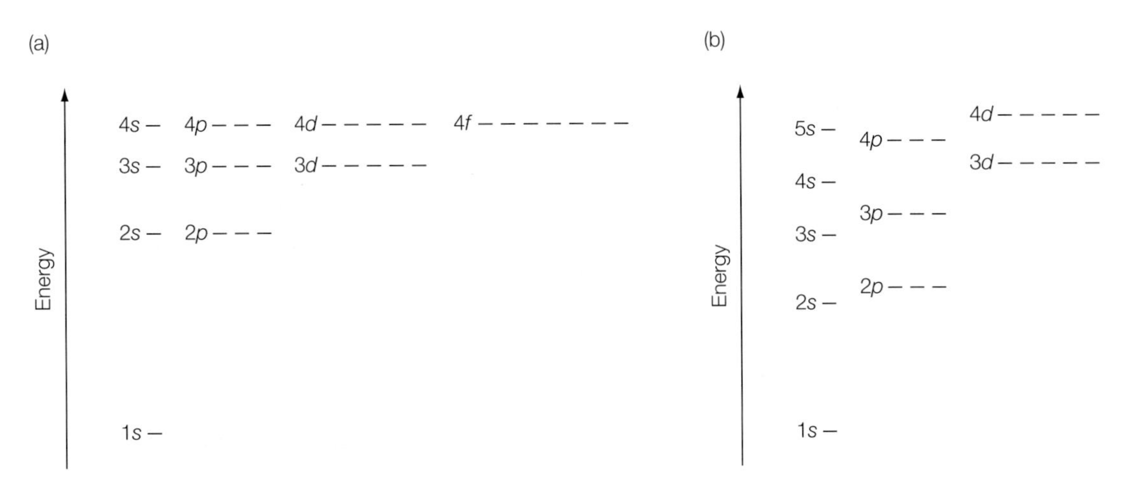

Figure 12.15 (a) The Bohr model predicts that many orbitals are degenerate; (b) including all interactions involving the electrons results in nondegenerate orbitals and an order for filling the orbitals with electrons.

four quantum numbers, with each orbital having one spin up and one spin down electron. Thus, for hydrogen, the single electron will be located in a 1s orbital. The two electrons present in helium will also be in the 1s orbital. The arrangement of the electrons in the orbitals, the electron configuration, is denoted for hydrogen and helium as $1s^1$ and $1s^2$, where the superscript identifies the number of electrons in the orbital. In general, only the outermost orbitals are shown as the inner orbitals are assumed to be filled. For example, lithium has three electrons. Two will be present in the 1s orbital and complete that shell and the third must be present in an $n = 2$ orbital. The electron configuration is denoted by $2s^1$, showing only the 2s state and not the inner 1s state. In the hydrogen model all of the $n = 2$ orbitals, the 2s and the three 2p orbitals, all have the same energy. However, the energies of these orbitals are modified due to the presence of more than one electron. As discussed above, one of the major effects is the shielding of the outer electrons, with the inner electrons modifying the effective nuclear charge. The extent of this shielding on the 2s and 2p orbitals differs. An electron in the 2s orbital has a greater presence near the nucleus; that is, a greater probability of being near the nucleus, than an electron in a 2p orbital. As a result of the greater presence, the electron in the 2s orbital experiences less shielding due to the 1s electrons. Consequently, the electron in the 2s orbital experiences a greater nuclear charge and is held more tightly than one in a 2p orbital. This corresponds to an electron in a 2s orbital having a lower energy than one in a 2p orbital. Consequently, the third electron will occupy the 2s orbital rather than one of the 2p orbitals.

The three p orbitals can hold up to six electrons. When more than one electron is present in a p orbital the electrons are expected to fill different p orbitals to minimize the repulsive electrostatic interaction by maximizing the distance between the electrons. For example, carbon has a total of six electrons, with two in the 1s orbitals, two in the 2s orbitals, and two in the 2p orbitals. The two electrons in the 2p orbitals are expected to occupy two different 2p orbitals of the three available. For nitrogen, the three electrons in the 2p orbitals will be in the each of the three orbitals. According to Hund's rule, the electrons in the 2p orbitals will preferentially prefer a configuration in which the spins are all aligned with the greatest number of unpaired electrons. The 2p electrons are filled in neon that has a total of 10 electrons with six in the 2p orbitals.

In potassium, the unpaired electron is located in a 4s orbital rather than a 3d orbital. Likewise, calcium has two electrons in the 4s orbital. The lower energy of the 4s orbital compared to the 3d orbital arises from the greater probability of an electron in the 4s orbital being near the nucleus and hence increasing its interactions with the nucleus. Finally with scandium, electrons begin to occupy the 3d orbitals. This begins the transition metals that are among the heaviest atoms found in proteins, including manganese, iron, nickel, copper, and zinc.

In general, for any given shell, the s orbitals will be filled preferentially compared to the p orbitals due to larger shielding experienced by electrons in the p orbitals. Likewise, the shielding ideal can be applied to argue that the p orbital is filled before the d orbital for the $n = 3$ and higher shells, and the d orbital before the f orbital in the $n = 4$ and higher shells. Electrons will preferentially occupy different p, d, or f orbitals is they are available before occupying the same orbital. This gives the following trend for increasing energy of orbitals:

$$1s < 2s < 2p < 3s < 3p < 4s < 3d < 4p < 5s < 4d \ldots$$

As the number of electrons increases, the detailed interactions between the electrons become increasingly important in defining the relative energies of the orbitals. For example, in the transition metals, the electron configuration of chromium is $4s^1 3d^5$ rather than the expected $4s^2 3d^4$ and copper has the configuration $4s^1 3d^{10}$ rather than $4s^2 3d^9$. The reason for the configuration of chromium is that by spreading the electrons over five d orbitals, the unfavorable electron repulsion is decreased. For copper, the completion of the d orbital provides a more stable configuration than the nearly complete configuration.

REFERENCES

Armstrong, F.A. (2004) Hydrogenases: active site puzzles and progress. *Current Opinion in Chemical Biology* **8**, 133–40.

Hoffert, M.I., Caldeira, K., Benford, G. et al. (2002) Advanced technology paths to global climate stability: energy for a greenhouse planet. *Science* **298**, 981–7.

US Department of Energy (2003) *Report of the Basic Energy Sciences Workshop on Hydrogen Production, Storage, and Use*. US Department of Energy, Argonne National Laboratory.

Vignas, P.M., Billoud, B., and Meyer, J. (2001) Classification and phylogeny of hydrogenases. *FEMS Microbiology Reviews* **25**, 455–501.

PROBLEMS

12.1 Write the potential for the hydrogen atom.

12.2 Calculate the energy of the (a) ground state and (b) first excited state.

12.3 Write the energy of an electron in a $3p_x$ orbital for the hydrogen atom in the Bohr model.

12.4 Calculate the position of the radial node of the 2s orbital.

12.5 Calculate the mean radius of the 1s state.

12.6 Calculate the most probable radius of the 1s orbital.

12.7 List the three quantum numbers for the hydrogen atom and their allowed values.

12.8 Calculate the wavelength of light absorbed for a transition of $n = 1$ to $n = 3$.

12.9 What is the physical interpretation of the three quantum numbers obtained for the hydrogen atom?

12.10 Calculate the wavelength of light absorbed when an electron makes a transition from an $n = 3$ state to an $n = 1$ state in copper.

12.11 Write in integral form (but do not solve) the expectation value of radius for the 2s state.

12.12 Write in integral form (but do not solve) the expectation value of momentum for the 2s state.

12.13 Write (but do not solve) the probability of finding an electron in the 2s state beyond a radius of $3a_0$.

12.14 How is the p_x orbital related to wavefunctions obtained using Schrödinger's equation?

12.15 Determine the probability of finding an electron in the 1s state beyond a distance of 1.058 Å.

12.16 Determine the probability of finding an electron in the 1s state between a distance of 0.529 Å and 1.058 Å.

12.17 Calculate the most probable radius for the 1s state using the wavefunction.

12.18 Calculate the expectation value of the radius for the 1s state.

12.19 Explain how the wavefunction for an electron in a 2s orbital can have a value of zero.

12.20 Schrödinger's equation can be divided into three parts using the separation of variables. The ϕ dependence is given by $\Phi(\phi) = Ae^{im_l\phi}$. Demonstrate why the value of m_l must be an integer.

12.21 Demonstrate that the function $\Phi(\phi) = Ae^{im_l\phi}$ is a solution to the phi dependence of Schrödinger's equation:

$$\frac{d^2}{d\phi^2}\Phi(\phi) = -m_l^2\,\Phi(\phi)$$

12.22 Demonstrate that the function $\Theta(\theta) = B\cos\theta$ is a solution to the theta dependence of Schrödinger's equation when $l = 1$ and $m_l = 0$:

$$\sin\theta\frac{d}{d\theta}\left[\sin\theta\frac{d\Theta(\theta)}{d\theta}\right] + [l(l+1)\sin^2\theta - m_l^2]\Theta(\theta) = 0$$

12.23 Determine the energy of the wavefunction $\Pi(r) = re^{-\alpha r}$ where $\alpha = \frac{m}{\hbar^2}\frac{e^2}{4\pi\varepsilon_0}$ by substituting this function into the radial form of the Schrödinger equation when $l = 0$:

$$\frac{d^2}{dr^2}\Pi(r) + \left(-\frac{l(l+1)}{r^2} + \frac{2m}{\hbar}\frac{e^2}{4\pi\varepsilon_0 r}\right)\Pi(r) = \frac{2m}{\hbar}E\Pi(r)$$

12.24 Explain how the wavefunction is related to the boundary surface of an orbital.

12.25 What transitions are forbidden due to the fact that photons have a spin of 1?

12.26 (a) Write Schrödinger's equation for the hydrogen atom if the classical potential energy between a positive and negative charge was given by

$$V(r,\theta,\varphi) = \frac{-e^2}{4\pi\varepsilon_0}\frac{\cos\theta}{r^3}$$

(b) What can be predicted about the energies of the wavefunctions for this modified potential?

12.27 Calculate the expectation (or average) value of r^2 for the 2s orbital.
12.28 How many orbitals are there in a shell with $n = 5$?
12.29 Explain the physical meaning of Z_{eff}.
12.30 Calculate the effective nuclear charge of helium using ionization energies of 2370 kJ mol^{-1} for helium and 1312 kJ mol^{-1} for hydrogen.
12.31 If Z_{eff} is 2.4 for the 1s orbital of lithium, what is the ionization energy? (For hydrogen the ionization energy is 1312 kJ mol^{-1}.)
12.32 Write Schrödinger's equation for the helium atom.
12.33 Explain qualitatively why the ionization energy for removing the first electron increases for beryllium compared to lithium.
12.34 Explain qualitatively why the ionization energy of removing the first electron decreases for boron compared to beryllium.
12.35 Explain why spin is not determined using Schrödinger's equation.
12.36 What is the interpretation of the four components obtained in the relativistic formulation of quantum mechanics?
12.37 Explain how electrons fill the orbitals of nitrogen.

13

Chemical bonds and protein interactions

In this chapter we first examine how the concepts of quantum mechanics can be extended to molecules in order to develop a quantitative description for chemical bonds. Schrödinger's equation is written for the hydrogen molecule, leading to expressions for valence bonds and the Hückel model. A central feature of proteins and other biological systems is building from repetitive units, such as amino acids or nucleic acids. The covalent geometry and other properties of the units are derived from the principles of quantum mechanics applied to molecules. However, the overall molecule is too large to be treated from a quantum-mechanical standpoint as many different interactions influence protein structures, including steric interactions, hydrogen bonds, electrostatic interactions, and hydrophobic effects. These interactions have pronounced influences on bond angles as the barriers to rotations are relatively low. For large molecules with many possible bond angles, there are many possible protein conformations. Although detailed quantum-mechanical descriptions of proteins are not feasible, the effects of the many interactions can be modeled as potentials that can then be used to describe the possible conformations of proteins. Whereas the thermodynamically favored conformations are closely related, the ability of theoretical calculations to predict only one stable conformation that matches the biologically observed structure is discussed.

SCHRÖDINGER'S EQUATION FOR A HYDROGEN MOLECULE

The simplest molecule, the hydrogen molecule, is used to introduce the concept of applying Schrödinger's equation to a molecule. For the hydrogen molecule, we must insert into Schrödinger's equation all of the interactions involving the nuclei and electrons. For H_2 there are two electrons, which are labeled 1 and 2, and two nuclei, labeled A and B (Figure 13.1). Only the electrons are considered to be moving, and so only the two electrons contribute to the kinetic energy:

$$\text{Kinetic energy} = \frac{p_1^2}{2m_e} + \frac{p_2^2}{2m_e} \quad (13.1)$$

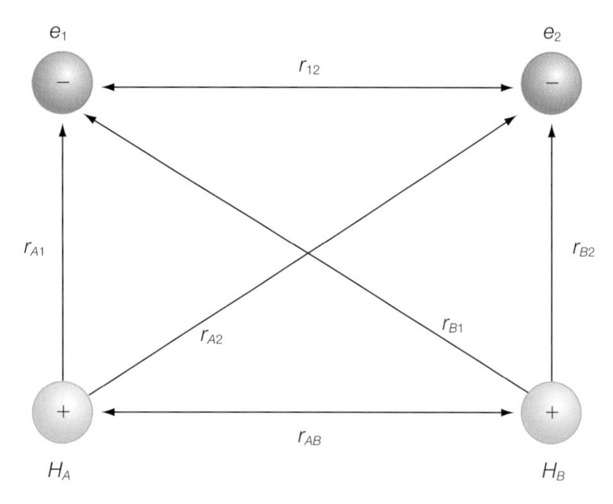

Figure 13.1
The potential of a hydrogen model is modeled as arising from electrostatic interactions involving the two electrons, e_1 and e_2, and two nuclei, H_A and H_B.

$$\text{Kinetic energy} = \frac{1}{2}mv^2 = \frac{1}{2m}(mv)^2 = \frac{p^2}{2m}$$

where p_1 and p_2 are the momenta of electrons 1 and 2, respectively. For four particles, there are a total of six interactions between individual particles. Each of the two nuclei can interact with the two electrons with a potential that decreases inversely with distance since the two charges have opposite signs. For example, the A nucleus and electron 1 have an electrostatic potential with a distance dependence of $-1/r_{A1}$. The other two electrostatic interactions are between electrons 1 and 2, with a distance dependence of $+1/r_{12}$, and two nuclei, with a distance dependence of $+1/r_{AB}$. In these latter two cases the sign of the interaction is positive as the two charges have the same sign. The total electrostatic potential for these four charges can be written as:

$$\text{Potential energy} = \frac{e^2}{4\pi\varepsilon_0}\left[-\frac{1}{r_{A1}} - \frac{1}{r_{A2}} - \frac{1}{r_{B1}} - \frac{1}{r_{B2}} + \frac{1}{r_{12}} + \frac{1}{r_{AB}}\right] \quad (13.2)$$

For two charges, q_1 and q_2, separated by a distance r_{12}

$$V(r) = \frac{q_1 q_2}{4\pi\varepsilon_0 r_{12}}$$

Now that the potential for the molecule has been established, we can write Schrödinger's equation. The wavefunction will be dependent upon six coordinates, the three that describe the first electron and the three that describe the second electron. For simplicity, this dependence of the wavefunction will be denoted as $\psi(r_1,r_2)$. Substituting the appropriate operators for the classical kinetic and potential energies yields Schrödinger's equation for the hydrogen molecule:

$$-\frac{\hbar^2}{2m}(\nabla_1^2 + \nabla_2^2)\psi(r_1,r_2) - \frac{e^2}{4\pi\varepsilon_0}\left[\frac{1}{r_{A1}} + \frac{1}{r_{A2}} + \frac{1}{r_{B1}} + \frac{1}{r_{B2}} - \frac{1}{r_{12}} - \frac{1}{r_{AB}}\right]\psi(r_1,r_2)$$
$$= E\psi(r_1,r_2) \quad (13.3)$$

Operators (Table 9.1)

$$p \rightarrow \frac{\hbar}{i}\nabla$$

$$r \rightarrow r$$

To solve this equation, we make some assumptions. The interactions between the two electrons, and the two nuclei, are assumed to be negligible compared to the interactions between the nuclei and electrons: this eliminates two of the six terms for the potential. With these assumptions, Schrödinger's equation for the hydrogen molecule can be written as:

$$-\frac{\hbar^2}{2m}(\nabla_1^2 + \nabla_2^2)\psi(r_1,r_2) - \frac{e^2}{4\pi\varepsilon_0}\left[\frac{1}{r_{A1}} + \frac{1}{r_{A2}} + \frac{1}{r_{B1}} + \frac{1}{r_{B2}}\right]\psi(r_1,r_2) = E\psi(r_1,r_2) \tag{13.4}$$

The wavefunction describing the two electrons is simplified as the product of wavefunctions for each electron $\psi(r_1,r_2) = \psi(r_1)\psi(r_2)$: the validity of this assumption will be addressed later. The equation can be rewritten by realizing that the derivative operators are specific to either electron 1 or 2, and so for each derivative operator part of the wavefunction can be considered to be a constant, yielding:

$$-\frac{\hbar^2}{2m}[\psi(r_2)\nabla_1^2\psi(r_1) + \psi(r_1)\nabla_2^2\psi(r_2)] - \frac{e^2}{4\pi\varepsilon_0}\left[\frac{1}{r_{A1}} + \frac{1}{r_{A2}} + \frac{1}{r_{B1}} + \frac{1}{r_{B2}}\right]\psi(r_1,r_2)$$
$$= E\psi(r_1,r_2) \tag{13.5}$$

This expression can be simplified by dividing the entire equation by $\psi(r_1)\psi(r_2)$ and grouping together the terms that depend upon r_1 and r_2:

$$-\frac{\hbar^2}{2m}\left[\frac{1}{\psi(r_1)}\nabla_1^2\psi(r_1) - \frac{e^2}{4\pi\varepsilon_0}\frac{1}{r_{A1}} - \frac{e^2}{4\pi\varepsilon_0}\frac{1}{r_{B1}}\right]$$
$$-\frac{\hbar^2}{2m}\left[\frac{1}{\psi(r_2)}\nabla_2^2\psi(r_2) - \frac{e^2}{4\pi\varepsilon_0}\frac{1}{r_{A2}} - \frac{e^2}{4\pi\varepsilon_0}\frac{1}{r_{B2}}\right] = E \tag{13.6}$$

These two equations can be separated by setting each of the individual parts equal to E_1 and E_2, which together add up to E. In this case, the association of the wavefunctions to the nuclei A and B are shown explicitly by the subscript.

$$-\frac{\hbar^2}{2m}\nabla_1^2\psi(r_1) - \frac{e^2}{4\pi\varepsilon_0}\left(\frac{1}{r_{A1}} + \frac{1}{r_{B1}}\right)\psi(r_1) = E_1\psi(r_1)$$

$$-\frac{\hbar^2}{2m}\nabla_2^2\psi(r_2) - \frac{e^2}{4\pi\varepsilon_0}\left(\frac{1}{r_{A2}} + \frac{1}{r_{B2}}\right)\psi(r_2) = E_2\psi(r_2) \qquad (13.7)$$

$$E = E_1 + E_2$$

Each of the two equations is dependent upon only a set of coordinates, allowing the wavefunctions to be solved explicitly. The resulting solutions will not be demonstrated here but the similarity of the equations to those for the hydrogen atom are consistent with the new solutions, behaving as modified forms of the hydrogen atom wavefunction due to the potential of the second proton included.

In solving this problem, there is always an ambiguity in identifying which electron is which. When the two atoms are far apart, the electrons are each localized around one nucleus. When the two atoms come within bonding distance, then the electrons can be found on either of the nuclei. These two electrons have identical properties and there is no identifiable property that would indicate that the two electrons were switched. Therefore, we must consider the two electrons to be identical. To reflect this property, the allowed wavefunctions are written as the linear combination of two solutions, with the second simply reflecting the possible switch:

$$\psi(r_1,r_2) = \psi_A(r_1)\psi_B(r_2) \pm \psi_B(r_1)\psi_A(r_2) \qquad (13.8)$$

These wavefunctions are substituted into Schrödinger's equation with one of the terms previously neglected: the interaction between the nuclei. The general dependence of this interaction on the separation between the nuclei, r_{AB}, arises from two contributions, the bonding and electrostatic interactions (Figure 13.2). As discussed for the multi-electron atoms, the two electrons will tend to be paired together, with opposite spin direction rather than having the same spin direction. Thus, the pairing of the electrons naturally results in a bond.

The bonding properties are a result of contributions of two factors. When two uncharged atoms are brought close together, their electrons interact. Random variations in the distributions of the electrons of one atom give rise to a transient dipole in a neighboring atom. The fluctuation-induced interaction becomes more negative as the separation decreases and results in a weak interaction, termed a London dispersion interaction, that has an approximate dependence of $1/r^6$ (Figure 13.2).

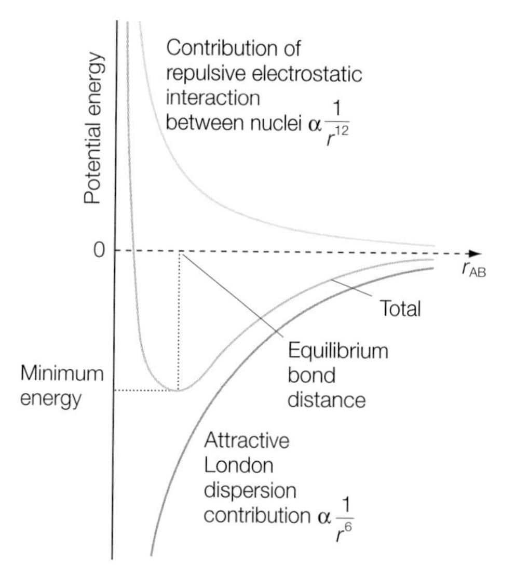

Figure 13.2 Potential energy for two atoms separated by a distance r, with both the attractive and repulsive contributions shown.

The second contribution is due to the repulsion between atoms at very short distances. The repulsion can be calculated using quantum mechanics as arising from trying to insert more than two electrons into one orbital. For the purpose of such calculations, simple models of electron-overlap repulsion are used. At long distances, the interactions are very weak and so the energy is considered to be zero. As the two atoms move closer together, the bonding interaction is favorable and the energy of the system decreases. When the two atoms are too close together, the electrostatic interaction becomes large and the overall energy becomes positive with a dependence of $1/r^{12}$. The use of these two potentials to describe the interactions between atoms is often termed the Lennard-Jones potential after Sir John Edward Lennard-Jones who first proposed its use. The combination of the attractive and repulsive effects results in an optimal separation distance between any two atoms and represents the ideal distance for a bond between the atoms. This net interaction has a minimum energy value at the equilibrium distance for the bond length.

Although this form of Schrödinger's equation can be solved, we have already been able to argue what the features of these wavefunctions should be. Likewise, we can argue that the energies for the molecular wavefunctions should have three terms. The first is the *energy of the isolated hydrogen atom, E_H*. The second and third energy terms arise from the bonding and the electrostatic contributions. The energies of these wavefunctions can be written as:

$$E_{\pm} = 2E_H + \frac{J \pm K}{1 \pm S^2} + \frac{e^2}{4\pi\varepsilon_0 r_{\mathrm{AB}}} \tag{13.9}$$

where J, K, and S are three constants representing the *bonding term*. The terms J and K reflect interaction energies involving the electrons and nuclei where the constant S is given by:

$$S = \int \psi_{\mathrm{A}} \psi_{\mathrm{B}} \, \mathrm{d}\tau \tag{13.10}$$

This parameter is a measure of the extent to which two atomic orbitals on different atoms overlap with each other. In order for the overlap to have a large value, the two wavefunctions must have large values at the same locations in space. The largest value of S is one, which occurs when the same wavefunction is substituted for both ψ_{A} and ψ_{B}. When two independent wavefunctions are substituted, such as the wavefunctions for a σ and π orbital, the value of S is the minimal value of zero, corresponding to no overlap.

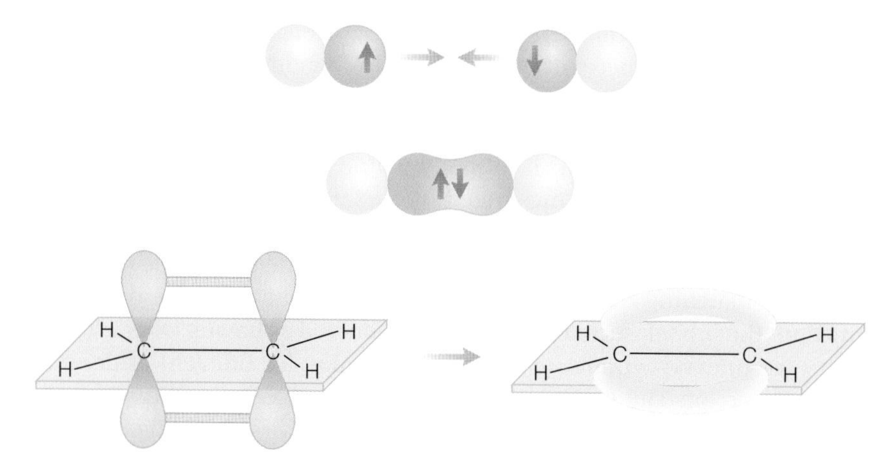

Figure 13.3 Molecular σ and π bonds arise from the overlap of atomic p and s orbitals.

VALENCE BONDS

In valence bond theory, a bond is formed when an electron in one of the atomic orbitals pairs its spin with the spin of an electron associated with another nucleus. Since the two electrons are identical, the overall wavefunction is written as a linear combination of individual wavefunctions (eqn 13.8). Since the molecular wavefunctions are written in terms of the atomic orbitals, the resulting molecular orbitals will also reflect the atomic orbitals. The lowest-energy solution of the hydrogen atom was the 1s orbital. The lowest-energy molecular orbital will be formed by a linear combination of two 1s orbitals and is called a σ bond (Figure 13.3). The individual 1s orbitals each have spherical symmetry. When these two atomic orbitals are combined the spherical symmetry is replaced by an axial symmetry. The molecular wavefunction also has an electron distribution reflecting these two states. For the higher-energy p atomic orbitals, the resulting molecular orbital will also have a distribution reflecting the individual orbitals forming a π bond.

The use of the linear combinations of wavefunctions results in the molecular wavefunctions having two different energies, with the extent of the energy difference being inversely proportional to $1 \pm S^2$. As the extent of the overlap for the two wavefunctions (S) increases, the energy difference between the two states also increases. The presence of the two energetic states (eqn 13.9) corresponds to what is commonly referred to as bonding and antibonding orbitals. The lower-energy bonding state is the bonding orbital and the higher-energy state is the antibonding orbital. A bonding orbital, such as a σ or π orbital, contributes to the overall strength of the bond when occupied. However, occupation of an antibonding orbital, such as a σ^* or π^* orbital, causes a decrease in the strength of a bond between two atoms. The symmetry of bonding and antibonding orbitals reflects the symmetry of the atomic orbitals and may possess an inversion symmetry. An orbital has inversion symmetry when after inversion through the center of the molecule, which formally would be the center of inversion, the orbital appears to be unchanged. For example, a σ_g orbital has such symmetry, as denoted by a subscript g to represent *gerade* symmetry (or even symmetry; *gerade* is the German word for even). When the same operation results in matching orbital values but with opposite signs for the wavefunction, the orbital, such as a σ_u orbital, is denoted as having *ungerade* symmetry (or uneven symmetry; *ungerade* is German for uneven) as denoted by a u subscript. The symmetry of these orbitals has a significant impact on their involvement in optical transitions as discussed in Chapter 14.

THE HÜCKEL MODEL

The concept of molecular orbital theory can be extended to large molecules of atoms. In 1931, Erich Hückel developed an approach that could be used for conjugated systems, for which the electron is considered free to travel throughout the length of a conjugated system. The molecular σ and π orbitals were treated as linear combinations of the atomic s and p orbitals, respectively. Several approximations are made to simplify the calculations. All carbon atoms are treated as being identical so that the contributing atomic orbitals are identical. The interactions between non-neighbors are neglected and the interactions between neighbors are reduced to a single parameter, β. The calculated energies can then be shown to split into two states that are separated in energy by the coupling parameter β. The resulting molecular orbitals are filled by the electrons of the conjugated system, starting from the lowest-energy state. The two critical states are the highest occupied molecular orbital, or HOMO, and the lowest unoccupied molecular orbital, or LUMO. These states are sometimes referred to as frontier orbitals of the molecule and are largely responsible for many of the chemical properties.

INTERACTIONS IN PROTEINS

The calculation of the electronic wavefunctions for molecules in proteins and other biological systems is difficult because proteins have heterogeneous surroundings. Also, the interactions with the surrounding amino acid side chains require an analysis of the protonation state of each group, which can be highly shifted from ideal solutions (Chapter 5). For this reason, the factors that give rise to a well-folded protein structure are complex and still impossible to predict based solely upon the protein structure alone.

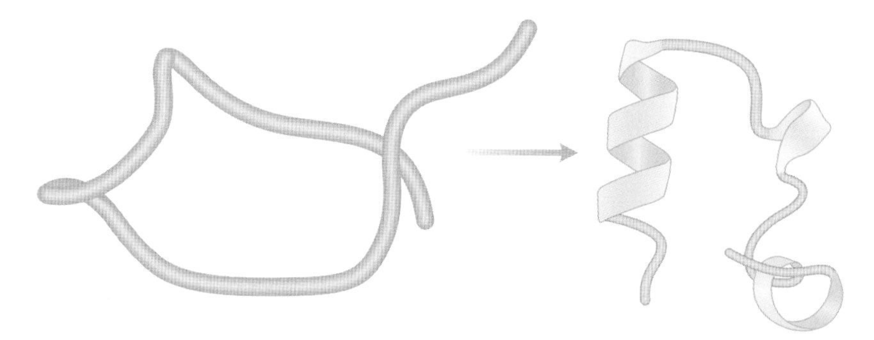

Figure 13.4 Proteins are expressed as polypeptide chains that fold into well-defined structures due to a large number of complex interactions involving the amino acid residues and the surrounding solvent.

The polypeptide chain interacts with the surrounding solution and begins to adopt a folded shape (Figure 13.4). A protein will always adopt a compact structure, forming a specific three-dimensional shape with a hydrophobic interior and hydrophilic exterior. After proper folding, the protein will have an active site that can perform a specific function. For example, enzymes have binding sites for substrates that are highly selective in, as well as being extremely efficient in, performing a catalytic reaction. For some proteins, additional polypeptide chains or cofactors must come together before the protein becomes active.

Although the structure of a protein cannot be predicted based upon its sequence, the properties of a protein can be understood once the structure has been determined, by examining the different interactions that would stabilize the structure and contribute to its function (Figure 13.5). The structure of a protein can be understood in terms of a conceptual hierarchy, which is influenced by different interactions. The sequence of amino acid

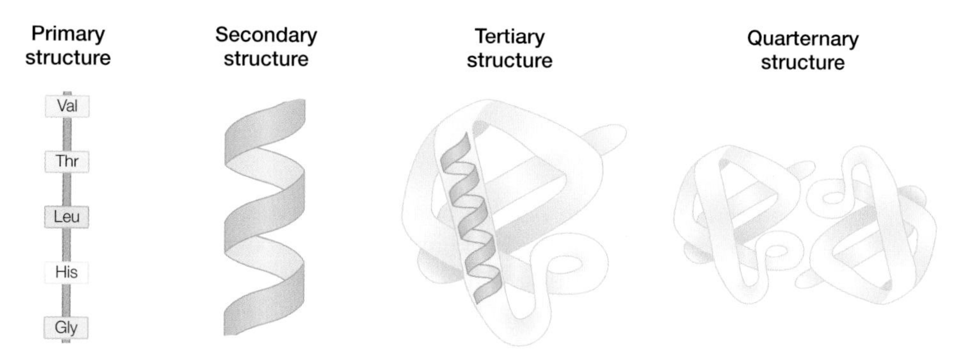

Figure 13.5 The structure of a protein can be understood in terms of different levels. The primary structure consists of the sequence of the amino acid residues linked by peptide bonds, the secondary structure can be organized into α helices and β sheets, the secondary structure folds into the tertiary structure, and the organization of different peptide chains describes the quaternary structure.

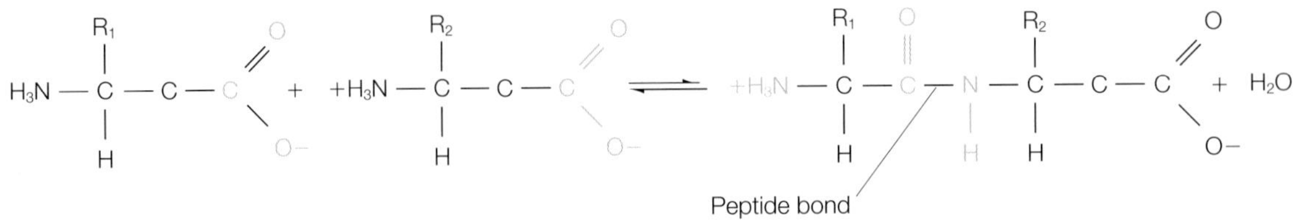

Figure 13.6 Formation of the peptide bond and the resulting polypeptide chain.

residues linked together by peptide bonds describes the primary structure of the protein. Localized regions of the protein fold into well-defined and stable arrangements of amino acid residues giving rise to structural patterns that are recurring in proteins, namely α helices and β sheets. The overall fold of these secondary-structure elements describes the tertiary structure of the protein. The quaternary structure is the arrangement of different polypeptides of the protein. In addition, proteins can also be described as having motifs; that is, arrangements of elements that are common in different proteins.

Peptide bonds

Critical for amino acids is their ability to form peptide bonds in which the carboxyl group of one amino acid is joined to the amino group of another amino acid, resulting in the loss of a water molecule (Figure 13.6). Many amino acids joined by peptide bonds form a polypeptide chain. This chain consists of the regularly repeating main chain or backbone and the side chains. The peptide unit is always rigid and planar with the hydrogen of the amino group opposite to the oxygen of the carbonyl group, with the exception of bonds involving proline. The bond between the carbonyl carbon atom and the nitrogen atom is not free to rotate, owing to the presence of the partial double-bond character. However, the bonds at the end of the peptide unit are free to rotate, allowing polypeptide chains to form a wide range of three-dimensional protein structures.

Steric effects

The nature of the peptide bond plays a critical role in the conformation of the protein. Of the backbone atoms, the oxygen has a partial negative charge and the nitrogen has a partial positive charge establishing a dipole moment. The six atoms of the peptide group lie in a plane and the chain can be considered to consist of a series of planes that can rotate relative to each other (Figure 13.7). By convention, the bond angles arising from rotations at the C_α position are identified as ϕ rotations and ψ rotations are those centered around the N–C_α bond. Both angles are defined to have a value of 180° when the polypeptide is in its fully extended conformation.

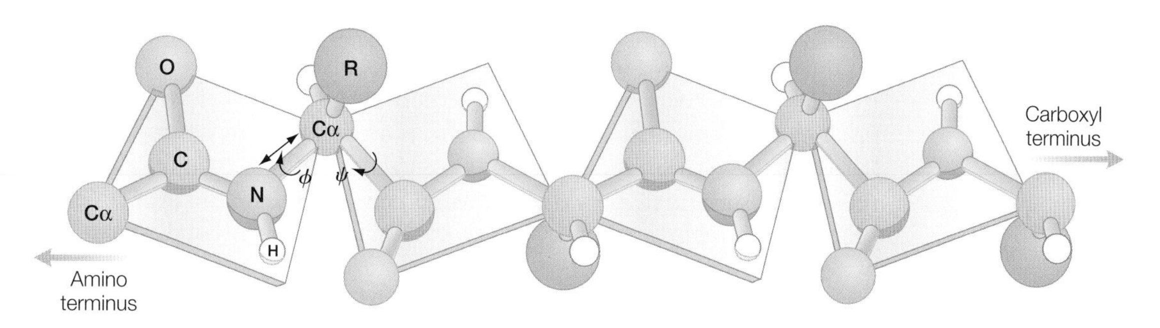

Figure 13.7 The planar peptide groups of a polypeptide chain. Each plane can rotate about the ϕ and ψ angles. Many combinations of angles are forbidden, including $\phi = \psi = 0°$.

Many rotation angles are prohibited due to steric hinderence between atoms in the polypeptide backbone and amino acid side chains. For example, the conformation in which both angles are zero is not allowed, owing to a steric overlap between the α-carbonyl oxygen and an α-amino hydrogen atom. Allowed values for these angles are graphically represented on a Ramachandran plot, which also identifies regions of secondary structure. Since proteins are usually compact structures, steric interactions between different parts of the protein that are spatially close also play a crucial role in the folding of the polypeptide.

Hydrogen bonds

Hydrogen bonding is an interaction between a proton donor and acceptor, typically involving oxygens and nitrogens in proteins (Figure 13.8). Hydrogen bonds arise from the electrostatic interaction between the partial negative charge of the acceptor and the partial positive charge of the donor. The strength of the hydrogen bond will depend upon the relative angles and distances. A typical distance between a hydrogen-bond donor and acceptor is 1.8 Å from the hydrogen to the acceptor, or about 2.8 Å between the nuclei of the donor and acceptor. In most protein structures the presence of hydrogen bonds can only be inferred from the structure. However, both nuclear magnetic resonance (NMR) and infrared spectroscopy provide a means to unambiguously identify hydrogen bonds (Chapters 15 and 16).

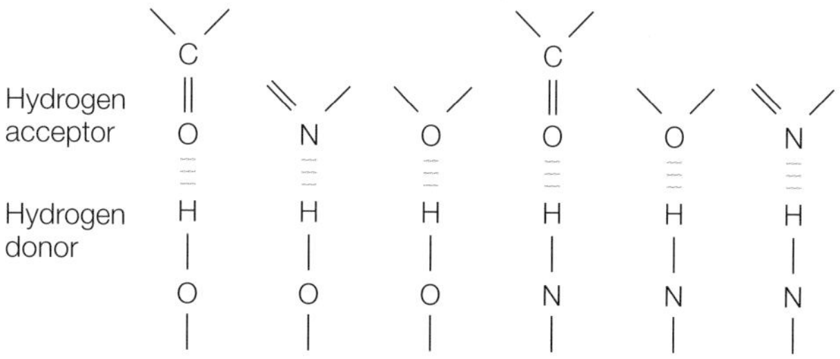

Hydrogen acceptor	C=O	N	O	C=O	O	N
Hydrogen donor	H–O	H–O	H–O	H–N	H–N	H–N

Figure 13.8 Common hydrogen bond donors and acceptors.

Hydrogen bonds play a critical role in many biological systems. Hydrogen bonds involving the main-chain atoms establish the stability of the secondary structures of proteins. In an α helix, the backbone is tightly wound around a central axis with a full turn for every 3.6 amino acid residues. For each amino acid residue forming the helix, except for the end residues, the hydrogen attached to the nitrogen of the backbone forms a hydrogen bond with the carbonyl oxygen of the amino acid residue four positions along the chain. Likewise, β sheets are stabilized by hydrogen bonds between the nitrogens and carbonyl oxygens of the backbone, although there is no specific sequence relationship between the amino acid residues forming each hydrogen bond. The recognition of hydrogen-bonding patterns by James Watson and Francis Crick were key to understanding the DNA structure. The double helix of DNA is stabilized by hydrogen bonds, with two hydrogen bonds found between every adenine and thymine and three hydrogen bonds between guanine and cytosine. In addition the model in which DNA could break the hydrogen bonds between the base pairs, as well as the more nonspecific base-stacking interactions and subsequent reform, was key to understanding the replication process.

Electrostatic interactions

The side chains of the amino acid residues lysine, arginine, glutamate, aspartate, and histidine are ionizable and hence electrostatic interactions contribute to protein stability and function. In a typical electrostatic model, the potential between *two charges*, q_1 and q_2, that are separated by a *distance*, r, is given by:

$$V(r) = \frac{q_1 q_2}{4\pi\varepsilon r} = (1389\ \text{kJ mol}^{-1})\frac{q_1(e)q_2(e)}{\varepsilon r(\text{Å})} \tag{13.11}$$

where ε is the *dielectric constant*. The dielectric constant of a vacuum is defined as 1.0 and its value in different solvents ranges from 80 in a polar solvent such as water to 2 for a nonpolar solvent such as benzene. The large value for water plays a critical role as it results in a decreased interaction between charges compared to a nonpolar solvent. In a protein, calculation of the value is made complex by the nonuniformity of both the composition and distribution of atoms in a protein. In addition, the surroundings will respond to the presence of a charge, leading to the reorientation of nearby dipoles and hydration of charges on the surface.

Hydrophobic effects

The distribution of amino acid residues in a protein usually ranges from very hydrophobic, with residues such as phenylalanine and isoleucine in the interior, to hydrophilic, with residues such as glutamate and aspartate on the surface. Hydrophobic factors also play a role in protein folding and stability. For membrane proteins, not only are the interior amino acid residues hydrophobic but also those residues are in contact with the lipid bilayer. The strength of hydrophobic interactions does not arise from direct forces involving nonpolar molecules but rather from the thermodynamics

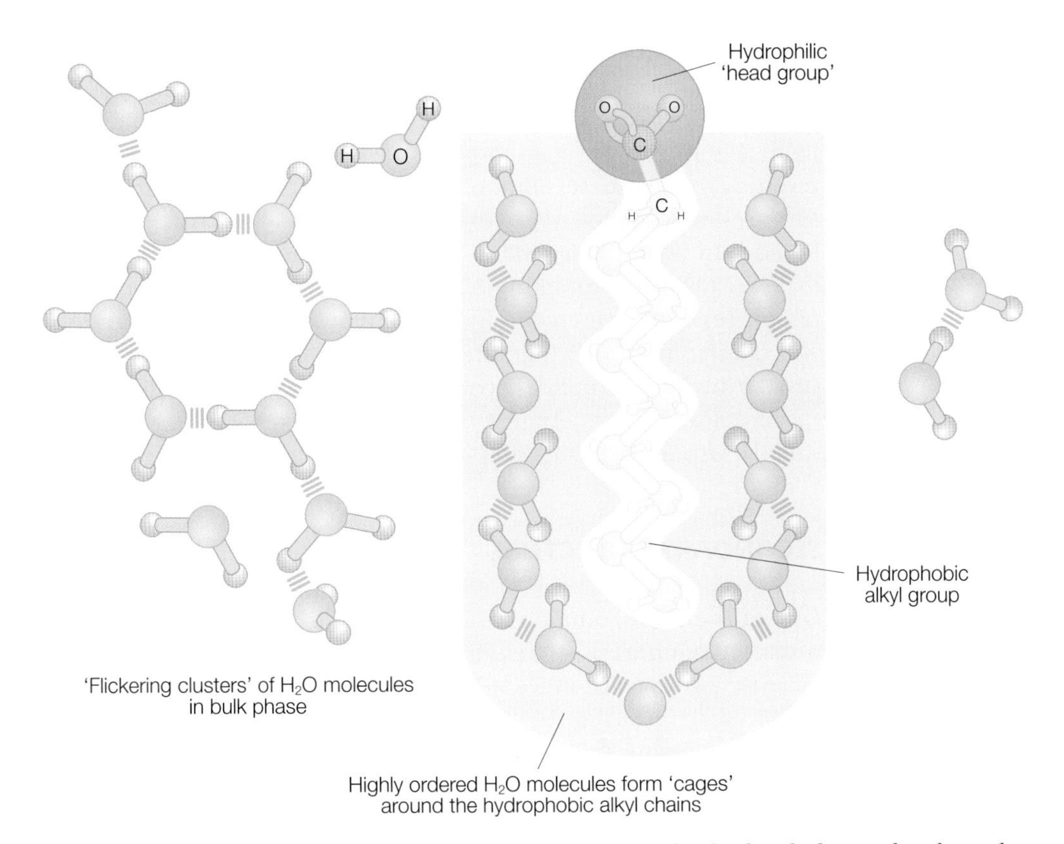

Figure 13.9 Highly ordered water molecules will form around a hydrophobic molecule such as a lipid.

involved in the introduction of a nonpolar molecule in a polar solvent. The Gibbs energy change has contributions from changes in both the enthalpy and entropy:

$$\Delta G = \Delta H - T\Delta S \tag{13.12}$$

The introduction of an ion into water will disrupt the hydrogen-bonding interactions of the water molecules in the vicinity of the ion (Figure 13.9). The loss of bonds will be compensated for by interactions between the ion and water, making the change in the Gibbs energy small. The introduction of a nonpolar molecule, such as a lipid, will also disrupt the hydrogen bonds but there will be no compensation for the loss of bonds. In addition, water molecules near the lipid will be constrained by the presence of the lipid and will form a highly ordered cage. For hydrocarbons in water, the change in enthalpy may be positive or negative depending upon how the newly formed hydrogen bonding compensates for the loss of bonding. However, the entropy is always more negative due to the ordering. The large decrease in the entropy dominates, and the change in the Gibbs energy is positive. Thus the hydrophobic effect of nonpolar molecules in water is a process driven by entropy rather than by enthalpy.

Only when the number of nonpolar molecules increases above a critical point will the tendency to disperse the nonpolar molecules be overcome, resulting in phase separation, with the lipids forming structures such as micelles (Chapter 4). Hydrophobic effects also participate in protein folding. In an unfolded state, the hydrophobic amino acid residues are exposed to the surrounding water. Since the sequestering of the hydrophobic residues is energetically more favorable, the protein conformation will change into a configuration with a hydrophobic interior in a process termed a hydrophobic collapse. Since the hydrophobic effect is not specific, this conformation is not unique but represents an intermediate state before other interactions stabilize the final state.

SECONDARY STRUCTURE

One of the most common arrangements for proteins is a compact structure consisting of α helices. In these proteins, the α helices are packed pairwise against each other so that one side of each helix provides a hydrophobic surface facing the interior, and the other side is hydrophilic and faces the aqueous solution. The classic examples of globular proteins with α helices are myoglobin and hemoglobin, which were the first proteins to have their structures solved in the 1950s by John Kendrew and Max Perutz (Figure 13.10). These proteins also contain a heme cofactor, which serves as the binding site for oxygen as the proteins transport oxygen in the muscles and cardiovascular system. Hemoglobin serves to bind oxygen under high oxygen levels in the lungs, carry the oxygen through the bloodstream, and release it to myoglobin in the muscles. The oxygen is bound to hemoglobin until it is needed for aerobic work. The binding of oxygen to the four hemes in hemoglobin is regulated allosterically by interactions among the four polypeptide chains.

Another common protein structure is the β structure, formed by antiparallel β strands. The simplest arrangement is obtained when each successive β strand is added adjacent to the previous strand, with a small twist yielding a β barrel if the protein is cylindrical, or a β sandwich, as found for more elongated proteins such as the Fenna–Matthews–Olsen, or FMO, protein (Figure 13.11). These proteins can bind a cofactor in the center, which is well sequestered from the surrounding environment. For example, the Fenna–Matthews–Olsen protein binds seven bacteriochlorophylls that are part of the light-harvesting pathway in photosynthetic green bacteria. Notice that, in addition to the β sheets, there are several α helices; the presence of both types of secondary structure is very common in proteins. The role of the seven bacteriochlorophylls in transferring light energy is discussed in Chapter 14.

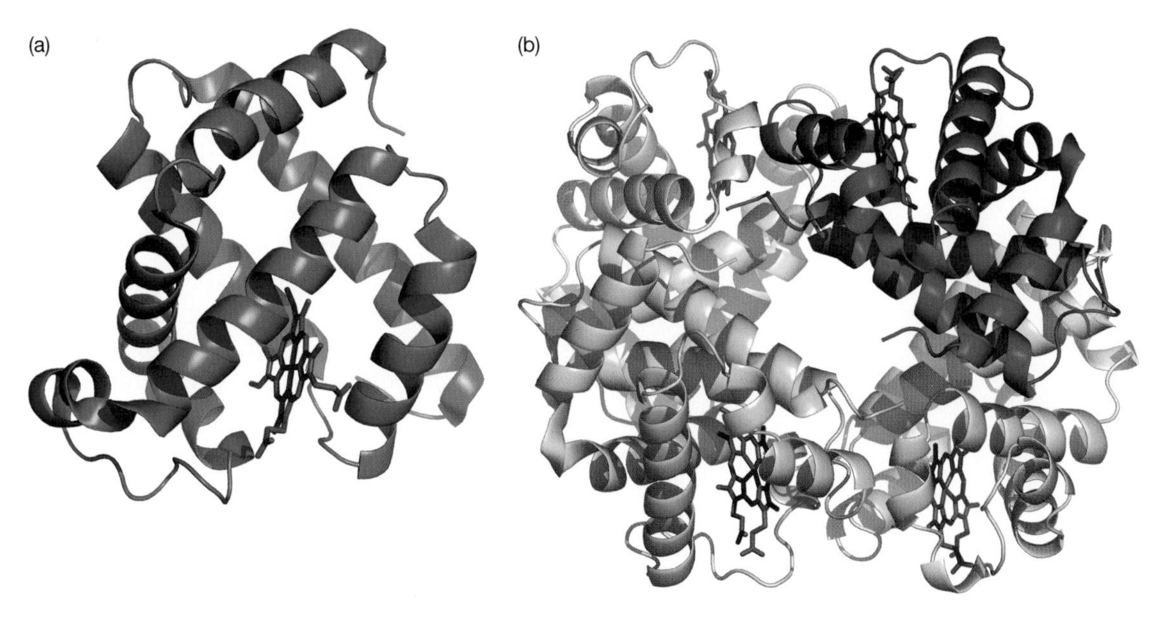

Figure 13.10 Three-dimensional structures of (a) myoglobin and (b) hemoglobin showing the extensive presence of α helices surrounding the heme groups. These were the first protein structures solved; for their efforts Max Perutz and John Kendrew received the Nobel Prize in Chemistry in 1962.

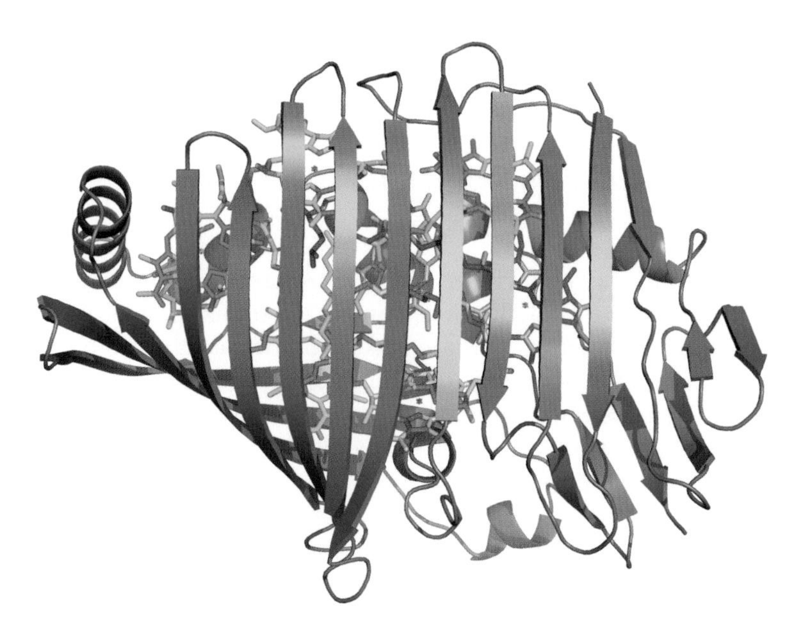

Figure 13.11 Fenna–Matthews–Olsen (FMO) protein with the β sandwich surrounding seven bacteriochlorophylls.

DETERMINATION OF SECONDARY STRUCTURE USING CIRCULAR DICHROISM

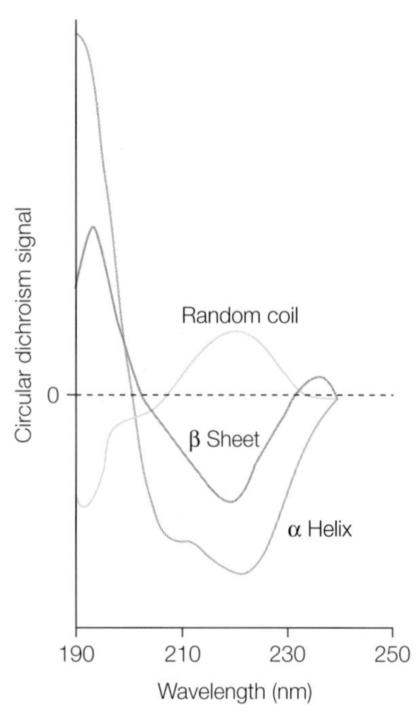

Figure 13.12 Circular dichroism spectra of α helices, β sheets, and random coils.

Chiral molecules can be distinguished based upon their interaction with polarized light. Light can be linearly polarized, in which case the electromagnetic fields oscillate back and forth along a line. Alternatively, light can be polarized circularly when the field rotates as the light propagates, so that when the light is viewed down the path of travel, the electromagnetic fields are observed to rotate in a circle in either a right-handed or left-handed direction. For chiral molecules, right- and left-handed circularly polarized light travels differently in an effect termed circular birefringence. As a result, chiral molecules show *circular dichroism*, namely *a difference in their absorbance for left- and right-handed polarized light*. In a circular dichroism spectrum, the wavelength dependence is measured for the difference in the absorbance for the right- and left-handed polarized light. For proteins, the effect of this optical activity is evident in a circular dichroism spectrum that is largely determined by the secondary structure of the protein (Figure 13.12). For example, α helices show a much stronger peak near 190 nm than β strands or random coils. By fitting a spectrum of a protein, it is possible to estimate the relative contributions of secondary structure. In addition to being used for structure prediction, circular dichroism is a useful measure of the extent to which a previously unfolded protein has been folded, and thus can be used in thermal or chemical denaturation studies.

RESEARCH DIRECTION: MODELING PROTEIN STRUCTURES AND FOLDING

Identification of the interactions that stabilize protein structures has provided the framework for the development of computational models of protein structure. Such models are becoming increasingly more sophisticated and are now routinely run for protein-structure determination (Chapter 15). To provide an accurate representation of the protein, these models include terms that reflect bond stretching, bending, and rotation. Although bond lengths and angles are formally determined by interactions of electrons and nuclei as described by quantum mechanics, these interactions can be treated by simple physical models. For example, the *bond-stretching potential*, $V(r)$, is determined by calculating the distance for each covalent bond, r, and comparing that distance to an *ideal value*, $r_{standard}$ (eqn 13.13). A similar expression can be written for the bending involving each angle θ that can be defined in terms of two neighboring bonds. Bond rotation through an angle φ is modeled as having a periodic dependence on the angle that corresponds to the different orientations of the substitutents. For example, the dependence for two tetrahedral carbons will show three minima corresponding to the substitutents being *trans* ($\varphi = 180°$), *gauche*$^+$ ($\varphi = 60°$), or *gauche*$^-$ ($\varphi = -60°$), when modeled with a $\cos(3\varphi)$ dependence (eqn 13.13). Torsion angles can also be defined in terms of three consecutive bonds, with a much larger potential that is associated with rotation around a double bond compared to a single bond. In each case, suitable constants must be determined that result in the potentials having similar values. The constants will always be positive as the potential will increase as the distances and angles are perturbed away from the standard values.

$$
\begin{aligned}
V(r) &= k_r(r - r_{standard})^2 \\
V(\theta) &= k_\theta(\theta - \theta_{standard})^2 \\
V(\phi) &= \frac{V_\phi}{2}(1 - \cos 3\phi)
\end{aligned}
\tag{13.13}
$$

Initial models of proteins made use of the ideal that the polypeptide could be modeled as a series of angles along the backbone with the individual bond distances held fixed, but mathematically these were not robust as rotations at the beginning of the polypeptide chain result in much larger effects than equivalent rotations at the terminus region. Instead, the calculation involves using N atoms that can move independently but are constrained by various interactions, including the potentials due to bonds. Thus, each atom is uniquely identified, has specific interactions with every ith atom, and experiences a potential given by an expression such as:

$$
\begin{aligned}
V_i = &\sum_j k(r_{bond_{ij}} - r_{eq})^2 + \sum_j k(\theta_{bond_{ij}} - \theta_{eq})^2 + \sum_j \frac{V_\phi}{2}(1 - \cos 3\phi_{bond_{ij}}) \\
&+ \sum_j \frac{B_{ij}}{r_{ij}^{12}} - \sum_j \frac{A_{ij}}{r_{ij}^{6}} + \sum_j \frac{Cq_iq_j}{r_{ij}}
\end{aligned}
\tag{13.14}
$$

The expectation for a protein is that the structure will adopt a conformation that represents the lowest-possible-energy state as given by the total potential energy. For a protein consisting of N atoms, each atom will be described by its position, generating *three parameters*, x, y, and z, and usually by at least one more parameter reflecting the motion of the atom, yielding a total of $4N$ parameters that must be determined to model the structure. For large proteins, the resulting possible values of the potential, termed an energy landscape, will have a large number of false minima but ideally only one true minimum (Chapter 8).

If the potential expression is subjected to a simple least-squares approach, then the potential will decrease until it reaches one of the false minima. Once in the false minimum, the calculated potential will never change as that would require allowing an increase in the potential. To reach the true minimum, the calculations have been modified to allow for an increase in the potential for a short time. One such approach is through a molecular-dynamics calculation in which the atoms are subject to motion that allows the potential to temporarily increase. In these models, atoms are free but constrained by forces. The *force on the ith atom, F_i*, is given by the product of the mass, m_i, and the acceleration, a_i, which can be written in terms of the position, r_i, or the potential V_i:

$$\vec{F}_i = m_i\vec{a}_i = m_i\frac{\mathrm{d}^2\vec{r}_i}{\mathrm{d}t^2} = -\frac{\partial V_i}{\partial \vec{r}_i} \tag{13.15}$$

$$a = \frac{\mathrm{d}r^2}{\mathrm{d}t^2}$$

$$V = -\int F\,\mathrm{d}r \text{ or } F = -\frac{\mathrm{d}V}{\mathrm{d}r}$$

The atoms are assigned velocities derived by assigning a distribution of velocities. Typically a Maxwellian distribution is used at a certain temperature (Figure 13.13):

$$f(v_i) = \left(\frac{m_i}{2\pi k_B T}\right)^{3/2} e^{-\frac{3m_iv_i^2}{2k_BT_i}} \tag{13.16}$$

This distribution has an average velocity of zero, and equal proportions of positive and negative velocities. As the assigned temperature is increased, the average velocity remains zero, but the distribution broadens with a larger percentage of higher velocities.

Once the forces and velocities are established, the system is allowed to evolve for a short time period of about 1 ps and the process is repeated again using the new positions. During each time step, once the velocities are assigned, atoms are allowed to move in *small time steps*, Δt:

$$\Delta x_i = v_{Ii}(\Delta t) - \nabla_{x_I}(V_{total})\frac{\Delta t^2}{2m_i} \tag{13.17}$$

After a specified number of time steps, the velocities are rescaled and the potentials are recalculated. At certain time points, the geometry may be optimized to avoid any

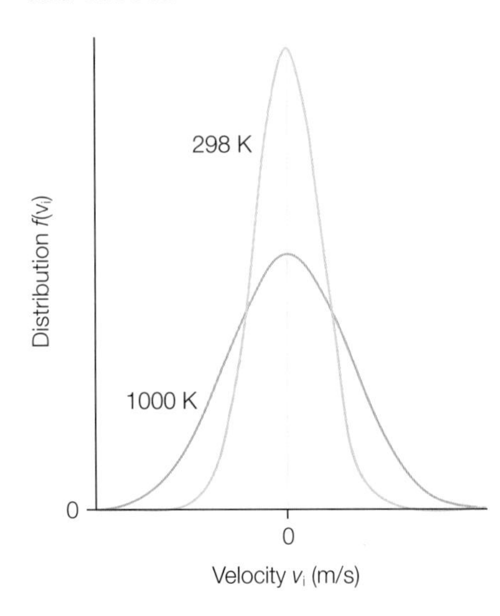

Figure 13.13 Maxwellian distributions of velocities at 298 and 1000 K.

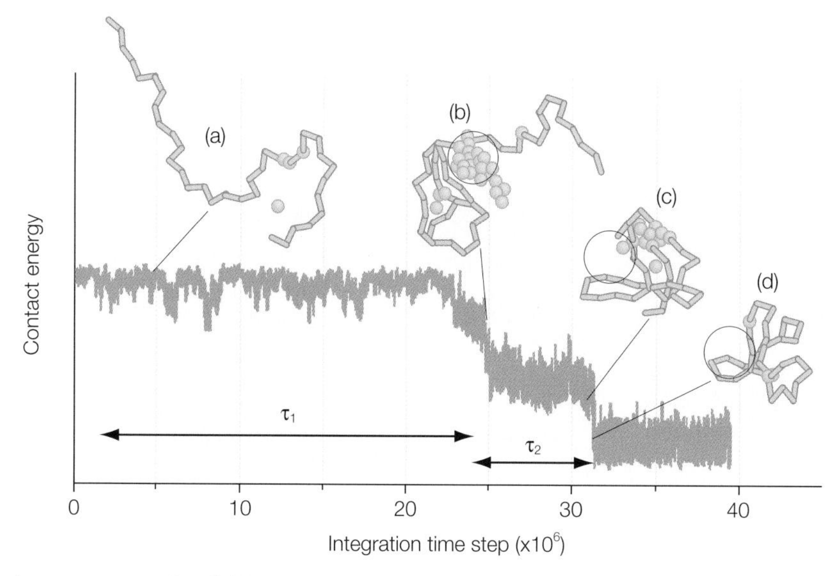

Figure 13.14 The folding trajectory for a protein from an unfolded to a folded configuration, (a) to (d). Modified from Onuchic and Wolynes (2004).

atoms from moving beyond realistic bond distances and angles. For refinement of X-ray data, the positions are also adjusted to agree with the diffraction data, by effectively adjusting the position of each atom so that it is located in a position of local electron density. For NMR data, distance constraints can be similarly imposed. Then the process repeats, typically in 100 increments for refinement and much longer for dynamics. As the trajectory progresses, the protein changes from an unfolded configuration to a globular structure (Figure 13.14).

In a molecular-dynamics calculation, a very high temperature is poised in order to provide enough energy to overcome local barriers. After a period involving hundreds to thousands of steps the temperature is cooled, allowing the structure to approach a true equilibrium. The temperatures are not intended to be realistic. For example, a temperature of 3000 K may be used, but this only provides a parameter to adjust the overall allowed degree of motion. Once the dynamics have been completed, the temperature is adjusted to 295 K, either instantly or very slowly. During refinement, in order to avoid the possibility of incorrect motion due to incorrect assignment of the charges of the amino acid side chains, the electrostatic term is not normally used.

In addition to being of use in refining structural models of proteins and probing their dynamics, computer models allow scientists to address one of the outstanding questions in structural biology, namely how the information necessary for proper folding is stored in the primary sequence of a protein (Daggett & Fersht 2003; Kuhlman et al. 2003; Clark 2004; Onuchic & Wolynes 2004; Lindorff-Larsen et al. 2005; Schueler-Furman et al. 2005). Proteins are thought to fold into intermediate states before reaching their

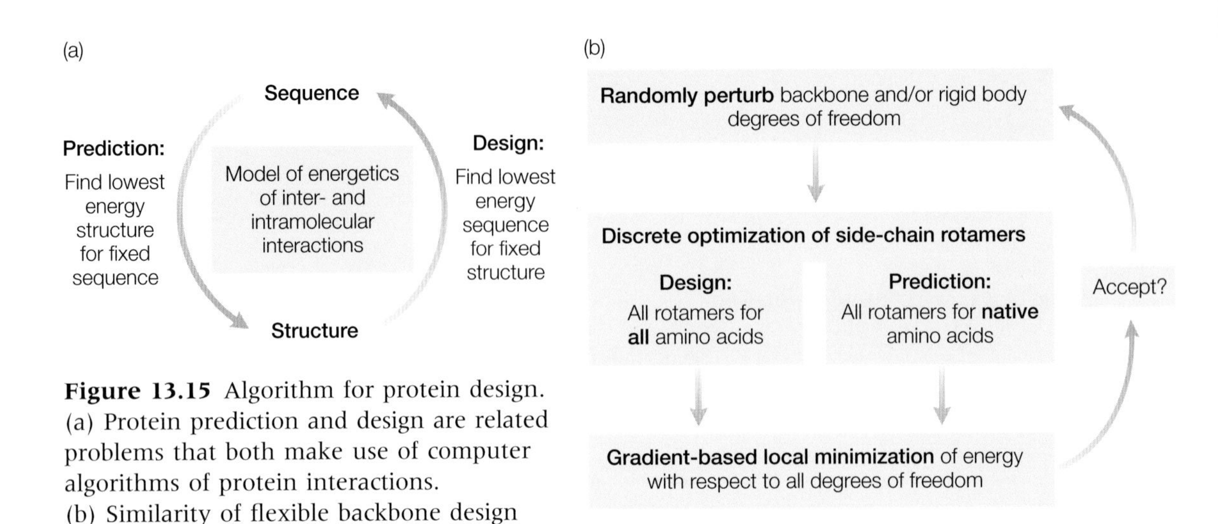

Figure 13.15 Algorithm for protein design. (a) Protein prediction and design are related problems that both make use of computer algorithms of protein interactions. (b) Similarity of flexible backbone design and structure prediction. Modified from Schueler-Furman et al. (2005).

final fold. Two general interactions influence the intermediate state. The overall globular shape reflects the sequestering of hydrophobic residues in the interior away from water molecules that minimize the unfavorable hydrophobic interactions. The presence of secondary structures, in particular α helices, reflects the formation of hydrogen bonds among neighboring amino acid residues. The intermediate state effectively provides the template for the subsequent formation of the fully folded conformation.

The prediction and design of the three-dimensional structures of proteins provides the opportunity to test the accuracy of our understanding of the interactions that underlie protein folding. Many of these approaches have focused on the interactions involved in the formation of the intermediate state, hydrophobic and hydrogen-bonding interactions, and packing interactions. Whereas the prediction algorithms identify the lowest-energy state of a given sequence, and in some cases an initial model, the design of a fold must involve the determination of the sequence with the lowest energy for the given structure (Figure 13.15). The successful design of new proteins has demonstrated the power of this methodology. The sequence with the lowest energy, computed for the fold of a small naturally occurring protein called a zinc finger, was found to adopt a structure that was found by NMR (Chapter 16) to be highly homologous to the starting sequence (Dahiyat & Mayo 1997). By iterating between the design and prediction algorithms, a new protein was designed that had a novel topology which proved to be very stable (Figure 13.16).

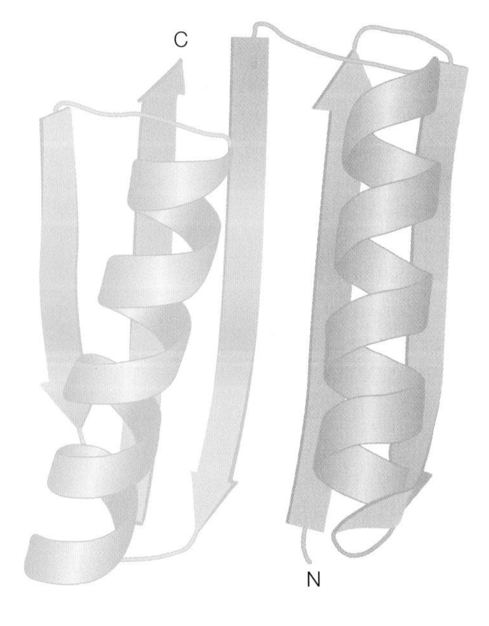

Figure 13.16 Three-dimensional structure of a novel design protein. Modified from Kuhlman et al. (2003).

Due to the complexity of proteins, challenges remain in the accurate modeling of protein interactions and their use in protein design and folding. As these algorithms become increasingly accurate, protein folding should become predictive, with the potential of designing enzymes that are capable of reactions that are not catalyzed by naturally occurring proteins.

REFERENCES

Clark, P.L. (2004) Protein folding in the cell: reshaping the folding funnel. *Trends in Biochemical Sciences* **29**, 527–34.

Daggett, V. and Fersht, A.R. (2003) Is there a unifying mechanism for protein folding? *Trends in Biochemical Sciences* **28**, 18–25.

Dahiyat, B.I. and Mayo, S.L. (1997) De novo protein design: fully automated sequence selection. *Science* **278**, 82–7.

Kuhlman, B., Dantas, G., Ireton, G.C. et al. (2003) Design of a novel globular protein fold with atomic-level accuracy. *Science* **302**, 1364–8.

Lindorff-Larsen, K., Rogen, P., Paci, E., Vendruscolo, M., and Dobson, C.M. (2005) Protein folding and the organization of the protein topology universe. *Trends in Biochemical Sciences* **30**, 13–19.

Onuchic, J.N. and Wolynes, P.G. (2004) Theory of protein folding. *Current Opinion in Structural Biology* **14**, 70–5.

Schueler-Furman, O., Wang, C., Bradley, P., Misura, K., and Baker, D. (2005) Progress in modeling of protein structures and interactions. *Science* **310**, 638–42.

PROBLEMS

13.1 Write the classical expression for kinetic energy for a molecule with two nuclei and two electrons, assuming that the nuclei are stationary.

13.2 Write the classical expression for potential energy for a molecule with (a) two nuclei, A and B, and two electrons, and (b) two nuclei, A and B, and three electrons.

13.3 Write Schrödinger's equation for a H_2 molecule. Assume that the nuclei do not interact.

13.4 Write Schrödinger's equation for a He_2^+ molecule.

13.5 Electrons must be considered as identical particles: how does this influence the wavefunctions describing molecules?

13.6 Why does the interaction between two atoms have an attractive $1/r^6$ dependence at large distances?

13.7 In a Hückel model, what effect does an increase in coupling have on the energetics?

13.8 Qualitatively, how are σ and π molecular orbitals related to atomic orbitals?

13.9 How is a peptide bond formed?

13.10 In a solvent such as benzene the dielectric constant is 4 compared to 78.5 for water. In which system would an electrostatic interaction between two charges be more pronounced?

13.11 Since the interactions, such as hydrogen bonds, found in proteins, are weak compared to covalent bonds, what stabilizes the folding of a protein?

13.12 Why are the (ϕ,ψ) coordinates of proteins found in only certain regions of a Ramachandran plot?

13.13 Since hydrogen bonds to water are not possible in the membrane, how can hydrogen bonds be formed in the cell membrane for membrane proteins?

13.14 How is the secondary structure influenced by the presence of hydrogen bonds?
13.15 What happens when a charge is generated in a protein?
13.16 Give two examples of proteins with α helices.
13.17 Why does hemoglobin have four polypeptide chains?
13.18 How is an energy landscape generated from consideration of the interactions in proteins?
13.19 How are molten globular states described by energy landscapes?
13.20 How is a distribution of velocities generated during a molecular-dynamics calculation?

14

Electronic transitions and optical spectroscopy

THE NATURE OF LIGHT

What is light? Michael Faraday and James Maxwell realized in the 1800s that light is an electromagnetic wave and they were able to develop a set of equations, known as Faraday's law and Maxwell's equations, that describe the properties of light. Key to the development of these equations was the ideal that changing electric fields give rise to changing magnetic fields and conversely changing magnetic fields give rise to changing electric fields. Light can be thought of as an oscillation that has a magnetic field and an electric field and can propagate through space (Figure 14.1). The peak-to-peak distance is the wavelength, and the number of times that the light wave oscillates per second is the frequency. Light has two waves, representing the electric-field and the magnetic-field components, which are perpendicular to one another and 90° out of phase. Since light is nothing more than an exchange between an electric and a magnetic field, it is a form of pure energy with no mass that must travel at a specific speed, c, of 2.988×10^8 m s^{-1} in a vacuum.

Because light is an oscillating field, it has a frequency, ν, of oscillation. Since its speed is constant, light will always cover a certain distance within one oscillatory cycle, which is the wavelength, λ. A high frequency corresponds to a short wavelength as the wavelength and frequency are inversely related according to:

$$\lambda \nu = c \tag{14.1}$$

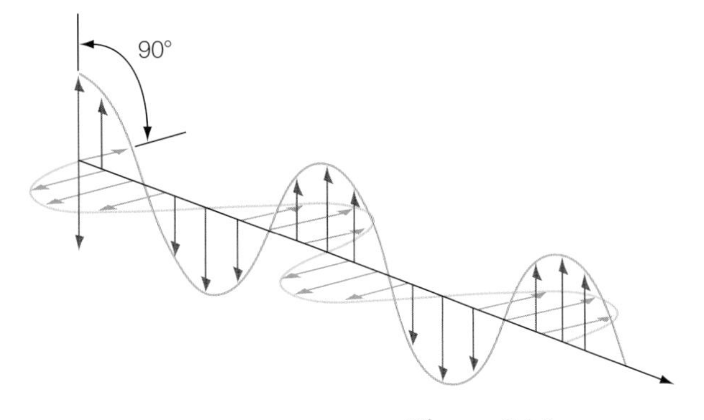

Figure 14.1
A schematic view of the light propagation of an electromagnetic field.

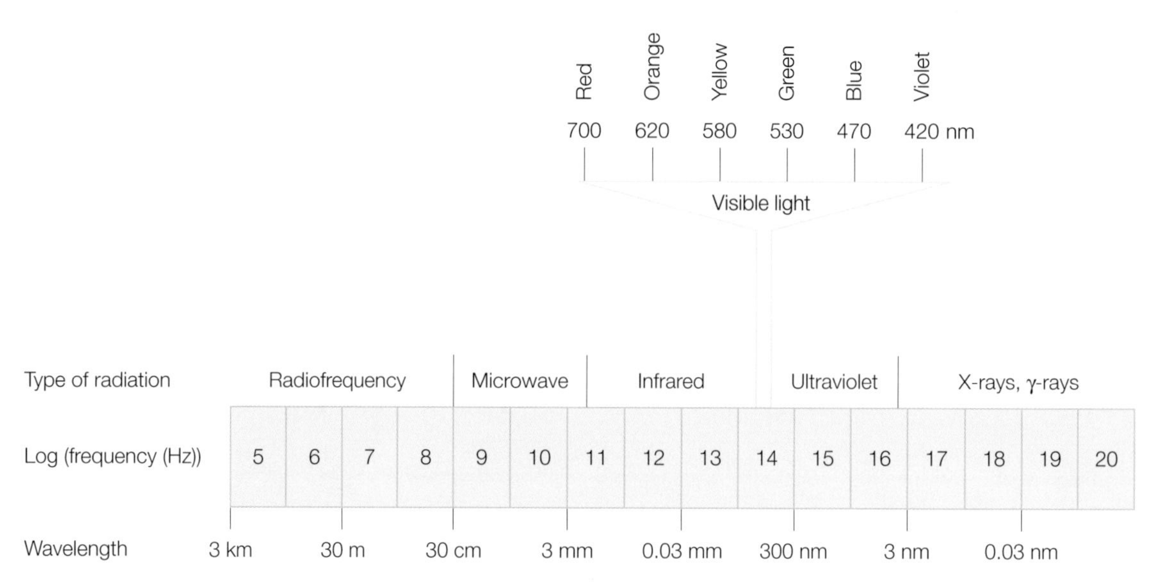

Figure 14.2 The electromagnetic spectrum of light.

Due to the quantization of energy (Chapter 9), the amount of energy in a wave of light is proportional to its frequency, with higher frequencies of light having higher energies:

$$E = h\nu \tag{14.2}$$

The number h is called Planck's constant and has the value of 6.626×10^{-34} Js.

Whereas the speed of light is restricted, light can have almost any energy (Figure 14.2). Radiowaves are low-energy electromagnetic radiation that travels through space between the radio station and your radio. Radiowaves are also used when performing nuclear magnetic resonance (NMR) experiments (Chapter 16). The frequencies for this type of radiation are in the order of 1–100 MHz; that is, 10^6 Hz or millions of times per second. Light with frequencies in the gigahertz range, or billions of oscillations per second, lies in the microwave region. Microwaves are used to cook foods in microwave ovens, to detect objects at long distances by radar, and to measure the properties of electrons in electroparamagnetic resonance (EPR; Chapter 16). At somewhat higher energies we have infrared light, with frequencies of roughly 10^{14} Hz that are characteristic of molecular vibrations (Chapter 11), then visible red light, visible orange, yellow, green blue violet, and the ultraviolet spectrum. At higher frequencies yet, there are X-rays at 10^{17} Hz, gamma rays, and other very high-energy photons. Of these frequencies, X-rays are useful for determining the structures of proteins since their wavelengths are about 1 Å and so are comparable with the size of atoms (Chapter 15).

THE BEER–LAMBERT LAW

The amount of light that is absorbed by a sample depends upon several factors including the concentration of the absorber, *c* (not the speed of light in this context), the pathlength of the sample, *l*, the intensity of the incident light, *I*, and the intrinsic ability of the sample to absorb light. The relationship among these parameters is given by the Beer–Lambert law, named after the astronomer Wilhelm Beer and the mathematician Johann Heinrich Lambert.

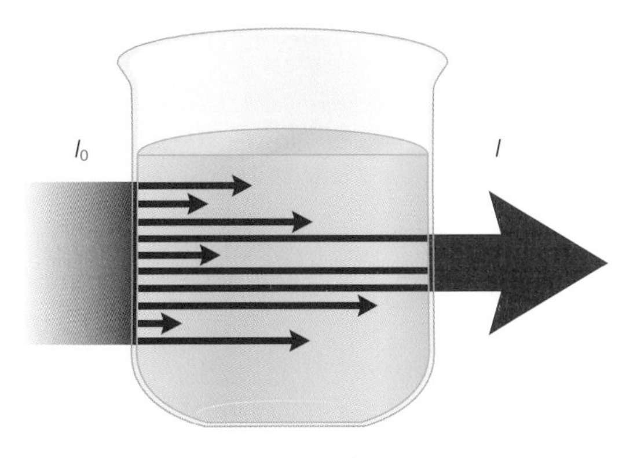

Figure 14.3 The intensity of light passing through a sample, I, can be considered as being absorbed at different distances from the initial point where the intensity is I_0.

The sample is considered to consist of a stack of infinitesimal slices, each of which has a *thickness*, d*x* (Figure 14.3). The *change in intensity*, d*I*, that occurs after light passes through the slice is proportional to the thickness, the *concentration of the absorber*, *c*:

$$dI = -\kappa cI\,dx \qquad (14.3)$$

where κ is the proportionality constant. Dividing this expression by I gives:

$$\frac{dI}{I} = -\kappa c\,dx \qquad (14.4)$$

To determine the total change in light after passing through all of the slices, this expression is integrated from 0 to l:

$$\int_0^{I_0} \frac{dI}{I} = -\kappa \int_0^{l} c\ dx \qquad (14.5)$$

$$\int \frac{dx}{x} = \ln x$$

Assuming that the concentration of the absorber is constant throughout the sample, then the expression can be rewritten as:

$$\ln \frac{I}{I_0} = -\kappa c \int_0^{l} dx = -\kappa cl \qquad (14.6)$$

$$\ln x = (\ln 10)(\log x) = 2.3 \log x$$

Converting the natural logarithm to base 10 logarithm gives:

$$\log \frac{I}{I_0} = -\varepsilon cl \qquad (14.7)$$

where the proportionality constant ε, known as the molar absorption coefficient, or extinction coefficient, is related to κ by:

$$\varepsilon = \frac{\kappa}{\ln 10} \qquad (14.8)$$

If we define the transmittance T to be the ratio of the intensities then the expression becomes:

$$\log T = -\varepsilon cl \quad T = \frac{I}{I_0} \qquad (14.9)$$

The absorbance or optical density is defined as:

$$A = -\log T \qquad (14.10)$$

yielding

$$A = \varepsilon cl \qquad (14.11)$$

MEASURING ABSORPTION

Notice that T must have a value between zero and one since the amount of light going through the sample will generally be less than the amount going through the reference. The absorbance A, therefore, will always be a positive number. The reason for defining the absorbance in this fashion is so that A is directly proportional to the concentration of the molecule in question. For proteins or nucleic acids, the absorbance provides an easy method to determine the sample concentration accurately. For this reason, perhaps the most common of all analytical tools in chemistry is the ultraviolet–visible absorbance spectrophotometer (Figure 14.4). The basic idea of the spectrophotometer is that a light source, usually either an ultraviolet deuterium lamp or a visible tungsten or tungsten/halogen lamp, is used to create two equal beams of light. A single wavelength of the light is selected by passing the light through a monochromator, which uses a dispersive element such as a grating or a prism to spatially separate the colors of light, followed by a slit to select one of the colors. One of these beams passes through the sample, while the second beam passes through

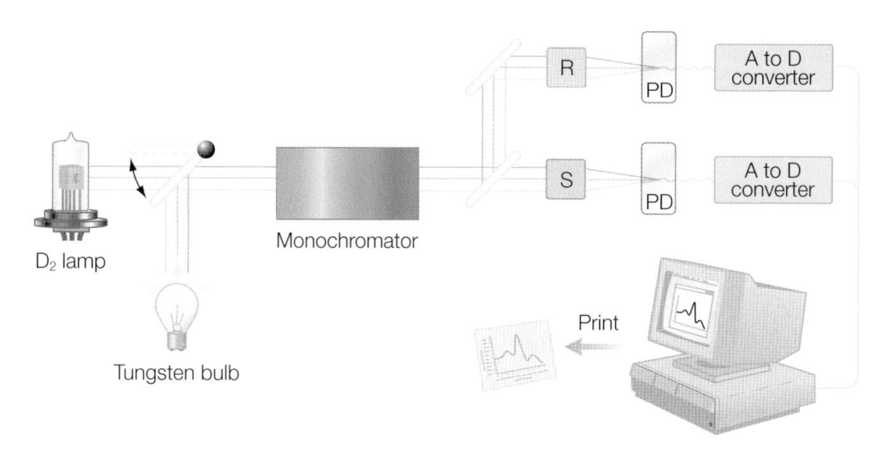

Figure 14.4 A sketch of a simple version of a double-beam ultraviolet–visible absorbance spectrophotometer with a sample compartment (S), a reference compartment (R), and photodetector (PD).

a reference sample cell that contains everything that is in the sample to be measured, except for the molecules being measured. For example, if you are measuring the absorbance of myoglobin in a phosphate buffer, the reference would be an identical cell with the phosphate buffer but no myoglobin. The advantage of using two beams is that the contribution of everything but the molecule of interest can be subtracted, yielding the signal of only that molecule.

Typically, the sample will be in a cuvette with a pathlength of 1 cm, so usually ε is provided with units of $mol^{-1}\ cm^{-1}$. The absorption spectrum of a biological sample will consist of a number of bands spread through the spectral region, as seen for bacteriochlorophyll in bacterial reaction centers (Figure 14.5). Many different types of molecular system have widely spaced electronic orbitals and therefore absorb in the ultraviolet region of the spectrum, including aromatic rings of proteins, the bases of nucleic acids, and many small aromatic molecules. Molecules with large conjugated systems, including bacteriochlorophylls (Figure 14.5), carotenoids, and hemes, have transitions corresponding to the visible and near-infrared regions as predicted by the particle-in-a-box model (Chapter 13). Remember that the color that you see is what the molecule does not absorb. For example, a flower is red when it absorbs blue light and reflects red light.

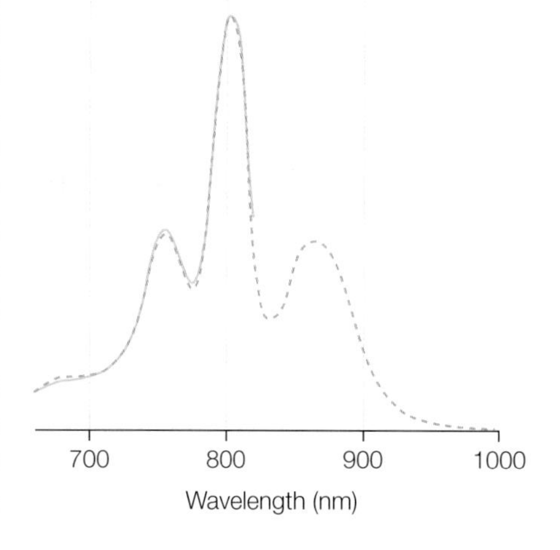

Figure 14.5 The absorbance spectrum of a bacterial reaction center (see Chapter 20).

In some cases, the absorption spectrum may contain more than one contribution. This may occur if a reaction is occurring in the sample. Alternatively, if a protein has a metal cofactor that can be in different redox states, it

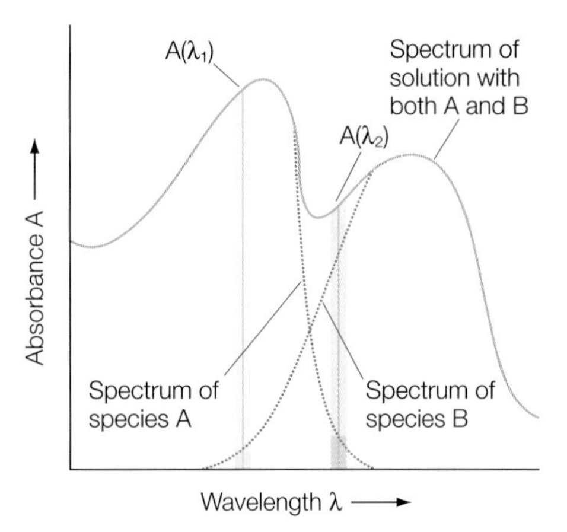

Figure 14.6 The contributions of different components (A and B) in an optical spectrum can be separated with measurements at different wavelengths. At an isosbestic point the absorption does not change as the relative concentrations of the components change.

may be possible to identify the relative amount of each state if they have different spectra. If the spectrum of each pure component is known then it is possible to determine the amount of each component by measuring the absorption at multiple wavelengths.

Assume that there are two components, A and B, wih concentrations [A] and [B], respectively (Figure 14.6). The total absorption at a given wavelength is given by:

$$A = A_A + A_B = (\varepsilon_A[A] + \varepsilon_B[B])l \tag{14.12}$$

If the spectrum is measured at two different wavelengths, λ_1 and λ_2, then the absorption at each wavelength is given by:

$$\begin{aligned} A(\lambda)_1 &= (\varepsilon_A(\lambda)_1[A] + \varepsilon_B(\lambda)_1[B])l \\ A(\lambda)_2 &= (\varepsilon_A(\lambda)_2[A] + \varepsilon_B(\lambda)_2[B])l \end{aligned} \tag{14.13}$$

Since there are two unknowns, [A] and [B], and two equations, it is possible to solve these equations for the unknowns. Some algebra yields:

$$\begin{aligned} [A] &= \frac{\varepsilon_B(\lambda)_2 A(\lambda)_1 - \varepsilon_B(\lambda)_1 A(\lambda)_2}{\varepsilon_A(\lambda)_1 \varepsilon_B(\lambda)_2 - \varepsilon_A(\lambda)_2 \varepsilon_B(\lambda)_1} \\ [B] &= \frac{\varepsilon_A(\lambda)_1 A(\lambda)_2 - \varepsilon_A(\lambda)_2 A(\lambda)_1}{\varepsilon_A(\lambda)_1 \varepsilon_B(\lambda)_2 - \varepsilon_A(\lambda)_2 \varepsilon_B(\lambda)_1} \end{aligned} \tag{14.14}$$

There may be a point in the spectrum at which the spectrum does not change as the relative concentrations of A and B change because the extinction coefficients are equal at this point; this is known as an isosbestic point. Having an isosbestic point demonstrates that the sample consists of two components in equilibrium with no intermediates and provides a useful reference point for spectral measurements.

TRANSITIONS

What determines the wavelength at which a molecule absorbs light? The absorption of light is associated with a transition of an electron from one state to another. Because the energies of the electronic states are quantized, the energy of the photon that can be absorbed is also quantized (Figure 14.7). Based upon the conservation of energy, the frequency of the light emitted when an electron makes a transition from state 1 to state 2 is:

$$\nu = \frac{1}{h}(E_2 - E_1) \tag{14.15}$$

where E_2 and E_1 are the energies of the states 2 and 1 respectively. Correspondingly, light is emitted when an electron starts in an excited state and makes a transition to a lower-energy state.

The energy difference for either absorption or emission can also be directly related to the wavelength of the absorbed or emitted photon according to:

$$\lambda = \frac{c}{\nu} = \frac{ch}{E_2 - E_1} \tag{14.16}$$

In some applications the wavelength is reported in terms of the wavenumber, $\tilde{\nu}$:

$$\tilde{\nu} = \frac{1}{\lambda} = \frac{\nu}{c} \tag{14.17}$$

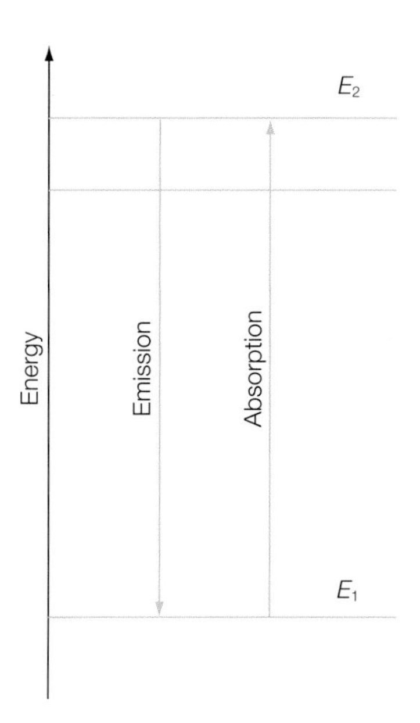

Figure 14.7 Absorption and emission correspond to the transition of an electron between two electronic states.

For the particle in a box, harmonic oscillator, and hydrogen atom, we have calculated the energies of every electronic state. Thus we were able to calculate the absorption spectrum for these cases. What does not arise directly from Schrödinger's equation is the amplitude of each absorption band. However, this can also be estimated from the wavefunctions, as we shall show below.

The absorption of a photon results in the transition of the electron from the lower state to the upper state (Figure 14.8) at a rate:

$$w = B\rho \tag{14.18}$$

where B is a constant called the *Einstein coefficient* and ρ is the *energy density*. The energy density is needed as the transition requires the presence of a photon and the energy density is a measure of how many photons are available with the correct frequency.

The total rate of absorption is determined by both the transition rate and the *total number of electrons in the lower state, N*, according to:

$$W = Nw = NB\rho \tag{14.19}$$

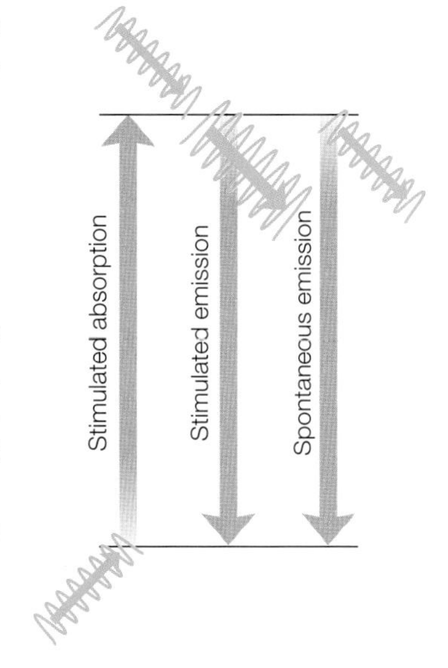

Figure 14.8 Three possible processes that can occur for absorption and emission: stimulated absorption, stimulated emission, and spontaneous emission.

For emission there are two different processes. Spontaneous emission starts with the electron in the higher-energy state and the electron makes a transition "spontaneously" to the lower state with the emission of a photon. For stimulated emission there must be a photon present in the initial state. The presence of this photon interacts with an electron and induces the transition of the electron down to the lower state. By conservation of energy, there are two photons of equal energy in the final state.

The *rate of stimulated emission, w'*, can be written in a form very similar to that of stimulated absorption:

$$w' = B'\rho \tag{14.20}$$

where B' is the *coefficient for stimulated emission*. In contrast, the *rate of stimulated emission, w'*, does not depend upon the presence of the initial photon so is given by:

$$w' = A \tag{14.21}$$

where A is a constant called the *Einstein coefficient of spontaneous emission*.

The *overall rate of emission, W'*, is given by the sum of the two rates multiplied by the *number of electrons in the upper state, N'*:

$$W' = N'(A + B'\rho) \tag{14.22}$$

Derivation box 14.1

Relationship between the Einstein coefficient and electronic states

It is possible to relate the Einstein coefficients to the wavefunctions of electrons. The intensity of the transition is determined by the coefficient B, which can be expressed as:

$$B = \frac{|\mu_{fi}|^2}{6\varepsilon_0\hbar^2} \tag{db14.1}$$

where μ_{fi} is called the *transition dipole moment*. The transition dipole moment can be calculated directly from the wavefunctions determined by Schrödinger's equation:

$$\mu_{fi} = \int \psi_f^* \mu \psi_i \, \mathrm{d}\tau \tag{db14.2}$$

where ψ_f and ψ_i are the *final* and *initial wavefunctions*, respectively. The operator μ is electric dipole moment operator and is given by the product of the charge and the distance between charges.

Sometimes, a related quantity called the *dipole strength, D_{fi}*, is used:

$$D_{fi} = \left[\int \psi_f \mu \psi_i \, \mathrm{d}\tau\right]^2 \tag{db14.3}$$

We can then write:

$$B = \frac{D_{fi}}{6\varepsilon_0 \hbar^2} \quad \text{(db14.4)}$$

This term has the advantage that it can be related directly to the experimental measured absorption band:

$$D_{fi} = \frac{10^3(3hc)}{8 \log e N_A \pi^3} \int \frac{\varepsilon(\upsilon)}{\upsilon} \, d\upsilon \quad \text{(db14.5)}$$

This relation can be derived using the Beer–Lambert law:

$$N \, dz = -\frac{1}{\sigma} \frac{dI'}{I'} \quad \text{(db14.6)}$$

This can be rewritten as:

$$\int_{I_0}^{I} \frac{dI'}{I'} = -\sigma N \int_0^l dz \quad \text{(db14.7)}$$

In these equations σ is the absorption cross-section and is associated with each molecule in the sample. To change the units to moles per liter we introduce Avogadro's number N_A and can write:

$$\log \frac{I_0}{I} = A = (-\log e) \ln \frac{I}{I_0} = \sigma N l \log e = \varepsilon l c \quad \text{(db14.8)}$$

and

$$\varepsilon = \sigma N_A 10^{-3} \log e \quad \text{(db14.9)}$$

This gives:

$$N \rho B = \frac{10^3 \varepsilon N I \, dz}{N_A \log e} \quad \text{or} \quad B = \frac{c}{h\upsilon} 10^3 \frac{1}{N_A \log e} \frac{\varepsilon}{\upsilon} \quad \text{with} \quad \rho = \frac{h\upsilon}{c} I \quad \text{(db14.10)}$$

These values are specific frequencies, so to obtain the total values we integrate:

$$B = \frac{10^3 c}{h N_A \log e} \int \frac{\varepsilon}{\upsilon} \, d\upsilon = \frac{D_{fi}}{6\varepsilon_0 \hbar^2} \quad \text{(db14.11)}$$

Albert Einstein used statistical mechanics (Chapter 8) to show that:

$$B' = B \quad \text{and} \quad A = \frac{8\pi h\upsilon^3}{c^3} B \tag{14.23}$$

Thus the rate of stimulated absorption and stimulated emission are equal and there is a net rate of absorption because there are more electrons in the lower states:

$$W_{net} = (N - N')B\rho \tag{14.24}$$

LASERS

The design of lasers is critically based upon the balance of the number of electrons in two electronic states. By establishing a larger number of electrons in the higher-energy state, it is possible to create a net emission of photons, all of which have the same frequency. The creation of a larger number in the higher-energy state is called population inversion. To achieve a population inversion it is necessary to provide energy into the system that can specifically change the energies of the electrons. In the simplest version, consider a total of four electronic levels (Figure 14.9). The goal

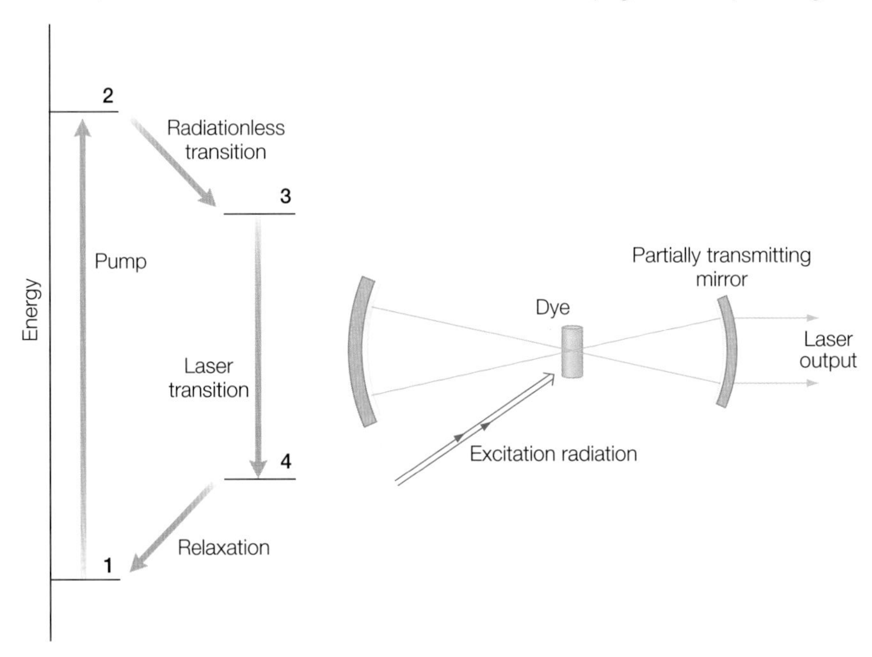

Figure 14.9 A schematic diagram of the transitions that occur in a laser and the physical layout of a laser. After pumping of electrons from level 1 to level 2, the excited electrons can make a radiationless transition to level 3, followed by the laser transition to level 4 and relaxation back to level 1.

is to create a population inversion of level 3 compared to level 4. Rather than trying to create this directly, two new states, 1 and 2, are introduced. An electron is first pumped from level 1 to level 2. The electronic levels are poised such that an electron in level 2 can make a radiationless transition to level 3. The laser transition is from level 3 to level 4. The electrons in level 4 are not stable but relax back to level 1. By repeated pumping, the electrons are cycled in order to maintain the desired population inversion. Lasers take advantage of population inversions to perform Light Amplification using Stimulated Amplification of Radiation. A population inversion in a dye, which is created by a pumping radiation, experiences repeated exposure to a light that bounces between two mirrors and has the energy matching the transition between levels 3 and 4. As the stimulated processes occur, the light intensity at that frequency increases until a equilibrium is reached. A small portion of the light is allowed to escape to be used for an experiment. Since the original designs, many other configurations have been developed including laser diodes that can be used for small-scale electronics. In addition to the continuous wave design, lasers can be operated in a pulse mode that allows the excitation of a sample at a specific wavelength for times of less than 1 fs.

SELECTION RULES

When an electron undergoes a transition from one state to another the quantum numbers describing the electronic state change. Since a photon carries angular momentum ($s = 1$), only certain changes in quantum numbers, representing the angular momentum of the electrons, are allowed to change. For an electron in the hydrogen atom, the electron is uniquely described by the four quantum numbers, n, l, m_l, and m_s. When an electron makes a transition, the change in angular momentum must exactly compensate for the angular momentum of the photon. Since the photon has $s = 1$, then the change in the quantum number l must be also 1. Thus some transitions are forbidden, such as a change from a d orbital ($l = 2$) to an s orbital ($l = 0$).

These selection rules can be derived explicitly from the wavefunctions derived from Schrödinger's equation and the definition of the transition dipole moment:

$$\mu_{fi} = \int \psi_f^* \mu \psi_i \, \mathrm{d}\tau$$

$$\mu_{fi} = \int \psi_f^* (-ez) \psi_i \, \mathrm{d}\tau \tag{14.25}$$

$$\mu_{fi} = \int_0^\infty R_{nflf} \left(-e\sqrt{\frac{4\pi}{3}}r\right) R_{nili} r^2 \, \mathrm{d}r \int_0^\pi \int_0^{2\pi} Y^*_{lfm_lf} Y_{10} Y_{li,m_li} \sin\theta \, \mathrm{d}\theta \, \mathrm{d}\phi$$

According to the properties of spherical harmonics:

$$\int_0^{\pi}\int_0^{2\pi} Y^*_{lfm_if} Y_{10} Y_{li,m_i i} \sin\theta \, d\theta \, d\phi = 0 \quad \text{unless } lf = li \pm 1 \tag{14.26}$$

Complete integration of these terms will of course give the exact dipole moment for any transitions. For example, for a transition from the 1s orbital to the 2pz orbital, the calculation gives the number 0.74 atomic units along the z direction and 0 along the x and y axes.

For molecules, there are also selection rules reflecting the symmetry of molecules (Figure 14.10) that arise from the conservation of momentum. Homonuclear molecules may possess parity. The parity of a molecular orbital is even, denoted by g from the German word *gerade* for even, if the sign of the orbital is unchanged for a vector passing through the center. Likewise, the parity is odd, denoted by u for *ungerade*, if the sign changes. Since the dipole operator has odd parity, the dipole moment will be exactly equal to zero if the wavefunctions have the same parity (because the integrand will be odd and the integral of an odd function is always zero if the limits are equal but of opposite sign).

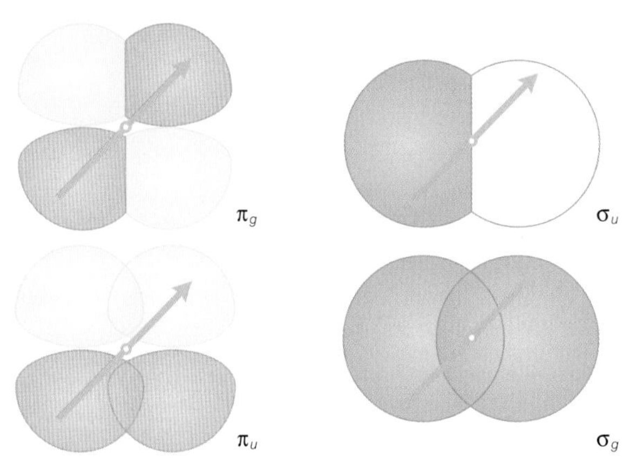

Figure 14.10 The parity of an orbital may be even (g) or odd (u) for homonuclear molecules.

THE FRANCK–CONDON PRINCIPLE

Biological systems have a large number of vibrational states and the transitions can be to any of these states. The Franck–Condon principle states that only certain transitions are observed. During a transition the nuclei can be considered to be not moving; thus the electron can be pictured as making a transition between two electronic states without any involvement of contributions from nuclei. The electrons are pictured as existing in states that are formed by vibronic oscillators that represent the bonds between atoms in the molecule. The transition occurs for a precise configuration of the nuclei without any alteration of the nuclear geometry. This is represented as a strictly vertical transition from the lower state to the higher state (Figure 14.11). The level marked with the asterisk is the most probable transition state because of the matching of the wavefunctions.

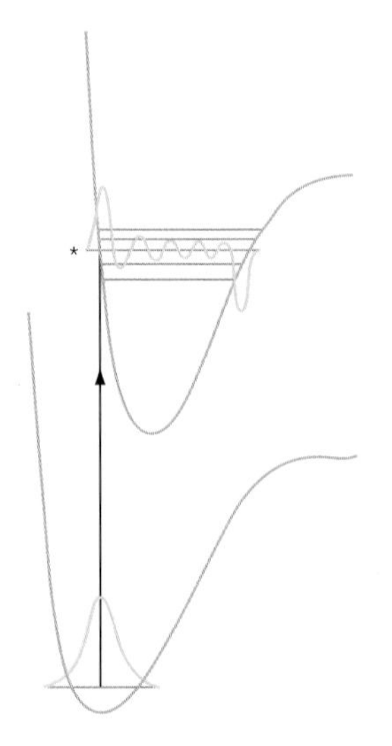

Figure 14.11 The Franck–Condon principle states that optical transitions correspond to vertical transitions between electronic states with no change of nuclear coordinates.

THE RELATIONSHIP BETWEEN EMISSION AND ABSORPTION SPECTRA

The key features of absorbance spectra are the wavelength of the peak of the band, the width, and intensity (Figure 14.12). The same features are used to describe fluorescence spectra. In comparing fluorescence and absorbance, the peak of the fluorescence is always red-shifted, or equivalently at a lower energy, with respect to the absorbance spectrum. Second, the fluorescence spectrum looks rather like a mirror image of the absorbance spectrum in terms of width and intensity.

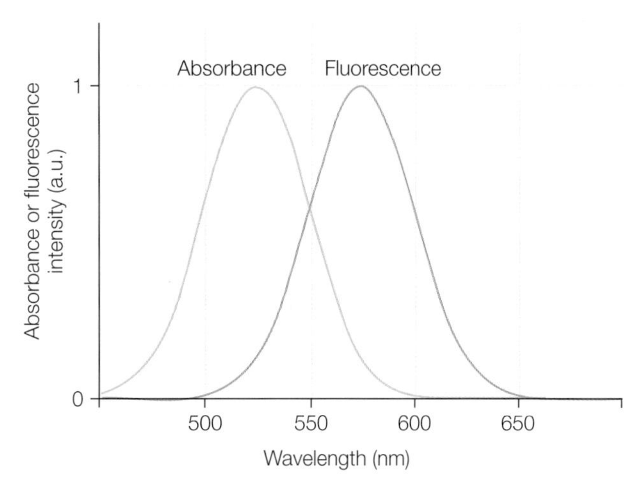

Figure 14.12 Comparison of the absorption and fluorescence spectra of a biological molecule.

The reasons for this correlation can be understood from a diagram of the electronic states (Figure 14.13). The majority of the molecules absorb energy from the lowest vibrational level of the ground state. However, the electrons can make transitions into any of a number of vibrational levels of the excited state. Thus, several different transitions are possible, and the absorption spectrum will have a width that reflects these different possibilities but they are unresolved in the spectrum. The transition to the lowest vibrational level of the excited state is often referred to as the zero–zero transition energy. Fluorescence arises from electrons in the excited state. Electrons in higher vibrational levels of the excited state relax very rapidly, in picoseconds or less, to the lowest vibrational level by losing heat to the surroundings. This means that fluorescence occurs from the lowest vibrational level of the excited state to any of a number of vibrational levels in the ground state. On average, this transition energy is less than the zero–zero transition energy; in other words, the emitted photons are at lower energy or to the red of the absorbed photons used to generate the excited state. Assuming that the vibrational bands in the ground and excited states have the same spacing, the two spectra will be shifted but have comparable widths, and hence appear to be mirror images.

The width evident in the spectra of biological molecules arises from different contributions (Figure 14.14). A major contribution to the width of the spectra is then from the presence of the vibrational states. It is observed in some cases that changes in vibrational states accompany an electronic transition, resulting in additional bands. In large organic molecules, these vibrational bands often are unresolved and increase the width of the band. A molecule may also have a distribution of structures at any given time that broadens the spectrum. In other words, different molecules in the sample have slightly different spectra because of the nuclear conformation they happen to have at the time of light absorption. Such

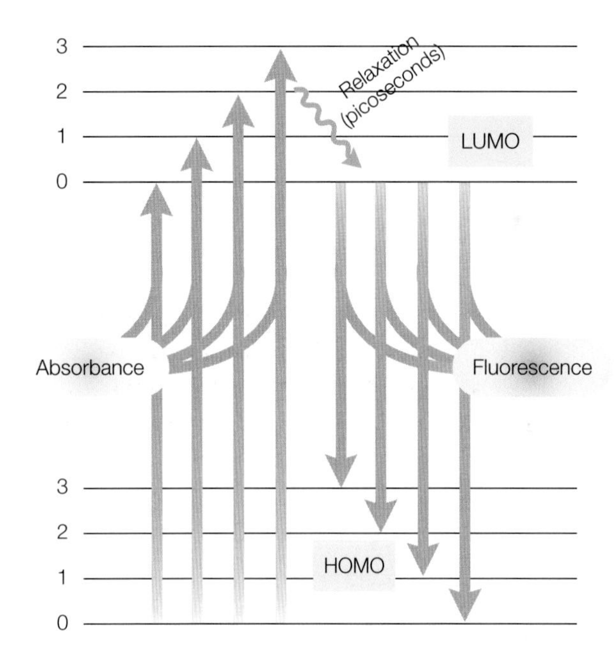

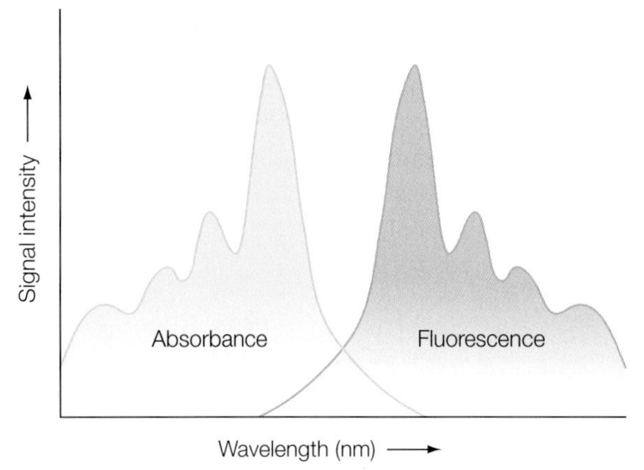

Figure 14.13 The absorbance is from the lowest vibration level of the highest occupied molecular orbital (HOMO) to one of several different vibrational states of the excited state, which is the lowest unoccupied molecular orbital (LUMO). Fluorescence is from the lowest vibrational level of the LUMO to any number of different levels of the HOMO. This results in a red-shifted spectrum for the fluorescence compared to the absorption spectrum.

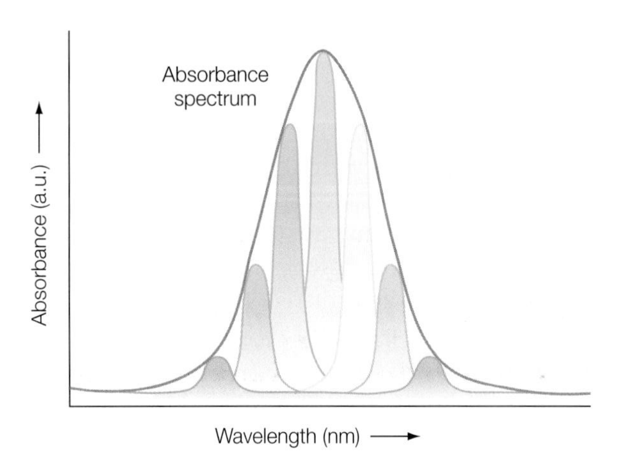

Figure 14.14 The large widths of optical spectra usually arise from the presence of unresolved bands.

differences are small and not resolved, resulting in a broad spectrum. Also, lines may be broadened due to the uncertainty principle but this is a significant contribution only for molecules with extremely short-lived excited states.

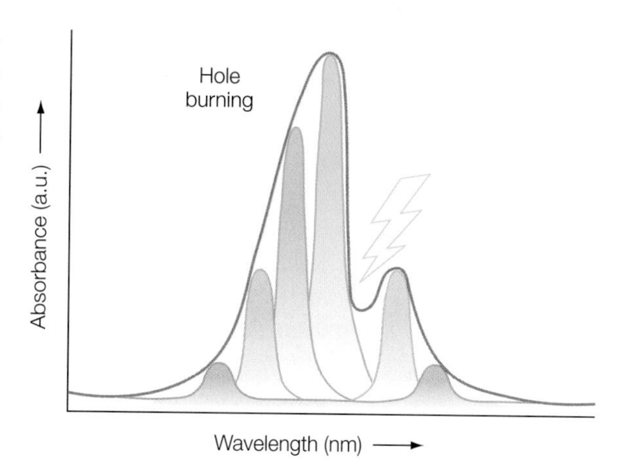

Figure 14.15
A hole-burning experiment triggered by a short light pulse.

The technique of hole-burning spectroscopy can be used to reveal how nuclear motion affects absorbance spectroscopy (Figure 14.15). These experiments are performed at temperatures near 1 K, where almost all nuclear motion stops. A very short laser pulse at a well-defined wavelength strikes the sample, causing all of the electrons associated with that specific wavelength to saturate; that is, the number of electrons in the ground and excited states match. The saturation results in loss of absorbance since the amplitude is given by the population difference that is now zero. The spectrum then has a "hole burned" at a specific part of the band. The hole will have a width equivalent to what would be seen if there were no significant heterogeneity in nuclear positions in the system. With time, the electrons will decay back to their normal distribution and the absorption recovers. By analyzing these types of spectra one can identify the vibrations are associated with the absorbing cofactor.

THE YIELD OF FLUORESCENCE

The emission of a photon is not the only possible mechanism by which an electron can undergo transition from an excited state to the ground state, because other nonradiative pathways, which usually involve generating heat instead of light, are also possible. The quantum yield of fluorescence is a measure of the fraction of transitions that occur by fluorescence compared to other pathways. The quantum yield is usually defined in terms of rates. Excited electronic states typically decay with first-order kinetics. The transition of the excited state to the ground state can be considered as being a competition between two first-order processes: fluorescence and nonradiative decay. If the fluorescence rate constant is k_f, and the nonradiative rate constants are each k_i, then the yield of fluorescence is just given by:

$$\phi_f = \frac{k_f}{k_f + \sum_i k_i} \tag{14.27}$$

When other processes are present, the observed rate that the excited state decays, k_{obs}, will be faster than the rate that is due to fluorescence only.

Since the rates are first order, the observed rate is given by just the sum of the rates for all possible decay pathways:

$$k_{obs} = k_f + \sum_i k_i \quad (14.28)$$

Sometimes the processes other than fluorescence involve mechanisms that are important to biology, such as photosensing processes like phototaxis, vision, and photosynthesis (Chapter 20). In these cases, the measurement of fluorescence provides a sensitive spectroscopic means of probing these processes.

FLUORESCENCE RESONANCE ENERGY TRANSFER (FRET)

Many applications of fluorescence to biological systems make use of energy transfer between a donor and acceptor. Because such transfer processes are highly dependent upon the distance, the transfer can serve as a measure of the donor–acceptor distance. One common method is to perform a FRET experiment, or fluorescence resonance energy transfer. The electron donor is excited and, if there is no nearby acceptor, fluorescence from the donor is observed. By if an acceptor is nearby, then the energy can be transferred from the donor to the acceptor and fluorescence is observed from the acceptor. The efficiency of transfer depends upon many factors. For example, the fluorescence spectrum of the donor should overlap with the absorption spectrum of the acceptor. In many applications, these factors are considered to be fixed and the efficiency of transfer depends on the *donor–acceptor distance*, r_{DA}, according to:

$$Efficiency = \frac{r_o^6}{r_o^6 + r_{DA}^6} \quad (14.29)$$

where r_o is a parameter that contains the various other factors. The value of r_o can be estimated but is usually determined experimentally with values of 10–60 Å being common.

MEASURING FLUORESCENCE

Fluorescence can be measured with an emission spectrophotometer. In this case, a second light source is used for excitation of the sample. This excitation light is passed through a monochromator before entering the sample compartment. Excited molecules in the sample fluoresce and the fluorescence is collected, passed through a second monochromator, and then passed onto a detector where the amount of fluorescence is measured. In the case of fluorescence spectrophotometers, one typically scans the monochromator associated with the detector. Alternatively, the excitation monochromator can be scanned if one wants to measure the spectrum of the material absorbing the light that results in fluorescence.

PHOSPHORESCENCE

As discussed above, there are certain types of electronic transition that are not allowed. One such case is a transition that involves both a change in electronic orbital and a flip of electron spin. When the spins of the two electrons in the highest orbitals are antiparallel, we refer to them as being in a singlet state (Chapter 12). If their spins are aligned (and in different orbitals, since you cannot have two electrons with the same spin in the same orbital), we refer to them as being in a triplet state. Sometimes, the spin of an electron in the excited state will flip, usually due to interactions with the magnetic moments of the surrounding nuclei. The electrons are then in a triplet state and a direct transition to the ground state, which is usually a singlet state, is not allowed, resulting in the triplet state having a long lifetime as short as microseconds but as long as seconds. When the transition does occur, light is given off and this is called phosphorescence. Some molecules readily undergo phosphorescence and are used in glow-in-the-dark toys.

RESEARCH DIRECTION: PROBING ENERGY TRANSFER USING TWO-DIMENSIONAL OPTICAL SPECTROSCOPY

In photosynthetic bacteria, light is captured by the light-harvesting complexes and funneled to the reaction center, which performs the primary photochemistry of converting the light energy into chemical energy (Chapter 20). Green sulfur bacteria have a large peripheral antenna complex, known as a chlorosome, which collects the energy of the light, using up to 10,000 bacteriochlorophylls. After light excitation, the light energy is efficiently transferred to the reaction center through a water-soluble pigment–protein complex that is termed the FMO protein after Richard Fenna, Brian Matthews, and John Olson, who led the early efforts to characterize the structure and function of this complex (Matthews & Fenna 1980; Olson 1998). The structure of the complex has been determined from two organisms using X-ray crystallography (Li et al. 1997). The protein consists of three identical subunits that form a trimer with a central symmetry axis (Figure 14.16). Each subunit has the distinctive feature of a large β sheet that is folded into a "taco shell" around seven bacteriochlorophylls. The relative distances between pairs of the bacteriochlorophylls range from 4

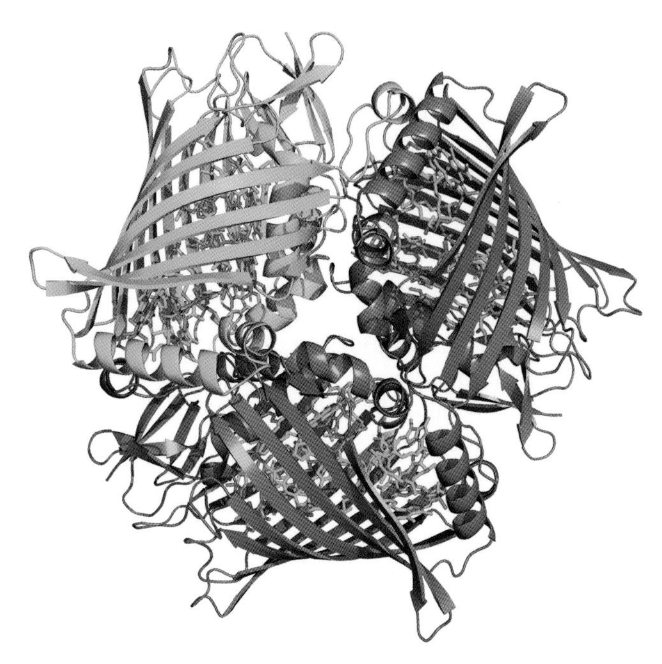

Figure 14.16
A schematic representation of the FMO complex, showing the extensive presence of β strands organized into a β sheet, which surround seven bacteriochlorophylls.

to 11 Å. Not only does each bacteriochlorophyll experience different interactions with the other bacteriochlorophylls but the protein environment surrounding each bacteriochlorophyll is unique.

One goal of both experimental and theoretical research on this complex was to understand the contribution of each bacteriochlorophyll in the process of energy transfer in the cell. Despite the difference observed in the structural features for each bacteriochlorophyll, the optical absorption spectrum observed in the FMO complex is relatively featureless at room temperature, with all of the bacteriochlorophylls absorbing light at the same wavelength. When the samples are cooled to low temperature, the contributions of the different bacteriochlorophylls is more distinctive, with the broad band observed at room temperature having some resolved underlying bands. Several theoretical calculations have been performed with the goal of assigning these partially resolved absorption bands to individual bacteriochlorophylls, but the calculations consistently show that the pigments are highly interacting and hence difficult to distinguish.

Time-resolved optical spectroscopy has helped to distinguish the relative contributions, at each point in time, of the different underlying bands but a unique temporal assignment could not be made unambiguously as the interactions could only be inferred indirectly. These experimental difficulties have been overcome with the development of two-dimensional (2D) femtosecond infrared spectroscopy (Brixner et al. 2005). The diagonal peaks in the 2D traces correspond to the positions of the peaks in the absorption spectrum. The off-diagonal peaks reveal which molecules are interacting and hence the couplings (Figure 14.17). The technique of 2D optical spectroscopy makes use of a series of ultrashort laser pulses in a manner analogous to the pulse sequences used in NMR (Chapter 16). The initial pulse excites the sample and all of the transitions of interest. After a delay time corresponding to the coherence time in NMR, the sample is illuminated with a second laser light pulse. Depending upon the delay time used, the signal at each frequency either increases or decreases. After another time interval corresponding to the population time, a third pulse probes the optical state of the sample. The signal arises from a fourth spontaneous pulse known as the photon echo that arises from the interactions of the three previous pulses with the transition dipole moments of the pigments. Each 2D spectrum shows the couplings a specific time after

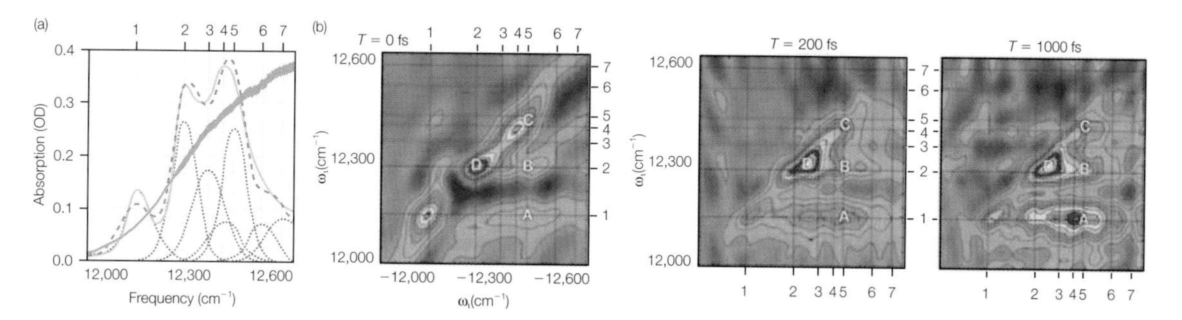

Figure 14.17 Experimental spectra of the FMO complex. (A) The linear absorption spectrum of the complex, showing the presence of several overlapping contributions at 77 K. The other three panels show the initial 2D spectrum and the spectra after 200 and 1000 fs. From Brixner et al. (2005).

excitation, and the use of successive delays provides the time-evolution profile of the couplings.

The ability of this technique to provide a direct mapping of the couplings as the energy-transfer process occurs within the complex is illustrated by the experimental 2D traces for the FMO complex. Before excitation, the positions of the three main diagonal peaks of the 2D spectrum match the positions of the peaks observed in the low-temperature absorption spectrum (Figure 14.18). In addition to the diagonal peaks, several off-diagonal features, both positive and negative, are evident in the 2D spectrum. These off-diagonal peaks arise from transitions involving excited states. The peak labeled A indicates that the bacteriochlorophylls identified with peaks 1 and 5 of the absorption spectrum are coupled. The presence of the off-diagonal peak labeled B reveals a coupling for the pigment's contribution to peaks 2 and 5. After light excitation the energy is transferred and the 2D spectrum is seen to evolve. The main diagonal peak shifts to lower energies, and the changes in the off-diagonal peaks indicate a change in the couplings among the bacteriochlorophylls. The results can be interpreted in terms of specific pathways of energy transfer for the different bacteriochlorophylls, rather than being simple stepwise energy transfer from the highest energy bacteriochlorophyll to the lowest. In the two possible pathways, certain states that are spatially close are skipped when they are energetically distinct.

These results for the FMO complex demonstrate that 2D time-resolved spectroscopy can overcome the problem of overlapping absorption bands in conventional spectra just as 2D NMR is needed to interpret spectra with many overlapping contributions. Scientists have performed 2D experiments with NMR for many years as the required manipulations of the radio-frequency waves have proved to be quite tractable. In contrast, such experimental control of light in the optical spectrum has proved to be more difficult until recent times. Recent efforts, including this work by Brixner

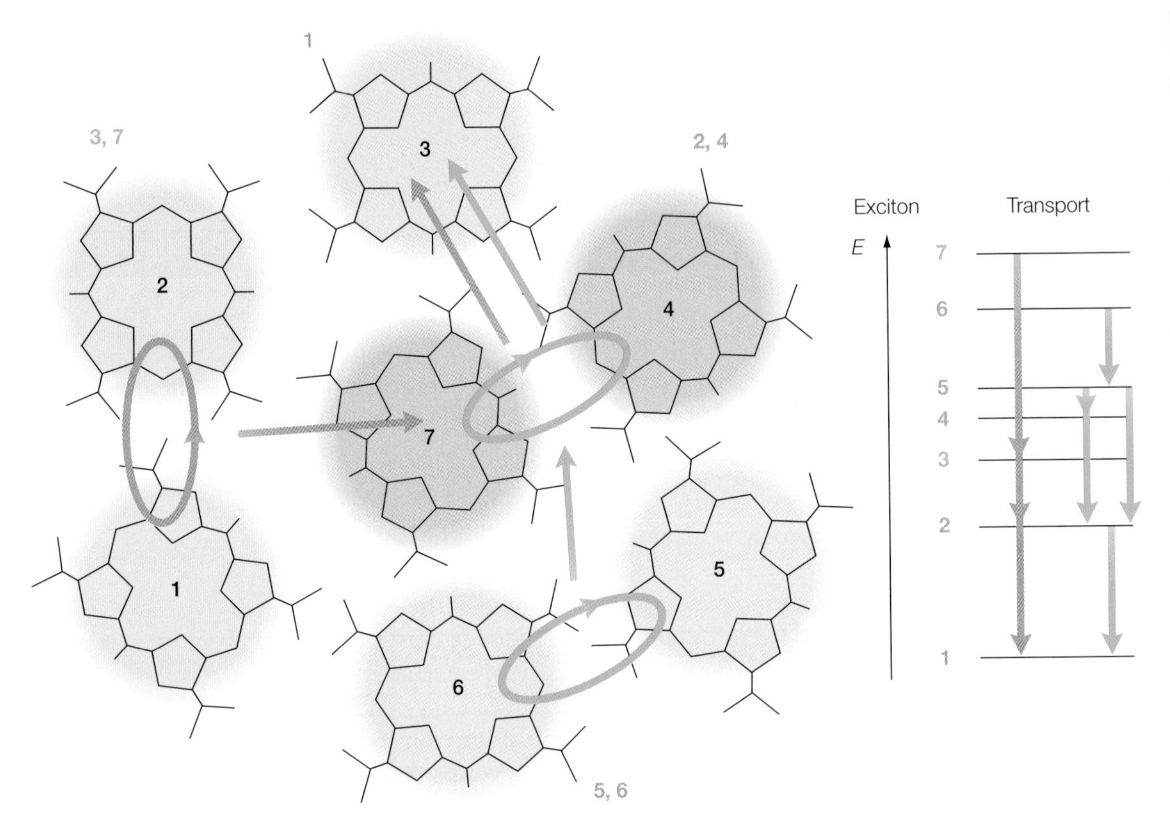

Figure 14.18 The structure of the seven bacteriochlorophylls in the complex. Two pathways for energy transfer are identified in the 2D experiments with the figure on the right showing the relative energies of the seven pigments. Modified from Brixner et al. (2005).

et al. (2005) on the FMO complex, demonstrate the feasibility of applying 2D optical spectroscopy to complex biological systems.

RESEARCH DIRECTION: SINGLE-MOLECULE SPECTROSCOPY

Most spectroscopic measurements are performed on the order of 10^{23} or a larger number of molecules and so represent an average over the entire ensemble. As a result, rare events cannot be identified from the average property. Also, a series of temporally distinguishable events, such as binding of polymerase to DNA followed by attachment of an amino acid and release, cannot be distinguished unless the process can be made to start simultaneously for all molecules. With technical advances in instrumentation, individual molecules can be detected optically and their time evolution can be measured (Nie & Zare 1997; Weiss 1999; Mollova 2003; Peterman et al. 2004).

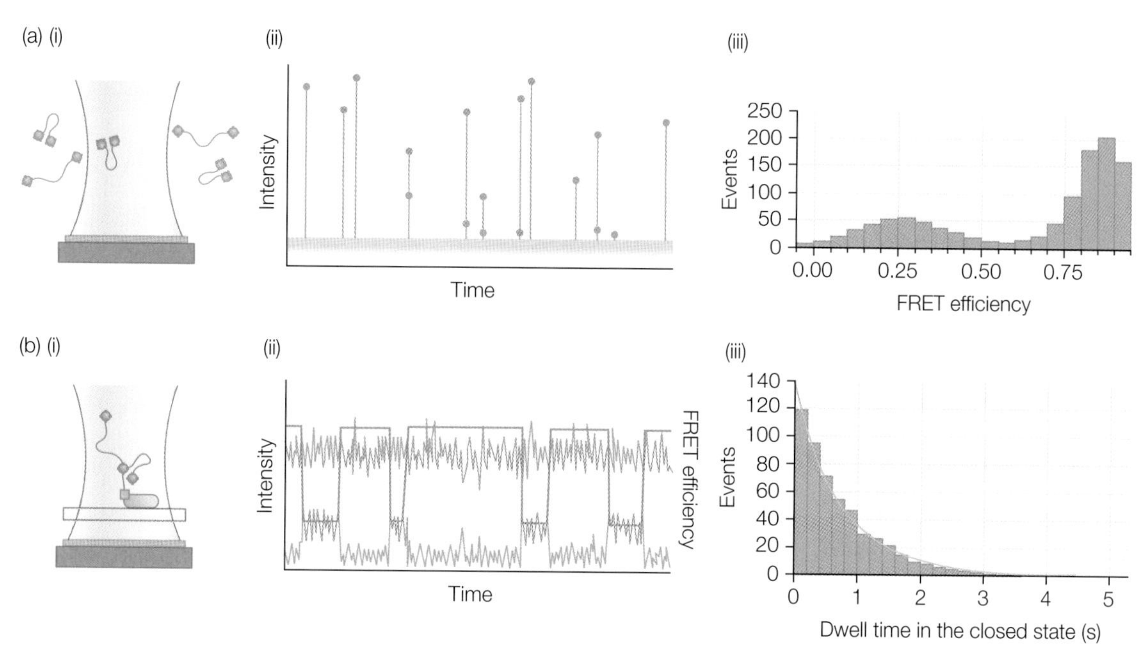

Figure 14.19 Illustration of the two basic configurations for observation of single-molecule fluorescence. (a) Molecules are allowed to freely diffuse and burst of fluorescence results in FRET efficiencies of 0.3 and 0.9, reflecting molecules in open and closed conformations, and (b) molecules are restricted in space allowing a time profile of the changes between the open and closed states. Modified from Mollova (2003).

In single-molecule spectroscopy, a key concept is that a single molecule can be repetitively cycled between the ground and excited states, yielding a large enough number of photons to be measured by averaging for about 1 ms. In particular, fluorescence is an ideal measurement since the system can be poised such that only molecules of interest fluoresce and background contributions can be minimized. Molecules can either be set to low concentrations and detected when they diffuse into the probe volume or spatially trapped at a specific location (Figure 14.19). In this case, the fluorescence is detected in bursts and is limited by the diffusion time of the molecule. Alternatively, the molecule can be spatially restricted by being tethered to a smooth surface, or trapped in a matrix or optical trap. Spatial restriction allows for extended measurements but can be limited by loss of the signal due to over-excitation of the molecule. Excitation is usually performed with a laser as its high intensity provides a high rate of excitation and consequently a measurable rate of emitted photons at the detector. In these measurements, two dyes are used to allow FRET measurements between an open state, with the two markers far apart, and a closed state, with the two markers close together. The single-molecule measurements can be interpreted in terms of the number of observations in different molecular configurations, corresponding to different FRET

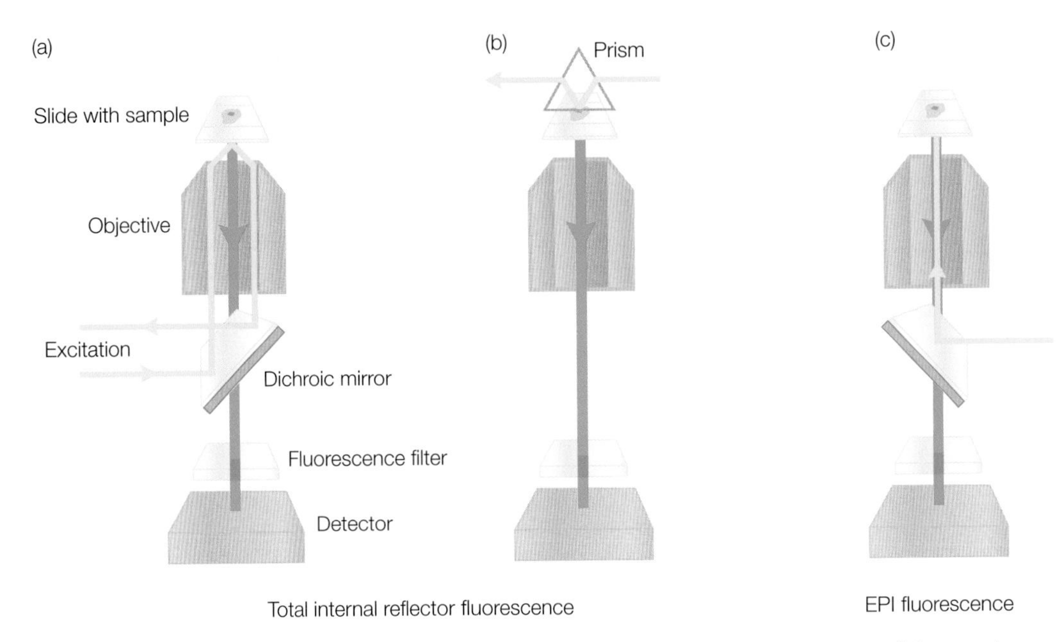

Figure 14.20 Different configurations for single-molecule measurements: (a) total internal reflection through the objective, (b) prism-based total internal reflection, and (c) epifluorescence illumination. Modified from Peterman et al. (2004).

efficiencies, or in terms of the time dependence of the FRET efficiency, corresponding to the rate of change between the open and closed molecular configurations.

There are several different fluorescent markers that can be attached to proteins for these studies. The marker may be a simple dye molecule such as tetramethyl rhodamine functionalized with malemide, the relatively stable dye cy3-malemide, dyes such as bis-[(*N*-iodacetyl)piperazinyl] sulforhodamine that have two attachment groups, or a label that can be incorporated genetically, for example green fluorescent protein (Chapter 19). Special optical configurations must be employed to detect single molecules in a very small area. The sample concentration is kept very low so that, at any given time, only one molecule is being probed by the beam. Several optical configurations have been used to measure fluorescence from single molecules involving different microscopic configurations, including total internal reflection (Figure 14.20a, b) and epifluorescence (Figure 14.20c).

HOLLIDAY JUNCTIONS

Genetic diversity is required for long-term survival of a species. Exchange of genetic material between homologous chromosomes provides a means of generating diversity. In homologous recombination the parent DNA aligns

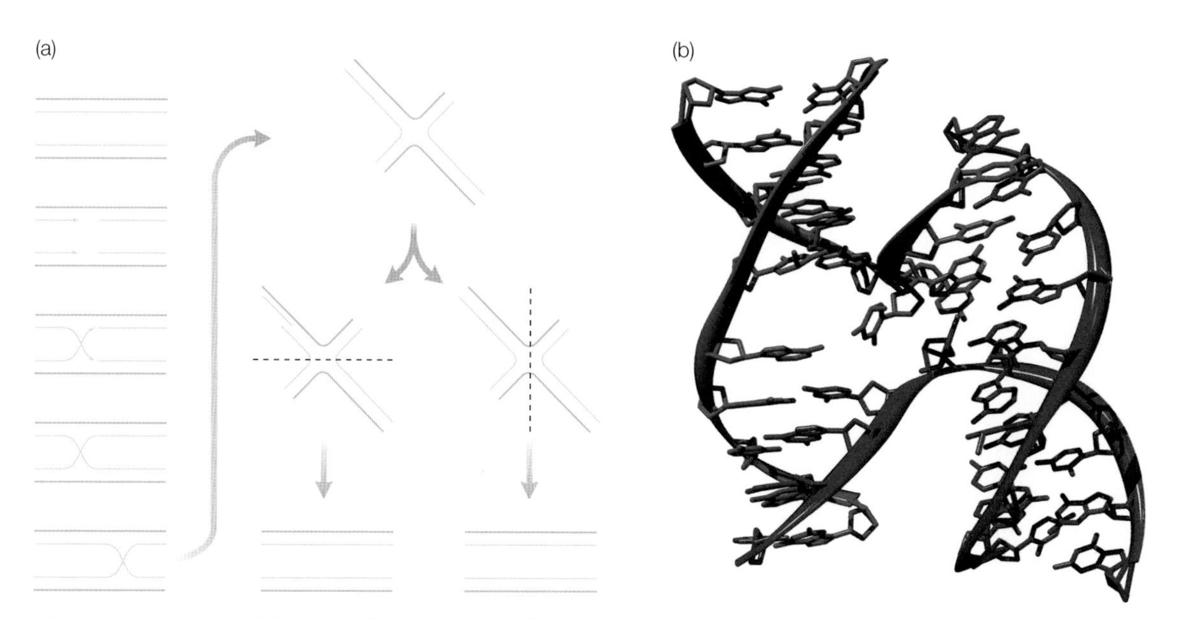

Figure 14.21 Model for a Holiday junction from two homologous duplexes. (a) The proposed pathway involving alignment, cleavage, invasion, sealing, and migration followed by different views of the cleavage process. (b) Three-dimensional view showing alignment of duplexes and minor and major groves. Modified from Berg et al. (2007).

and new DNA molecules are formed by breakage and joining of the homologous segments. This process is facilitated by the protein RecA and is used by cells to repair damaged DNA. The process has been seen by use of electron microscopy of DNA undergoing homologous recombination.

In genetic recombination, a DNA molecule is formed with a sequence derived from two strands. In 1964 Robert Holliday proposed a model that involved the formation of an intermediate state that is now termed a Holliday junction. The first step is the alignment of two homologous duplexes (Figure 14.21). A strand from one duplex and the corresponding strand are nicked, leading to a cross-over in which strands from different duplexes are joined. The junction point can move along the strands through a process called branch migration. The strands are then nicked again, resulting in mixed strands. The structure of the Holliday junction has been determined primarily using NMR, although some X-ray three-dimensional structures are available.

To study the properties of the intermediate states, the DNA is first immobilized by a 5′ biotin tag to a streptavidin-coated glass slide (Churchill 2003; McKinney et al. 2003). The motions of the DNA strands are observed by placing fluorescent markers at the ends of the DNA (Figure 14.22). For this particular set of experiments, a dye is attached at the end of one strand and serves as the energy donor and a different dye is attached to the end of another strand and serves as the energy acceptor. The Holliday junction

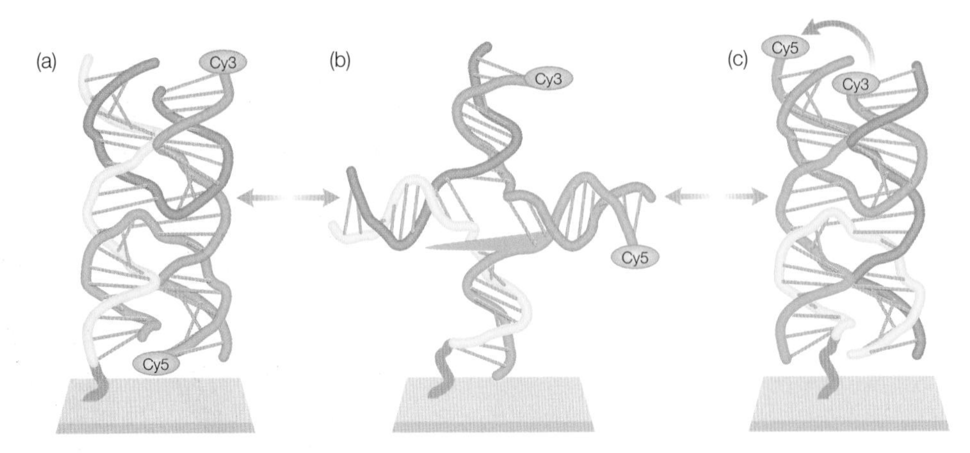

Figure 14.22 The different conformations of the Holliday junction showing the different relative positions of the energy donor and acceptor. (a) A conformation with a limited FRET signal, (b) a proposed intermediate, and (c) an alternate conformation with a strong FRET signal. Modified from Churchill (2003).

is a dynamic structure that can adopt two different conformations that are in equilibrium with each other. In one conformation the distance between the energy donor and acceptor is large and energy transfer cannot occur, but in the other conformation the distance is small and energy transfer does occur.

In a bulk measurement of the Holliday junction the presence of the different conformations cannot be distinguished since all states are sampled simultaneously and averaged. Single-molecule FRET measurements allow the measurement of the distance between the donor and acceptor from a single molecule. As the conformation switches back and forth, the fluorescence will be measured from either the donor, indicating that the donor–acceptor distance is large, or the acceptor, indicating that the donor–acceptor distance is small (Figure 14.23). In these experiments, a variable time interval is evident between conformational switching because the process is stochastic. That is, the switching occurs with a certain probability only, and so each time the period is different, and only the average time must agree with bulk measurements of the switching rate.

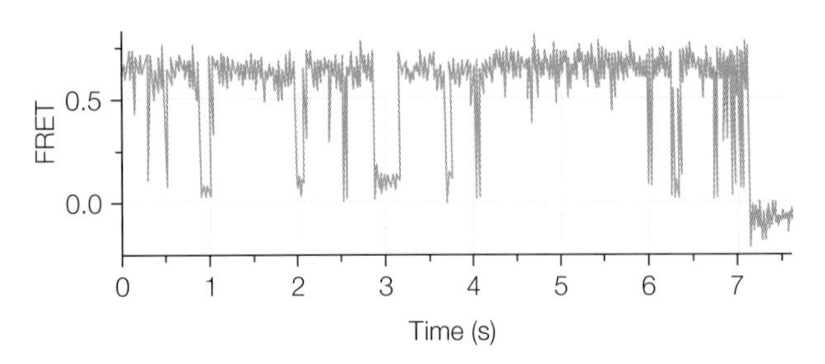

Figure 14.23 The measurements of the Holliday junction give a signal from either the energy acceptor, with a high FRET value, or from the energy donor, with a low FRET value. Modified from Churchill (2003).

FRET can also be used to study interactions between different parts of DNA due to protein binding (Katilitene et al. 2003). For example, some DNA restriction endonucleases, such as *Ngo*MIV, function by binding simultaneously at two specific recognition sites, making a DNA loop, and then cleaving the site. By labeling different ends of the DNA it is possible to get profiles for the distances between the two ends as a function of parameters such as the enzyme concentration, and to derive possible models for the arrangement of the enzyme on the DNA strands.

REFERENCES

Berg, J.M., Tymoczko, J.L., and Styer, L. (2007) *Biochemistry*, 6th edn. W.H. Freeman, San Francisco.

Brixner, T., Stenger, J., Vaswani, H.M. et al. (2005) Two-dimensional spectroscopy of electronic couplings in photosynthesis. *Nature* **434**, 625–8.

Churchill, M.E.A. (2003) Watching flipping junctions. *Nature Structural Biology* **10**, 73–5.

Katiliene, Z., Katilius, E., and Woodbury, N.W. (2003) Single molecule detection of DNA looping by *NgoMIV* restriction endonuclease. *Biophysical Journal* **84**, 4053–61.

Li, Y.F., Zhou, W., Blankenship, R.E., and Allen, J.P. (1997) Crystal structure of the bacteriochlorophyll *a* protein from *Chlorobium tepidum*. *Journal of Molecular Biology* **271**, 456–71.

Matthews, B.W. and Fenna, R.E. (1980) Structure of a green bacteriochlorophyll protein. *Accounts of Chemical Research* **13**, 309–17.

McKinney, S.A., Declais, A.C., Lilley, D.M.J., and Ha, T. (2003) Structural dynamics of individual Holliday junctions. *Nature Structural Biology* **10**, 93–7.

Mollova, E.T. (2003) Single-molecule fluorescence of nucleic acids. *Current Opinion in Chemical Biology* **6**, 823–8.

Nie, S. and Zare, R.N. (1997) Optical detection of single molecules. *Annual Review of Biophysics and Biomolecular Structure* **26**, 567–96.

Olson, J.M. (1998) Chlorophyll organization and function in green photosynthetic bacteria. *Photochemistry and Photobiology* **67**, 61–75.

Peterman, E.J.G., Sosa, H., and Moerner, W.E. (2004) Single-molecule fluorescence spectroscopy and microscopy of biomolecular motors. *Annual Review of Physical Chemistry* **55**, 79–96.

Weiss, S. (1999) Fluorescence spectroscopy of single biomolecules. *Science* **283**, 1676–83.

PROBLEMS

14.1 Write Beer's law and define all of the terms.

14.2 What is the frequency of light with a wavelength of (a) 1 nm, (b) 500 nm, and (c) 3 cm? Give the spectral classification of the radiation of each frequency.

14.3 What is the wavelength of light with a frequency of (a) 10^{18} Hz, (b) 10^{6} Hz, and (c) 10^{12} Hz. Give the spectral classification of the radiation of each wavelength.

14.4 A 0.1% solution of lysozyme gives an absorption of 2.7 at 280 nm. (a) If the concentration is 0.01%, what absorption would be measured? (b) If an absorbance of 1.0 is measured, what is the concentration?

14.5 Adenosine 5′-phosphate has an extinction coefficient of $1.54 \times 10^{4}\ M^{-1}\ cm^{-1}$ at 260 nm. (a) If the concentration of the sample is 1 μM, what absorbance would be measured in a 1-cm cuvette? (b) If an absorbance of 0.1 is measured, what is the concentration?

14.6 If the extinction coefficient of a molecule is $1000\ M^{-1}\ cm^{-1}$ and you measure an absorption of 0.1 in a 1-cm cuvette, what is the concentration of the molecule?

14.7 For a sample in a 1-cm cuvette, if the extinction coefficient of a molecule is $1 \times 10^4\ M^{-1}\ cm^{-1}$ at 500 nm, what is the absorbance of a sample at 500 nm if the concentration is (a) 1 μM and (b) 0.2 mM?

14.8 A protein sample is measured to have absorbance values of 0.717 and 0.239 at 240 and 280 nm, respectively. Assume that the spectrum is due to the presence of the aromatic amino acid residues tryptophan and tyrosine and determine the concentrations of these residues using the following extinction coefficients.

Wavelength (nm)	**Extinction coefficient ($M^{-1}\ cm^{-1}$)**	
	Tyrosine	**Tryptophan**
240	11,300	1960
280	1500	5380

14.9 For a given electron how is the rate of stimulated emission related to the light energy density?

14.10 What is the physical basis for the restriction of $\Delta l = \pm 1$ for an electronic transition?

14.11 Describe the difference between a stimulated and spontaneous process.

14.12 Why can a stimulated emission process be used to make a laser and not spontaneous emission?

14.13 For a given molecule, how is the fluorescence spectrum related to the absorption spectrum?

14.14 What factors give rise to the linewidth in biological samples?

14.15 What is the FRET efficiency if the value of the parameter r_o is 10 Å and the distance between the energy donor and acceptor is (a) 1 Å, (b) 10 Å, and (c) 20 Å?

14.16 If the observed lifetime of an excited state is 1.0 ns^{-1} and the natural radiative rate constant is 0.1 ns^{-1}, what is the nonradiative rate constant?

14.17 For bacterial reaction centers (Chapter 20), the quantum yield of fluorescence is 10^{-4}. The natural radiative decay rate is $10^8\ s^{-1}$. Assuming that there are no other nonradiative processes, what is the rate constant for the initial transfer of an electron from the excited state?

14.18 Why do some photosynthetic organisms have chlorosomes attached to the cell membrane?

14.19 What is the interpretation of an off-diagonal peak in a two-dimensional optical spectrum?

14.20 What can be measured in a single-molecule experiment that can only be inferred in a bulk experiment?

14.21 Why does the measured switching time between conformations vary in a single-molecule experiment?

14.22 Explain the signals shown for a FRET measurement of a Holliday junction; be sure to relate the signals to the three equilibrium states of the Holliday junction. Cy3 is the donor and Cy5 is the acceptor.

15

X-ray diffraction and extended X-ray absorption fine structure

Crystallography is the most common experimental method for obtaining a detailed picture of a protein or protein complex. Such experiments involve interpretation of the diffraction of X-rays from many identical molecules in an ordered array commonly referred to as a crystal. This experimental technique provides information on the positions of individual atoms within a biological complex. Having a detailed structure of a macromolecule aids greatly in understanding its function. Determining a protein's structure by X-ray crystallography consists of growing crystals of the purified protein, measuring the directions and intensities of X-ray beams diffracted from the crystals, and using computers to transform the X-ray measurements. This method produces an image of the crystal's contents. This image must be interpreted, which involves computer graphics to display the electron density of atoms in the molecule and the construction of a consistent molecular model.

X-rays are used in diffraction experiments because they have a wavelength of about 1 Å, which matches the atomic bond. The basic principles of diffraction are the same for X-rays as for visible waves, which have much larger wavelengths. Waves are described by their amplitude, A, wavelength, λ, frequency, ν, and velocity, v (Figure 9.1). When two waves come together in space they can add together constructively or destructively (Figure 15.1). When the position of two peaks matches they will add up constructively, but when a peak is added to a trough the two waves cancel.

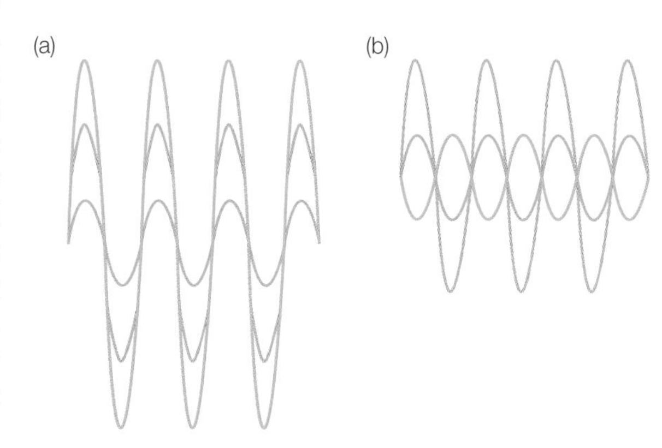

Figure 15.1 Waves can add together either (a) constructively or (b) destructively.

Diffraction is the property of many waves coming together. As a simple example, consider

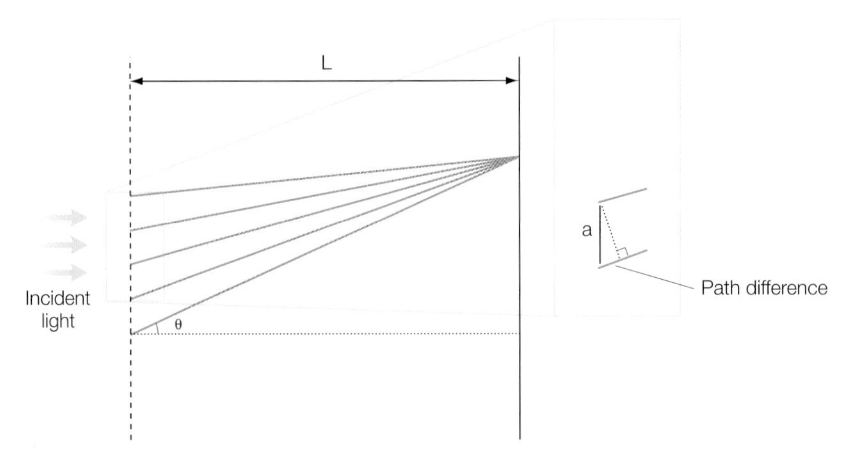

Figure 15.2 Diffraction from multiple slits.

a wave passing through a series of slits that are separated by a distance, a (Figure 15.2). Passing through a slit does not change the wavelength or amplitude. After passing through the slits, the waves all travel to a certain point in space, at a distance, D, and an angle, θ, and add together either constructively or destructively. Considering two waves that pass through neighboring slits, the only difference between the two waves is the distance that each traveled. To determine this difference in distance, a line is drawn from one slit that is perpendicular to a line formed by the path of the second wave (Figure 15.2). Assuming that the distance D is much larger than the separation a, the difference in pathlength can then be seen to be equal to $a \cos \theta$. Whether these two waves will add up constructively or destructively will depend only upon this pathlength difference. The two peaks will add up constructively when d is equal to λ, or correspondingly when:

$$a \sin \theta = \lambda \tag{15.1}$$

As the angle is increased or decreased, the path difference between neighboring slits is no longer equal to λ and the peaks of the two waves no longer match. The amplitude of the combined waves decreases until they reach a minimum when the peak from one wave corresponds to the trough from the second. This occurs when the path difference is equal to $\lambda/2$.

Extending the argument for all slits shows that the waves passing through all of the slits will combine constructively when the angle is:

$$\sin \theta = \frac{\lambda n}{a} \tag{15.2}$$

where n is an integer. For example, when n equals 1, the path distance between any two neighboring slits is equal to l, with the waves from the other slits having larger path differences of 2λ, 3λ, 4λ, etc., so that all of the waves will add together constructively. Thus, this angle corresponds to a diffraction peak.

The same arguments can be made for a line of equally spaced molecules forming a one-dimensional crystal. A wave striking each of the molecules can be considered to be a source of new waves. Assuming that there is no absorption, the wavelength remains the same for each wave. Normally, proteins in solutions will have different relative positions and orientations so that there is no net scattered wave. However, when every protein is aligned the same way, with the same relative distance, then the waves from each protein can combine into a diffraction pattern. Since each protein is aligned identically, each of the scattered waves will have the same amplitude and the only difference between waves is the relative path difference. As was done for the slits, the difference in pathlength can be related to the distance between the molecules and angle, yielding again eqn 15.2. So, for a line of molecules, the positions at which the diffraction points appear depend only upon the relative separation of the molecules. The composition of the molecule determines how much of the incident wave gets scattered; that is, the intensity of the wave.

BRAGG'S LAW

These ideas can be used to consider the interaction of X-rays with three-dimensional crystals through an analysis that was developed by the family team of William Henry Bragg and William Lawrence Bragg. They were able to use this formulism to solve a number of structures of simple inorganic crystals in the early 1900s. The crystal is considered to consist of lattice planes formed by the molecules of each layer of the crystal. Each lattice plane is treated as a mirror reflecting the X-rays back at the same angle, θ, at which they struck the mirror (Figure 15.3). Whether the X-rays reflected from two different lattice planes add together constructively or destructively depends upon the relative path difference for the two waves. For neighboring planes, the pathlength difference $AB + BC$ by geometry is:

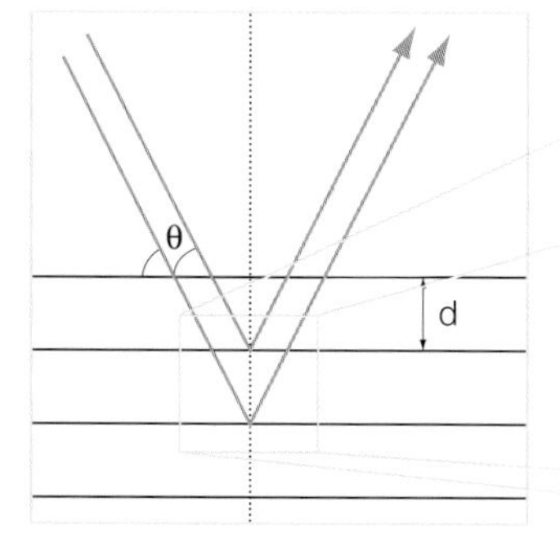

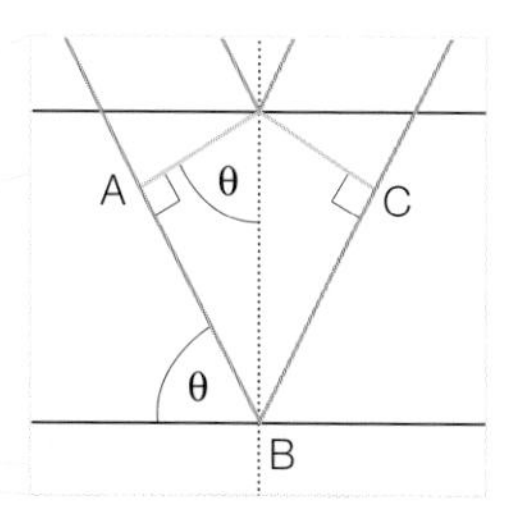

Figure 15.3
A diagram showing Bragg's law.

$$AB + BC = 2d \sin \theta \tag{15.3}$$

Constructive interference occurs when this pathlength difference is an integer number of wavelengths, or:

$$n\lambda = 2d \sin \theta \tag{15.4}$$

where n is a positive integer; this equation is known as Bragg's law. At this angle, each of the waves from the other lattice planes will also add together constructively since those relative pathlengths will be a multiple of the wavelength. Destructive interference occurs when the pathlength equals $n\lambda/2$.

As a detector is rotated through the angle Θ, it will alternatively measure strong X-rays followed by weak ones. Notice that this analysis makes a prediction that if a lattice plane is inserted midway between each of the existing planes then the additional waves will have a pathlength equal to a multiple of half a wavelength, resulting in destructive interference. As discussed below, this corresponds to the difference between primitive lattices and centered lattices and shows that centered lattices will have reflections systematically missing compared to primitive lattices.

To measure all of the diffracted waves, the crystal is rotated relative to the X-ray beam. As the crystal rotates, the distance between the lattice planes will change depending upon the specific packing of the molecular protein in the crystal. Therefore, for most angles, it is necessary to consider all three dimensions of the crystal, which can be grouped into certain classes, known as Bravais lattices, as described below.

BRAVAIS LATTICES

A crystal is composed of molecules that are packed such that there is an exact separation distance between the molecules in all three directions x, y, and z. Thus, a crystal can be considered to be built from regularly repeating structural motifs which may be atoms, molecules, or proteins. Each molecule or protein is packed in a crystal such that it is possible to move a certain length from an atom of the protein and come back to the same atom. The packing of the molecules is represented by a lattice that specifies the location of a structural motif (Figure 15.4). The unit cell is the smallest box that contains one of these repeating patterns, and is commonly formed by joining neighboring lattice points (Figure 15.5).

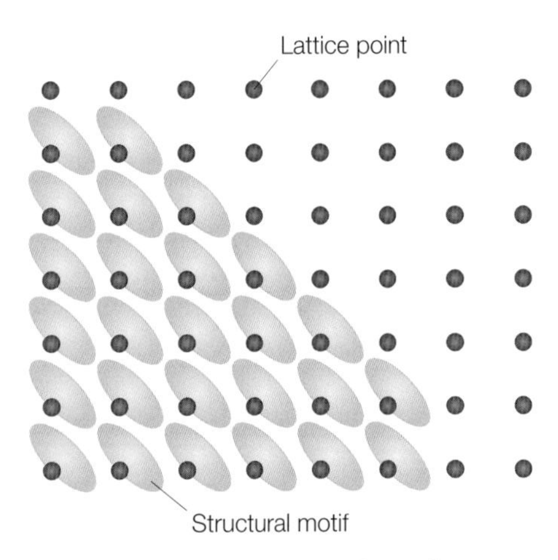

Figure 15.4 Every crystal can be considered as consisting of a structural motif that repeats in space, forming a lattice.

In 1850, it was shown by Auguste Bravais that the different repeating units, termed unit cells, can be classified into seven crystal systems that reflect the rotational symmetry of the cell (Table 15.1). For example, a cubic unit cell has four 3-fold axes, denoted by C_3, in a tetrahedral array while a monoclinic unit cell has one 2-fold axis and a triclinic cell has no rotational symmetry.

Bravais was able to show that there are only 14 distinct crystal lattices in three dimensions (Figure 15.6). The

Table 15.1
The seven Bravais crystal systems.

System	Symmetry*
Triclinic	None
Monoclinic	One C_2 axis
Orthorhombic	Three perpendicular C_2 axes
Rhombohedral	One C_3 axis
Tetragonal	One C_4 axis
Hexagonal	One C_6 axis
Cubic	Four C_4 axes in tetrahedral arrangement

*The rotation axes are denoted by C_n, where $360°/n$ is the rotational symmetry.

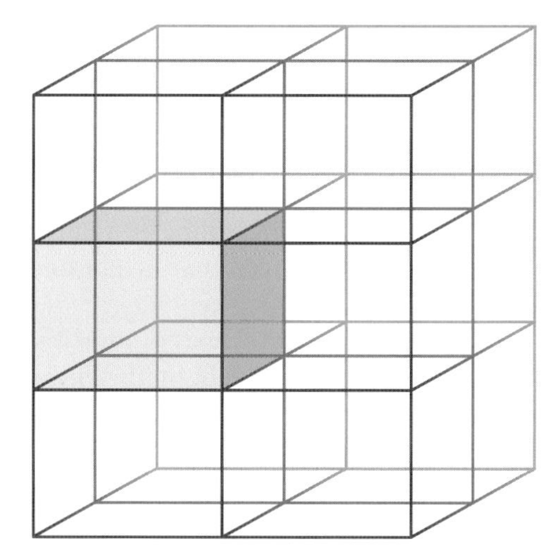

Figure 15.5 Crystals are formed by repeating units termed unit cells.

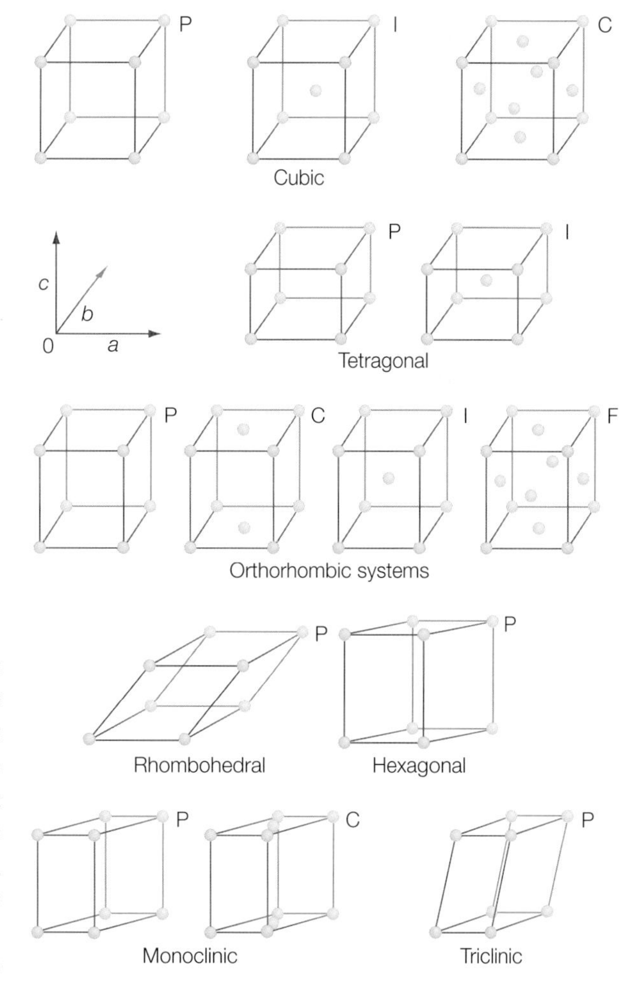

Figure 15.6 The Bravais lattices showing primitive (P), body-centered (I), face-centered (F), and side-centered (C) lattices.

simplest form of these cells are the primitive cells, denoted as P, which have lattice points at each corner. The non-primitive cells have lattice points, not only at the corners, but also at additional locations. A body-centered unit cell, denoted by I, has a lattice point at the center of the cell, a side-centered cell, denoted by C, has two lattice points on opposite faces, and a face-centered cell, denoted by F, has lattice points at the center of each of the six faces.

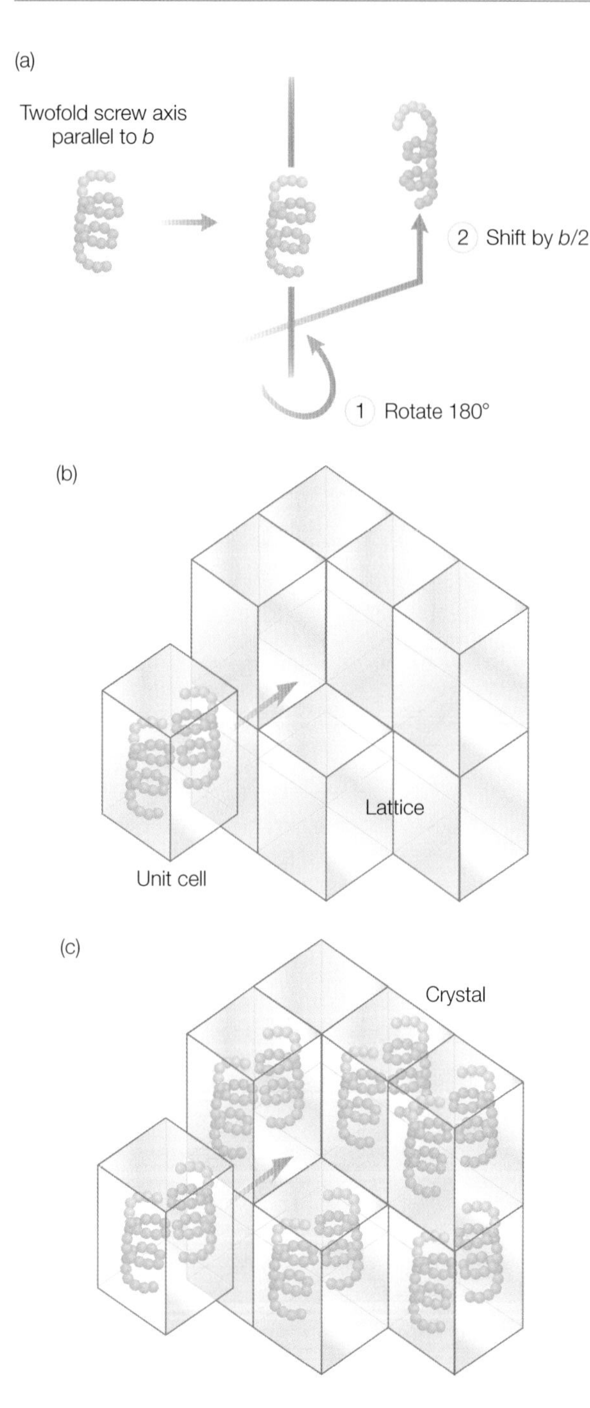

Figure 15.7 The packing of a protein in a crystal with 2-fold symmetry: (a) the symmetry relationships, (b) the packing of the unit cell, and (c) the packing in the crystal.

PROTEIN CRYSTALS

Proteins are only stable in the presence of the proper buffer; therefore, the crystals are always grown from solutions. Proteins can pack into lattices but the choice of symmetries is limited because of the intrinsic asymmetry of the protein backbone. There are two mirror-image forms of amino acids, called the L isomer and the D isomer, but in proteins only the L isomer is found (see Figure 1.7). Thus, no symmetry operation that involves a mirror symmetry is allowed. This restricts the number of possible space groups to 65 for which proteins, or any other asymmetric molecule, can crystallize.

The packing of protein crystals is illustrated by the simple helical molecule (Figure 15.7). In this case, the molecules are packed according to a 2-fold symmetry that is common in protein crystals. The two symmetry-related molecules form the unit cell that is repeated in the crystal.

Because of the irregular shapes of proteins, they pack in crystals relatively inefficiently, leaving a large amount of space unoccupied by protein. As a result, a large fraction of a protein crystal, anywhere from 20 to 80%, is filled not by protein but rather by the crystallization solution. This makes protein crystals much softer than simple salt crystals.

Crystals of proteins are obtained by adding a salt to a protein solution. The presence of the salt reduces the protein's solubility and forces it to either form crystals or precipitate. Fast salting out favors precipitation whereas slow salting out favors the formation of highly ordered crystals. One approach to crystallizing a protein is to slowly make a supersaturated solution and allow it to sit undisturbed until crystals are formed after a few days (Figure 15.8). Alternatively, the crystallization can be slowed down even further by the use of vapor diffusion in which the water from a protein solution is removed slowly by vapor exchange with a large reservoir solution that is initially poised at a higher salt concentration. Since the container is a closed system, the protein drop will concentrate slowly with time and hopefully produce crystals once the protein concentration exceeds the solubility.

Protein precipitate has no particular shape but protein crystals have well-defined facets and edges despite their large solvent content (Figure 15.9). The overall morphology of the crystal will usually reflect the space group of the crystal. Proteins will often crystallize in more than one space group as the contacts that stabilize the crystal involve hydrogen bonds, and salt bridges can be formed by different combinations of surface resides.

Some proteins are crystallized easily whereas others require much effort because the proper conditions, such as pH, temperature, salt concentration, and protein concentration, can be found only by trial and error, which requires considerable time, effort, and protein. Sometimes the crystallization is restricted by limitations in the stability of a protein. All of these problems are tremendously multiplied in proteomics, which has the goal of determining the structure of every protein in a given organism. Despite these difficulties, efforts are underway in a number of laboratories to achieve this goal by developing expression systems that can yield large quantities of each gene product and large-scale robotic control of crystallization.

DIFFRACTION FROM CRYSTALS

The diffraction pattern from a complex molecule cannot be solved directly from Bragg's law, but requires a more

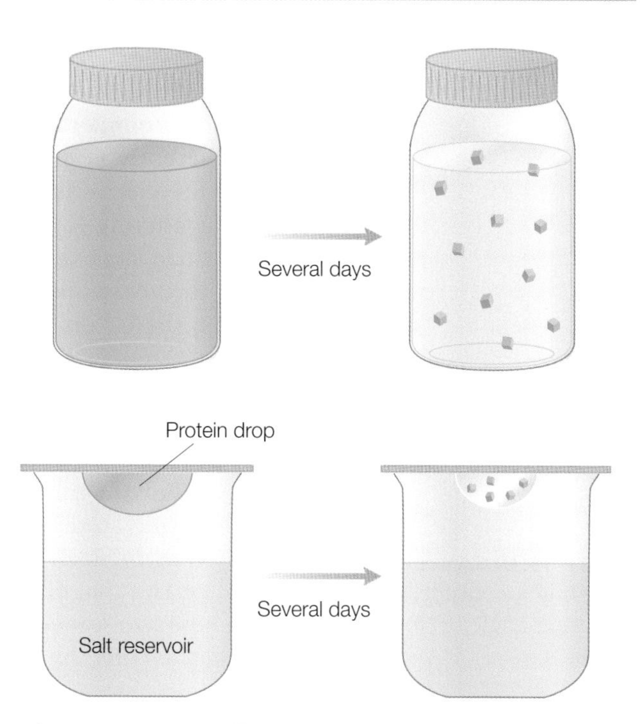

Figure 15.8 Crystallization using batch- and vapor-diffusion methods. The protein drop in the vapor-diffusion method is usually very small, at about 1 μl, to minimize the amount of protein.

Figure 15.9 Crystals of the protein phenoxazinone synthase. See Smith et al. (2004).

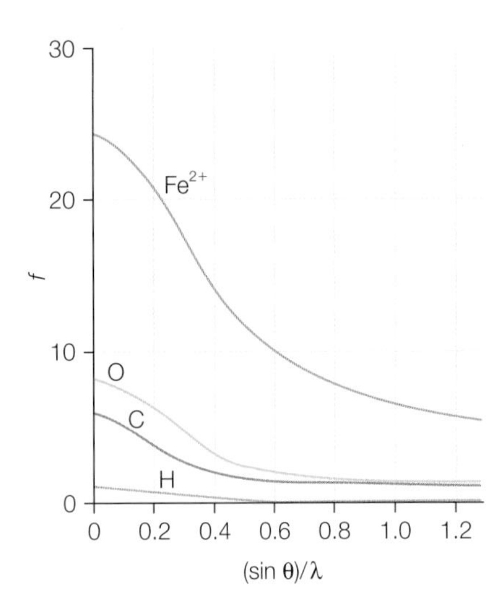

Figure 15.10 Dependence of the scattering amplitude, f, on the scattering angle, θ, for different atoms.

detailed analysis since the contribution from each atom of the molecule must be considered in the analysis. Each type of atom present in a molecule will scatter the X-rays differently (Figure 15.10). Since the X-rays are scattered by the electrons of the atoms, the primary effect is that the scattering is larger for atoms with more electrons, so that at an angle of 0°, the amplitude of the scattering, f_j, increases as the atomic number increases. For every atom, there is a similar drop off in the amplitude as the angle between the initial beam and the measuring point increases.

A molecule is considered to be composed of atoms, each of which has a certain scattering amplitude, f_j, at a point in space denoted by x_j, y_j, and z_j. These molecules pack inside the unit cell of a Bravais lattice. For any two neighboring atoms, A and B, whether the scattered waves add constructively or destructively depends upon the relative phases that differ according to the pathlength difference of the two scattered waves. If the phase of each scattered wave is denoted by ϕ_j, then the sum of the two waves **F** is:

$$\mathbf{F} = f_A e^{i\phi_A} + f_B e^{i\phi_B} \qquad (15.5)$$

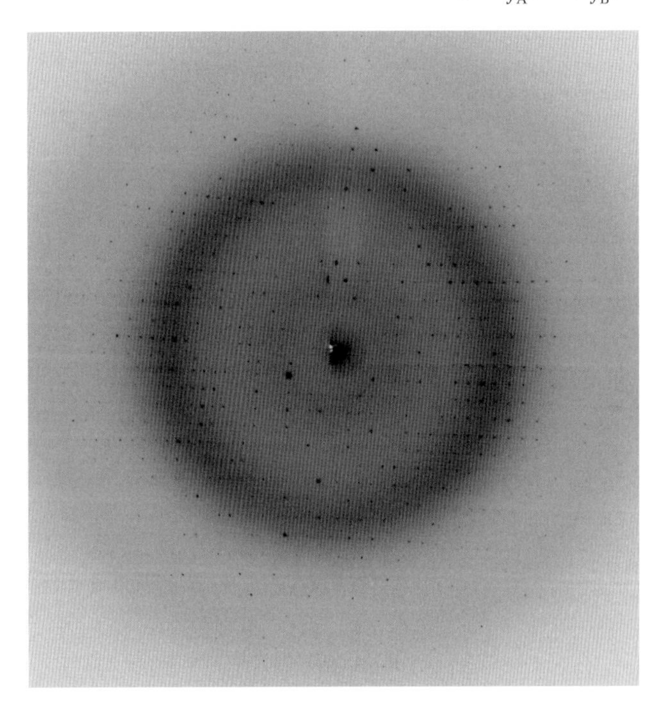

Figure 15.11 Oscillation photograph of the diffraction from crystals of the small metal-binding protein. The dark ring is due to solvent present in the crystal.

To determine the scattering from a molecule, the scattering from each atom is added together as the product of the amplitude and phase factor according to:

$$\mathbf{F}(hkl) = \sum_j f_j e^{i\phi_j(hkl)} \qquad (15.6)$$

The quantity $\mathbf{F}(hkl)$ is called the structure factor, with the indices hkl being used to specify the diffraction peaks. In a typical measurement, the protein crystal is oscillated through a small angle and the diffraction peaks form a series of circles (Figure 15.11). The position of each peak is determined by the space group. By measuring the diffraction for a series of different angles it is possible to identify the space group by comparing the positions of the peaks at the different angles. After the space group is found, each peak is identified with a hkl value and the intensity of each spot is measured.

Because protein crystals have such a significant amount of the crystallization solution, there is a dark ring due to the solvent. To correct for this and other contributions, the intensity measurement includes a subtraction of the average darkness of the region surrounding each spot. For a given protein crystal, tens to hundreds of thousands of peaks are measured, yielding a listing of intensities for each *hkl* value.

In a diffraction experiment, information concerning the molecules forming the crystals is found by analyzing the intensity of each diffraction point. The amplitude of each diffraction point, $\mathbf{F}(hkl)$, is related to the position of each atom according to eqn 15.6. This equation can be reversed by use of a Fourier transform that yields:

$$\rho(r) = \frac{1}{V}\sum_{hkl}\mathbf{F}(hkl)\, e^{-2\pi i(hx+ky+lz)} \tag{15.7}$$

In this equation, $\rho(r)$ is the *electron density* at the position r in the crystal, since in practice the discrete atomic positions are not observed. The intensity of the scattered waves decreases with increasing angle (Figure 15.10), so measurement of the higher-angle diffraction peaks can be difficult, especially for proteins crystals that typically diffract X-rays weakly. The greatest angle for which the diffraction can be measured is identified by the resolution limit, with the larger angles corresponding to the smaller resolution limit. To solve the structure of a protein, the resolution limit should be at least 3 Å. For a lower value of 5 Å, the backbone can be traced but the positions of the side chains cannot be determined, whereas a resolution limit of 1 Å allows the determination of the positions of the protons that are the weakest-diffracting atoms.

Derivation box 15.1

Phases of complex numbers

The diffraction from a crystal is described in terms of structural factors that are complex numbers with both an amplitude and phase. A complex number, $\mathbf{F}$, can be considered to consist of a real component, A, and an imaginary component, iB:

$$\mathbf{F} = A + iB \quad \text{where } i = \sqrt{-1} \tag{db15.1}$$

The complex conjugate of a complex number, $\mathbf{F}^*$, is defined as:

$$\mathbf{F}^* = A - iB \tag{db15.2}$$

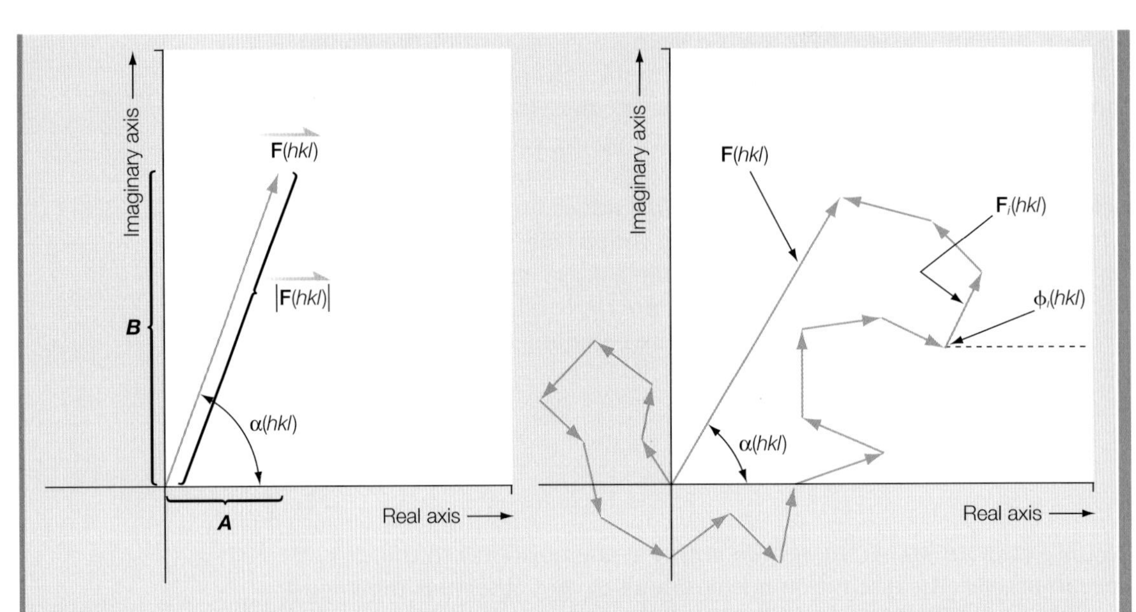

Figure 15.12 Representation of structure factors as arising from a summation of vectors in a comple plane. Each structure factor is represented by a length $\mathbf{F}(hkl)$ and phase $\alpha(hkl)$. The scattered wave from each atom gives rise to the structure factor for each reflection.

Complex numbers can be considered to be vectors in a complex plane with the component along the real axis being A and the component along the imaginary axis being B (Figure 15.12). The phase angle, α, is defined as the angle between the vector and real axis. The components A and B can then be expressed in terms of the vector amplitude and the phase angle:

$$A = F\cos\alpha \quad \text{and} \quad B = F\sin\alpha \tag{db15.3}$$

The vector amplitude or length F is sometimes termed the magnitude or modulus of the vector. From the geometry of the vector, the amplitude and angle can be expressed in terms of the two components A and B:

$$F = \left|\mathbf{F}\right| = \sqrt{A^2 + B^2} = \sqrt{\mathbf{FF^*}} \quad \alpha = \tan^{-1}\frac{B}{A} \tag{db15.4}$$

Combining eqns db15.1 and db15.3 yields:

$$\mathbf{F} = A - iB = F\cos\alpha + iF\sin\alpha = F(\cos\alpha + i\sin\alpha) = Fe^{i\alpha} \tag{db15.5}$$

In X-ray diffraction, the intensity of each reflection differs as the contributing atoms scatter with different phases, hence the sum of all contributions differ. Each structure factor $\mathbf{F}(hkl)$ can be considered as the sum of all contributions of the X-rays scattered from all atoms within the unit cell (Figure 15.12). For each $\mathbf{F}(hkl)$, each ith atom contributes to the sum

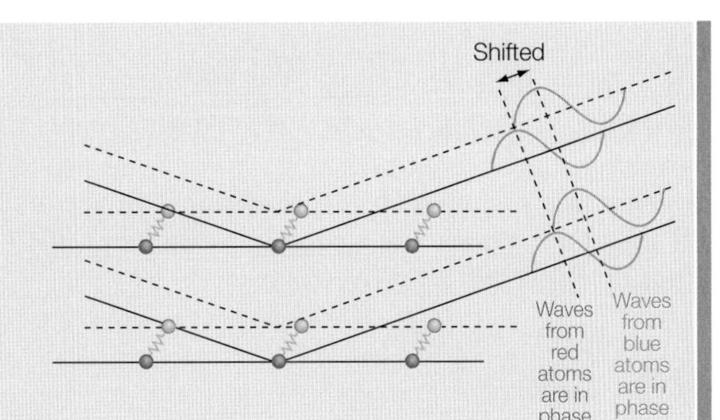

Figure 15.13 The scattering from the lattice of each atom type is in phase but the waves from the two lattices are not in phase.

according to its atomic scattering factor, $\mathbf{f}_i(hkl)$, which has both an amplitude $f_i(hkl)$ and a phase $\varphi_i(hkl)$:

$$\mathbf{F}(hkl) = F(hkl)e^{i\alpha(hkl)} = \sum_{i=1}^{N} \mathbf{f}_i(hkl) = \sum_{i=1}^{N} f_i(hkl)e^{i\phi(hkl)} \quad \text{(db15.6)}$$

The amplitude $f_i(hkl)$ depends upon the type of atom (Figure 15.10) and the phase $\varphi_i(hkl)$ depends upon the position of the atom in the unit cell. Notice that $\varphi_i(hkl)$ is the phase for the scattering of the *i*th atom and is distinct from $\alpha(hkl)$, which is the phase of the *hkl* reflection (Figure 15.12).

To derive an expression for the phase $\varphi_i(hkl)$, consider the simple example of a diatomic molecule that forms an orthorhombic crystal (Figure 15.13). Atom 1 is positioned at the origin and atom 2 is positioned in the *xy* plane at (*x*,*y*,0). Each atom will scatter the X-rays, with the waves scattered from all atom 1 being in phase and the waves from all atom 2 being in phase, but the waves from the two sets of atoms are not necessarily in phase. The phase of the *hkl* reflection will shift by $2\pi h$ if atom 2 is positioned a distance of *a* away from atom 1. Since atom 2 is shifted in the *x* direction by a distance *x* relative to atom 1, the phase difference for this shift is given by this factor reduced by x/a:

$$\phi_x(hkl) = 2\pi h\left(\frac{x}{a}\right) \quad \text{(db15.7)}$$

Considering all three dimensions yields a total phase difference of:

$$\phi(hkl) = 2\pi\left(\frac{hx}{a} + \frac{ky}{b} + \frac{lz}{c}\right) \quad \text{(db15.8)}$$

The expression for the structure factor can now be simplified by introducing the *diffraction vector*, **S**, and the *position vector*, **r**:

$$\mathbf{S} = h\mathbf{a}^* + k\mathbf{b}^* + l\mathbf{c}^*$$
$$r = (x/a)\mathbf{a} + (y/b)\mathbf{b} + (z/c)\mathbf{c} \qquad \text{(db15.9)}$$

where $(\mathbf{a},\mathbf{b},\mathbf{c})$ and $(\mathbf{a}^*,\mathbf{b}^*,\mathbf{c}^*)$ represent unit vectors in real and reciprocal space, respectively. The expression for the structure factor (eqn db15.6) can now be written in terms of the position and diffraction vectors:

$$\mathbf{F}(\mathbf{S}) = \sum_{i=1}^{N} f_i(hkl)e^{2\pi i(r_i \cdot S)} \qquad \text{(db15.10)}$$

A more general form of this relationship treats the electron density of the diffracting object as a continuous object rather than as a sum of discrete atoms. In this case, the summation is replaced by an integral and the discrete atomic scattering factors are replaced by the electron density at the location:

$$\mathbf{F}(\mathbf{S}) = \int \rho(r)e^{2\pi i(\mathbf{r}\cdot\mathbf{S})}\,\mathrm{d}V \qquad \text{(db15.11)}$$

This last equation represents a Fourier transform from real space to reciprocal space. Jean Baptiste Joseph Fourier lived in France during the French Revolution and discovered Fourier transforms as a method of relating two functions by integral transforms. These transforms are used in many applications, including the conversion of the temporal response of timed nuclear magnetic resonance (NMR) pulses into a chemical-shift spectrum (Chapter 16).

PHASE DETERMINATION

In a diffraction experiment, the intensity of each diffraction peak, that is $|\mathbf{F}(hkl)|^2$, is measured. To determine the electron density, the value of the amplitude and phase of $\mathbf{F}(hkl)$ is needed. Unfortunately, the phase cannot be determined directly from the intensity because the structure factor is a complex number and the phase associated with the structure factor is not determined in the experiment. There are three approaches that can be used to overcome this phase problem.

Molecular replacement

In the case that the structure of the protein crystallized is unknown, but the structure of a related protein has already been determined, then the approach of molecular replacement can be used. The structure of the related protein is first modified to make a structural model that resembles the unknown protein. Based upon a comparison of the sequences, regions are removed that are found in the related protein but not in the unknown protein. Also, amino acid residues that are not conserved can be replaced with alanine. Computers are used to place the resulting structural model in the unit cell of the unknown protein crystal. For any given position and orientation of the model, the resulting diffraction pattern can be calculated by using eqn 15.6.

Isomorphous replacement

In this case, the phase problem is overcome by use of special modifications of the protein using a biochemical approach. Heavy metal compounds are added to the solution containing the crystals. These metal compounds are designed to bind to specific amino acid groups such as cysteine. Crystals of the protein with the heavy metal are measured using diffraction, and a difference approach can be used to determine the phase.

Consider one specific diffraction peak at a given *hkl* value. The native protein will have a structure factor that can be represented as a vector with an amplitude F_P rotated by an angle α_P (Figure 15.14). The protein with a bound heavy metal will have a different amplitude F_{PH} and different angle α_{PH}. The difference between these two vectors is due to the heavy metal that has an amplitude F_H and angle α_H. Since the angles associated with F_P and F_{PH} are unknown, the tips of their vectors can lie anywhere along two circles. However, since the difference between these two vectors is the vector $\mathbf{F_H}$, if the circle associated with F_{PH} is displaced by the vector $\mathbf{F_H}$, then the correct solution must lie at the points where the two circles intersect, and the correct angle is determined. Since two independent intersection points are found, at least two derivatives are needed to determine the correct one. Normally, the phases are found using several derivatives to minimize the error.

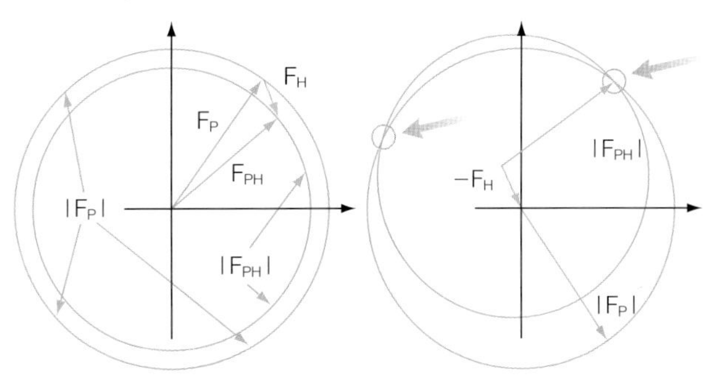

Figure 15.14 Diagram used to calculate phases from the native protein, F_P, heavy metal, F_H, and derivative, F_{PH}, scattering factors.

Anomalous dispersion

An assumption of the analysis of the X-ray diffraction is that the only effect of the atom is to scatter the X-ray beam. However, if the energy of the X-rays is at a value corresponding to a transition energy for a certain atom, then the phase will include a contribution associated with the transition of electrons for that specific atom, termed the anomalous dispersion. For proteins, the anomalous dispersion is large for any bound transition metals. By comparing the diffraction at different wavelengths, with energies near and away from the transition energies, the differences in the diffraction arising from the anomalous dispersion can be identified and the location of the metal can be identified. This approach is now commonly used to provide phase information by incorporating selenium at positions normally occupied by methionine by replacement of that residue with selenomethionine.

In addition to use of anomalous dispersion for phase information, for proteins with metal cofactors, the anomalous dispersion provides an unambiguous identification of the metal as each metal has a different wavelength dependence. This identification is particularly useful for the assignment of individual atoms in complex metal clusters or new cofactors. For example, the protein phenoxazinone synthase (Figure 15.9) was known to be a multicopper oxidase with four copper atoms. The electron density calculated for the protein showed the presence of a fifth metal cofactor (Figure 15.15). The diffraction was measured at different wavelengths at the Advanced Light Source synchrotron and the electron-density map calculated for copper showed a peak at the same position, demonstrating that the metal was a fifth copper atom. The role of this copper atom has not been determined and is under investigation (Smith et al. 2006).

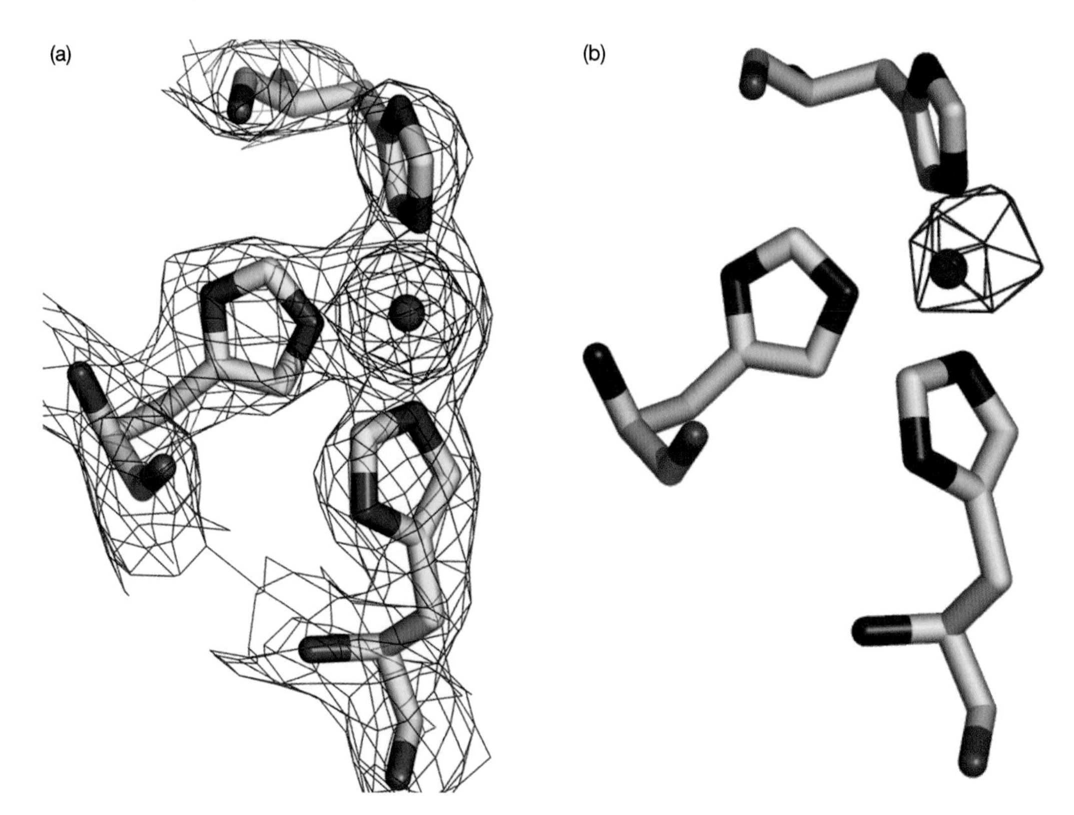

Figure 15.15 Electron-density maps showing for one region the positions of (a) all atoms of the protein phenoxazinone synthase and (b) only the copper atoms. Modified from Smith et al. (2006).

MODEL BUILDING

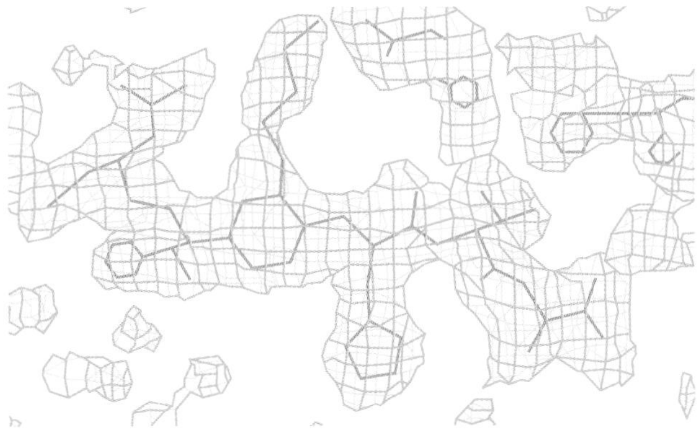

Once the phases have been determined then it is possible to calculate the electron density. The atoms present in a protein, C, N, and O, cannot be distinguished but rather a continuous electron density is observed that must be modeled as arising from a polypeptide chain. Although the density may appear to be difficult to interpret, it can be done because the polypeptide chain is uniquely defined and the arrangement of the side chains can be determined using gene or protein sequencing (Figure 15.16).

Figure 15.16 The electron density for a polypeptide chain.

The placement of the protein in the electron density is still done manually in most cases, although computer algorithms are becoming increasingly more powerful and useful for identifying the possible positions of atoms. This work is done using special programs in which both the structural model and the electron density are displayed and the operator can move the model using a mouse. These programs also have the capability of lengthening the polypeptide chain and it is placed in the density. Normally, the chain is built as a polyalanine peptide until the positions of critical amino acid residues, namely those near cofactors or the active site, are located. The positions of all of the atoms are then refined by computer programs that minimize the difference between electron density calculated using the structure factors for each atom (Figure 15.16) and the density calculated using the measured structure factors, F_{obs}.

For proteins, the electron density of itself is not sufficient to actually identify the positions of all of the atoms. However, the positions of individual amino acid residues are highly restricted since they must be part of the polypeptide chain. The orientation of each residue can be represented by two angles, ϕ and ψ, and the allowed values of these angles in proteins is shown on a Ramachandran plot. In addition, the identity of each side chain is determined by the sequence of the gene encoding the protein. Thus, after the backbone is built, the computer operator simply replaces each alanine with the proper residue and rotates the side chain into density. Likewise, cofactors are first built using chemical models and then placed into density. The *accuracy* of the resulting model is measured by the *Rfactor*, which is given by:

$$Rfactor = \sum_{hkl} \frac{F_{obs} - F_{calc}}{F_{obs}} \quad (15.8)$$

For proteins, this typically will have a value of between 0.15 and 0.20.

EXPERIMENTAL MEASUREMENT OF X-RAY DIFFRACTION

X-rays can be generated in the laboratory by bombarding a metal with high-energy electrons (Figure 15.17). As the electrons decelerate they generate bremsstrahlung radiation, which is approximately independent of wavelength. Superimposed upon this background are a few high-intensity sharp peaks that arise from the collision of the high-energy electrons with the electrons in the inner shell of the metal. The collision expels an electron from the inner shell and an electron from an upper energy level drops into the vacancy. This transition is associated with the emission of a photon. For copper, the wavelength of this photon is 1.54 Å for the L-to-K transition (Kα radiation).

For many protein diffraction experiments, measurements are also performed using synchrotrons. Synchrotron radiation is generated by electrons that circulate in a large storage ring (Figure 15.18). As the electrons circulate in the ring they generate bremsstrahlung radiation (Figure 15.17). Whereas the intensity of the laboratory X-ray beam is restricted by the efficiency of the cooling system, synchrotron radiation is limited by the strength of the magnets that bend the electrons to produce the radiation.

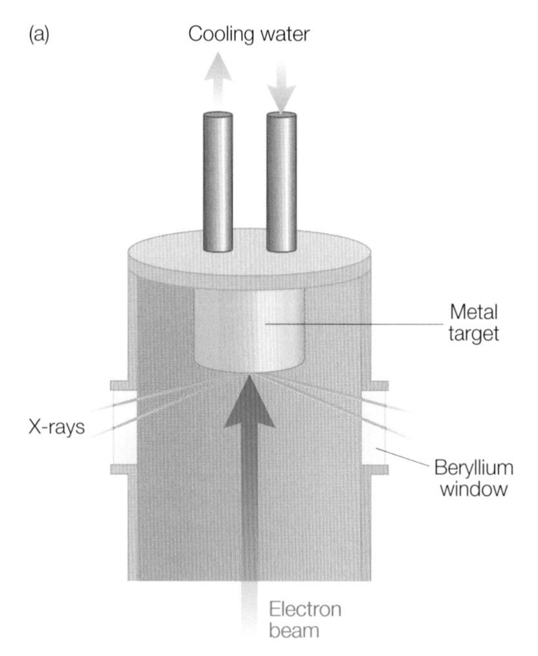

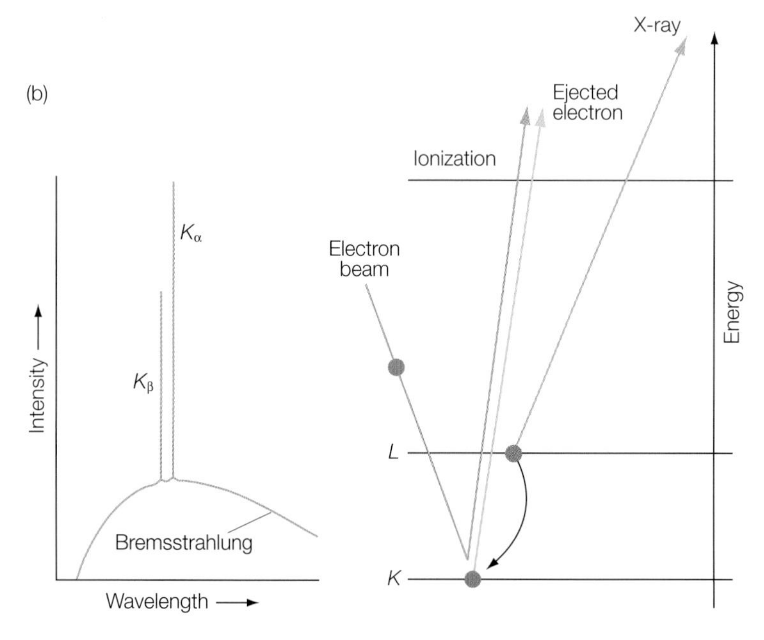

Figure 15.17 Experimental production of X-rays. (a) X-rays are generated in the laboratory by striking a target, usually copper, with a beam of electrons. (b) This results in a broad distribution of X-rays, known as bremsstrahlung radiation, and two sharp peaks corresponding to Kα transitions.

(a)

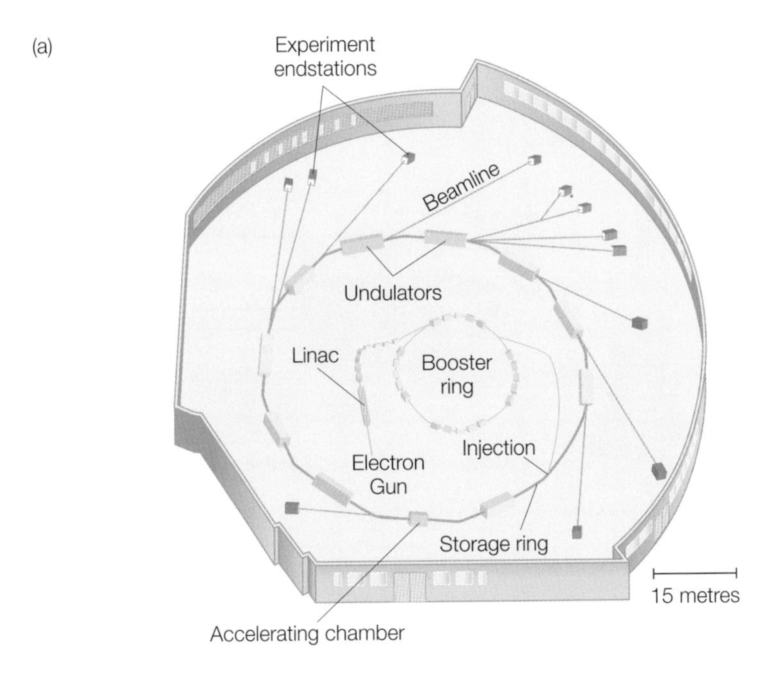

(b)

Figure 15.18 The Advanced Light Source. (a) A schematic view of the synchrotron. (b) Overhead view of the synchrotron. Figures courtesy of the Advanced Light Source.

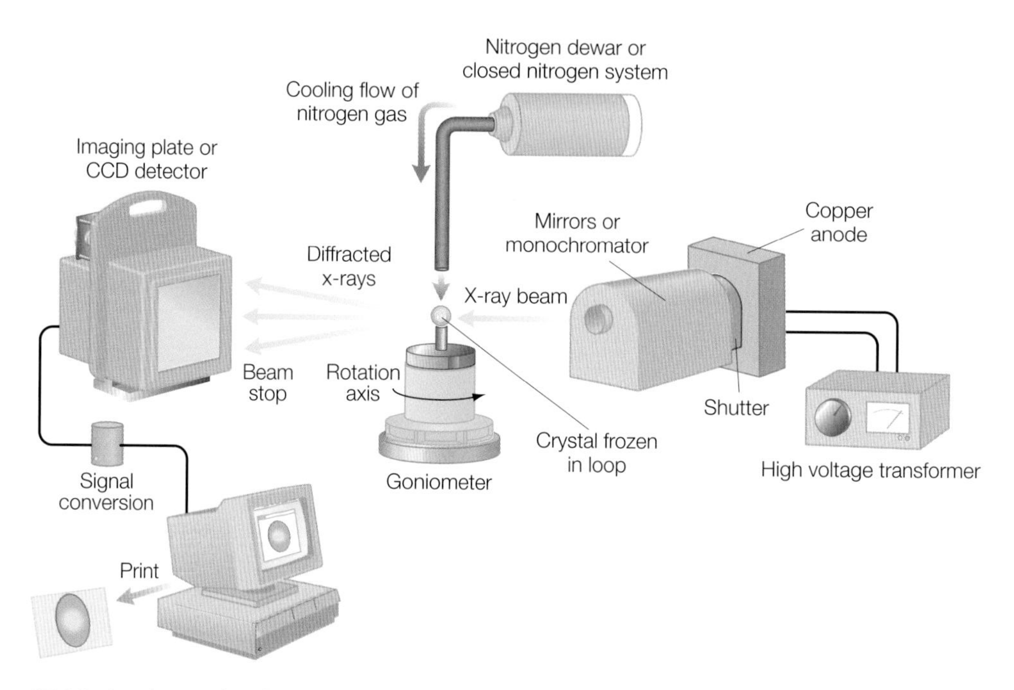

Figure 15.19 A scheme for the measurement of X-ray diffraction data.

The bremsstrahlung radiation is 100–1000 times more intense than what is available in the laboratory, with significant improvements planned for the future.

The structures of proteins are solved using single-crystal X-ray diffraction techniques. In this technique the intensity of every diffraction point should be measured. The crystal is positioned between the X-ray source and a detector (Figure 15.19). Usually, the protein crystal is frozen to minimize damage from the beam and cooled constantly by a stream of nitrogen gas. The electronic detector is designed to simultaneously measure the diffraction over a large area. Once measured, the diffraction is digitized and displayed on a computer monitor. With a few measurements, the space group is identified and orientation of the crystal is determined. The crystal is then rotated in increments with the diffraction measured for each angle. Electronic detectors are used, allowing the intensity of each spot to be integrated as the data for the next angle are measured so that, at the end of data collection, a full data-set is ready for analysis. For a laboratory source, a full data-set can be measured in 1 or 2 days while only an hour would be needed for the same crystal at a synchrotron. More importantly, the much greater intensity of the synchrotron source allows the accurate measurement of much smaller crystals than is possible using the laboratory source.

EXAMPLES OF PROTEIN STRUCTURES

All of this work is done so that the three-dimensional structure of proteins can be determined. Proteins are involved at all levels of cellular function and an understanding of their three-dimensional structure is required to model their function at a molecular level. For example, proteins have the capacity to specifically bind virtually any molecule. Enzymes play a critical role in cellular regulation and have the capacity to specifically stabilize the transition state, which gives them tremendous catalytic function. Myoglobin and hemoglobin have an oxygen-binding capacity that can be very sensitive to the conditions of the circulatory system and muscles.

Pepsin is an enzyme that digests proteins in the highly acidic environment of the stomach. Pepsinogen is a precursor that contains a segment of 44 residues that are proteolytically removed in the formation of pepsin. Diffraction analysis has shown that the active site of the precursor form is blocked by residues of the precursor segment at neutral pH (Figure 15.20). Note that a positively charged lysine side chain interacts electrostatically with a pair of negatively charged aspartate residues. When the pH is lowered, the salt bridges are broken and the exposed catalytic site hydrolyzes the peptide bond between the precursor and pepsin moieties.

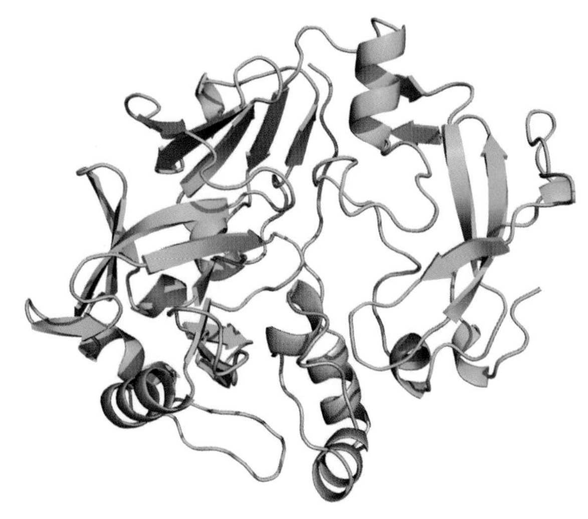

Figure 15.20 The backbone structure of pepsinogen (light blue) showing the stabilization of a precursor segment (brown).

Myoglobin and hemoglobin are oxygen-carrying proteins in the circulatory system and muscles. Hemoglobin is present in the blood and binds oxygen in the lungs, travels through the circulatory system, and transfers the oxygen to myoglobin in the muscles, where it is stored until needed. In both proteins the binding of oxygen to the heme group is regulated by the partial pressure of oxygen, as well as by pH and other factors. The ability of proteins such as these to bind molecules can be studied by X-ray diffraction by preparing the protein in different states. The conformational changes that myoglobin undergoes as it binds molecules is shown by the panels of Figure 15.21, which show the structures of the heme site (seen edge on) with a bound CO molecule and the protein after dissociation of the CO molecule. The location of the bound CO is shown as a dark blue cylinder and a light purple cylinder after dissociation. For more details on this series of experiments see Ostermann et al. (2000).

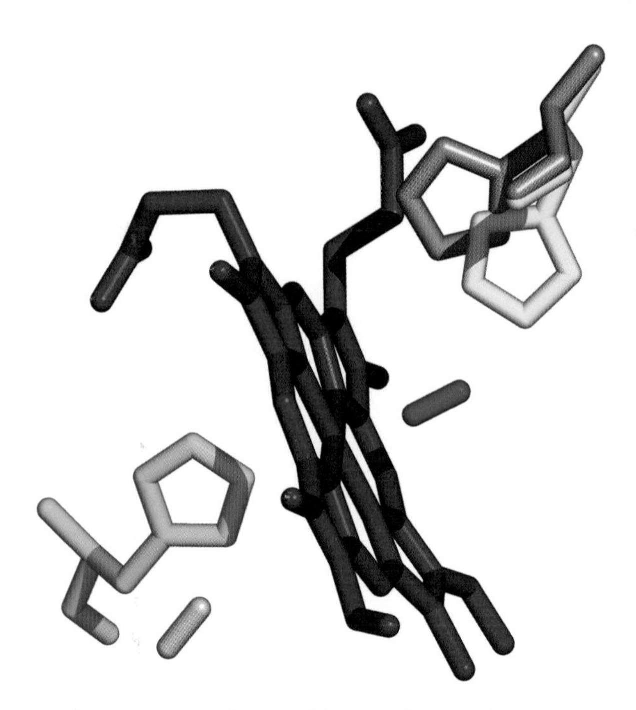

Figure 15.21 The structure of myoglobin with CO (shown as a small rod) bound to the heme (dark blue) and after dissociation (light purple). Notice that one histidine shifts in response to binding while the other does not appreciably move.

In addition to trapping proteins in different functional states it is now possible to monitor protein structure as it undergoes changes in time. Since data collection requires at least 1 h at a synchrotron, the experimental approach must be modified to observe subsecond structural changes. In considering the Bragg law (eqn 15.4), instead of measuring the diffraction for different angles of the crystal, it is also possible to use a range of wavelengths using the Laue method. For a fixed orientation of the crystal, the Bragg condition for different reflections will be satisfied simultaneously for different diffraction peaks for different combinations of wavelengths. In this case, a full data-set can be measured, ideally with only one orientation of the crystal, and the exposure time can be as short as a few milliseconds.

RESEARCH DIRECTION: NITROGENASE

As discussed in Chapter 3, nitrogen fixation is a fundamental biological process that is part of the overall nitrogen cycle for the Earth. The formation of ammonia from nitrogen and hydrogen gas is, overall, exothermic but kinetically is extremely slow due to the initial step of breaking the triple bond of N_2. Despite this unfavorable step, the overall reaction takes place readily in nature due to facilitation of the process by the nitrogenase enzyme system. Nitrogenase consists of two essential metalloproteins, the iron protein and the molybdenum–iron protein, which are named after their cofactor composition. The three-dimensional structure of the molybdenum–iron protein was originally reported at a modest resolution limit of 2.8 Å (Kim & Rees 1992). Subsequently, the quality of the diffraction data significantly improved to a resolution limit of 1.6 Å (Einsle et al. 2002). The major outcome of the high-resolution structure was the addition of a new atom in the cofactor (Figure 15.22).

The presence of the additional atom in the higher-resolution structure illustrates the difficulty in modeling the structures of proteins at limited resolutions. These limitations have a significant impact on proteins with complex metal clusters as the details concerning the metal coordination are often uncertain and difficult to model. One impact of any resolution limit is that the lack of measurable diffraction past a certain angle results in truncation errors in the Fourier analysis. To illustrate this point, the electron density at a point adjacent to one of the iron atoms of the cofactor was expressed as a function of the *resolution limit*, d_{max}:

$$\rho(r) = \int_0^{1/d_{max}} 4\pi s^2 f_{Fe}(s) \frac{\sin 2\pi sr}{2\pi sr} \, ds \qquad (15.9)$$

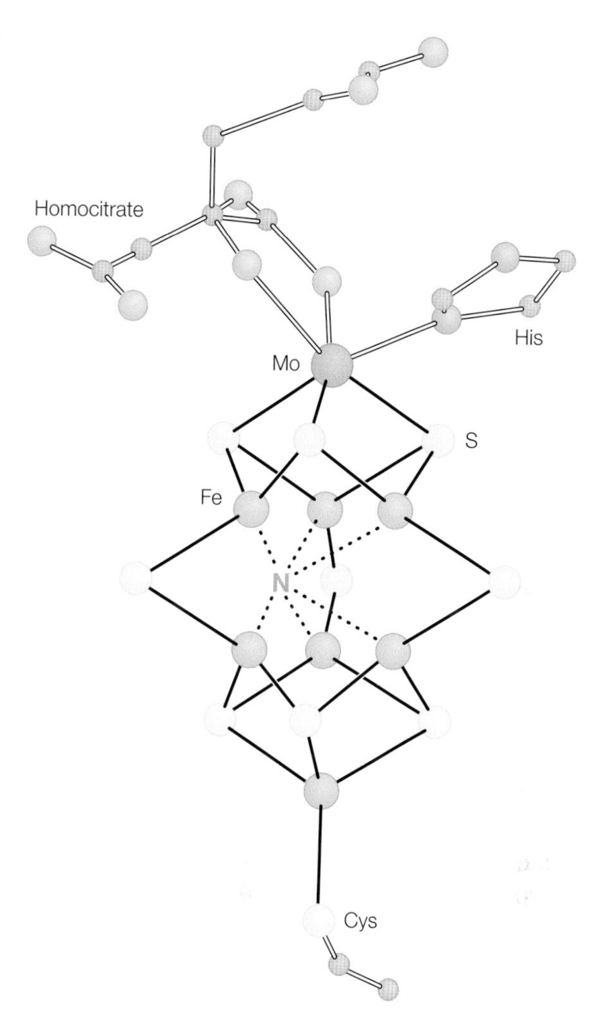

Figure 15.22 Model of the molybdenum–iron cofactor showing the presence of a central atom, tentatively identified as nitrogen, that was not present in the original model. Modified from Einsle et al. (2002).

where f_{Fe} is the atomic form factor for iron and $s = 1/d$ with d being the resolution (Einsle et al. 2002). When a finite resolution limit of 2.0 Å is inserted, an artificial negative density, sometimes termed a ripple, is calculated at the position where the putative nitrogen atom is located (Figure 15.23). The effects associated with the six central iron atoms and nine sulfur atoms of the cofactor combine to create a significant

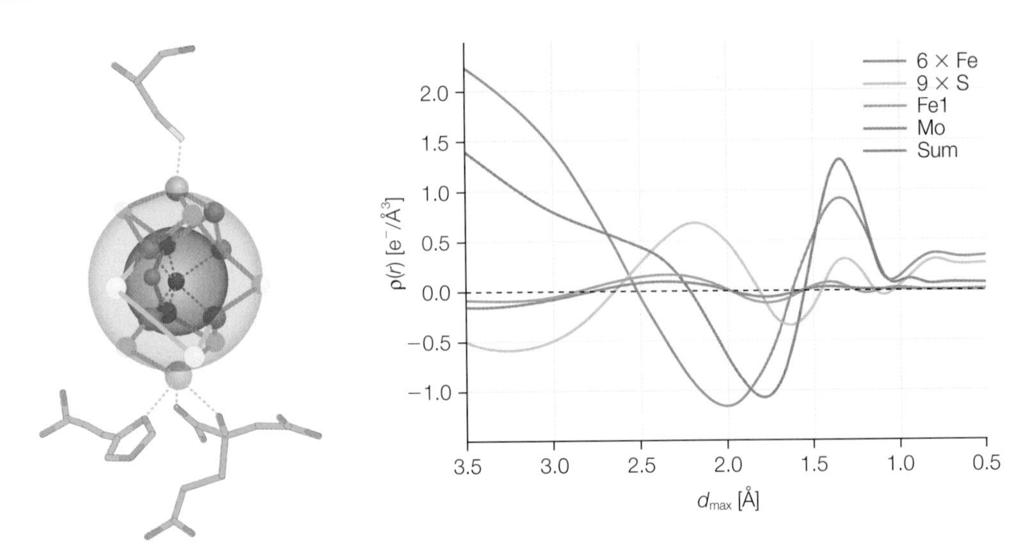

Figure 15.23 The effect of a series of termination errors on the resolution-dependent electron-density profile around the iron atom of the cofactor results in the calculation of negative density. Contributions of individual atom types to the resolution-dependent profile show a pronounced negative density at 1.8 Å. Modified from Einsle et al. (2002).

negative ripple that obscures the electron density of any light atom. In the higher-resolution data, the truncation effects are less pronounced and the calculated density no longer shows the ripple of negative density. Unambiguous identification of the new atom based solely upon the electron density is problematic as all of the light atoms – carbon, nitrogen, oxygen, and sulfur – are chemically plausible and consistent with the density, although the interaction of nitrogenase with dinitrogen and ammonia suggests the assignment of nitrogen. The need to insert an interstitial nitrogen adds to the complexity of the cluster and questions arise concerning the biosynthesis of the cofactor and assembly into the protein.

The mechanism of this complex enzyme system is under active investigation (Barney et al. 2005, 2006; Texcan et al. 2005; Peters & Szilagyi 2006). As part of the dynamics of the overall process, the iron protein docks to the molybdenum–iron protein, transfers an electron, and dissociates in a repetitive process until the eight-electron transfer is completed. Coupled to the docking-and-release gating process is the hydrolysis of MgATP. Crystal structures have been determined in different nucleotide states that identify conformational changes in the nitrogenase complex during ATP turnover (Figure 15.24). These structures show that the arrangement of the two proteins in the complex is dependent upon the nucleotide state. For each conformational state, different distances are observed between the redox cofactors involved in the electron-transfer process implying differences in the electron-transfer rates (Chapter 10). The spatial arrangement of these docking geometries may reflect large movements of the two proteins during the cycle of nucleotide binding and hydrolysis. In addition to such structural studies, work is ongoing to delineate the individual steps, in particular to locate where the dinitrogen binds, and to identify all intermediate states. Spectroscopic studies have been performed on intermediate states that are trapped by rapid freezing in liquid nitrogen, showing the sequential formations of a series of intermediate states, including diazene, hydrazine, and amine intermediates (Figure 15.25).

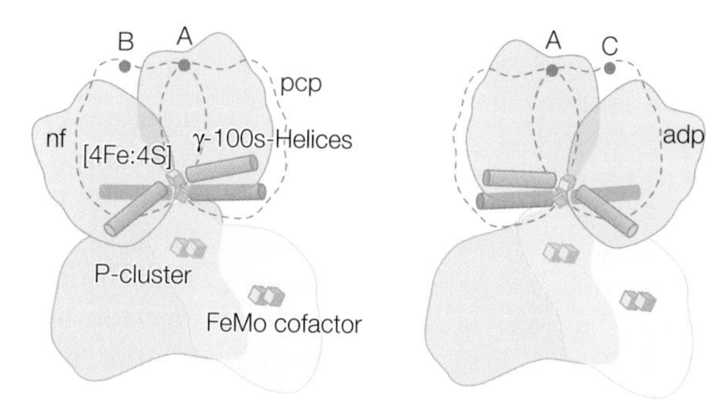

Figure 15.24 A schematic representation of the docking geometry for nf, the nucleotide form, pcp, the complex with an ATP analogue, and adp, the complex with ADP. Modified from Tezcan et al. (2005).

EXTENDED X-RAY ABSORPTION FINE STRUCTURE

In the X-ray diffraction experiments, the X-rays are assumed to only scatter from the atoms without any other changes. The exception to this assumption occurs when the energy of the X-rays is near a transitional energy of some of the atoms in the protein. The changes arising associated with this spectral region can be used to provide phase information through the anomalous dispersion effect discussed previously. Alternatively, this spectral region can also be used to probe the environment of metals.

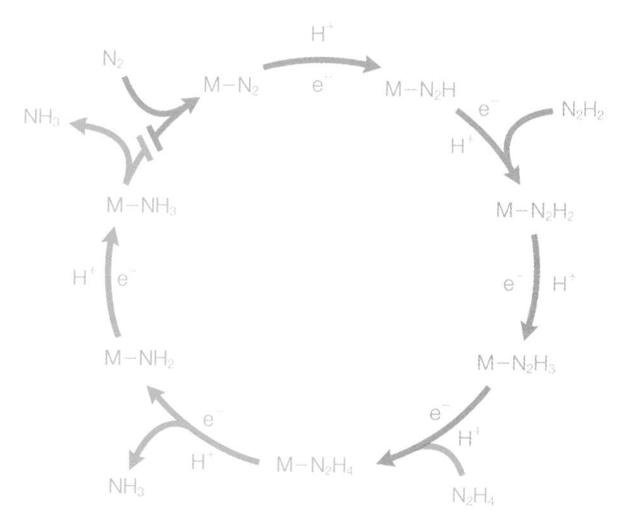

Figure 15.25 Putative process of N_2 reduction to NH_3 following a series of intermediate states. Modified from Barney et al. (2006).

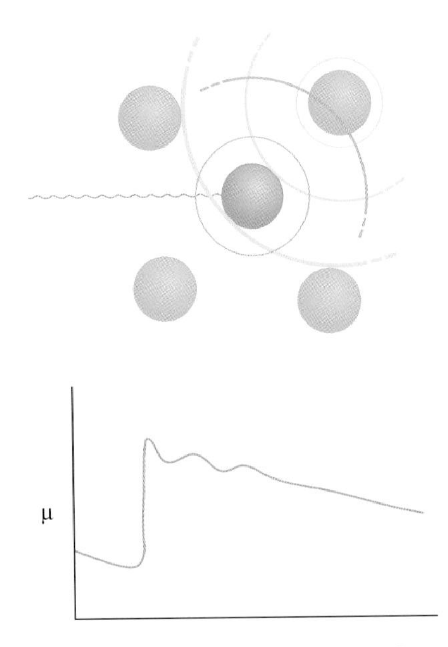

Figure 15.26 The absorption of X-rays shows a transition at a metal site followed by oscillations due to reflections from nearby atoms.

For a free metal, as the energy of the X-rays is varied, the intensity of the beam undergoes an abrupt shift when it matches a transition energy (Figure 15.26). If another atom is nearby then the transmission of the X-rays immediately after the transition is found to have features that are identified as extended X-ray absorption fine structure (EXAFS). The absorption of the photon leads to the emission of an electron from the atom, which has wavelength determined by the de Broglie relationship. This wave can be backscattered from a nearby second atom, and the backscattered wave consequently contributes to the initial wave in either a constructive or destructive manner that is dependent upon the pathlength, $2d$, where d is the distance between the atoms, relative to the wavelength:

$$\text{Phase} = 2\pi\left(\frac{2d}{\lambda}\right) \tag{15.10}$$

In a typical EXAFS experiment, the angles are fixed and the energy of the incident photon is varied. As the photon energy is varied the energy of the emitted electron can be calculated according to:

$$E = h\nu - E_{ion} = \frac{p^2}{2m_e} \tag{15.11}$$

where E_{ion} is the energy required to remove the electron from the atom. The phase will change as the energy of the photon changes due to the changes in the wavelength of the electron, which are found by using the de Broglie relationship (Chapter 9) and eqn 15.11:

$$\text{Phase} = \frac{(4\pi d)}{\lambda} = (4\pi d)\frac{p}{h} = \frac{(4\pi d)}{h}\sqrt{2m_e(h\nu - E_{ion})} \tag{15.12}$$

The change in phase as the energy of the photon, $h\nu$, changes gives rise to the ripples in the EXAFS signal. The phase is proportional to the bond distance d, so a Fourier transform can be used to determine the bond distance.

This technique is sensitive only to those atoms that have transition energies in the appropriate energy region, namely transition metals. Thus, for a protein, the signals will provide a direct measure of the coordination of the metal cofactors. As an example, there are many iron-containing

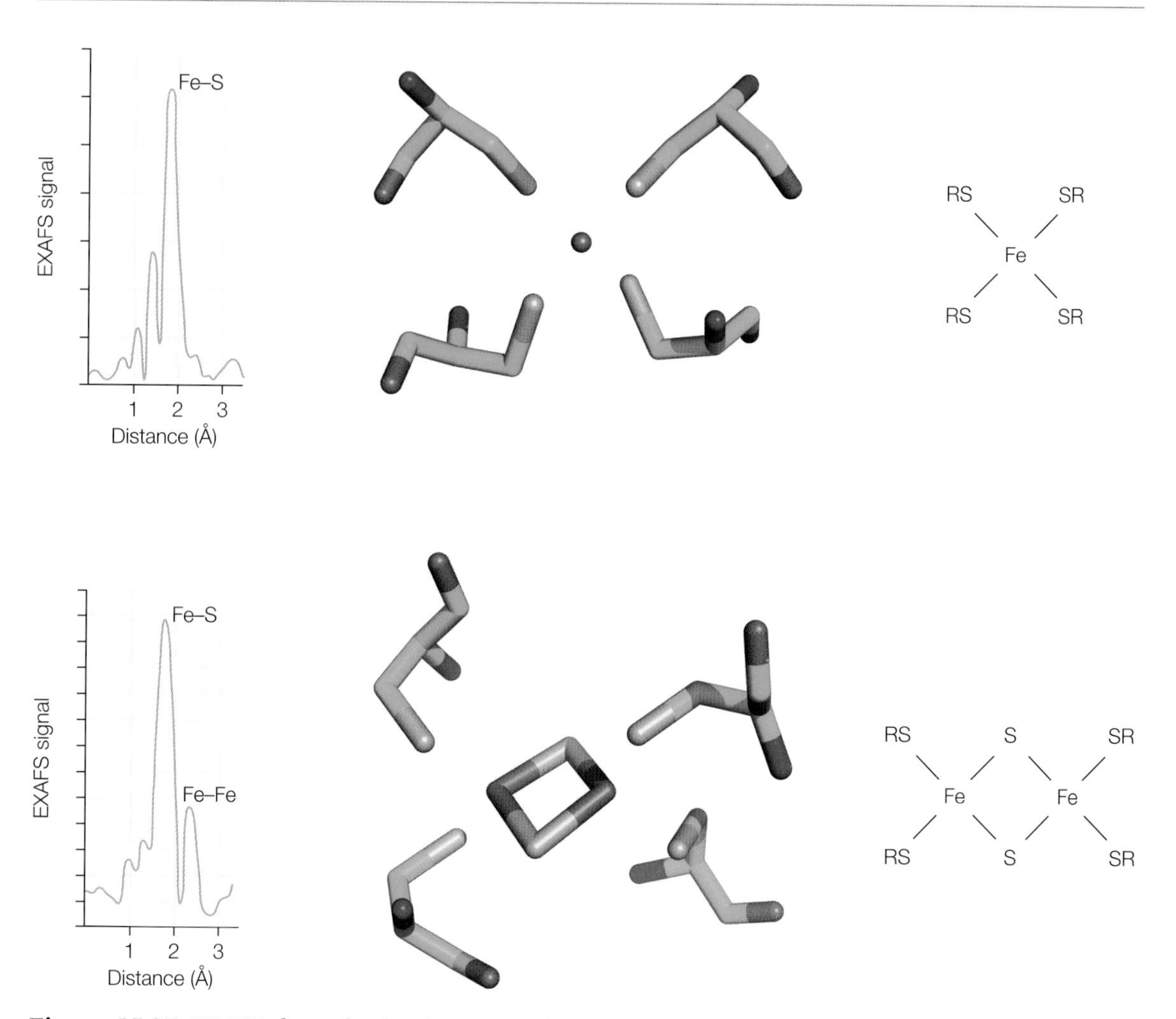

Figure 15.27 EXAFS data of rubredoxin and ferredoxin. Also shown are the three-dimensional and chemical structures of the iron–sulfur clusters. Modified from Yachandra (1995).

proteins that also contain sulfur. Typically, the metal cofactor of iron–sulfur proteins can be either a single iron, a Fe_2S_2 cluster, or a Fe_4S_4 cluster with the ligands all being provided by cysteine residues. For a single-iron protein, such as rubredoxin, only one peak is seen at 2.26 Å in the EXAFS spectrum, associated with the nearly identical Fe-S distances of the iron and the sulfur atoms of the coordinating residues (Figure 15.27). For a Fe_2S_2 cluster, two peaks are observed, typically at 2.25 and 2.73 Å, arising from the Fe–S and Fe–Fe distances respectively.

REFERENCES

Barney, B.M., Lee, H.I., Dos Santos, P.C. et al. (2006) Breaking the N_2 triple bond: insights into the nitrogenase mechanism. *Dalton Transactions* **2006**, 2277–84.

Barney, B.M., Yang, T.C., Igarashi, I. et al. (2005) Intermediates trapped during nitrogenase reduction of *N≡N, CH_3–N=NH,* and *H_2N=NH_2*. *Journal of the American Chemical Society* **127**, 14960–1.

Einsle, O., Tezcan, F.A., Andrade, S.L.A. et al. (2002) Nitrogenase MoFe-protein at 1.16 Å resolution: a central ligand in the FeMo-cofactor. *Science* **297**, 1696–1700.

Kim, J. and Rees, D.C. (1992) Crystallographic structure and functional implications of the nitrogenase molybdenum-iron protein from *Azotobacter vinelandii*. *Nature* **360**, 553–9.

Ostermann, A., Waschiphy, R., Parak, F.G., and Nienhaus, G.U. (2000) Ligand binding and conformational motions in myoglobin. *Nature* **404**, 205–8.

Peters, J.W. and Szilagyi, R.K. (2006) Exploring new frontiers of nitrogenase structure and mechanism. *Current Opinion in Chemical Biology* **10**, 101–8.

Smith, A.W., Camara-Artigas, A., Olea, C., et al. (2004) Crystallization and initial X-ray analysis of phenoxazinone synthase from *Streptomyces antibioticus*. *Acta Crystallographica D* **60**, 1453–5.

Smith, A.W., Camara-Artigas, A., Wang, M., et al. (2006) Structure of phenoxazinone synthase from *Streptomyces antibioticus* reveals a new type 2 copper center. *Biochemistry* **45**, 4328–87

Tezcan, F.A., Kaiser, J.T., Mustafi, D. et al. (2005) Nitrogenase complexes: multiple docking sites for a nucleotide switch protein. *Science* **309**, 1377–80.

Yachandra, V.K. (1995) X-ray absorption spectroscopy and applications in structural biology. *Methods in Enzymology* **246**, 638–75.

PROBLEMS

15.1 Why are X-rays needed to determine the structure of proteins in diffraction?

15.2 Why are protein structures not directly observable in an X-ray microscope?

15.3 How is the X-ray scattering factor dependent on atomic number and angle?

15.4 Using Bragg's law, calculate the angle at which the $n = 1$ diffraction peak is observed, $\lambda = 1$ Å and (a) $d = 5$ nm and (b) $d = 0.5$ nm.

15.5 What is a protein crystal?

15.6 Explain how the composition of a protein crystal differs from that of a simple salt crystal.

15.7 Explain what the 'phase problem' is for solving protein structures.

15.8 Explain how Figure 15.14 is used in protein crystallography, including a description of the symbols in the figure.

15.9 Explain the difference between isomorphous replacement and anomalous dispersion.

15.10 Explain how a Ramachandran plot is used in model building.

15.11 Why are mirror symmetries excluded for protein crystals but not inorganic crystals?

15.12 X-rays can be generated using copper targets. Calculate the wavelength of a photon that is emitted when an electron makes a transition from the $n = 3$ level to the $n = 1$ level in copper.

15.13 X-rays can be generated using molybdenum targets. Calculate the wavelength of a photon that is emitted when an electron makes a transition from the $n = 3$ level to the $n = 1$ level in molybdenum.

15.14 Explain why synchrotrons are useful for solving structures of proteins.

15.15 Explain how the lack of all reflections in the diffraction data analysis changes the calculated electron density.

15.16 What are the possible advantages for the unusual homocitrate ligand of nitrogenase?

15.17 How do the different docking conformations facilitate the enzymatic function of nitrogenase?

15.18 Suppose that an EXAFS experiment shows the presence of one peak at 2.3 Å for a metalloprotein with a metal cluster. What is the interpretation of the peak?

16

Magnetic resonance

Magnetic resonance is widely used for spectroscopic measurements of biological systems. The concept of magnetic resonance can be applied to two probes: with the nuclei, as is done in nuclear magnetic resonance (NMR), or with the electrons, as done in electron paramagnetic resonance (EPR). The basic concepts of NMR are discussed, including the use of different pulse techniques for improvements in the signal quality. The use of NMR to determine protein structures is presented with the example of a protein associated with neurological disease. Also discussed is the relationship between NMR and the technique of magnetic resonance imaging (MRI) that is commonly used in hospitals. The presentation switches to the basic concepts of EPR followed by the use of this technique to understand the function of proteins, including heme proteins and ribonucleotide reductase, a critical enzyme that catalyzes the conversion of nucleotides into deoxynucleotides.

NMR

Every particle that has a spin also has a magnetic dipole moment, with protons, neutrons, and electrons having a spin of 1/2 (Chapter 17). The protons and neutrons forming a nucleus will tend to pair up their spins. In general, nuclei with even numbers of protons and neutrons will have zero spin, those with an odd number of either protons or neutrons will have a half-integer spin, and those with both protons and neutrons being odd will have an integer spin (Table 16.1). Since the isotopes of any given atom have different numbers of neutrons, the nuclear spin of isotopes will necessarily be different. For proteins, most work is done on the NMR signals arising from the protons that have a spin of 1/2. The NMR signals arising from protons can be eliminated by substitution with deuterium. The non-zero spins of the isotopes ^{13}C and ^{17}O provide the opportunity to perform specific isotopic substitutions and to measure the NMR signals from those isotopes.

Table 16.1
Nuclear-spin properties of atoms commonly found in proteins.

Nucleus	Natural abundance (%)	I	$\gamma(10^7\ T^{-1}\ s^{-1})$
^{1}H	99.98	1/2	26.75
^{2}H	0.02	1	4.107
^{12}C	98.89	0	–
^{13}C	1.11	1/2	6.727
^{14}N	99.64	1	1.938
^{15}N	0.36	1/2	0.37
^{16}O	99.96	0	–
^{17}O	0.04	1/2	−3.627

Electromagnetic radiation interacts with dipole moments. For example, a radio and television antenna responds to a broadcast signal through the electrons of the antenna moving up and down in the antenna. To understand how the nuclear spin interacts with the electromagnetic radiation in an NMR experiment, we need to know the relationship between the nuclear spin and the magnetic dipole moment. Consider a nucleus with a spin angular momentum I. Along the z direction, the angular momentum is quantized such that the z projections m_I can range from $+I$ to $-I$. This nucleus will have a *magnetic dipole moment*, μ, given by:

$$\vec{\mu} = \gamma \vec{I} \quad \text{and} \quad \mu_z = \gamma \hbar m_I \tag{16.1}$$

The *coefficient* γ, the proportionality constant between the magnetic dipole moment and the spin, is called the *magnetogyric ratio* and can be measured experimentally. The magnetic moment can also be expressed as a *nuclear g-factor*, g_I, *and the nuclear magneton*, μ_n:

$$\gamma \hbar = g_I \mu_N \quad \text{and} \quad \mu_N = \frac{e\hbar}{2m_p} = 5.051 \times 10^{-27}\ \mathrm{JT^{-1}} \tag{16.2}$$

In a *magnetic field*, B, the dipole moment of the nucleus will try to align along the direction of the field, just as a compass needle will align along the direction of the Earth's magnetic field. For a given orientation the *energy*, E, is given by the dot product of the field and dipole moment:

$$E_{m_I} = -\vec{\mu} \times \vec{B} \tag{16.3}$$

Normally the magnetic field is defined to be along the z direction, so this reduces to:

$$E_{m_I} = -\mu_z B = -\gamma \hbar B m_I \tag{16.4}$$

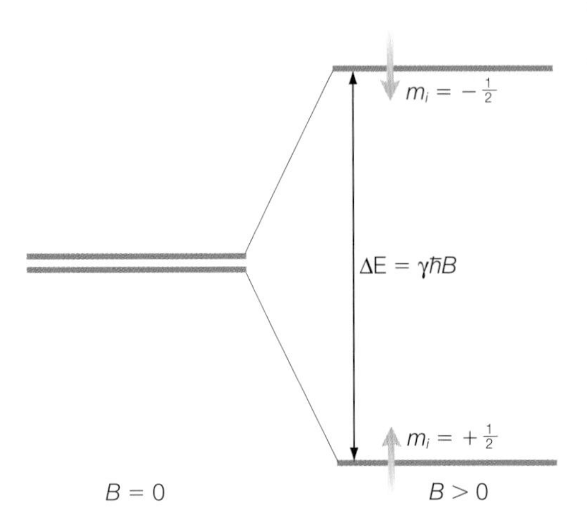

Figure 16.1 The nuclear-spin energy levels in the presence of a magnetic field.

These energies can be expressed in terms of the *Larmor frequency*, ν_L, by:

$$E_{m_I} = -m_I h\nu_L \quad \text{or} \quad \nu_L = \frac{\gamma B}{2\pi} \tag{16.5}$$

For a nucleus with a spin of 1/2, this interaction with the magnetic field will result in a splitting of the energies for the two spin states (Figure 16.1). For a spin 1/2 system, there is only one transition between the spin down and up states. If the spin is larger there are more states and more transitions. For spin 1 there are three states (+1, 0, −1) and two possible transitions of equal energy with $\Delta m_I = 1$. In general there are $2I + 1$ states for a given spin.

The experimental system needed to measure an NMR signal is in principle very simple, with just a radiofrequency transmitter coupled to the sample, in a loop in a magnet followed by a receiver (Figure 16.2). Although signals can be measured by simple systems, accurate measurements require very precise equipment. The magnetic field must be uniform throughout the sample, which is often rotated. The most resolved signals make use of very high field magnets. Whereas routine studies of the products of a synthesis reaction can be performed on a system of 300 MHz, for proteins spectrometers operating at higher frequencies of 600 MHz and above are required to obtain separation of closely lying peaks. Such spectrometers operate with a field strength requiring magnetic fields of at least 12 T, which is generated using a superconducting magnet. Superconductors are materials that change their properties dramatically at very low temperatures, typically at liquid helium temperatures or 1–10 K. For these materials, the resistance normally encountered by the electrons in the material is not present at temperatures of a few degrees Kelvin. The superconducting material can maintain electrical current forever and so superconducting magnets can produce extremely homogeneous fields. For many proteins, the spectra are complex and difficult to interpret, so many NMR experiments are now performed on systems that operate at 900 MHz to maximize the resolution of the spectra.

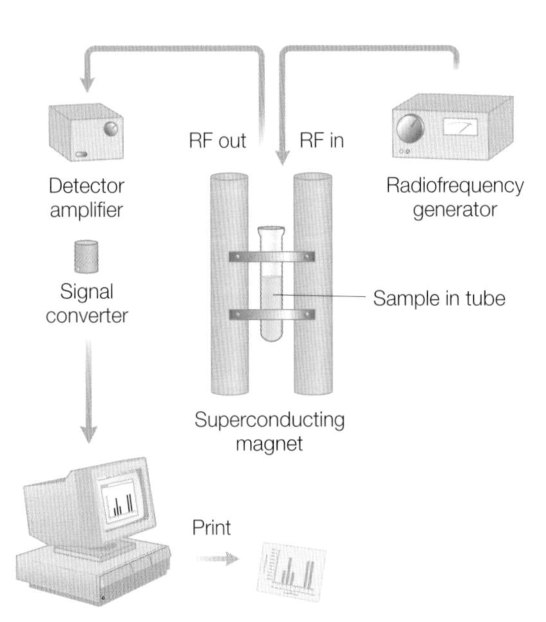

Figure 16.2 A simple scheme of an NMR spectrometer.

Chemical shifts

Every dipole of a protein will interact with the external magnetic field. The magnetic field experienced by each dipole will be slightly different than the applied field due to the contributions of these dipoles. The additional contribution is proportional to the applied field and is usually written as:

$$\delta B = -\sigma B \tag{16.6}$$

Here the parameter σ is the *shielding constant*. The field at the nucleus is then given by the sum of the two terms:

$$B_{loc} = B + \delta B = (1 - \sigma)B \tag{16.7}$$

and the resulting Larmor frequency is

$$\nu_L = \frac{\gamma B_{loc}}{2\pi} = (1 - \sigma)\frac{\gamma B}{2\pi} \tag{16.8}$$

To make the frequency shift independent of the specific experimental conditions, the shift is expressed in terms of the *chemical shift*, δ:

$$\delta = \frac{\nu - \nu^0}{\nu^0} \times 10^6 \tag{16.9}$$

where ν^0 is a *standard reference* such as tetramethylsilane, $C_4H_{12}Si$, which has four methyl groups, CH_3, bonded to Si. Because all 12 hydrogens are equivalent, the NMR spectrum has a single peak that can be used for reference.

The factor of 10^6 is included because the shifts are in the range of 1–10 ppm. Note that as the shielding increases, the chemical shift decreases:

$$\delta = \frac{(1 - \sigma)B - (1 - \sigma^0)B}{(1 - \sigma^0)B} \times 10^6 = \frac{\sigma^0 - \sigma}{1 - \sigma^0} \times 10^6 \approx (\sigma^0 - \sigma) \times 10^6 \tag{16.10}$$

The calculation of the chemical shift for even small molecules is difficult and normally not done for large molecules such as proteins. However, qualitative estimations can be done as the chemical shifts normally are due to certain interactions. The dominant contribution is the electron density of the group. A high electron density corresponds to a large shielding and a low chemical shift. The electrons surrounding the nuclei respond to the applied field, causing an induced magnetic field. A greater density of electrons corresponds to a larger induced dipole and a larger

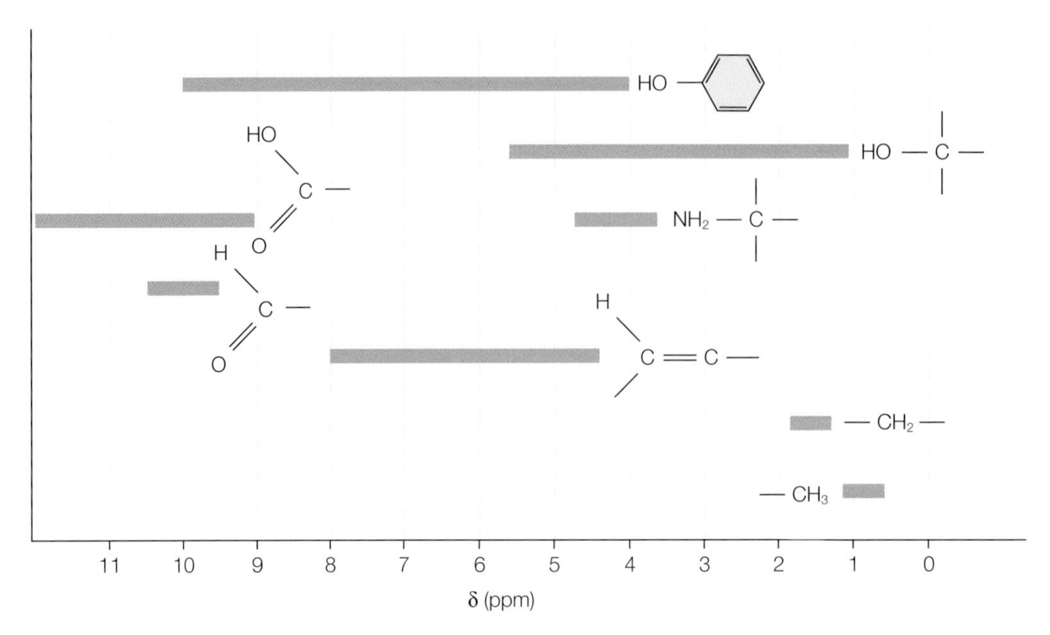

Figure 16.3 Chemical shifts for protons in different groups.

shielding. Electronegative groups withdraw electron density from the protons, give less shielding, and hence have larger shifts. Every functional group has a specific characteristic chemical shift (Figure 16.3).

Other factors can also contribute to the observed chemical shifts. Neighboring atoms may increase the shielding experienced by a proton. Ring currents can occur in delocalized π electrons of aromatic rings. The contribution of these currents is strongly dependent upon the geometry as the shift will be smaller for protons above the ring but will increase for protons on the side of the ring. In proteins, any bound metals may also produce large local magnetic fields.

Spin–spin interactions

NMR spectra usually exhibit splittings of certain lines. The splitting is independent of the magnetic field strength and due to interactions between neighboring spins. Consider the effect of one proton on another. The magnetic field experienced by a spin depends upon whether the neighboring spin is up or down. Thus, the local field is either slightly higher or lower due to the presence of the neighboring proton, and the original line is split into two lines, one slightly higher and the other slightly lower, as shown in Figure 16.4. The strength of this interaction is described by a *scalar coupling constant, J*.

As the number of protons interacting with a given proton increases, the number of lines also increases. Consider a CH_2 group interacting with a proton. The spins of the H_2 may be up/up, up/down, down/up, or down/down. Since the up/down and down/up configurations lead to the same energy, these two lines overlap and what is observed are three lines of which the middle is twice as large as the two side lines. This is called a 1:2:1 triplet and is shown in Figure 16.4. Three interacting protons will lead to a 1:3:3:1 pattern and larger numbers can be determined easily using Pascal's triangle.

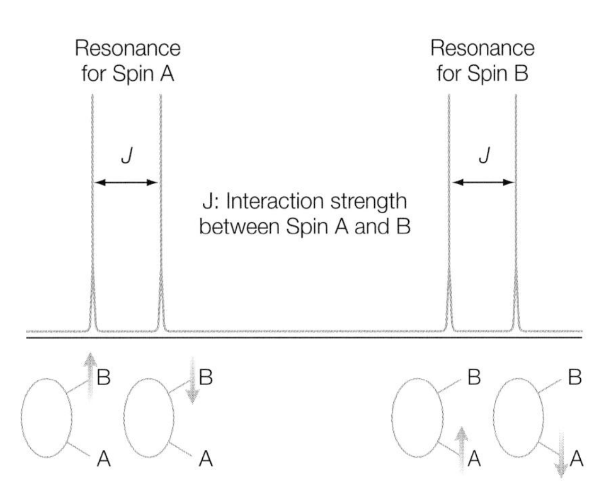

Figure 16.4 An NMR spectrum with each resonance, due to spin A or B, split into two lines due to interactions with the nearby spin.

In an NMR spectrum, the presence of two close peaks may arise from spin–spin coupling but it may also be accidental and simply due to the complexity of the spectrum. These two cases can be distinguished experimentally by changing the field strength. If the two lines arise from the spin–spin interaction then altering the field will not change the splitting. If they are two separate lines, their separation will be proportional to the field strength. However, this is normally not done as the use of two-dimensional NMR will also yield the same information as described below. Instead, another means of identifying whether peaks arise from a single spin but are split is to measure the total area of the peaks, as this should be directly proportional to the number of spins, although broadening of a peak due to a factor such as proton exhange may make assignment of the areas problematic.

Pulse techniques

NMR experiments can be performed using a conventional continuous-wave instrument but modern instruments use radiofrequency pulses due to improved signal-to-noise features and versatility. The use of the pulse techniques arises from the ability to align magnetic dipoles due to the presence of an external magnetic field. In the absence of the external field, there is no preferred direction for the dipoles and they will lie at random angles on a cone such that their projection is +1/2 or −1/2 (Figure 16.5). In the presence of the field, there is a net magnetization, M, along the z direction and the spins start precessing along that direction.

Now consider the effect of a radiofrequency pulse that has a field strength B_1. If this field is given a frequency that matches the Larmor frequency, the magnetization vector will begin to precess around the direction of B_1 (Figure 16.6). As the spins precess their direction changes from along z to a cone around B_1. The duration of the pulse then will determine the final direction of the spins, with a 90° pulse resulting in magnetization moving from the z direction to the x–y plane. After the 90° pulse, the magnetization vector is in the x–y plane. As time passes, the individual spins precess

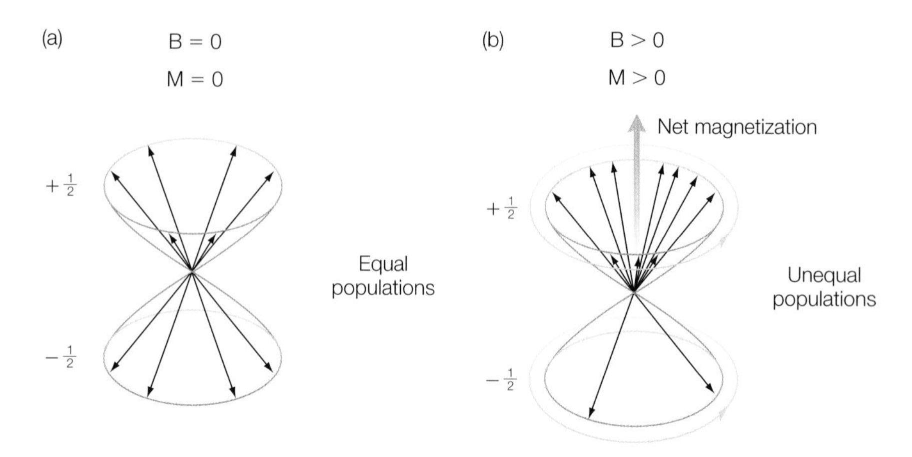

Figure 16.5 The presence of a magnetic field, B, shifts the relative populations of the spins from (a) equal populations to (b) unequal populations and a net magnetization, M.

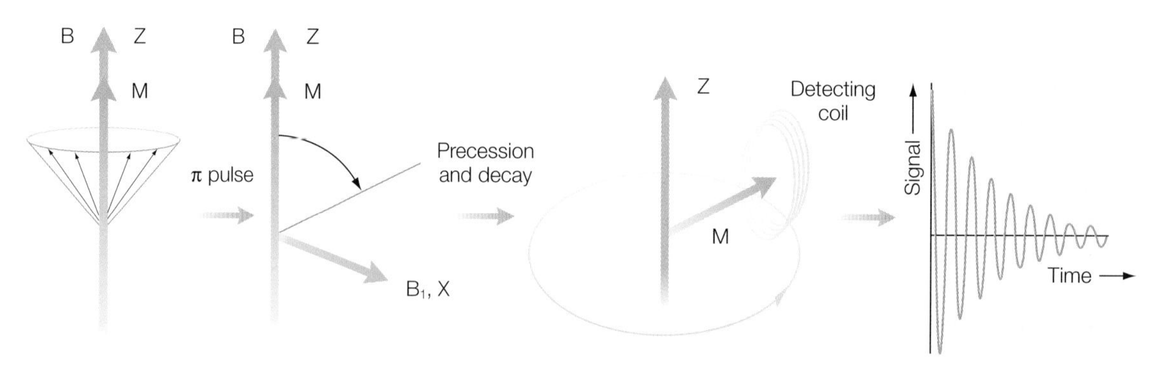

Figure 16.6 After a 90° pulse the net magnetization M is turned from the z-direction to the x-direction and it begins to precess, yielding an oscillating signal.

at different frequencies and start to separate. The magnetization vector decreases with a time constant, T_2, resulting in an exponentially decreasing signal. The form of this signal is called the free-induction decay. Each spin makes a different contribution to the decay depending upon the specific frequency for the spins that can be separated out by performing a Fourier transform on the signal.

A normal NMR spectrum is measured using a pulse as described. Modern NMR makes use of multiple pulses to generate two-dimensional spectra. Depending upon the sequence of pulses, the resulting spectra will be sensitive to different types of interaction between protons. For example, correlation spectroscopy (COSY) uses two 90° pulses and is sensitive to standard through-bond interactions between protons. Use of an alternate pulse sequence allows determination of through-space interactions through the nuclear Overhauser effect, as described below.

Two-dimensional NMR: nuclear Overhauser effect

The nuclear Overhauser effect is a specific NMR tool for the determination of which nuclei are near each other. The tool is based on the idea that the intensity of an NMR peak will change if the peak associated with another nearby proton is simultaneously irradiated at the resonance frequency of that proton. The nearby proton can interact through a dipole–dipole interaction that has a strong distance dependence of r^{-6}, and so this interaction is effective only for distances of 5 Å. This interaction with the nearby proton causes a change in the relative spin population of the original proton and, consequently, the NMR intensity of the peak corresponding to the original proton is altered.

Consider the nuclear Overhauser effect for a polymer with five protons (Figure 16.7). Protons 2 and 5 are close in proximity, within 4 Å, whereas the other pairs are farther apart. The diagonal shows the five peaks corresponding to the five protons and two off-diagonal peaks corresponding to the interaction between protons 2 and 5.

The nuclear Overhauser effect on the NMR spectrum can be understood by considering the populations of two different spins. The presence of the magnetic field splits the energy of each of two spins, creating four energy levels. The level corresponding to both spins up is the lowest energy level, the two up/down levels have equal energy, and the level with both spins down has the highest level (Figure 16.8). At thermal equilibrium, most of the spins are in the lowest energy level with fewer in the higher energy levels. Consider a simple situation where there are initially 14 spins occupying the levels (Figure 16.8). When a saturating radiofrequency

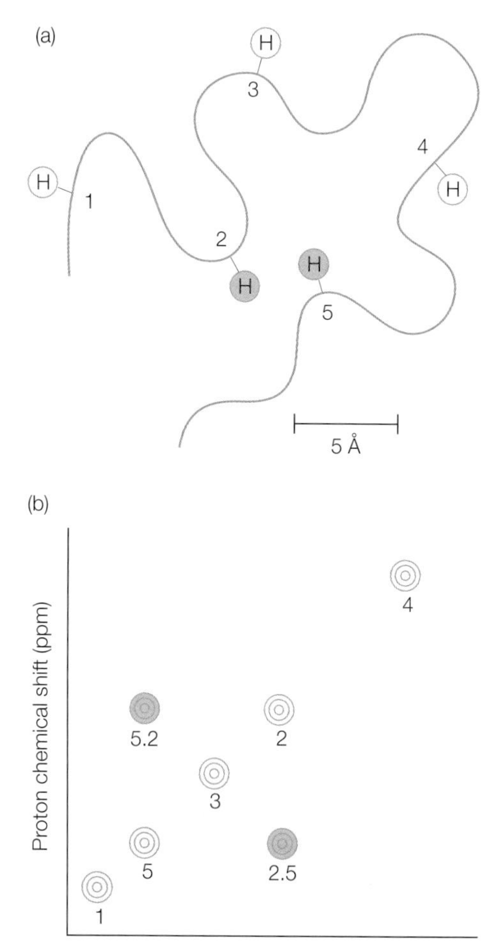

Figure 16.7 The two-dimensional NMR spectrum of (a) a simple polymer containing a total of five spins, resulting in (b) five diagonal resonances and two off-diagonal resonances.

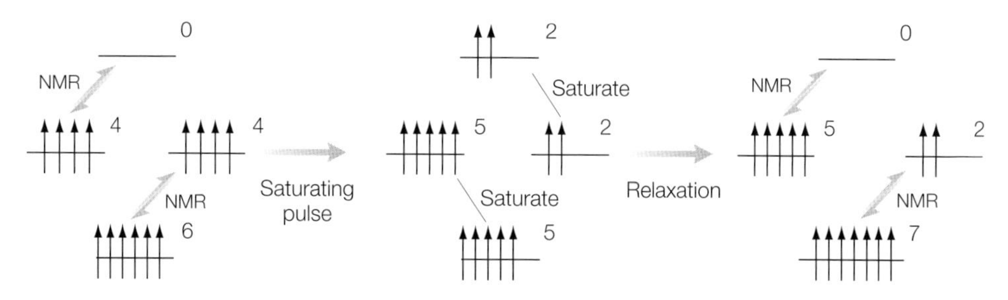

Figure 16.8 The Overhauser effect on the spin populations of different energy levels.

pulse is applied to the system that has an energy matching the energy difference for one of the spins, the populations of those spins will equilibrate. After the pulse, the spins in the uppermost level will relax down to the lowest level. When a second radiofrequency pulse probes the second spin, the population levels are now greater and the NMR signal is enhanced. Thus, the pulse sequence uses the dipole–dipole interactions between two coupled spins to enhance the NMR signals of those spins, allowing their detection as cross peaks in the two-dimensional spectrum.

NMR spectra of amino acids

An NMR experiment is usually a COSY measurement in which the cross peaks arise from nuclei that interact through chemical bonds. For individual amino acid residues, the resulting spectrum results in a pattern that is characteristic of the bonding arrangements. Consider the idealized COSY spectra of several amino acid residues (Figure 16.9). The assignment of the peaks begins with a peak in a region that is free of overlapping peaks, with the others following from the COSY connectivity. In assigning the spectrum of the amino acid residues, the protons associated with the carboxyl and amino groups are not included, as in a polypeptide chain they would not contribute to the spectrum.

The simplest amino acid residues are Gly and Ala. Glycine has one $C_{\alpha}H$ proton present and so shows a single peak at 3.97 ppm. Alanine has one proton at the C_{α} position and three at the C_{β} positions. The three protons of the methyl position are all magnetically equivalent and contribute one peak that has an integrated area 3-fold larger than the peak from the C_{α} position. Since these protons are bonded together there is a cross peak showing that connectivity. For threonine, the methyl protons are magnetically equivalent as was found for alanine. Since the methyl protons are connected only to the $C_{\beta}H$ there is only one cross peak involving the methyl protons. Likewise, the $C_{\alpha}H$ proton is connected only to the $C_{\beta}H$ proton, resulting in only one cross peak involving the $C_{\alpha}H$ proton. Thus, the two peaks near 4 ppm can be assigned through the cross peaks. For valine, the spectrum has contributions from four sets of protons. The methyl groups each contribute near 1 ppm whereas the $C_{\alpha}H$ proton is at 4.18 ppm and the $C_{\beta}H$ proton is at 3.13 ppm. The connectivity gives a cross peak between the $C_{\alpha}H$ proton and only the $C_{\beta}H$ proton whereas $C_{\beta}H$ proton is also coupled to each of the methyl groups.

RESEARCH DIRECTION: DEVELOPMENT OF NEW NMR TECHNIQUES

The primary use of NMR is to determine the structures of molecules. For biological systems, NMR provides the means to determine protein

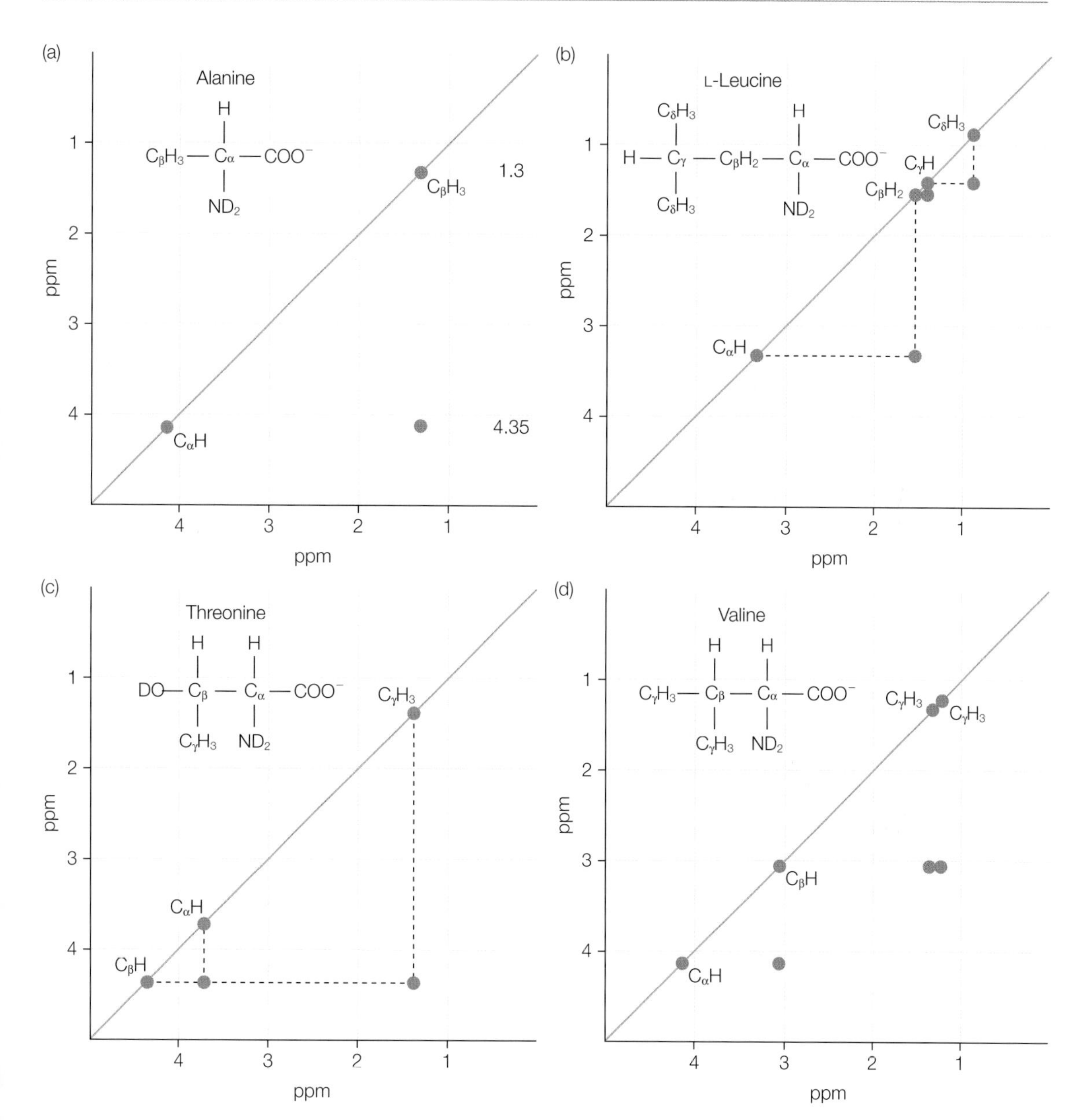

Figure 16.9 Examples of NMR spectra of (a) alanine, (b) leucine, (c) threonine, and (d) valine.

structures if the structures are not too large. Although NMR techniques are being developed to overcome this limitation, the upper boundary for normal NMR spectroscopy is a molecular mass of about 50,000 Da. Therefore, an active area of research is the development of novel techniques that can be used for larger structures (Riek et al. 2000; Hakumaki & Brindle 2003; Tugarinov et al. 2004; Vaynberg & Qin 2006). In conventional two-dimensional NMR, the spectra of large proteins typically display broad lines and the poor resolution limit imposes a fundamental limitation. In the new approaches, such as transverse relaxation-optimized spectroscopy, or TROSY, the pulse sequences are designed to suppress harmful spin-relaxation effects that cause decay of the NMR signals. Among the scientists who developed these techniques was Kurt Würthrich, who was awarded a Nobel Prize in 2002 in recognition of his efforts.

The analysis of NMR spectra is often limited by the width of the individual peaks. One major source of linewidth is anisotropy of the chemical shift. In solution, if the protein is tumbling rapidly only the average value is observed. However, if the motion is slow, then the entire anisotropy contributes and the lines are broad. Such reduced motion is a problem with large proteins or proteins that aggregate or form large complexes. Special pulse techniques can be used to reduce linewidths. For example, the dipolar fields of protons may be reduced by the use of so-called decoupling procedures. The use of TROSY partially alleviates the effect of relaxation, which is suppressed by adjusting the pulse sequence to create field strengths where the transverse relaxation rate is minimal. A comparison of the spectra from TROSY and the more conventional COSY is shown (Figure 16.10). In addition, these efforts have benefited from new labeling strategies.

As is evident in the TROSY spectra, NMR spectra are commonly obtained with the use of two different types of nuclear spin, such as protons and ^{15}N or ^{13}C. For proteins isolated from expression systems, the protein can be labeled with these isotopes using enriched media. For ^{15}N, the natural abundance of the isotope is sufficient, although the protein concentration must be poised sufficiently, high. In such a heteronuclear system, the spectrum contains a peak for each unique proton associated with the nonproton nuclear spin. One of the most common experiments for proteins is to measure the heteronuclear single quantum correlation (HSQC) spectrum using ^{15}N. Every amino acid residue, excluding proline, has an amide proton attached to nitrogen in the peptide bond spin in addition to any side chains that contain a nitrogen-bound proton. Thus the HSQC spectrum ideally provides a straightforward approach for identifying a contribution for every residue, provided that the protein is folded properly.

As an example, TROSY experiments have been performed on nucleic acids to provide direct experimental evidence for hydrogen-bonding interactions, which are usually only inferred from X-ray structures. These experiments made use of ^{15}N, which has a nuclear spin of 1/2, substituted for the predominant isotope ^{14}N, which has no nuclear spin (Table 16.1). For DNA, the ^{15}N–^{15}N couplings across Watson–Crick hydrogen bonds can be identified in oligonucleotides while simultaneously monitoring the ^{1}H couplings (Figure 16.11). These experiments are intrinsically low-sensitivity measurements and therefore require the use of the better-resolution TROSY approach. While TROSY is useful in enhancing the resolution of amide

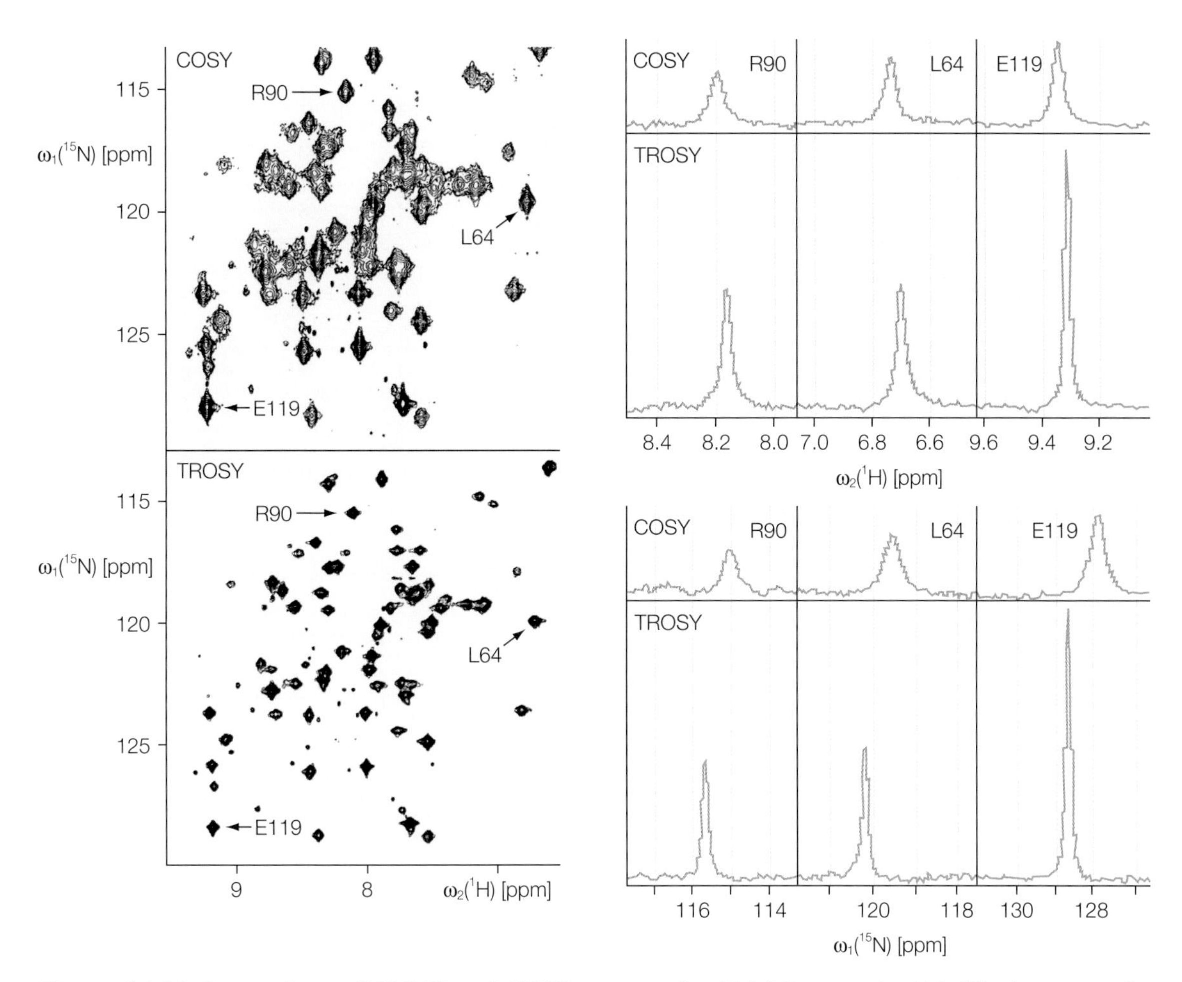

Figure 16.10 Comparison of TROSY and COSY spectra of a 110-kDa protein, 7,8-dihydroneopterin aldoase. From Riek et al. (2000).

bonds in proteins, the approach has been extended to methyl groups resulting in sharp NMR signals for very large proteins exceeding 1000 kDa in size. Since amino acid side chains spin rapidly about their axes, the magnetic interactions are effectively decreased. This application for methyl groups has been used to examine the structure of the 20 S proteasome CP (Spranger & Kay 2007). The NMR structure confirms the previously determined X-ray diffraction structure showing the presence of two heptameric rings stacking together to form a barrel. The use of the methyl-based TROSY technique provided information about 100 methyl groups found in the isoleucine, leucine, and valine amino acid residues. The successful determination was aided by the overall symmetry of the protein as many spins were magnetically equivalent, thus reducing the total number of unknown interactions.

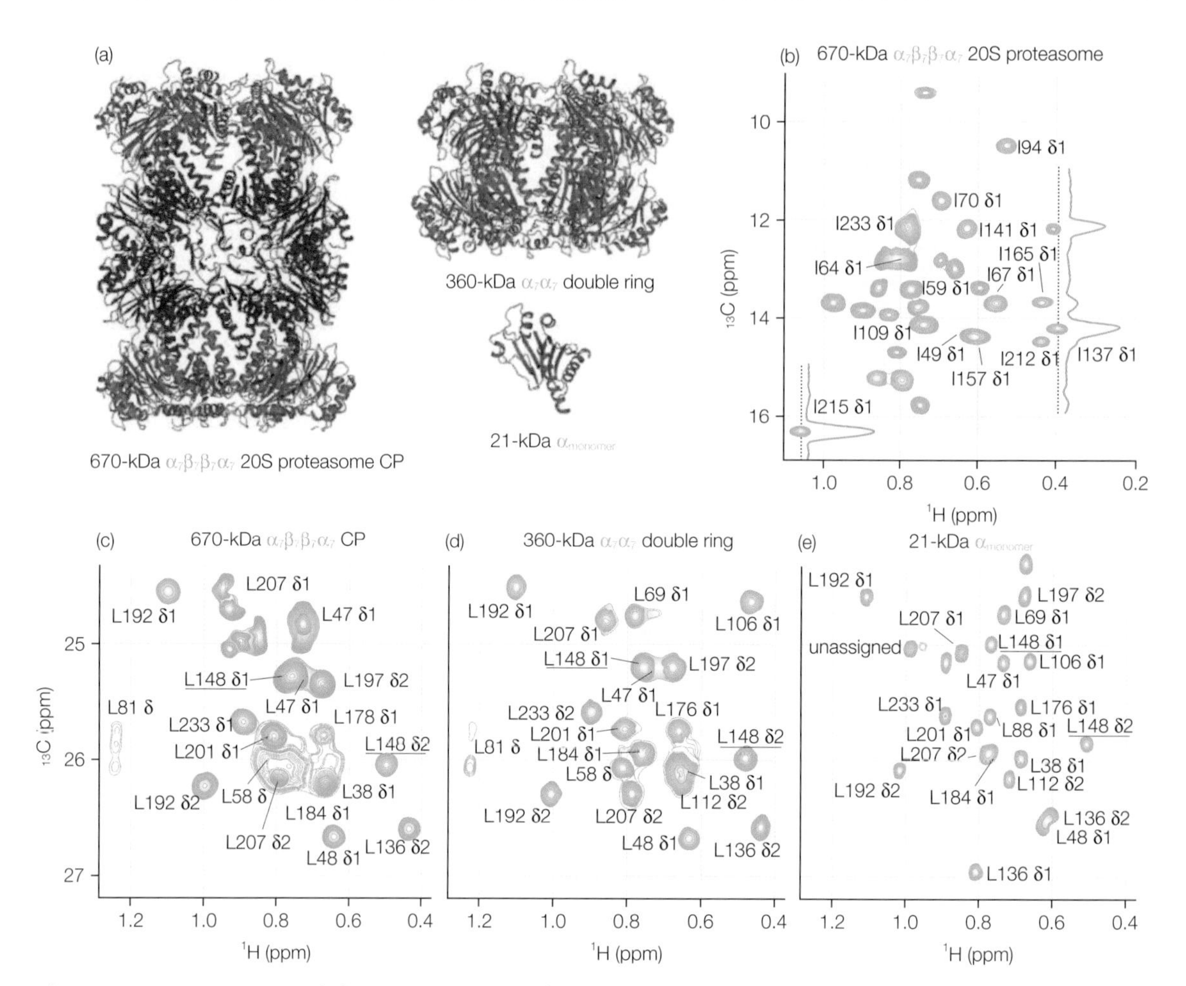

Figure 16.11 Structure of the proteasome and TROSY spectra. (a) The structure of the proteasome shows a overall barrel shape formed from rings of two different subunits. (b–e) The methyl-based TROSY spectra, which are color-coded according to the assigned region, show very sharp peaks that allow assignments despite the large size of the complex. From Sprangers and Key (2007).

A second approach is to use solid-state NMR (McDermott 2004). The protein is no longer in solution and so does not tumble, and a broad distribution would be seen due to the anisotropy of the chemical shift. However, the contribution can be eliminated with spinning of the sample at the magic angle of 54.74° (Figure 16.12). At this angle, the dipole–dipole dependence of $1–3\cos^2\theta$ gives an average value of zero and so there is no contribution to the spectrum. The application of this technique requires that the rotation frequency be not less than the width of the spectrum. This requires the use of kilohertz rotation frequencies for the sample in the magnet, which are achieved with the use of gas-driven spinners.

Together, the development of these techniques for NMR provides the opportunity to address a major question in proteomics, namely how to determine the structures of the proteome; that is, all the proteins that are found in an organism (Stauton et al. 2003). In addition, with its application in the determination of protein structures, NMR is also used widely as a tool for drug discovery. Many of the technical developments have proven useful for such efforts, in particular the screening of low-molecular-mass molecules with novel ligands to the desired target molecule (Villar et al. 2004).

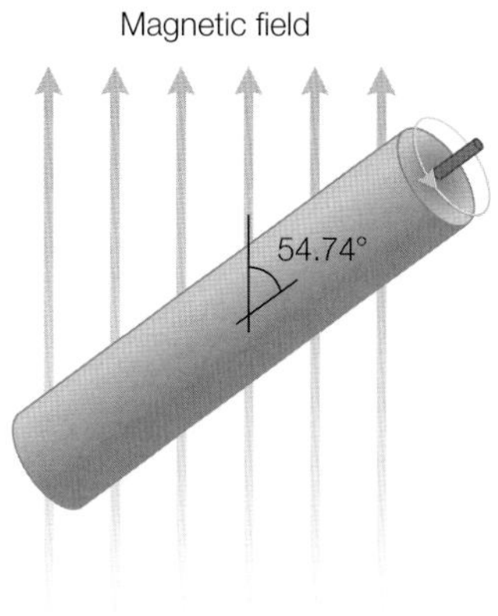

Figure 16.12 Magic-angle spinning with the spinning at 54.74° relative to the applied magnetic field.

Determination of macromolecular structures

The chemical shifts of most protons in proteins lie in the range of 1–9 ppm. The chemical shifts for all 20 amino acid residues have been measured (Table 16.2). These values provide guidelines as to the expected values for the different amino acid residues: in particular, how the protons are coupled within an amino acid residue. Determination of the structure requires that all of the peaks must be assigned individually to specific protons in the protein. To make this assignment the primary structure of the protein must be known as each of the 20 amino acids, with their different side chains, has a characteristic NMR pattern. NMR experiments are used to determine the structure of a protein by providing constraints on the distances between backbone protons, neighboring side chains, and spatially close side chains. Programs use this information together with the known geometric constraints, using energy minimization to generate possible structures.

RESEARCH DIRECTION: SPINAL MUSCULAR ATROPHY

Spinal muscular atrophy is a genetic, motor neuron disease caused by progressive degeneration of motor neurons in the spinal cord. The disorder causes weakness and wasting of the voluntary muscles. Weakness is often more severe in the legs than in the arms. Childhood spinal muscular atrophies are all autosomal recessive diseases. This means that they run in families and more than one case is likely to occur in siblings or cousins of the same generation. Parents usually have no symptoms, but carry the gene. The gene for spinal muscular atrophy has been identified and accurate diagnostic tests exist.

The *SMN1* gene encodes a protein consisting of 294 amino acid residues (Lefebvre et al. 1995; Lorson et al. 1998; Holzbaur 2004). The SMN protein has several sequence motifs: a highly basic region in residues 28–91, two proline-rich regions in residues 195–248, and a tyrosine/glycine-rich region in residues 261–278. The protein sequence does not share homology with any other proteins, except for residues 92–144, which were identified as being homologous to a Tudor domain. How the protein functions to ensure the survival of motor neurons remains uncertain. It is

Table 16.2

^{1}H Chemical shifts of the amino acid residues (see Figure 1.8) in the random-coil conformation*.

	Chemical shift (ppm)			
Residue	**NH**	**$C^{\alpha}H$**	**$C^{\beta}H$**	**Others**
Gly	8.39	3.97		
Ala	8.25	4.35	1.39	
Val	8.44	4.18	2.13	$C^{\gamma}H_3$ 0.97, 0.94
Ile	8.19	4.23	1.90	$C^{\gamma}H_2$ 1.48, 1.19 $C^{\gamma}H_3$ 0.95 $C^{\delta}H_3$ 0.89
Leu	8.42	4.38	1.65, 1.65	$C^{\gamma}H$ 1.64 $C^{\delta}H_3$ 0.94, 0.90
Pro (*trans*)		4.44	2.28, 2.02	$C^{\gamma}H_2$ 2.03, 2.03 $C^{\delta}H_2$ 3.68, 3.65
Ser	8.38	4.50	3.88, 3.88	
Thr	8.24	4.35	4.22	$C^{\gamma}H_3$ 1.23
Cys	8.31	4.69	3.28, 2.96	
Asp	8.41	4.76	2.84, 2.75	
Glu	8.37	4.29	2.09, 1.97	$C^{\gamma}H_2$ 2.31, 2.28
Asn	8.75	4.75	2.83, 2.75	$N^{\gamma}H_2$ 7.59, 6.91
Gln	8.41	4.37	2.13, 2.01	$C^{\gamma}H_2$ 2.38, 2.38 $N^{\delta}H_2$ 6.87, 7.59
Met	8.42	4.52	2.15, 2.01	$C^{\gamma}H_2$ 2.64, 2.64 $C^{\varepsilon}H_3$ 2.13
Lys	8.41	4.36	1.85, 1.76	$C^{\gamma}H_2$ 1.45, 1.45 $C^{\delta}H_2$ 1.70, 1.70 $C^{\varepsilon}H_2$ 3.02, 3.02 $N^{\varepsilon}H_3^+$ 7.52
Arg	8.27	4.38	1.89, 1.79	$C^{\gamma}H_2$ 1.70, 1.70 $C^{\delta}H_2$ 3.32,3.32 NH, NH_2^+ 7.17, 6.62
His	8.41	4.63	3.26, 3.20	$C^{\delta 2}H$ 7.14 $C^{\varepsilon 1}H$ 8.12
Phe	8.23	4.66	3.22, 2.99	$C^{\delta}H$ 7.30 $C^{\varepsilon}H$ 7.39 $C^{\zeta}H$ 7.34
Tyr	8.18	4.60	3.13, 2.92	$C^{\delta}H$ 7.15 $C^{\varepsilon}H$ 6.86
Trp	8.09	4.70	3.32, 3.19	$C^{\delta 1}H$ 7.24 $C^{\varepsilon 3}H$ 7.65 $C^{\zeta 3}H$ 7.17 $C^{\eta}H$ 7.24 $C^{\zeta 2}H$ 7.50 $N^{\varepsilon 1}H$ 10.22

*Measured at pH 7.0 and 35°C as peptide Xaa in tetrapeptide Gly-Gly-Xaa-Ala. Modified from Wüthrich (1986).

Table 16.3
Structural statistics for the Tudor domain. From Selenko et al. (2001).

Number of distance restraints	**1402**
Coordinate precision N, Cα, C	0.42 ± 0.09
Ramachandran plot	
Most favored	80.0 ± 4.1
Additional allowed	19.8 ± 4.1

found in many different cell types but the mutations lead to termination of growth only for motor neurons. The protein plays a role in RNA metabolism as it has essential interactions with spliceosomes that participate in pre-mRNA processing (Meister et al. 2002; Gubitz et al. 2004).

The determination of the SMN structure has not been possible, owing to a number of difficulties with sample preparation. In structural analysis, a common approach towards overcoming such difficulties is to work with a smaller portion that is biochemically more tractable. This strategy proved useful for the Tudor domain of SMN, which was solved using NMR. The Tudor domain was expressed in *Escherichia coli* and purified, and distance restraints were derived primarily from nuclear Overhauser enhancement spectroscopy. The resulting structure was well defined with a high number of restraints for each amino acid residue (Table 16.3).

Programs use the distance constraints with the known geometric constraints, using energy minimization to generate possible structures. Because the final answer is not unique, a series of structures is usually presented, each of which is possible given all of the information (Figure 16.13). Differences reflect both uncertainties in the assignments and disorder in the structure and provide a measure of the coordinate precision in the structure determination (Table 16.3).

The Tudor domain of SMN was found to fold as a strongly bent antiparallel β sheet with the β strands connected by short loops (Figure 16.14). The structure has a number of features that could not be identified easily from the sequence, such as a negatively charged surface that is proposed to interact with another class of proteins, termed Sm proteins. One of the amino acid residues on the

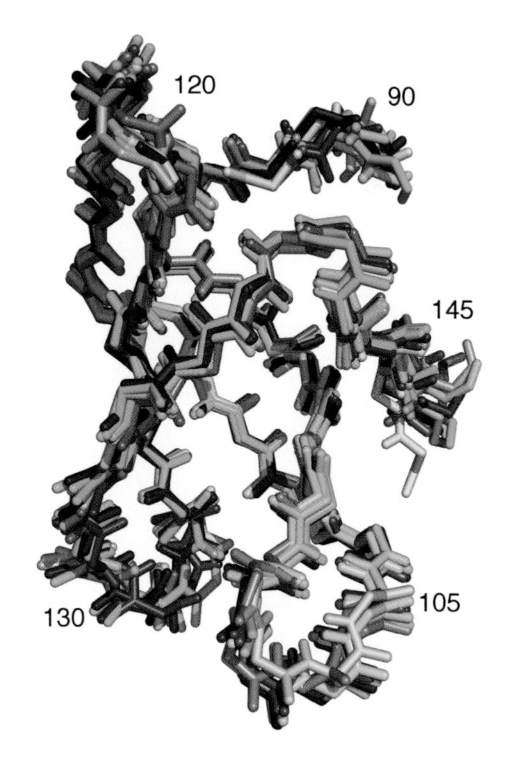

Figure 16.13 A superposition of 20 allowed structures of the Tudor domain.

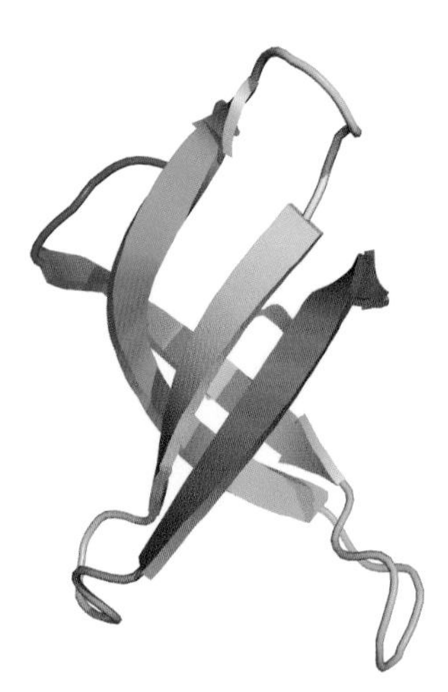

Figure 16.14 The three-dimensional structure of the Tudor domain of SMN showing the extensive β sheet.

surface of the structure is Glu-134. The mutation Glu to Lys at 134 had been found previously in a patient with spinal muscular atrophy and the structure of the protein with this mutation was also studied. The structure with this mutation was found to not disrupt the wild-type structure but replacement of the negatively charged Glu with the positively charged Lys alters the charge distribution on the surface. This suggests that the SMN protein interacts with other proteins through electrostatic interactions involving surface residues.

MRI

The technical developments resulting from NMR have proven to be useful for medical applications. For both NMR and MRI, information is obtained about the presence of nuclear spins. In the NMR experiment, a very homogeneous external magnetic field is applied and the response of the nuclear spins to radiofrequency pulses is measured. Normally, one measures a molecule in which the resonance energy is slightly shifted for different spins of the molecule, due to the features near each spin (such as electron density), which slightly alter the local magnetic field as measured by the chemical shift. MRI makes use of the same fundamental ideas to map out the composition of whole organisms.

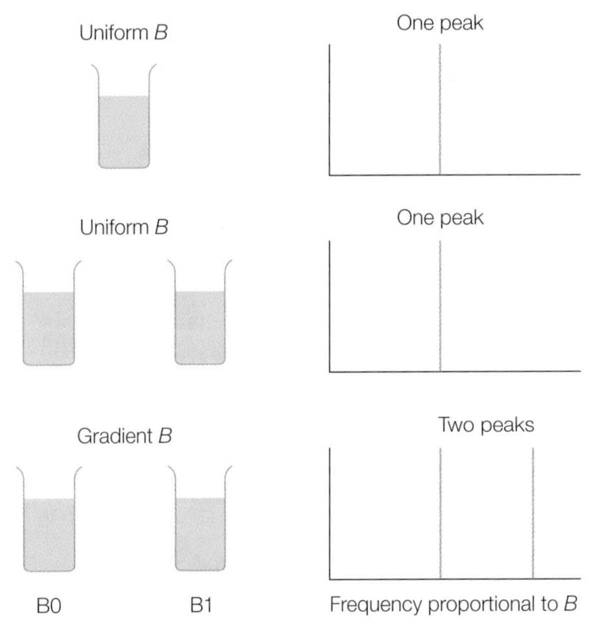

Figure 16.15 Expected spectra of capillaries containing water for different numbers of capillaries and magnetic-field configurations.

First, consider a case where a homogeneous sample, water in a small glass tube, is placed inside the magnetic field in an NMR spectrometer (Figure 16.15). Because the sample is homogeneous, the protons respond equally to the magnetic field and there is a single resonant peak in the NMR spectrum. Second, consider a situation where two glass tubes are placed side by side in the spectrometer. Normally, this is not done as the systems are designed for only one sample, but it can be performed as a *gedenken* (or thought) experiment. All of the spins in the two tubes experience the same magnetic field, so the NMR spectrum remains a single peak. Third, keep the same two tubes of water in the spectrometer but now replace the uniform magnetic field with a field that slowly increases with increasing distance from the left pole face of the magnet. In this case, the spins in each tube experience the same environment but not the same magnetic field strength, as the tube on the right-hand side has a much stronger magnetic field than the one on the left. The resonant condition for each protein is proportional to the strength of the magnetic field B (eqn 16.5). Due to the different strengths of the magnetic field, the resonant frequencies are now different, and two peaks are measured in the spectrum.

Eqn 16.5:

$$E_{m_I} = -m_I h\nu_L \quad \text{or} \quad \nu_L = \frac{\gamma B}{2\pi}$$

In general, if a large container of water is placed in a homogeneous field only a single peak is measured. When the same container is placed in a magnetic field that increases linearly with distance from one pole face, then the spectrum effectively maps out an image of the water. The container can be considered to consist of a series of planes, each at a certain distance from the left pole face. As the magnetic field increases from left to right, each plane will resonate at an increased frequency that is proportional to the distance from the left pole face. The number of protons in each plane will determine the amplitude of the signal, so the resulting spectrum will provide a count of the number of protons at every distance and the overall profile will match the image of the water in the container.

In an MRI experiment, the strength of the magnetic field is not fixed. Rather, the strength of the field increases steadily from one pole face to the other. As a result the protons will resonate at different frequencies depending upon their distance from the poles with an intensity proportional to the number of protons. This produces an image that can be made three-dimensional by rotation of the magnetic field by about the same amount as the structure being imaged (for example, your head) as shown in Figure 16.16. MRI is of great medical use as it is not invasive; that is, the pictures can be obtained without the introduction of any drugs or probes into the body.

In conventional MRI, protons from different tissues can be distinguished. These images can be sharpened by use of contrast agents. These agents sharpen the contrast between the protons of interest and those in their proximity by changing the spin relaxation time of protons in the different tissues. For example, contrast agents can be injected through the bloodstream and tissues to improve the contrast of the circulation system. Work is underway to improve the quality of the contrast between different tissues. For example, the contrast agent can be made inactive by attaching certain sugar groups to the agent until they are cleaved by specific enzymes, allowing monitoring of gene delivery and expression.

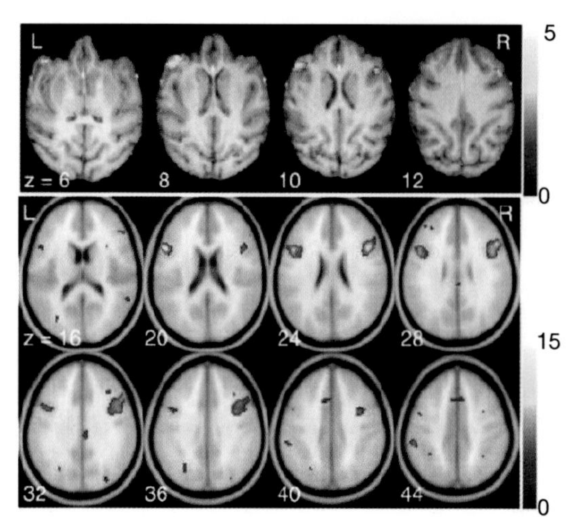

Figure 16.16 Comparison of a shift-activated response in prefrontal cortex in monkeys (upper panel) and humans (lower panel). From Nakahara et al. (2002).

Improvements in the resolution of the resulting images have led to their application as a tool for studying physiological responses to certain stimuli. Functional MRI, or fMRI, can be used to study changes during cognitive tasks. For example, the visual and task response needed to sort cards is associated with transient activation in regions of the prefrontal cortex that is different in monkeys and humans (Figure 16.16). These imaging techniques provide a new and promising aid in the development of therapeutics targeted towards specific cells.

ELECTRON SPIN RESONANCE

Electrons have a spin of 1/2 and so have magnetic dipole moments that can interact with electromagnetic radiation. Since electrons normally pair up, the net electron spin of a molecule is usually zero. Thus, most materials do not have EPR signals. Unpaired electrons occur when radicals are present or when metals are present. Many proteins have radicals and metals that are located at their active site: EPR is an excellent probe of the characteristics of the active site. The dipole moment is proportional to the spin according to:

$$\vec{\mu} = -\frac{g_e e}{2m_e}\vec{S} \tag{16.11}$$

where g_e is the *g factor* for a free electron and is equal to 2.0023. The proportionality factor has a very similar appearance as the dipole moment for the nuclear spin (eqn 16.12), except that the charge contributes a negative sign and the mass is now the electron mass. The projection of the electron spin along the z direction is quantized so the z projection of the magnetic dipole moment can be written as:

$$\mu_z = -\frac{g_e e}{2m_e} m_s \hbar = -g_e \mu_B m_s \quad \text{defining } \mu_B = \frac{e\hbar}{2m_e} \tag{16.12}$$

The parameter μ_B is called the *Bohr magneton* and has the same form as the nuclear magneton but is substantially larger by the ratio of the masses:

$$\frac{\mu_B}{\mu_N} = \frac{(e\hbar)/(2m_e)}{(e\hbar)/(2m_p)} = \frac{m_p}{m_e} = 1786 \tag{16.13}$$

Just as for the nuclear spins, the electronic spins will split the atomic energy levels in the presence of an external magnetic field, B (Figure 16.1):

$$E = -\mu_z B = g_e \mu_B m_s B \tag{16.14}$$

Note that because the electrons and protons have opposite charges, the lower-energy state for this case is the −1/2 spin.

Transitions will occur when the energy of the applied light is equal to this energy difference, giving:

$$\Delta E = g_e \mu_B B = h\nu \tag{16.15}$$

This resonance frequency is substantially different than that used for the NMR, primarily due to the difference in the dipole moments (eqn 16.14). Using typical values of $\nu_{NMR} = 100$ MHz, $B_{NMR} = 10$ T, and $B_{EPR} = 0.3$ T gives:

$$\nu_{EPR} = \nu_{NMR} \frac{m_N}{m_e} \frac{B_{EPR}}{B_{NMR}} \approx 100\ \text{MHz} \times 2000 \times \frac{0.3\ \text{T}}{10\ \text{T}} \approx 10\ \text{GHz} \tag{16.16}$$

This frequency is in the microwave region of light. Older systems typically used klystrons as the microwave source whereas newer systems use diodes (Figure 16.17). The microwaves are carried to the sample by use of waveguides whose size is matched to the wavelength of the microwave. For a 10-GHz frequency, the wavelength:

$$\lambda = \frac{c}{\nu} = \frac{3 \times 10^{10}\ \text{cm s}^{-1}}{10 \times 10^9/\text{s}} = 3\ \text{cm} \tag{16.17}$$

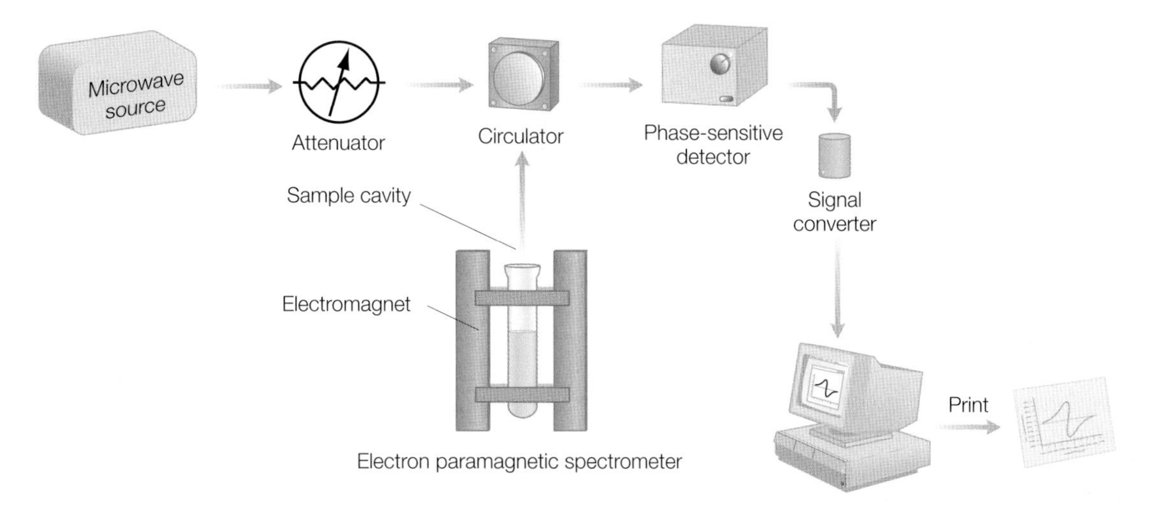

Figure 16.17 Schematic diagram of an EPR spectrometer.

Eqn 14.1:

$\lambda \nu = c$

Since the waveguides operate at specific frequencies (with adjustments possible only over a narrow region), the frequency for a typical EPR experiment is fixed and the magnetic field is varied. Because the magnetic fields are substantially smaller, conventional electromagnets can be used and the magnetic field strength can be varied by changing the applied current.

In an EPR experiment, the microwave field is monitored as the magnetic field is swept. When the magnetic field strength creates an energy difference that matches the energy of the microwaves (eqn 16.16) resonance absorption occurs. Experimentally the absolute change in the microwaves is susceptible to error from factors such as baseline drifts. For increased sensitivity, most experiments not only sweep the magnetic field but also apply a small alternating magnetic field simultaneously. Thus, the absolute EPR signal is not measured, only the difference in the signal as a result of the small alternating field; that is, the slope of the signal is measured. This technique, termed phasesensitive detection, is more sensitive as only those signals that oscillate with the proper frequency are allowed.

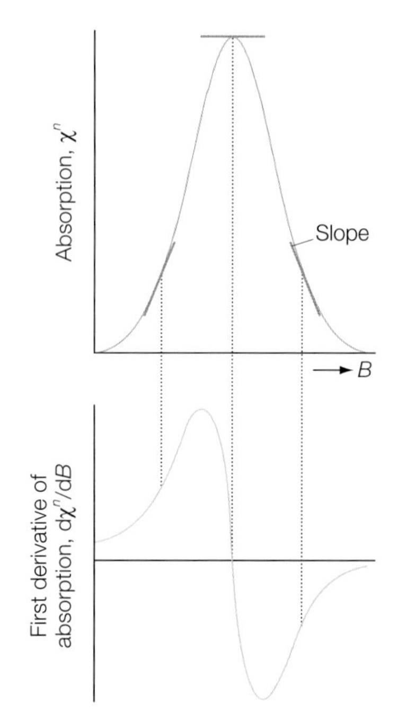

Figure 16.18 EPR spectra can be measured as absorption (χ'') but are usually measured using phase-sensitive detection that results in a signal representing the derivative of the microwave absorption, $d\chi''/dB$.

As the magnetic field is swept, the EPR signal is zero and so is the slope. When the magnetic field approaches the resonance condition, the EPR signal increases and the slope initially increases but then decreases until a slope of zero is reached (Figure 16.18). Increasing the magnetic field further results in a decrease of the EPR signal and negative slope until both return to zero. Thus, EPR spectra are normally reported in terms of the derivative spectrum.

As with NMR, the resonance frequency is proportional to the magnetic field and is not normally reported. For EPR, the samples are described by the g factor:

$$g = \frac{h\nu}{\mu_B B} \tag{16.18}$$

In terms of the derivative signal, the g factor corresponds to the crossing point in the middle of the signal. For some radicals, differences compared to the free electron value of 2.0023 are interpreted in terms of the interaction of the electronic spin with the local environment. However, many electron spins show large g values that arise due to sensitivity of electrons for the local environment, as discussed below.

Hyperfine structure

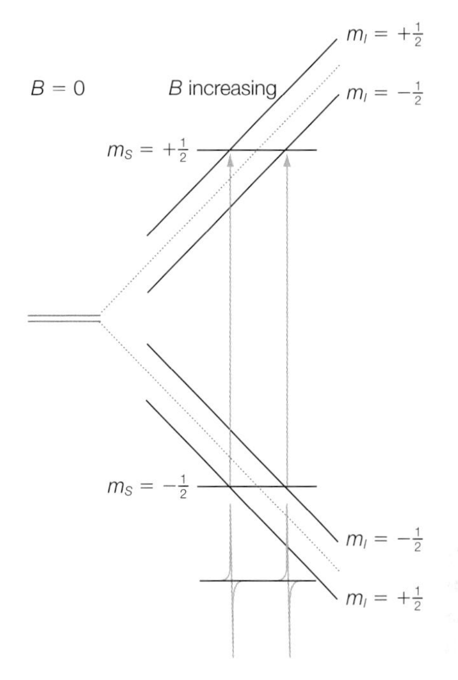

Figure 16.19 The interaction between an unpaired electron spin and a nuclear spin results in four energy levels and the splitting of the EPR signal.

The g value will be different in protein environments, due to differences in the local magnetic field compared to the applied field. In EPR, one of the primary effects is the contribution of the proton spins and their associated dipole moments on the field. In the presence of a nuclear spin, I, the local magnetic field, is:

$$B_{loc} = B + am_I \quad \text{where} \quad m_I = \pm 1/2 \quad \text{for a proton} \qquad (16.19)$$

where a is called the hyperfine coupling constant. For an interaction with a proton (m_I of $+1/2$ and $-1/2$) each of the two energy levels splits, resulting in four levels. Since m_I must be constant during the experiment, two of the transitions are possible in an EPR experiment at magnetic fields of:

$$B = \frac{h\nu}{g\mu_B} - \frac{1}{2} \quad \text{and} \quad B = \frac{h\nu}{g\mu_B} + \frac{1}{2} \qquad (16.20)$$

The presence of one line is then split equally into two lines (Figure 16.19). If the nuclear spin is larger, the splittings will accordingly become more complex with $2I + 1$ hyperfine lines being present, although not all of the lines may be resolved (See Chapter 20).

Electron nuclear double resonance

We can now consider a double-resonance condition as we did for NMR. The microwave frequency is fixed at resonance and a radiofrequency wave is applied. When the radiofrequency is at resonance for a nuclear spin nearby the electron spin, then the radiofrequency will flip the nuclear spin and be sensed by the electron (Figure 16.20). Writing the total energy of the system as:

$$E = \pm\frac{g_e\mu_B B}{2} \pm \frac{ha}{4} \mp \frac{g_N\mu_N B}{2} \qquad (16.21)$$

leads to the transitions occurring at

$$\nu_{RF} = \frac{A}{2} \pm \nu_0 \quad \text{where} \quad h\nu_0 = g_N\mu_N B \qquad (16.22)$$

Thus, the EPR signal changes when the radiofrequency is resonant with a nearby nuclear spin. The EPR effectively becomes a means to obtain the

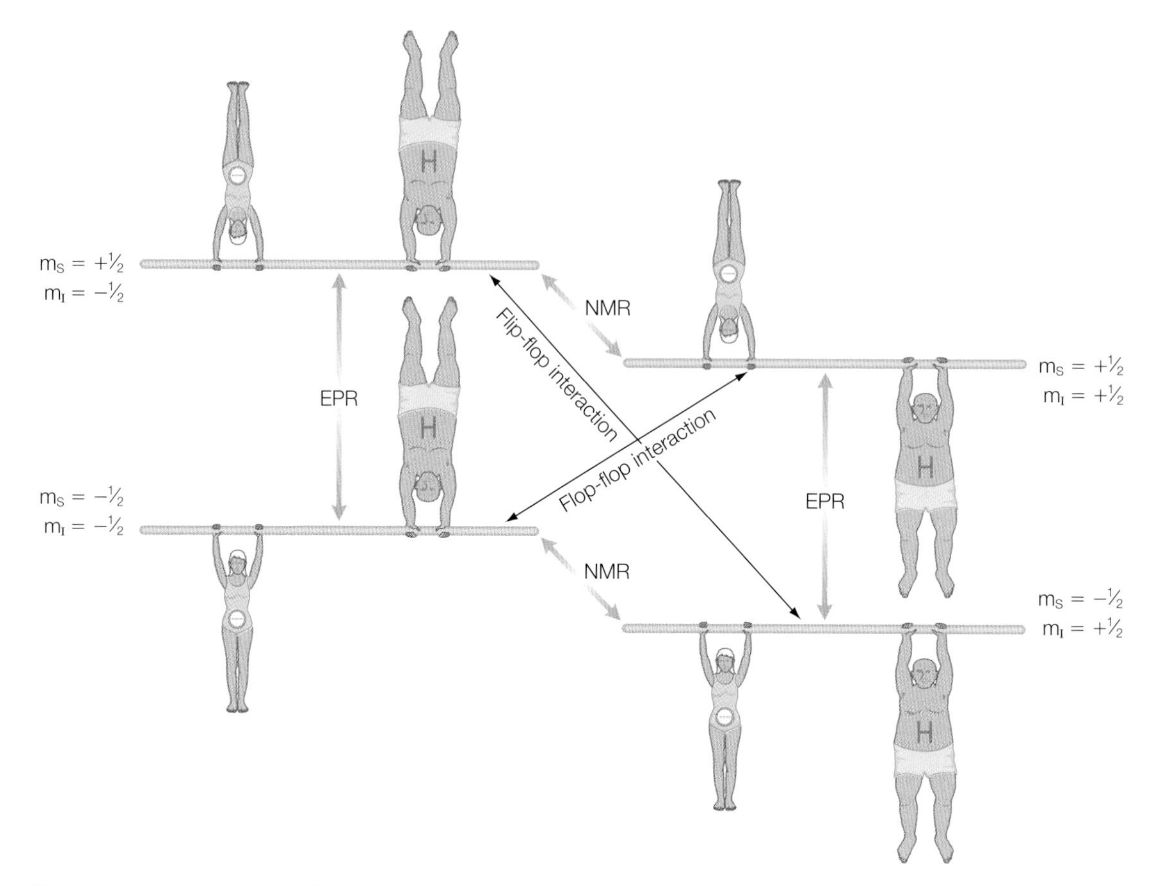

Figure 16.20 Energy-level diagram for one unpaired electron and one proton. EPR transitions correspond to vertical transitions and NMR transitions correspond to diagonal transitions.

NMR spectrum selectively for the environment of the unpaired electron. This technique, named ENDOR for electron nuclear double resonance, was developed by George Feher in 1956. Whereas ENDOR is largely applied to protons, any nuclei with a net spin can be studied using this technique.

Spin probes

In addition to studying metal- and radical-containing proteins, other biological components can be characterized by attaching spin probes; that is, small molecules that contain an unpaired electron spin. Spin labels have a stable free radical, which is usually a nitroxide derivative. Spin labels can be incorporated into the fatty-acid chains of lipids in both artificial and natural membranes where their EPR spectra give information concerning the mobility of lipid chains. Alternatively, spin labels have been designed to specifically bind to cysteine residues in proteins. By introducing reactive residues, such as cysteines, incorporated using mutagenesis, spin-probe binding sites can be placed throughout the surface of a protein.

By introducing more than one spin probe to a protein, the coupling between the spin probes can be measured to study conformational changes of the protein.

RESEARCH DIRECTION: HEME PROTEINS

Heme proteins such as myoglobin, hemoglobin, and cytochrome are globular proteins that have, as cofactors, hemes, which are planar molecules with central iron atoms (Figure 16.21). The central iron atom can have an EPR signal depending upon the state of ligation. There are four nitrogens in the heme plane, which always serve as ligands and two out-of-plane coordination sites (Figure 16.21). Myoglobin in the deoxy form has Fe^{2+} with two axial histidine ligands. Oxygen can bind to the sixth coordination position and the iron remains Fe^{2+}. Alternatively, with other ligands the oxidation state becomes Fe^{3+}. As the coordination of the iron atom changes, the electronic structure also changes. For example, the changes of the electronic states correspond to shifts in the allowed transition energies, resulting in a distinctive optical spectrum for each state (Figure 16.22).

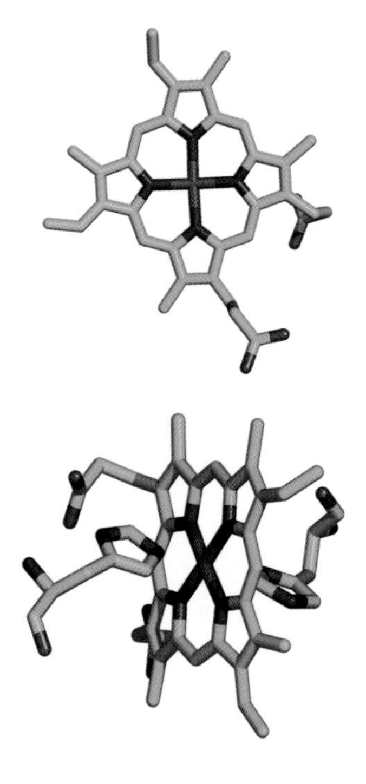

Figure 16.21 Coordination of the Fe in the heme group of myoglobin to (top) nitrogen atoms in plane and (bottom) amino acid residues.

These relationships between coordination of the metal and the electronic state can be understood in terms of ligand or crystal field theory. Whereas an understanding of the EPR spectra requires a detailed understanding of the properties of individual transition metals, the effect of different coordinations can be explained on a qualitative level. Consider as an example the iron of heme. The EPR signal of the iron is dependent upon the properties of the five and six electrons in the d orbitals for Fe^{3+} and Fe^{2+} respectively. The energies of the individual d orbitals can be described by considering the different types of electronic interaction using a hierarchical approach. In ligand field theory, the nature of the ligands is not considered, only the relative location and number. For heme, the iron can be considered to be coordinated by four equivalent nitrogens in the heme plane with different ligands along the heme normal, providing a tetragonal field.

The ferric state of iron has five electrons in the outer d orbitals. These five electrons will fill the five d orbitals to minimize the energy of the system. The d orbitals can be energetically grouped into the three orbitals, d*xy*, d*yz*, and d*zx*, which are more sensitive to the x–y coordination (these are termed the t_{2g} orbitals) and the two orbitals, d*xx*, which are more sensitive to the axial coordination (these are

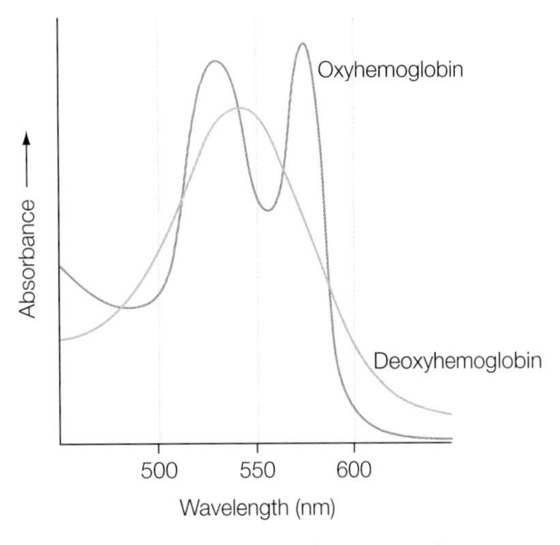

Figure 16.22 The optical spectra of different states of myoglobin.

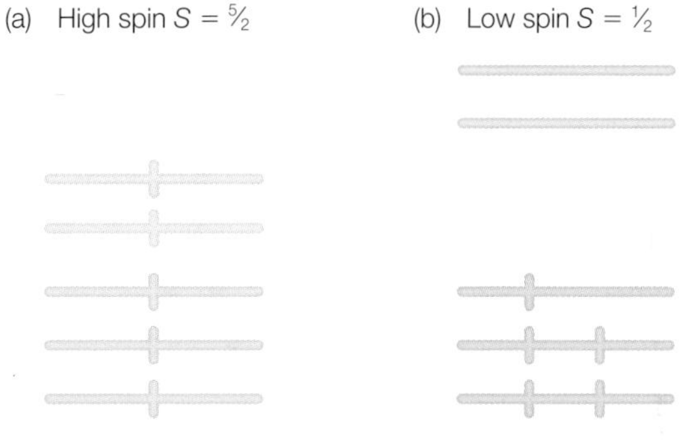

Figure 16.23 Energy-level diagram for (a) high- and (b) low-spin Fe^{3+}.

the e_g orbitals). As the axial coordination changes compared to the planar coordination, the two e_g orbitals will lie at increasingly higher energy than the other three orbitals (Figure 16.23). The five electrons will fill the lower three orbitals, resulting in two pairs of electrons and one unpaired electron, forming a low-spin state. If the axial coordination is similar then the five orbitals have similar energies. The five electrons will prefer to occupy different orbitals to minimize the electron–electron repulsion energy. As a result, all five electrons are unpaired and a high-spin state with an effective spin of 5/2 is formed. For the case of Fe^{2+}, the six electrons will have values of 2 and 0 for the high- and low-spin states respectively.

EPR has been used extensively to characterize the electronic structure of metals and relate any changes to functional features. For example, cytochrome cd_1 serves a respiratory nitrite reductase, with electrons delivered to the enzyme from the cytochrome bc_1 complex via electron carriers such as azurin, a small copper protein, in the nitrogen cycle (Allen et al. 2000; Wasser et al. 2002; Stevens et al. 2004). This enzyme catalyzes the one-electron reduction of nitrite to nitric oxide in the nitrogen cycle. The protein has two domains, one containing a heme *c* and the other containing a heme d_1, with nitric oxide being released from the d_1 heme after reduction. During this process both hemes undergo redox changes with nitric oxide being tightly bound for the Fe(II) state but not the Fe(III) state. Coupled with the redox changes are structural rearrangements of the protein, which involve alteration of the heme coordination.

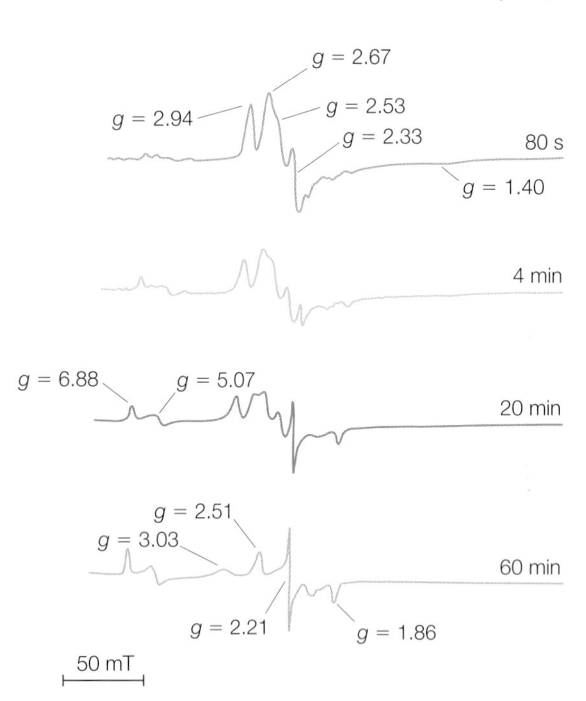

Figure 16.24 The EPR spectra of cytochrome cd_1 at different times after oxidation. Modified from Allen et al. (2000).

These events can be carefully monitored by use of EPR as six-coordinated low-spin ferric hemes are very sensitive to the chemical nature and precise coordination of the axial ligands. For example, the His–Fe–His coordination has an EPR spectrum that is distinctive from a His–Fe–Met coordination (Figure 16.24). In the presence of hydroxylamine, NH_2OH, cytochrome cd_1 is oxidized, thus providing a mechanism to examine the time evolution of the spectral features as the oxidation state changes. Initially, the d_1 heme has the spectrum of a low-spin heme, with *g* values of 2.94, 2.33, and 1.40, which is indicative of a His–Fe–Met coordination. As the protein becomes oxidized with time, large changes are evident in the EPR spectra. These spectra evolve with primary features of g values at 2.51, 2.20, and 1.86 that are consistent with a His–Fe–Tyr coordination. Also present as a minor component is a high-spin species with g values of 6.88 and 5.07. The combination of these spectroscopic studies leads to a model of the protein as having three states (Figure 16.25). Two of these states are active states: the initial protein conformation that binds the hydroxylamine and the resulting oxidized state. The third state is an inactive state that is stable for a prolonged time but can be returned to the active state by other regulated proteins.

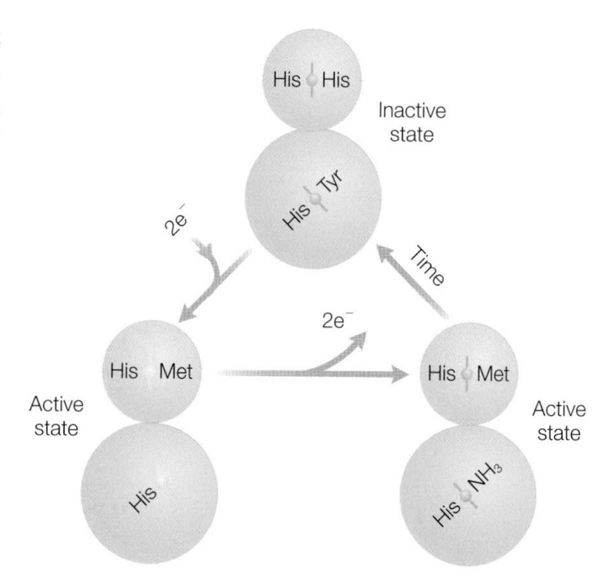

Figure 16.25 A model of the different states of cytochrome cd_1. Modified from Allen et al. (2000).

RESEARCH DIRECTION: RIBONUCLEOTIDE REDUCTASE

A critical step of DNA biosynthesis is the conversion of nucleotides into deoxynucleotides, which is catalyzed by several classes of enzyme known collectively as ribonucleotide reductase. Many of these enzymes, class I ribonucleotide reductases, possess a characteristic EPR spectrum centered around $g = 2.00$ with a pronounced hyperfine interaction (Figure 16.26). Although these enzymes contain two iron atoms that form a diferric cluster, the EPR spectrum does not arise from the metal cofactors. Rather, the spectrum arises from a third non-metal cofactor, a tyrosine radical.

The existence of amino acid radicals was first discovered in proteins in the 1970s and they have now been identified in a number of different enzymes (Stubbe & van der Dork 1998; Stubbe 2003). In each case, the amino acid radical can be identified primarily by the EPR spectrum, which should be centered near $g = 2.00$ and have hyperfine structure. In some cases, the radicals are transient species, but in the case of ribonucleotide reductase the tyrosine radical is stable. Surprisingly, the tyrosine radical and the diferric cluster are not near the active site of the protein that binds the nucleotide (Figure 16.27). Rather, the

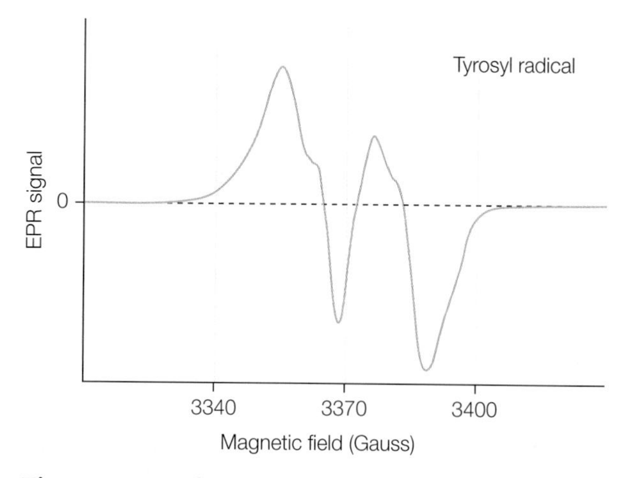

Figure 16.26 The EPR spectrum of ribonucleotide reductase. From Stubbe and van der Dork (1998).

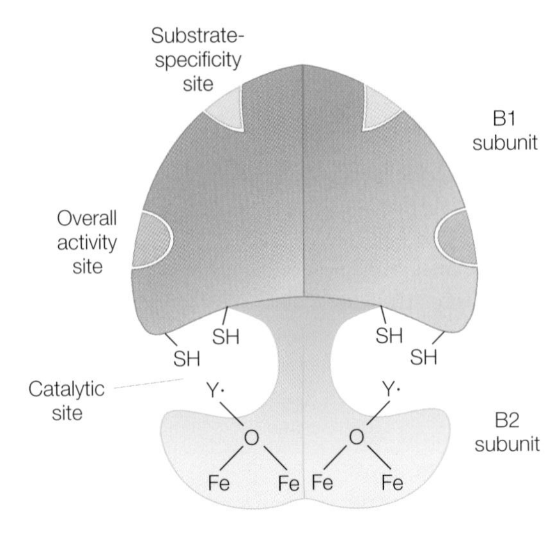

Figure 16.27 A model of ribonucleotide reductase showing the position of the tyrosine radical relative to the iron cofactor and active site.

nucleotide is bound at a distant location containing two cysteine residues, which that are oxidized as part of the enzymatic cycle, transiently forming thiyl radicals. These radicals can be observed only when the intermediate states are trapped.

REFERENCES AND FURTHER READING

Allen, J.W.A., Watmough, N.J., and Ferguson, S.J. (2000) A switch in heme axial ligation prepares *Paracoccus pantotrophus* cytochrome cd_1 for catalysis. *Nature Structural Biology* **7**, 885–8.

Gubitz, A.K., Feng, W., and Dreyfuss, G. (2004) The SMN complex. *Experimental Cell Research* **296**, 51–6.

Hakumaki, J.M. and Brindle, K.M. (2003) Techniques: visualizing apoptosis using nuclear magnetic resonance. *Trends in Pharmacological Sciences* **24**, 146–9.

Holzbaur, E.L.F. (2004) Motor neurons rely on motor proteins. *Trends in Cell Biology* **14**, 233–40.

Lefebvre, S., Burglen, L., Reboullet, S. et al. (1995) Identification and characterization of a spinal muscular atrophy determining gene. *Cell* **80**, 155–65.

Lorson, C.L., Strasswimmer, J., Yao, L.M. et al. (1998) SMN oligomerization defects correlates with spinal muscular atrophy severity. *Nature Genetics* **19**, 63–7.

McDermott, A.E. (2004) Structural and dynamic studies of proteins by solid-state NMR spectroscopy: rapid movement forward. *Current Opinion in Structural Biology* **14**, 554–61.

Meister, G., Eggert, C., and Fischer U. (2002) SMN-mediated assembly of RNPs: a complex story. *Trends in Cell Biology* **12**, 472–8.

Nakahara, K., Hayashi, T., Konishi, S., and Miyashita, Y. (2002) Functional MRI of macaque monkeys performing a cognitive set-shifting task. *Science* **295**, 1532–6.

Riek, R., Pervushin, K., and Wüthrich, K. (2000) TROSY and CRINEPT:NMR with large molecular and supramolecular structures in solution. *Trends in Biochemical Sciences* **25**, 462–8.

Selenko, P., Sprangers, R., Stier, G., Buhler, D., Kischer, U., and Sattler, M. (2001) SMN Tudor domain structure and its interaction with the Sm proteins. *Nature Structural Biology* **8**, 27–31.

Sprangers, R. and Kay, L.E. (2007) Quantitive dynamics and binding studies of the 20S proteosome by NMR. *Nature* **445**, 618–22.

Staunton, D. Owen, J., and Campbell, I.D. (2003) NMR and structural genomics. *Accounts of Chemical Research* **36**, 207–14.

Stevens, J.M., Daltrop, O., Allen, J.W.A., and Ferguson, S.J. (2004) C-type cytochrome formation: chemical and biological enzymes. *Accounts of Chemical Research* **37**, 999–1007.

Stubbe, J. (2003) Di-iron tyrosyl radical ribonucleotide reductases. *Chemical Opinion in Chemical Biology* **7**, 183–8.

Stubbe, J. and van der Dork, W.A. (1998) Protein radicals in enzyme catalysis. *Chemical Review* **98**, 705–62.

Tugarinov, V., Hwang, P.M., and Kay, L.E. (2004) Nuclear magnetic resonance spectroscopy of high-molecular-weight proteins. *Annual Review of Biochemistry* **73**, 107–46.

Vaynberg, J. and Qin, J. (2006) Weak protein-protein interactions as probed by NMR spectroscopy. *Trends in Biotechnology* **24**, 22–7.

Villar, H.O., Yan, J., and Hansen, M.R. (2004) Using NMR for ligand discovery and optimization. *Current Opinion in Chemical Biology* **8**, 387–91.

Wasser, I.M., de Vries, S. Moenne-Loccoz, P., Schroder, I., and Karlin, K.D. (2002) Nitric oxide in biological dentrification: Fe/Cu metalloenzyme and metal complex NO_x redox chemistry. *Chemical Review* **102**, 1201–34.

Wüthrich, K. (1986) *NMR of Proteins and Nucleic Acids*. Wiley Interscience, New York.

PROBLEMS

16.1 Calculate the resonance frequency of a proton in a magnetic field of (a) 7 T and (b) 20 T.

16.2 Calculate the resonance frequency of a ^{14}N nucleus in a magnetic field of (a) 20 T and (b) 9 T.

16.3 If the mass of a proton was only 4×10^{-29} kg, what would nuclear magneton become?

16.4 Explain what usually causes a large chemical shift.

16.5 Explain what a shielding constant is.

16.6 Why are high magnetic fields normally useful in NMR experiments?

16.7 For two coupled protons, what is the value of the lowest-energy state in the presence of a magnetic field?

16.8 How many lines are present when a proton interacts with four protons?

16.9 What type of magnetic field is used in an MRI experiment?

16.10 When a spin pulse is applied to the sample with the direction of its magnetic field perpendicular to the external magnetic field, what happens to the spins?

16.11 How many lines are present when a proton interacts with a ^{14}N nucleus?

16.12 How many lines are present when a proton interacts with another proton?

16.13 Describe the expected two-dimensional NMR spectrum of methionine.

16.14 Predict what the two-dimensional NMR would be for the newly found amino acid "carbonine 2", which has the following structure:

```
              CH3
                \
     CH3        CH2           H
       \         |            |
        CH  —  CH  —  CH2  —  Cα  —  COO⁻
       /                      |
     CH3                     ND2
```

16.15 What is a spin echo in an NMR experiment?

16.16 If three small beakers of water are lined up between the poles of a magnetic field, explain what signals would be observed for a homogeneous field compared with an inhomogeneous field.

16.17 How can MRI be used to reveal head trauma resulting from an accident?

16.18 Calculate the resonance frequency of a free electron in a magnetic field poised at (a) 0.4 T and (b) 0.3 T.

16.19 An EPR signal is observed at 0.3 T magnetic field for a 9 GHz microwave frequency. Calculate the g-factor.

16.20 Explain quantitatively why radiofrequencies are used in NMR experiments, whereas microwaves are used in EPR experiments.

16.21 An EPR spectrum is observed to be split into three equally spaced lines. What is the reason for this splitting?

16.22 How does an EPR signal from an unpaired electron change due to coupling to a proton?

16.23 How is the EPR spectrum of iron in a protein related to the oxidation state and coordination?

16.24 For Fe^{2+} what is the value of the high-spin state?

16.25 For Fe^{3+} what is the value of the low-spin state?

16.26 What EPR signals are observed from cytochromes?

16.27 How is the EPR spectrum of iron in a protein related to the oxidation state and coordination?

16.28 What EPR signals are expected for an amino acid radical?

16.29 Ferredoxins are proteins that serve as electron carriers in cells. Ferredoxins contain a Fe_2S_2 cofactor that is in a square configuration and coordinated to the protein by cysteine residues. Predict what the EPR spectrum would look like.

16.30 Explain the origin of the EPR spectrum of ribonucleotide reductase.

16.31 Briefly explain the basic mechanism for the ribonucleotide reductase and the role of amino acid radicals.

Part 3

Understanding biological systems using physical chemistry

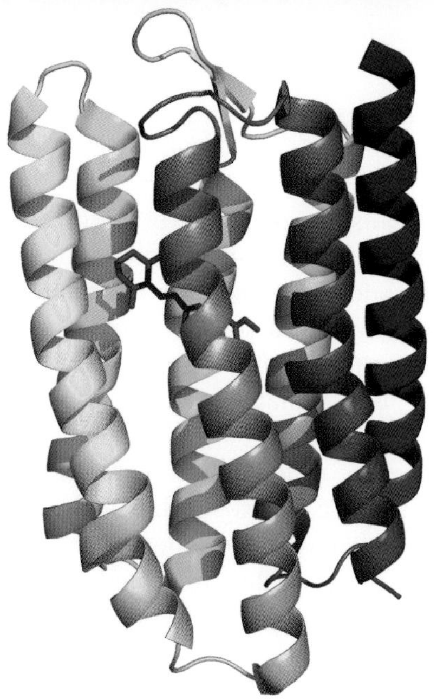

17

Signal transduction

Vision is one of the many signal-transduction processes in the body. In signal transduction, a signal in the form of light, or odor, or taste, is converted into a neurological signal that can be interpreted by the brain. Transduction processes occur in certain cellular membranes and involve protein complexes that are able to transform the signal with great sensitivity and specificity. The detection of light in vision occurs in the eye of vertebrates. The light is focused on to two types of light-sensing cells: rods, which sense low levels of light but cannot discriminate colors, and cones, which are less sensitive but are reactive to certain colors. Each human retina contains millions of these rods and cones, which individually are long, narrow neurons with an inner and an outer compartment (Figure 17.1). The outer compartment of a retinal rod cell contains stacks of discs, which are densely packed with the receptor for light, rhodopsin. The inner compartment generates ATP to power the transduction process and also synthesizes the necessary proteins. The plasma membrane contains proteins called ion channels (see Chapter 18), which control the flow of ions across the membrane. Sodium ions can flow rapidly through these channels in the dark but light blocks the channels. The decrease in the flow of sodium ions causes the plasma membrane to become more negative in the inside, or hyper-polarized. The change in membrane potential is sensed by the synapse and conveyed as a signal to the brain (Figure 17.2).

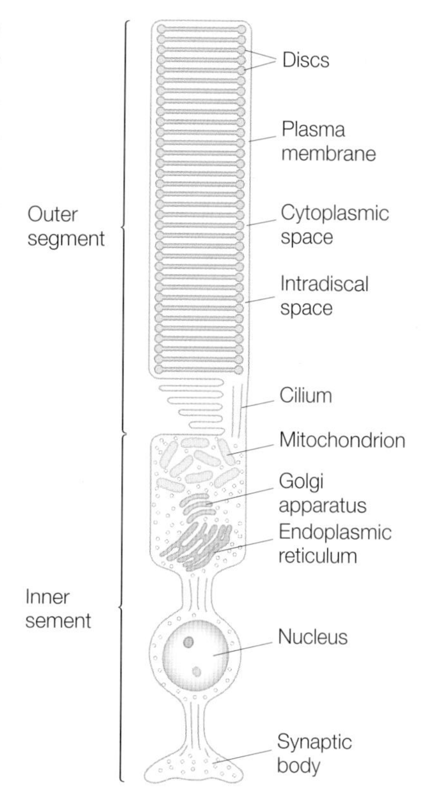

Figure 17.1 A schematic diagram of a retinal rod cell, with a width of approximately 1 μm and a length of about 40 μm.

BIOCHEMICAL PATHWAY FOR VISUAL RESPONSE

The visual process can be divided into four steps: recognition, conversion, amplification, and processing. The recognition step in vision is unusual for

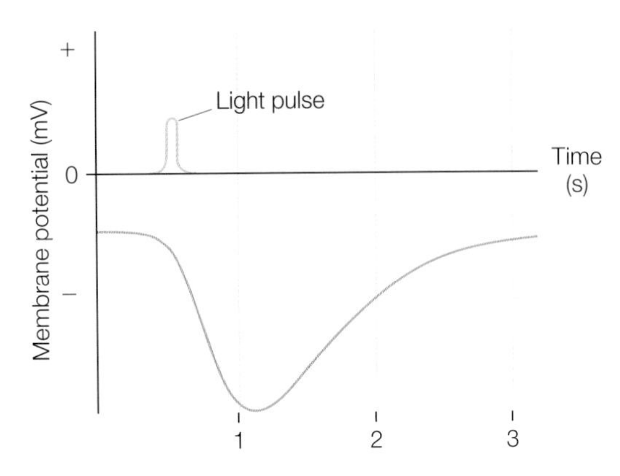

Figure 17.2 Light-induced hyperpolarization of a retinal rod cell.

biological signal transduction as it involves the absorption of light by a pigment buried inside of a protein called rhodopsin rather than the binding of a signal molecule to a protein receptor, as found in other signal-transduction processes. After recognition, there is a signal conversion, which requires a structural change of the protein and an associated molecule called retinal in response to light absorption. Once the signal has activated rhodopsin, it is amplified by many orders of magnitude, allowing the signal to be processed into a change in membrane potential and consequently a signal to the brain.

On a biological level, these four steps are achieved by coupling the action of rhodopsin to a G-protein cascade (where G-protein is short for guanine nucleotide-binding regulatory protein; Figure 17.3). Rhodopsin is located largely in the plasma membrane, and has an extramembraneous domain. Excitation of retinal leads to a conformational change of the rhodopsin, which facilitates binding of the protein transducin. Upon binding, transducin undergoes a change that results in the exchange of bound guanosine 5′-diphosphate (GDP) with

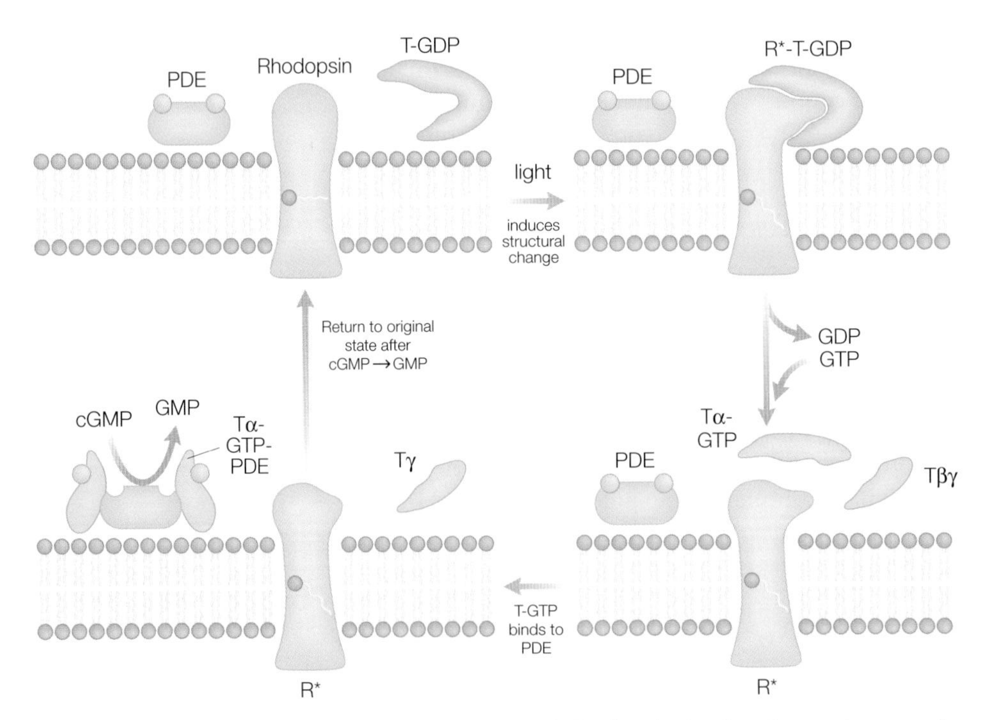

Figure 17.3 A schematic illustration of the coupling of rhodopsin in the G-protein cascade process.

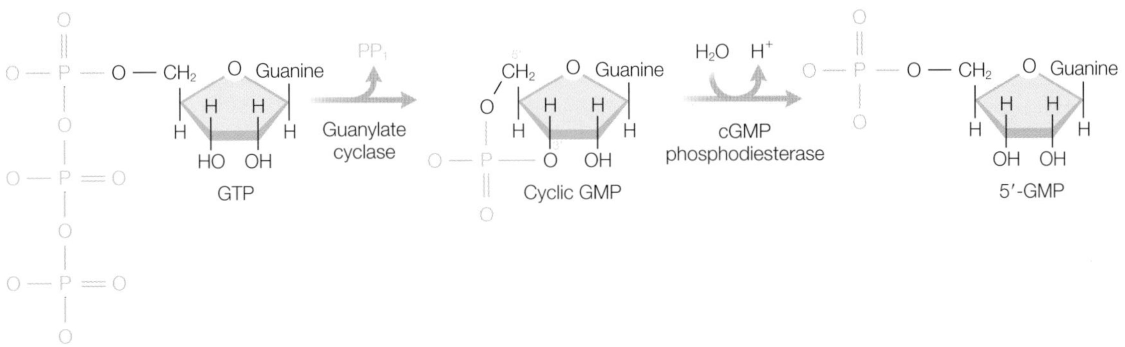

Figure 17.4 Structure of guanosine 3′,5′-cyclic monophosphate, or cyclic GMP, which serves as the messenger in the G-protein cascade process.

guanosine 5′-triphosphate (GTP). The transducin with bound GTP is released from the rhodopsin and one of the protein subunits, the α subunit, separates. This form of transducin, consisting of the α subunit and bound GTP, binds to phosophodiesterase and activates that enzyme, leading to a rapid conversion of cyclic GMP (guanosine 3′,5′-cyclic monophosphate, or cGMP) into GMP. The cycle is reset after the rhodopsin returns to the original dark state with the incorporation of retinal after the regeneration of *cis*-retinal by a retinal isomerase. Through this cycle, absorption of a single photon leads to a rapid loss of cyclic GMP through both depletion of the precursor GTP after transducin binding and by activation of the phosophodiesterase (Figure 17.4). In a sense, the overall biological process can be thought of as a two-stage electronic device that has a high-gain, low-noise character.

Why does this lead to a change in membrane potential? The membrane potential is controlled by ligand-gated ion channels that bind cyclic GMP. When cyclic GMP is bound, the channels are open and ions can flow through the membrane. The rapid loss of cyclic GMP leads to a loss of bound ligand, the channels close, ion flow stops, and the membrane potential becomes more negative. How can an animal's eye respond quickly to a sudden light in the eyes when the recovery by hydrolysis of GTP by transducin is relatively slow? There are additional protein regulators in the cell, which accelerate the hydrolysis to provide a rapid response.

SPECTROSCOPIC STUDIES OF RHODOPSIN

The role of retinal in the initiation of the visual process was discovered by George Wald who won the Nobel Prize for Medicine in 1967. Retinal is bound to rhodopsin through a protonated Schiff-base linkage, which is formed when the aldehyde group of retinal binds with the amino group

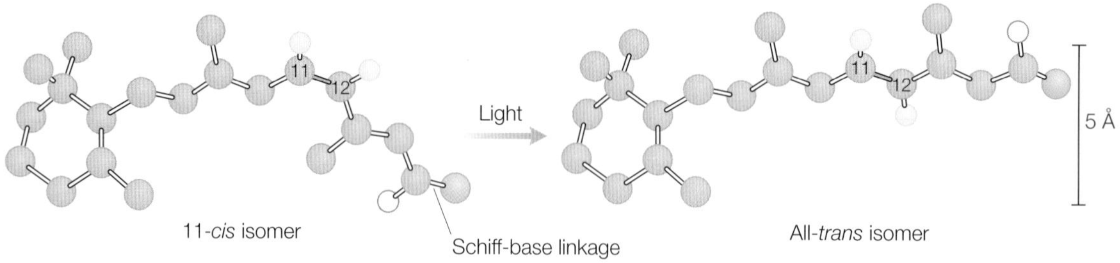

Figure 17.5 Upon light excitation, retinal is converted from a *cis* to *trans* isomer.

of Lys 296 in a hydrolysis reaction (Figure 17.5). The formation of the Schiff base shifts the wavelength maximum from 380 to 500 nm and produces a highly conjugated and strongly absorbing cofactor with an extinction coefficient of 40,000 mol^{-1} cm^{-1}.

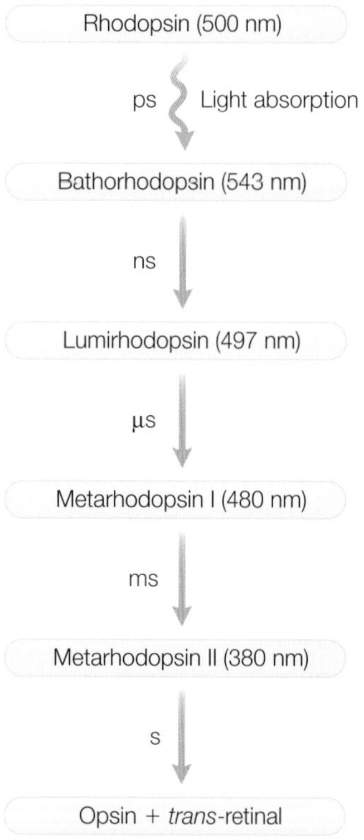

Due to the presence of the retinal, rhodopsin could be characterized by transient optical spectroscopy. In addition, the timing problem of spectroscopic studies of many enzymes, that is the difficulty in poising all of the enzymes in the same state, was easily overcome as an incident light pulse could be used to initialize the energy conversion. These studies showed that before the actual isomerization occurred, the protein changes into a number of intermediate states. These states were originally identified as optical states and so are named according to the respective wavelength maximum (Figure 17.6). Later these optical states were identified in terms of changes in the molecular configurations. The first state, bathorhodopsin, is formed in a few picoseconds and then converts to a series of other states on timescales ranging from nanoseconds to milliseconds until the all-*trans*-retinal is formed. The observed changes in the optical spectra cannot be predicted based upon a simple particle-in-a-box model (Chapter 10) as the length of the chromophore does not change. Rather, the optical shifts can be understood only in terms of detailed molecular models that show changes in the energy of the ground state of the retinal, arising from alteration of the relative position of the electron orbitals for the carbon atoms as the bond rotates in the isomerization.

Figure 17.6 The photocycle of rhodopsin showing the presence of many intermediate states before the formation of the all-*trans*-retinal.

BACTERIORHODOPSIN

The biochemical characterization of rhodopsin proceeded slowly due to the limited stability of the isolated protein and the irreversibility of the photoreaction. Therefore, much of the research in this area was centered on another related protein, bacteriorhodopsin. Bacteriorhodopsin is found in the bacterium *Halobacterium halobium*,

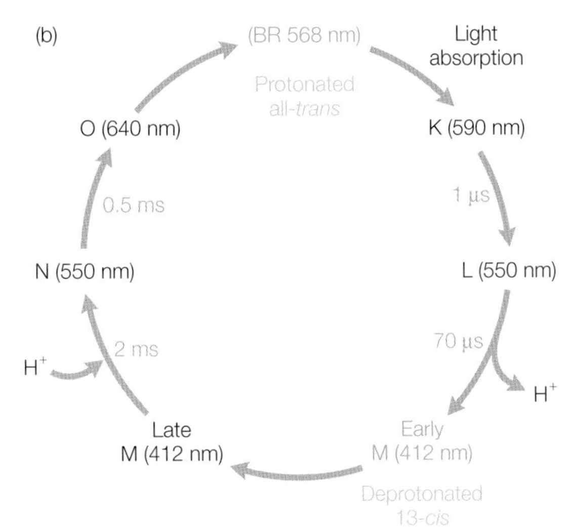

Figure 17.7 Photocycle of a bacteriorhodopsin. (a) In response to light the retinal of bacteriorhodopsin undergoes a *trans*-to-*cis* isomerization. (b) The isomerization process proceeds through a number of intermediate steps at different timescales. Unlike rhodopsin, the cycle is fully reversible and coupled to the transfer of a proton across the membrane.

which grows in briny waters such as the Dead Sea. This bacterium has a distinctive purple color due to the presence of the extremely high concentrations of retinal found in bacteriorhodopsin. The bacteriorhodopsin is present, not for a visual response such as phototaxis, but rather to convert light energy into chemical energy in the form of proton gradients across the cytoplasmic membrane.

Both rhodopsin and bacteriorhodopsin are proteins with a simple polypeptide chain formed by about 250 amino acid residues. Many of the amino acid residues of these polypeptides are conserved between rhodopsin and bacteriorhodopsin. As was found for rhodopsin, a retinal binds through a Schiff base to a lysine, which is residue 216 in bacteriorhodopsin. The photocycle of bacteriorhodopsin has similarities to that of rhodopsin, with the presence of different spectral states that are generated on varying lifetimes; however, there are two critical differences (Figure 17.7). First, the isomerization in bacteriorhodopsin proceeds from *trans* to *cis* in a reversible process rather than the irreversible *cis*-to-*trans* isomerization of rhodopsin that requires the reincorporation of a new *cis* isomer of retinal before the protein becomes active again. Second, the absorption of light by bacteriorhodopsin leads to the transfer of a protein across the cell membrane to be used for the generation of ATP from the buildup of a proton gradient.

The photocycles of rhodopsin and bacteriorhodopsin were established with the use of transient optical spectroscopy, Fourier transform infrared spectroscopy, and resonance Raman studies. Although the protein is large with many overlapping vibrational bands, the vibrations associated with the retinal could be distinguished by making use of light. Spectra were measured both before and after a laser flash. The differences between the two spectra were used to identify changes in vibrations, which arise from light-induced structural changes of the retinal and surrounding protein. These difference spectra were combined for comparison with isolated retinal and isotopic labeling of the retinal to assign the vibrations to specific bonds. For example, the spectrum shows a carbon–carbon stretch mode for the dark state at 1201 cm^{-1}, shifted to 1167 cm^{-1} in the light, consistent with the *trans*-to-*cis* isomerization. Also, the identification of the protonation state of the Schiff base could be made, based upon a shift of a bond from 1646 to 1629 cm^{-1} when the sample was deuterated.

STRUCTURAL STUDIES

The first indications of the structure of rhodopsin came from the sequence of the gene encoding bacteriorhodopsin. This sequence showed the presence of seven long stretches of hydrophobic amino acid residues connected by short segments of hydrophilic residues. If these residues were assumed to form an α helix, then each helix would be long enough to span the cell membrane. Thus, the sequence of bacteriorhodopsin was used to predict the presence of seven transmembrane helices. The arrangement of these helices was not determined by this analysis but mutagenesis studies showed that residues from the C and E helices influenced the functional properties.

A critical development was the elucidation of the three-dimensional structure using electron microscopy. Bacteriorhodopsin is highly concentrated in the cell membrane and forms a crystalline array. The availability of two-dimensional crystals provided the opportunity to use electron microscopy. When the electron beam probes a sample, the image mapped out is a projection of the protein onto a plane. To determine a three-dimensional image, images are obtained for the sample rotated through all possible angles. By knowing how much the sample has turned and how the image has changed for each angle, it is possible to infer the three-dimensional object.

A major limitation of images derived from biological samples by electron microscopy is that the atoms that compose proteins – carbon, nitrogen, and oxygen – have very comparable numbers of electrons. Not only can these atoms not be distinguished from each other but these atoms are low in contrast; that is, it is difficult to identify the protein from the surrounding matrix holding the sample. Samples can be labeled with heavy atoms, but then the images are limited by the quality of the labeling and are much too course for determination of a protein's structure. Increasing the strength of the signal by using intense electron beams does not help as this results in rapid radiation damage of the sample.

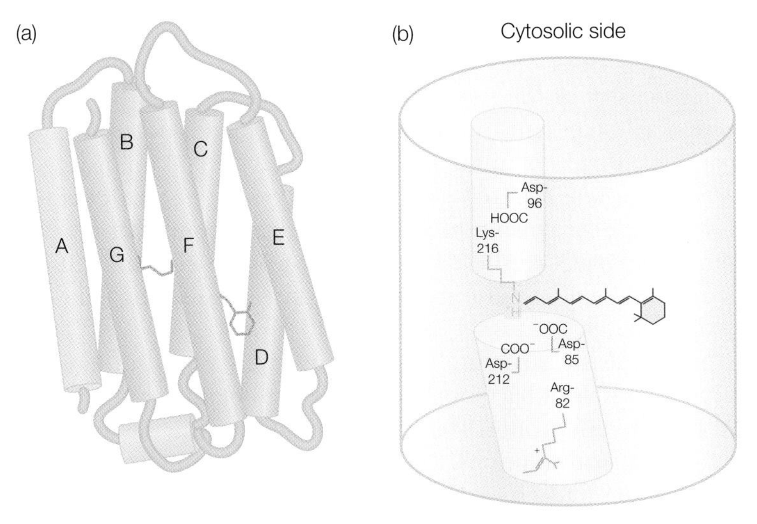

Figure 17.8 Three-dimensional structure and analysis of bacteriorhodopsin. (a) Three-dimensional structure of bacteriorhodopsin as determined by electron microscopy. The structure shows the presence of seven transmembrane helices (A–G) that enclose a series of protonatable residues forming a proton pathway. (b) The retinal is bound through a Schiff base to Lys-216 with the protonatable residues Arg-82, Asp-96, and Lys-216 nearby. The structure provided a molecular framework for the interpretation of the spectroscopic data. Modified from Henderson et al. (1990).

The structure of bacteriorhodopsin became a staging ground by Richard Henderson and coworkers for the development of new technical approaches to overcome these difficulties. To avoid radiation damage the intensity of the beam was decreased by limiting the focusing and the samples were measured at low temperature. A significant improvement in the resolution was achieved by not depending upon the real images that have a limited resolution but by measuring the diffraction of electrons from the two-dimensional crystal. Unlike protein crystallography, the phase is not a problem. The electron density can be determined from the images and hence the phases for the diffraction data can be found immediately (see Chapter 15).

These electron-microscopy studies produced a three-dimensional structure (Figure 17.8). The resolution was 2.8 Å in the plane but was more limited in the perpendicular direction due to the inability to tilt the sample to a very steep angle. Bacteriorhodopsin was seen to have seven long α helices with short connecting regions, as predicted by the analysis of the sequence. The retinal was bound to Lys-216 as predicted by the spectroscopic studies. The coupling of the structure with known mutagenesis studies led to identification of a proton pathway involving amino acid residues: Asp-96, Asp-85, Asp-212, and Arg-82 play critical roles in the proton transfer. Later electron-microscopy studies of rhodopsin showed that the structures of the two proteins are very similar.

Although achievement of the structure provided a detailed molecular model of the function, a number of questions were still unanswered. The spacing of the amino acid residues was too far apart to provide a uninterrupted proton pathway, and the light-induced structural changes were not determined. Answers to these questions required structural studies that provided more detail of not only the dark state but also the light-induced intermediate states.

The quality of the structure was improved after well-ordered, three-dimensional crystals of bacteriorhodopsin were obtained after many years of effort by different groups. The breakthrough was the inclusion of lipids in the crystallization solutions. The lipid concentration was poised so that phase separation occurred, with the protein being concentrated in the cubic phase along with the lipid (see Chapter 4). The protein is solubilized with lipids rather than detergents and was thus incorporated into the cubic lipid phase where the regular arrangement of the lipids promotes the formation of protein crystals.

The resulting structure determined from the crystals confirmed the earlier electron-microscopy results with the only major differences found in the loop regions, which were difficult to model in the electron micrographs. Because the crystal structure was determined at a much higher resolution, with the best crystals reaching a resolution limit of 1.55 Å, structural features were now resolvable (Figure 17.9). In particular, the structure showed the presence of water molecules bound inside the protein. The water molecules answered the question of how the proton could traverse the long 10-Å distances found between the amino acid residues forming the proton pathway. The pathway has many water molecules that provide the 2-Å steps needed for individual proton transfers.

The light-induced structural changes were identified by trapping the protein in different photochemical states and determining the structures in those states. The very early state shows only small changes near the retinal. The latter state shows not only the isomerization of the retinal but also the changes in the positions of the waters and amino acid residues forming the proton pathway. In particular, the bound water molecules near the *trans*-retinal are in different positions after the *cis*-retinal is formed, which is consistent with a movement of a proton along the pathway.

Because these structural states can be related directly to the spectroscopic states, the photochemical cycle of bacteriorhodopsin has been well established (Figure 17.10). Because the retinal is tethered, the isomerization leads to changes in the positions of the surrounding helices, which are relatively small – less than 1 Å – but sufficient to drive the proton transfer.

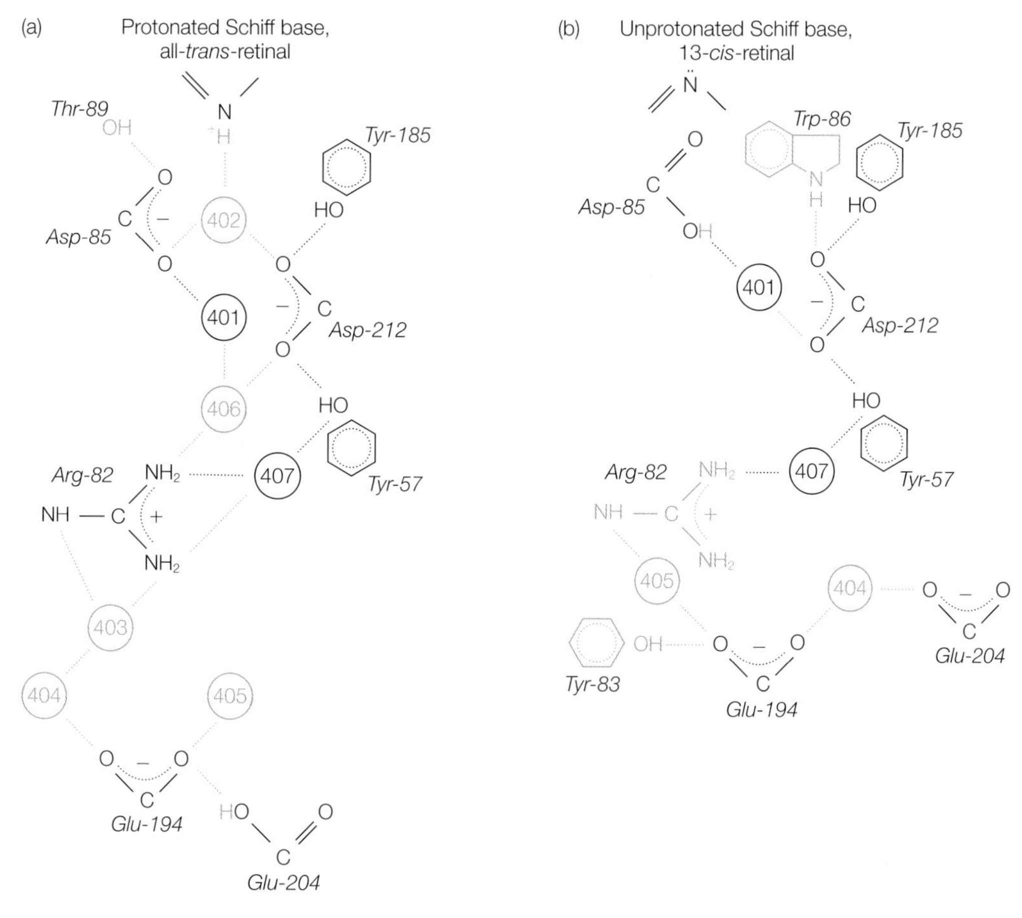

Figure 17.9 The structure of bacteriorhodopsin, as determined by X-ray diffraction, revealed many details about the proton pathway, including the presence of bound water molecules that complete the pathway. The positions of the amino acid residues and waters change in response to the isomerization. Modified from Belrhali et al. (1999).

The transition from the dark state to the initial K state happens when light is absorbed. During the transition from K to L, proton transfer occurs due to change in the retinal-binding site and a small movement of helix C that shifts Asp-85 toward the retinal. The proton associated with the retinal is transferred to Asp-85 as facilitated by Tyr-89 and Asp-212. In the transition from L to M, the deprotonation causes the retinal to shift, consequently moving helices F and G. During the M-to-N transition, the retinal is reprotonated by Asp-96 and a movement of helices F and G opens a channel that causes Asp-96 to be reprotonated. The protein then relaxes back to the dark state.

The structure then acts as a proton switch, with one side of the membrane well coupled to the retinal, leading to the early proton release

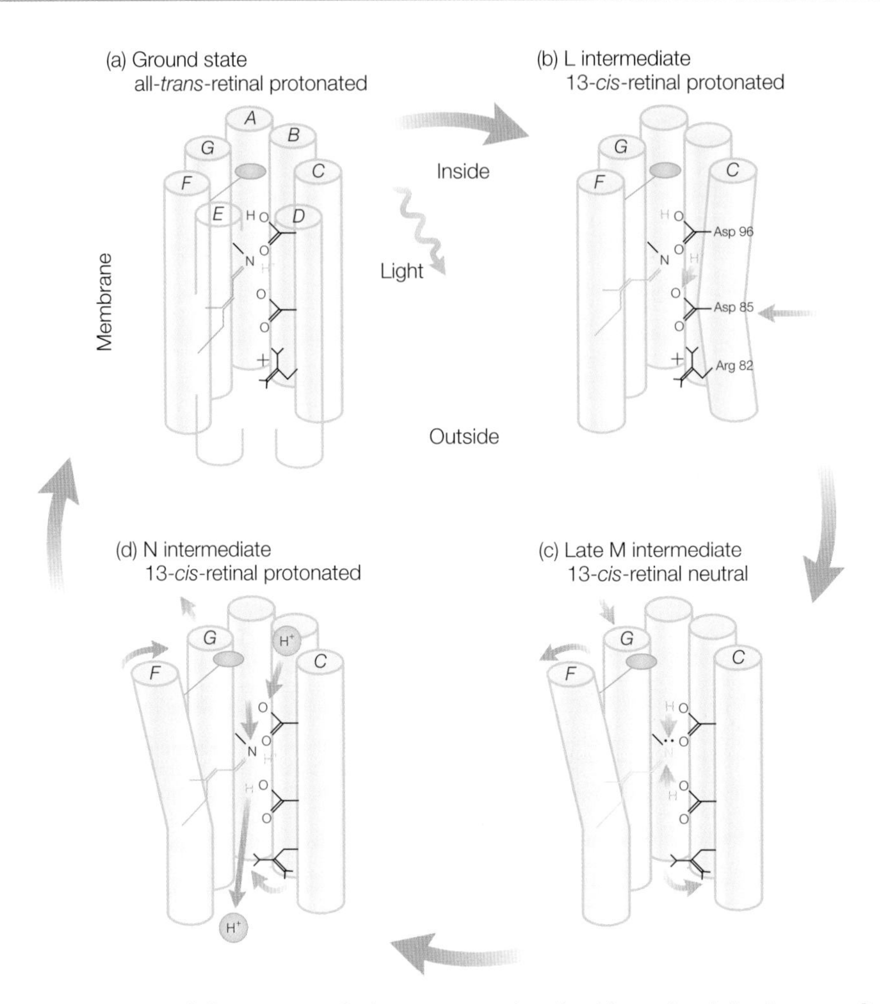

Figure 17.10 A summary of the structural changes associated with each of the intermediate states of bacteriorhodopsin. Modified from Kuhlbrandt (2000).

during steps 1 and 2 (Figure 17.11). The protein then switches, due to conformational changes, to be coupled to the other side of the membrane, leading to the proton uptake during steps 3 and 4. The system then resets with the reprotonation of Glu-194 and the re-isomerization of the retinal in step 5.

COMPARISON OF RHODOPSINS FROM DIFFERENT ORGANISMS

The three-dimensional structure of rhodopsin has also been solved using X-ray diffraction and was found to be closely related to that of bacteriorhodopsin, as expected based on the spectroscopic studies and sequence comparisons. Whereas the 348 amino acid residues form seven long helices located at very similar positions, the loops connecting the helices differ. The domains in the extramembraneous regions are thought to shift during the photocycle in a manner that is not fully established, to bind proteins such as tranducin during the signaltransduction process (Figure 17.3).

Members of the rhodopsin family are found in all three domains of life, the Archaea, Eubacteria, and Eukaryota. The strong structural homology between rhodopsin and bacteriorhodopsin is consistent with the concept that all opsins have similar folds with seven transmembrane helices surrounding a retinal molecule (Figure 17.12). Despite the structural homology, the different opsins have significantly different functions involving light-driven ion transport or photosensing signaling. Rhodopsin serves as a G-coupled protein whereas bacteriorhodopsin serves as a proton pump. Halorhodopsin is found in halophilic archaea and serves primarily as a chloride pump, although it can also transport other ions such as bromide. Sensory rhodopsins (SRI and SRII) are found in the membrane with other proteins, such as kinases, which serve to transmit the light-induced structural changes into cellular signals.

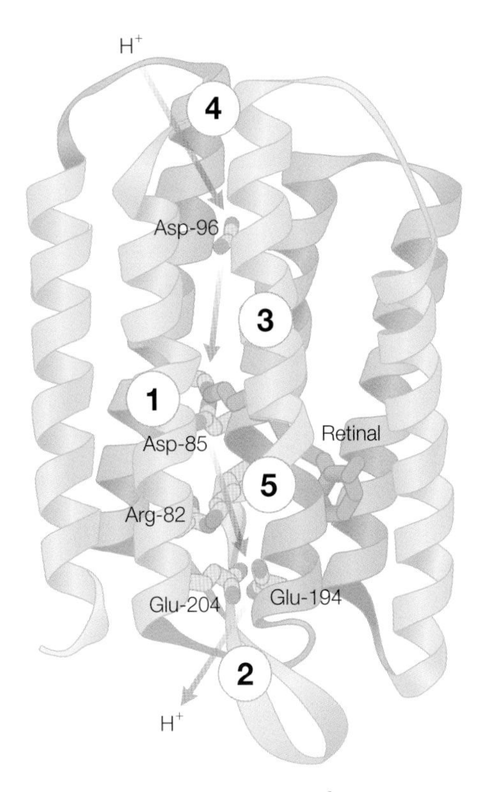

Figure 17.11 Proton transfer can be considered to be driven by a light-activated switch that activates a series of steps (1–5). 1, the sequential deprotonation of the Schiff base and protonation of Asp-85; 2, proton release; 3, reprotonation of the Schiff base and deprotonation of Asp-96; 4, reprotonation of Asp-96; 5, deprotonation of Asp-85 and reprotonation of the release site. Modified from Luecke et al. (1999).

The overall structure of sensory rhodopsin is very similar to that of bacteriorhodopsin, with the presence of the transmembrane helices. The loops connecting the helices are more extensive and serve as part of the binding domain for the G-proteins. Sensory rhodopsin contains an 11-*cis*-retinal chromatophore that isomerizes to an all-*trans*-configuration upon light absorption. The largest difference compared to bacteriorhodopsin and rhodopsin is the presence of additional subunits, the bound transducer proteins. In response to light-induced structural changes of the helices surrounding the retinal, which have not yet been delineated, these transducer proteins initialize a phosphorylation cascade that regulates the flagellar motors. Although the structures of the transducer proteins have not yet been determined, structures of soluble domains suggest that the proteins are remarkably long, with a total length of about 400 Å and the bulk of the protein on the cytoplasmic side of the membrane, formed by long helices that are responsible for the signaling activity.

Halorhodopsin transports chloride ions rather protons but still possesses many of the same structural features as bacteriorhodopsin. Both have seven transmembrane helices surrounding the retinal, although halorhodopsin is a trimer in the membrane rather than a monomer. Halorhodopsin has

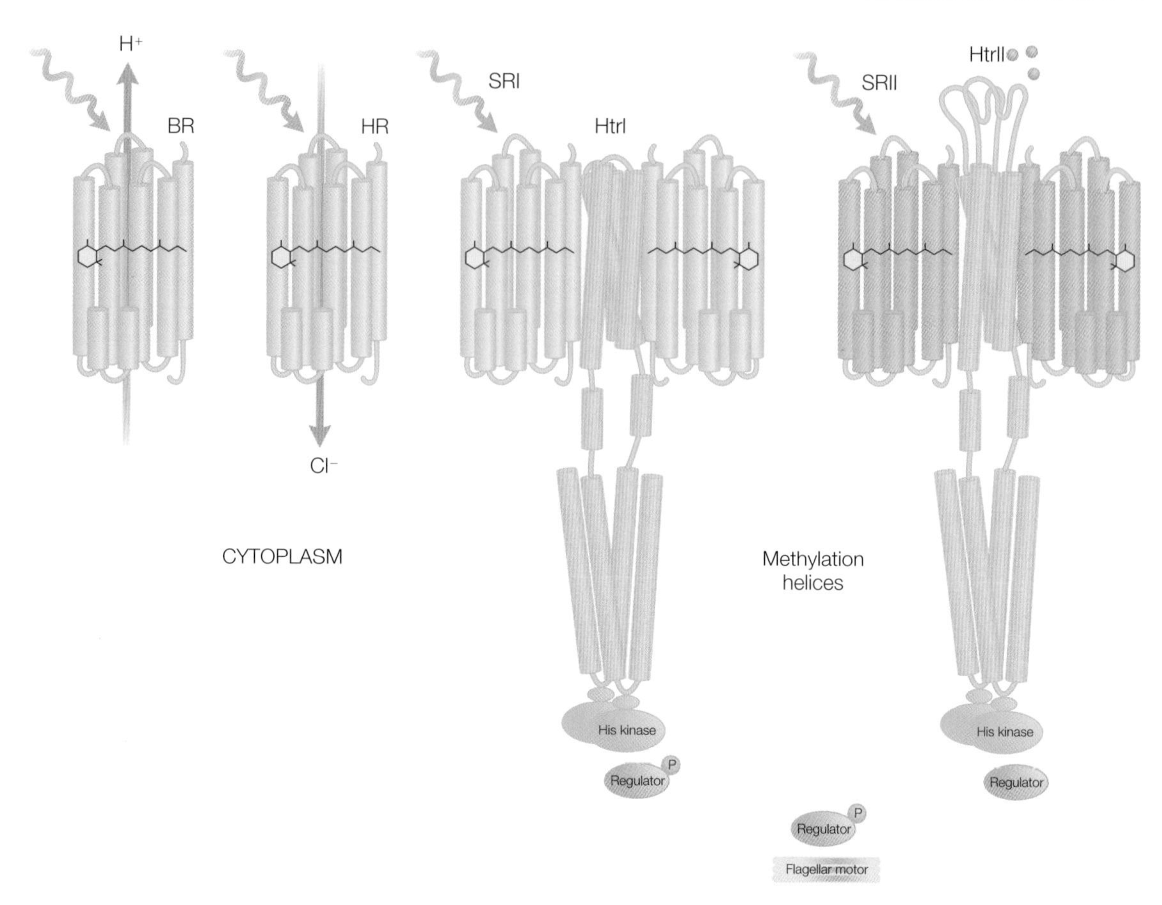

Figure 17.12 Four archad rhodopsins: bacteriorhodopsin (BR), halorhodopsin (HR), and the phototaxis receptors sensory rhodopsins (SRI and SRII). The sensory rhodopsins are complexed to the cognate transducer proteins (HtrI and HtrII). Modified from Spudrich et al. (2000).

an exterior surface that has significant electrostatic character, which serves to interact with a chloride ion, allowing putative identification of the chloride channel. Unlike ion channels, there is no open channel in the protein but rather the pathway opens during the photocycles.

Retinal binds halorhodopsin at a site very similar to that of bacteriorhodopsin (Figure 17.13). Nine of the surrounding residues are conserved and water is found, as was true for bacteriorhodopsin. A single chloride ion was found near the retinal at a location corresponding to that of Asp-85. Ion translocation is proposed to occur through a mechanism related to that of bacteriorhodopsin. The ion is proposed to be driven by ion–dipole interactions involving the NH group of the retinal before it is released toward the cytoplasm and a new chloride ion enters the transport site. Only the replacement of the negatively charged Asp-85 as a proton acceptor in bacteriorhodopsin by chloride in halorhodopsin changes the kinetic preference and therefore the ion specificity. Thus in both cases the retinal serves as a switch for the movement of the proton or ion.

RHODOPSIN PROTEINS IN VISUAL RESPONSE

Much of the characterization of the visual process has centered in rhodopsin in rod cells. Less is known about the structure of the photoreceptors in the cone cells. Color vision is provided by three receptors that are each responsive to a different color: blue, green, and red. These receptors have been sequenced and were found to be highly homologous to rhodopsin. These cone receptors are modeled as having a structure similar to rhodopsin, with a central retinal molecule bound to a Schiff base. The absorption is tuned to different wavelengths by differences in the amino acid residues in the retinal-binding site. Together the photoreceptors in the rod and cone cells provide the information for the initial stages of visual processing. The eyes also provide another function; namely, they serve as light sensors to regulate circadian rhythms of the body. Melanopsin is a protein that is thought to have a fold similar to rhodopsin and to bind *trans*-retinal, which has an absorption maximum at 480 nm. Light results in a *trans*-to-*cis* isomerization that triggers a signal cascade, as found for the other rhodopsins. Despite this similarity of the photocycle, the cellular response is probably much different as the protein is not in the rod or cone cells but rather is located in special ganglion cells that are coupled to the optical nerve. In addition to questions concerning the response mechanism, the reason why vertebrates respond to blue light remains an open question.

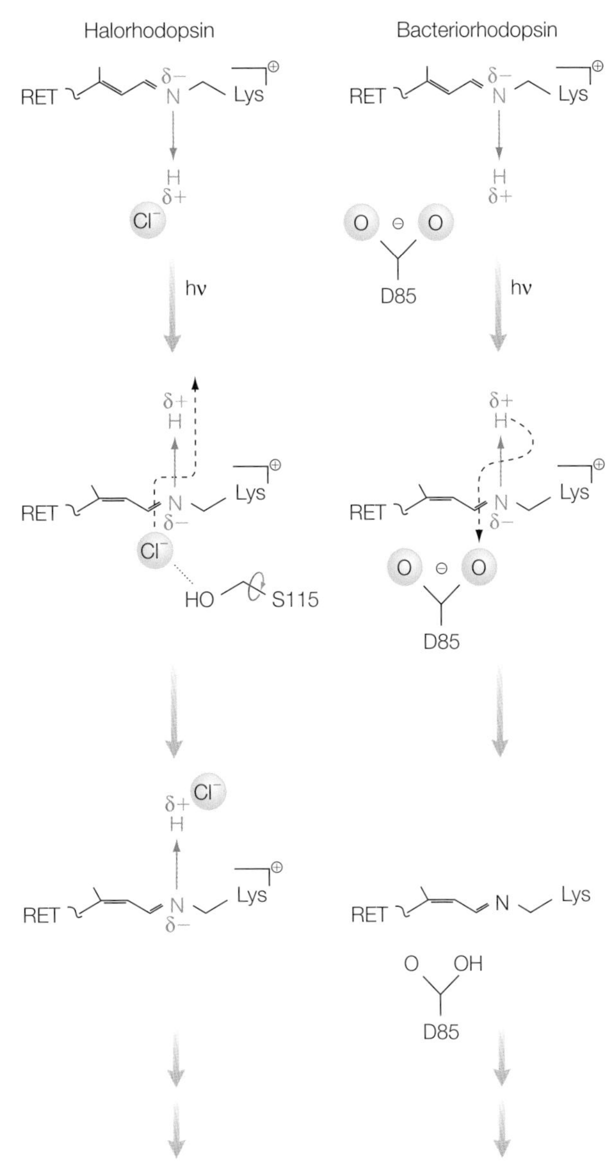

Figure 17.13 Comparison of the light-induced changes for bacteriorhodopsin and halorhodopsin. Modified from Kolbe et al. (2000)

REFERENCES AND FURTHER READING

Belrhali, H., Nollert, P., Royant, A. et al. (1999) Protein, lipid, and water organization in bacteriorhodopsin crystals: a molecular view of the purple membrane at 1.9 Å resolution. *Structure* **7**, 909–17.

Blurner, K.J. (2004) The need for speed. *Nature* **427**, 20–1.

Edman, K., Nollert, P., Royant, A. et al. (1999) High-resolution X-ray structure of an early intermediate in the bacteriorhodopsin photocycle. *Nature* **401**, 822–6.

Essen, L.O. (2002) Halorhodopsin: light-driven ion pumping made simple? *Current Opinion in Structural Biology* **12**, 516–22.

Filipek, S., Teller, D.C., Palczewski, K., and Stenkemp, R. (2003) The crystallographic model of rhodopsin and its use in studies of other G protein-coupled receptors. *Annual Review of Biophysics and Biomolecular Structure* **32**, 375–97.

Foster, R.G. (2005) Bright blue times. *Nature* **433**, 698–9.

Haupts, U., Tittor, J., and Oesterhelt, D. (1999) Closing in on bacteriorhodopsin: progress in understanding the molecule. *Annual Review of Biophysics and Biomolecular Structure* **28**, 397–9.

Henderson, R., Baldwin, J.M., Ceska, T.A., Zemlin, F., Beckmann, E., and Dowing, K.H. (1990) Model for the structure of bacteriorhodopsin based upon high-resolution electron cryo-microscopy. *Journal of Molecular Biology* **213**, 899–929.

Kolbe, M., Besir, H., Essen, L.O., and Oesterhelt, D. (2000) Structure of the light-driven chloride pump halorhodopsin at 1.8 Å resolution. *Science* **288**, 1390–6.

Kuhlbrandt, W. (2000) Bacteriorhodopsin – the movie. *Nature* **406**, 569–70.

Landau, E.M., Pebay-Peyroula, E., and Neutze, R. (2003) Structural and mechanistic insight from high resolution structures of archaeal rhodopsins. *FEBS Letters* **555**, 51–6.

Lanyi, J.K. (2005) Proton transfers in the bacteriorhodopsin photocycle. *Biochimica Biophysica Acta* **1757**, 1012–18.

Lanyi, J.K. and Luecke, H. (2001) Bacteriorhodopsin. *Current Opinion in Structural Biology* **11**, 415–19.

Luecke, H., Schobert, B., Lanyi, J.K., Spudrich, E.N., and Spudrich, J.L. (2001) Crystal structure of sensory rhodopsin II at 2.4 Angstroms: insights into color tuning and transducer interactions. *Science* **293**, 1499–1503.

Luecke, H., Schobert, B., Richter, H.T. et al. (1999) Structural changes in bacteriorhodopsin during ion transport at 2-Angstrom resolution. *Science* **286**, 255–60.

Oprian, D.D. (2003) Phototaxis, chemotaxis, and the missing link. *Trends in Biochemical Science* **28**, 167–9.

Palczewski, K., Kumasaka, T., Hori, T. et al. (2000) Crystal structure of rhodopsin: a G-coupled receptor. *Science* **289**, 739–45.

Pebay-Peyroula, E., Rummel, G., Rosenbusch, J.P., and Landau, E.M. (1997) X-ray structure of bacteriorhodopsin at 2.5 Å from microcrystals grown in cubic lipid phases. *Science* **277**, 1676–81.

Rummel, G., Hardmeyer, A., Widmer, C. et al. (1998) Lipidic cubic phases: new matrices for the three-dimensional crystallization of membrane proteins. *Journal of Structural Biology* **121**, 82–91.

Spudrich, J.L. (2000) A chloride pump at atomic resolution. *Science* **288**, 1358–9.

Spudrich, J.L. and Luecke, H. (2002) Sensory rhodopsin II: functional insights from structure. *Current Opinion in Structural Biology* **12**, 540–6.

Spudrich, J.L., Yang, C.H., Jung, K.H., and Spudrich, E.N. (2000) Retinylidene proteins: structures and functions from archaea to humans. *Annual Review of Cell and Developmental Biology* **16**, 365–92.

PROBLEMS

17.1 Why do rod cells respond to light?

17.2 What are the roles of GMP and transducin in vision?

17.3 What is the initial light-induced structural change in bacteriorhodopsin?

17.4 Why does the absorption spectrum of rhodopsin change as the photocycle proceeds?

17.5 How does the isomerization of retinal differ in rhodopsin and bacteriorhodopsin?

17.6 How was FTIR used to characterize the photocycle of bacteriorhodopsin?

17.7 Why was electron microscopy used for the initial determination of the structure of bacteriorhodopsin rather than protein crystallography?

17.8 How is phase determination different for electron microscopy compared to X-ray diffraction?

17.9 How were cubic lipid phases important for the determination of the structure of bacteriorhodopsin?
17.10 What questions remained after the determination of the structure of bacteriorhodopsin by electron microscopy?
17.11 Why is the presence of water molecules important in bacteriorhodopsin?
17.12 Why does bacteriorhodopsin function as a light-activated proton switch?
17.13 What similarities are found for rhodopsin, bacteriorhodopsin, halorhodopsin, and sensory rhodopsin?
17.14 Describe the functional differences between bacteriorhodopsin and sensory rhodopsin.
17.15 What is the role of the transducer proteins in sensory rhodopsin?
17.16 How can halorhodopsin transfer a chloride ion without the presence of a channel?
17.17 Explain the role of dipole interactions in triggering proton transfer in bacteriorhodopsin.
17.18 Contrast the role of dipole interactions in triggering proton transfer in bacteriorhodopsin compared to chloride transfer in halorhodopsin.
17.19 How is the cellular location of malanopsin different from that of the other photoreceptors of the eye?
17.20 Why does the mutation of certain amino acid residues, such as Asp-85, result in a structurally intact but nonfunctional bacteriorhodopsin?
17.21 Compare and contrast the proton-switch model of bacteriorhodopsin and halorhodopsin.

18

Membrane potentials, transporters, and channels

MEMBRANE POTENTIALS

The cell membrane is a lipid bilayer. The composition of cell membranes varies among cell types and growth conditions, but in general the major component is phospholipids. Phospholipids are amphipathic molecules; that is, they have both a polar head group and a hydrophobic fatty acid portion (Chapter 7). Because of the central hydrophobic core of the bilayer, the cell membrane is largely impermeable to ions and polar molecules. In contrast, water traverses membranes much more readily. The permeability of small molecules ranges over several orders of magnitude and is correlated with the solubility of each molecule in nonpolar solvents relative to water (Figure 18.1). The cell makes use of this ability to control the relative concentrations of ions to build ion gradients across the cell membrane for many metabolic processes.

The cell controls the balance of ions through certain proteins present in the membrane, called transporters and channels. For example, in

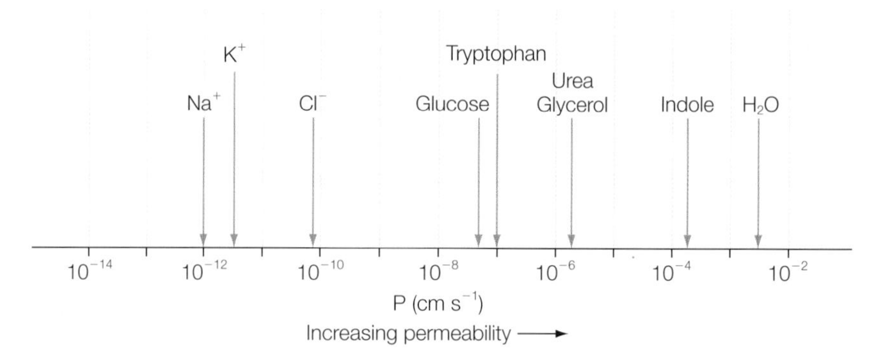

Figure 18.1 Permeability of a membrane for different types of molecules.

virtually every animal cell, the fluid surrounding the cells has ion concentrations that are different from those found in the cell. Typically, the concentrations outside the cell are poised at 140 mM for Na^+ and 5 mM for K^+. Yet the concentrations inside the cells are maintained at concentrations of 10 mM for Na^+ and 100 mM for K^+. This balance is maintained by various proteins, such as the Na^+/K^+ transporter, which pumps two potassium ions into the cell while transferring three sodium ions out of the cell.

ENERGETICS OF TRANSPORT ACROSS MEMBRANES

The transport of ions depends upon the change of free energy of the transported ion. Consider an uncharged solute molecule with *concentrations* c_1 and c_2 for the two sides of the membrane. The free energy change in transporting the species across the membrane is:

$$\Delta G = RT \ln \frac{c_2}{c_1} \tag{18.1}$$

For a charged species, the electric potential must also be considered:

$$\Delta G = ZF\,\Delta V \tag{18.2}$$

where Z is the electrical charge, F is the faraday constant, and ΔV is the potential difference across the membrane. This gives a total free energy change for the transport of ions across the membrane:

$$\Delta G = RT \ln \frac{c_2}{c_1} + ZF\,\Delta V \tag{18.3}$$

Active transport requires a coupled input of free energy. For example, consider the transport of an uncharged molecule from $c_1 = 10^{-3}$ mM to $c_2 = 10^{-1}$ mM:

$$\Delta G = 2.3RT \log \frac{10^{-1}}{10^{-3}} = 2.7 \text{ kcal mol}^{-1} \tag{18.4}$$

This process is energetically unfavorable and will not occur spontaneously. The transport could be coupled to ATP hydrolysis but alternatively it could be coupled to the transport of another molecule across the membrane. This often occurs in cell membranes, in proteins known as antiporters, which can simultaneously transport two different molecules in opposite directions. As an artificial example, consider the possible coupled transport of glucose and urea from the inside to the outside of a cell (this is not actually found in nature). The coupled reaction would be written as:

$$\text{Glucose}_{\text{out}} + \text{urea}_{\text{in}} \leftrightarrow \text{glucose}_{\text{in}} + \text{urea}_{\text{out}} \qquad (18.5)$$

Set the concentrations to 1 mM for both glucose and urea outside and 10 mM inside:

$$\Delta G = RT \ln \frac{(c_{\text{glucose}})_{\text{in}}(c_{\text{urea}})_{\text{out}}}{(c_{\text{glucose}})_{\text{out}}(c_{\text{urea}})_{\text{in}}} = RT \ln \frac{(10 \times 10^{-3}\ \text{M})(1 \times 10^{-3}\ \text{M})}{(1 \times 10^{-3}\ \text{M})(10 \times 10^{-3}\ \text{M})} = 0 \qquad (18.6)$$

Now set the outside concentrations to 1 mM for glucose and 10 mM for urea and both inside concentrations at 10 mM:

$$\Delta G = RT \ln \frac{(c_{\text{glucose}})_{\text{in}}(c_{\text{urea}})_{\text{out}}}{(c_{\text{glucose}})_{\text{out}}(c_{\text{urea}})_{\text{in}}} = RT \ln \frac{(10 \times 10^{-3}\ \text{M})(1 \times 10^{-3}\ \text{M})}{(10 \times 10^{-3}\ \text{M})(10 \times 10^{-3}\ \text{M})}$$
$$= 58\ \text{meV} \log(0.10) = -58\ \text{meV} \qquad (18.7)$$

The transport is now energetically favorable. Notice that when the units are converted to base 10 logarithms, a factor of 10 difference in concentration corresponds to an energy of 58 meV at room temperature.

In the intestines, glucose is cotransported with Na^+ into epithelial cells by a symporter, which is a channel that can transport two molecules in the same direction simultaneously. For the Na^+/glucose symporter, two sodium ions are transported with every glucose molecule:

$$2\text{Na}^+_{\text{out}} + \text{glucose}_{\text{in}} \rightarrow 2\text{Na}^+_{\text{in}} + \text{glucose}_{\text{out}} \qquad (18.8)$$

The energy to transport the glucose is provided by the simultaneous transport of the Na^+. Consider concentrations of 12 and 145 mM for the intracellular and extracellular sodium concentrations, respectively, and a typical membrane potential of –50 mV. For each Na^+ the change in free energy is calculated using eqn 18.3:

$$\Delta G = (2.3 \times 2.47\ \text{kJ mol}^{-1}) \log \frac{12\ \text{mM}}{145\ \text{mM}} + 1(96.5\ \text{kJ V}^{-1}\ \text{mol}^{-1})(-0.05\ \text{V})$$
$$= -11.0\ \text{kJ mol}^{-1} \qquad (18.9)$$

For every 2 mol of Na^+ moved, the energy to transport 1 mol of glucose is:

$$\Delta G = 2\ \text{mol}\ (11.0\ \text{kJ mol}^{-1}) = 22\ \text{kJ} \qquad (18.10)$$

This energy provides the opportunity to transport glucose against a large concentration gradient. The glucose does not build up in the epithelial cells because there is a passive glucose channel that allows the glucose to pass from the epithelial cells into the bloodstream. In this case, the transporter serves as a uniporter, which transports a single molecule.

The energy for the Na^+/K^+ pump is driven by the hydrolysis of ATP. As discussed above, the transport of Na^+ out of the cell moves the ion from a low concentration of 10 mM to 140 mM, so this term will be unfavorable. Since a charge is being moved, we need to include the contribution of the membrane potential, which is 70 mV, being more negative in the inside. Using eqn 18.3 we can calculate the energy change for the Na^+ as:

$$\Delta G = (2.3 \times 2.47\ \text{kJ mol}^{-1}) \log \frac{140\ \text{mM}}{10\ \text{mM}} + 1(96.5\ \text{kJ V}^{-1}\ \text{mol}^{-1})(+0.07\ \text{V})$$
$$= +13.6\ \text{kJ mol}^{-1} \quad (18.11)$$

The transport of K^+ is in the opposite direction. The ion moves from a low concentration of 5 mM to a high concentration of 100 mM, which is unfavorable, but the contribution of the membrane potential has changed sign and is favorable and the overall free energy change is nearly zero:

$$\Delta G = (2.3 \times 2.47\ \text{kJ mol}^{-1}) \log \frac{100\ \text{mM}}{5\ \text{mM}} + 1(96.5\ \text{kJ V}^{-1}\ \text{mol}^{-1})(-0.07\ \text{V})$$
$$= +1.0\ \text{kJ mol}^{-1} \quad (18.12)$$

Since 3 mol of Na^+ is transported with 2 mol of K^+ the net energy change is:

$$\Delta G_{\text{total}} = 3(+13.6\ \text{kJ mol}^{-1}) + 2(+1.0\ \text{kJ mol}^{-1}) = +42.8\ \text{kJ mol}^{-1} \quad (18.13)$$

To estimate the free energy change contributed by converting ATP to ADP, we must correct the standard free energy difference of $-31.3\ \text{kJ mol}^{-1}$ by the contribution of the high concentration of ATP in the cell. If there is a thousand-fold greater amount of ATP than ADP, the free energy difference is estimated to be:

$$\Delta G = -31.3\ \text{kJ mol}^{-1} + (2.3 \times 2.47\ \text{kJ mol}^{-1})\log(10^3)$$
$$= -49.1\ \text{kJ mol}^{-1} \quad (18.14)$$

The dependence of the free energy on concentration quantifies the tendency of molecules to have equal concentrations on both sides of membranes. However, if we consider a charged molecule, such as a proton, the free energy difference is zero when:

$$\Delta V = -\frac{RT}{FZ} \ln \frac{(c_p)_{\text{out}}}{(c_p)_{\text{in}}} \quad (18.15)$$

Under normal cellular conditions, it takes very little transfer of protons across the membrane before the electric potential ΔV that is generated is equal to the chemical potential due to the difference in the proton concentrations. Thus, the transfer of protons does not impact on the proton concentrations on either side. For example, if you have a 10-fold excess of protons on one side of the membrane and add a protonophore, a molecule that allows the transfer of protons, but nothing else, across the membrane, such as nigericin, you would assume that the protons transfer until the chemical and electric potentials balance without changing the proton concentrations.

In a mitochondrial membrane, the ratio of proton concentration is about 10:1, corresponding to a chemical potential of 60 meV. The electric potential is about 180 meV so the total free energy to drive protons across the membrane is 240 mV. The oxidation of NADH requires about 2.3 eV, so roughly 10 protons are pumped across the membrane per NADH oxidized. Remember that volt (V) is a unit of electrical potential whereas an electron volt (eV) is a unit of energy. One eV is the amount of free energy required to transfer 1 mol of a singly charged positive ion across a membrane with an opposing electric potential of 1 V. For example, the free-energy transfer of 1 mol of protons across a membrane with a 200 mV potential is 200 meV.

TRANSPORTERS

Transport proteins selectively mediate the passage of molecules across the membrane, which is otherwise impermeable. More than 360 transporter families have been identified, highlighting the critical role of transport processes in cells. For example, the major facilitator superfamily transfers its substrates against concentration gradients by coupling the transfer to a second, energetically favorable movement. These transporters have 12 transmembrane helices, which are seen to be organized into two domains in the structures of LacY, which mediates the coupled transport of lactose with protons, and GlpT, which transports glycerol 3-phosphate and phosphate (Figure 18.2). The two domains are related to each other by an approximate 2-fold symmetry axis. These structures support the idea that transport occurs through a process of alternating access, in which the protein undergoes domain changes that alternately provide substrate access to one side of the membrane or the other, but never to both simultaneously.

Glutamate transporters play a critical role in the control of neurotransmitters, small diffusible molecules that transmit nerve impulses across synapses. Neurotransmitters are used for communication between neurons, and many drugs, such as cocaine or Prozac, target proteins that inhibit neurotransmitter transporters. The predominant excitatory neurotransmitter

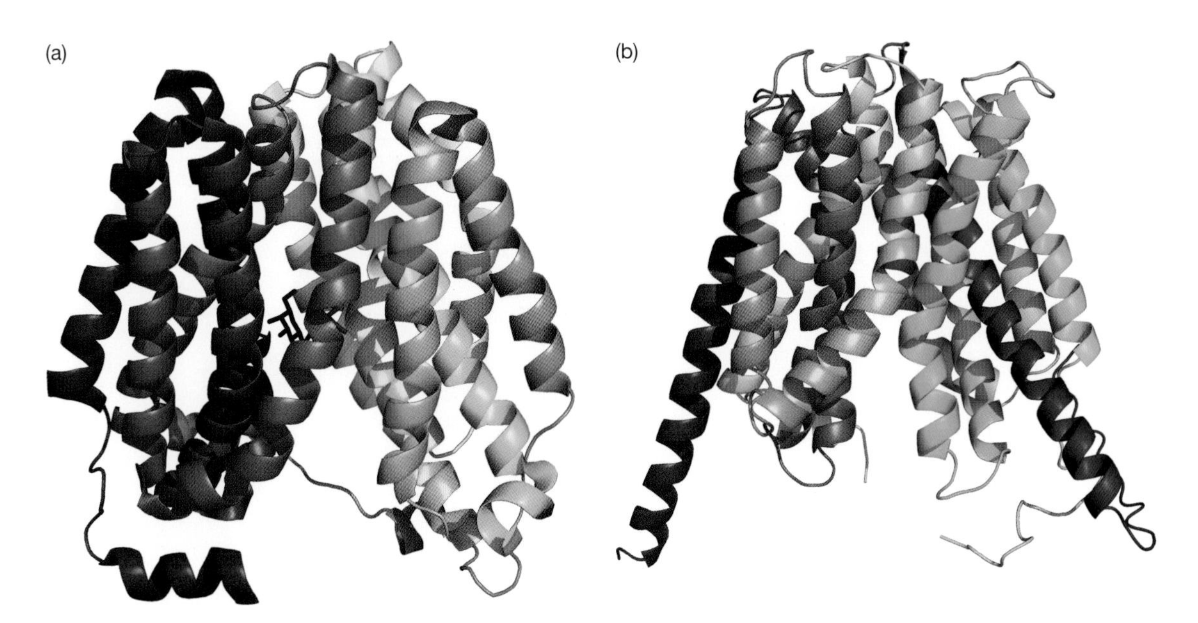

Figure 18.2 The three-dimensional structures of the transporters (a) LacY and (b) GlpT.

in the mammalian central nervous system is the amino acid glutamate. Glutamate transporters transport three Na^+ ions and one proton across the membrane while glutamate and one K^+ ion are transferred in the opposite direction. The glutamate-transporter family has properties that are distinct from those of the major facilitator superfamily, in particular only eight transmembrane helices. The fold of homolog of a glutamate transporter shows the presence of two segments that contain α helices, forming short hairpin loops, which could serve to gate the transport process (Figure 18.3). In addition to the ability of short helices to serve as a gate through a loop movement, the transmembrane helices are often bent and the flexing of a helix around its bend may provide a means of changing the opening of transporters (Figure 18.4).

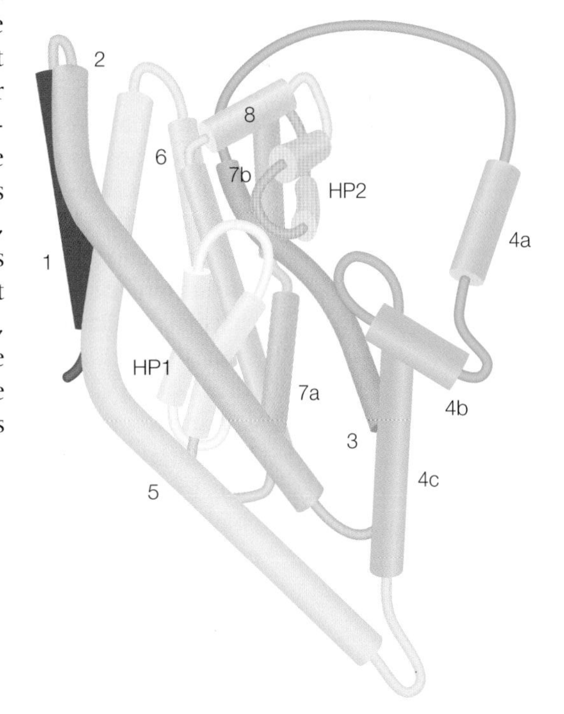

Figure 18.3 Representation of the glutamate transporter showing the α helices as cylinders. The hairpin helices that are proposed to serve as a gate are identified as HP1 and HP2. Modified from Yernool et al. (2004).

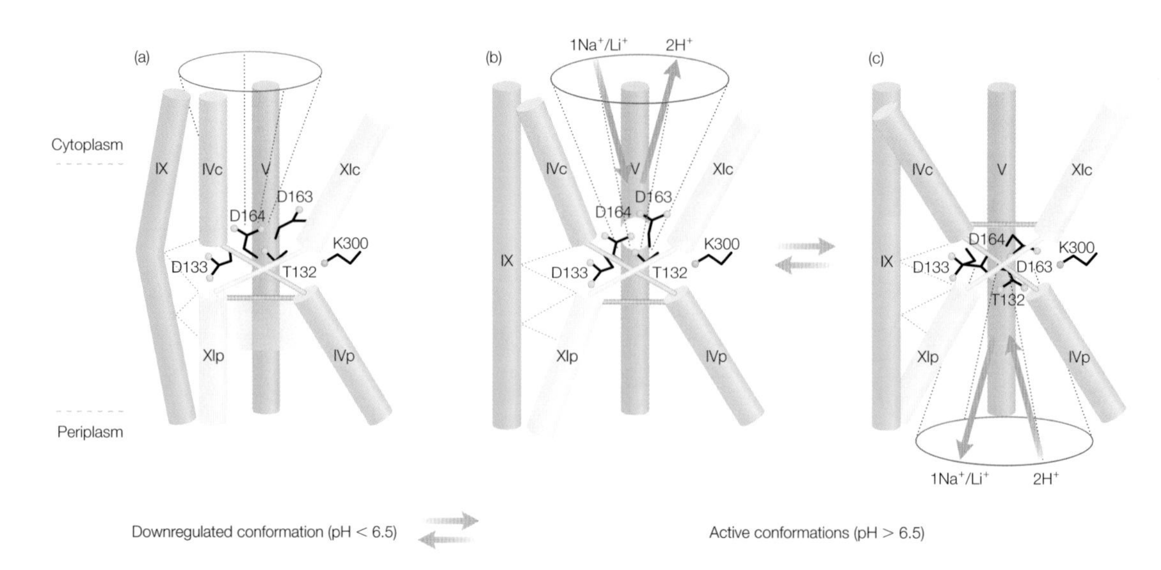

Figure 18.4 A possible mechanism for the regulation of translocation of molecules based upon the structure of the Na^+/H^+ antiporter. (a) Under acidic conditions, transport is blocked. (b) Activation by alkaline pH. (c) Na^+ (Li^+) results in opening on the periplasmic side. Modified from Hunte et al. (2005).

The availability of several different structures of porins has provided a molecular basis for the conductance of molecules in transporters. Porins are molecules that allow the passage of different molecules through the membrane. The porins from the outer membrane, such as FhuA, which is an iron transporter, are β-barrels with a large central channel and a domain that can cover the channel. The aquaporins are a family of 30-kDa proteins that are found throughout all kingdoms and in all species and regulate the movement of water and aliphatic alcohols across the cell membrane. When open, the rate of conductance is extremely rapid at about 10^9 s^{-1}, which is close to the diffusion limit of water molecules, without sacrificing selectivity, as the transport of other solutes is negligible. The structure shows eight transmembrane helices folded around a central channel, which is shaped like an hourglass and has a width that varies from 3.5 to 15 Å (Figure 18.5). In the structure of the glycerol transporter, three bound glycerol molecules were found occupying the central channel. Conduction through the channel is regulated by a number of interactions. The initial wide opening allows the water molecules or substrate molecule to maintain hydrogen bonding with the surrounding water. The channel has a narrow region to filter molecules based on their size. Proton conduction in the form of H_3O^+ is prevented by electrostatic interactions with amino acid residues that line the channel and the dipoles of short α helices. Computer simulations of the passage of water suggest that the water molecules rotate and become aligned through dipole interactions as they travel though the channel.

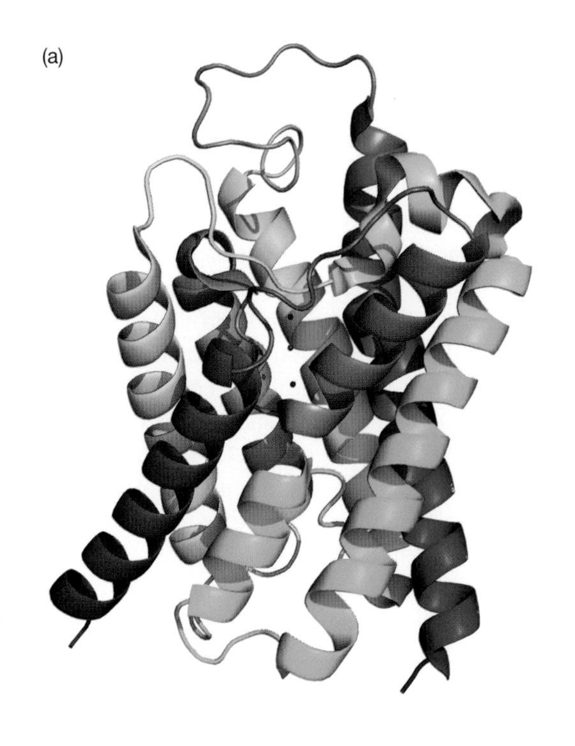

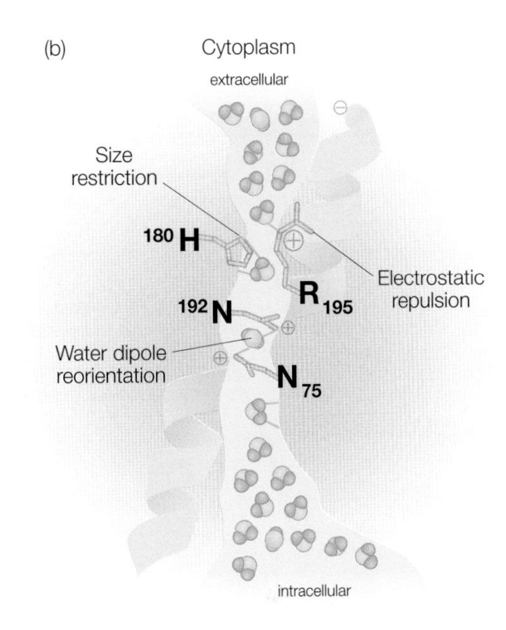

Figure 18.5 The aquaporins. (a) The three-dimensional structure of aquaporin. (b) A schematic representation of the water channel of aquaporin. Peter Agre won the Nobel Prize in Chemistry in 2003 for his work on aquaporin.

Whereas these structures provide insight into the molecular mechanism of transport, many questions remain. For example, how are the movements of the substrate and ions physically coupled? A hint is provided by the structure of a Na^{+}/Cl^{-}-dependent neurotransmitter transporter. In the structure one of the two sodium atoms bound in the transporter comes into contact with a bound leucine molecule (Figure 18.6). The binding site is formed by a partially unwound transmembrane helix devoid of water with main-chain atoms and helix dipoles providing binding interactions.

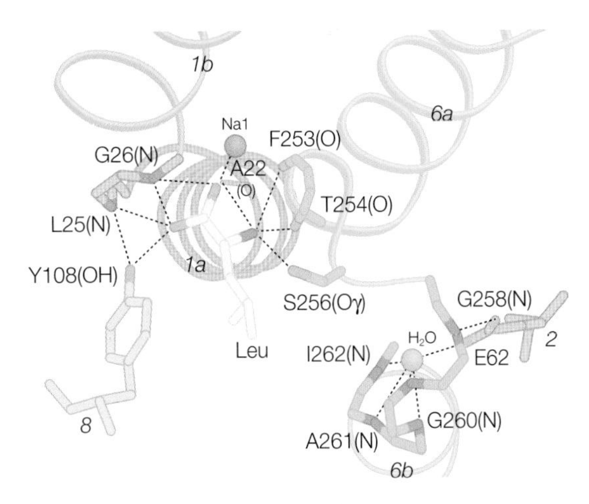

Figure 18.6 The leucine-binding site of the Na^{+}/Cl^{-} transporter showing the presence of a bound Na^{+} ion. Modefied from Yamashita et al. (2005).

ION CHANNELS

Cell membranes possess ion channels that are proteins designed to transport specific ions across the cell membrane. Ion channels can be distinguished from ion transporters by certain characteristics. Channels can transport ions at a significantly faster rate than transporters. Also, the rate of ion

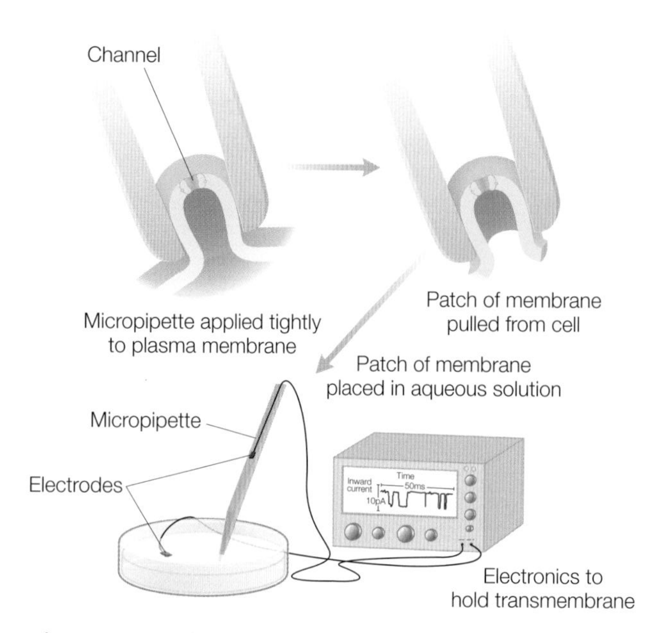

Figure 18.7 The patch-clamp technique for the measurement of an individual ion channel.

transport in a channel is usually gated; that is, the channel is open or closed depending on another factor. The controlling factors are the binding of a specific ligand for ligand-gated channels and the membrane potential for voltage-gated channels.

Gating is very fast, with a channel opening in microseconds for a short period of a few milliseconds, thus providing a rapid cellular response. Experimentally, ion flux is measured electrically as changes in voltage or current using special devices called patch clamps (Figure 18.7). This technique is very sensitive, allowing the measurement of small currents due to the flow of ions through a channel. Pulling a small region out of a membrane that contains the channel using a fine glass pipette isolates the channel. Once the electrical circuit is complete, the flow of approximately 10,000 ions through an ion channel in 1 ms is detectable. The properties of gated channels are characterized by the inclusion of the appropriate ligand or by setting the voltage at a specific value.

The three-dimensional structures have been determined for several classes of channel, including the mechanosensitive channel and the potassium channel (Figure 18.8). The mechanosensitive channel opens in response to a mechanical motion, namely the stretching of the cell membrane, to respond to the associated change in osmolarity. Potassium channels conduct K^+ ions across the cell membrane in many different cellular processes. Some potassium channels are ligand-gated whereas others are voltage-gated. Despite the differences in the response mechanism of these two channels, the structure of these channels show remarkable similarity, with the presence of a central channel surrounded by several transmembrane helices, as was also found for transporters.

The diversity of mechanosensitive channels is reflected in the lack of any identified sequence motifs associated with mechanosensitivity and the lack of any homology between the structures of the MscL and MscS channels. The structure of the MscS channel suggests a mechanism for the gating of this channel (Figure 18.9). Two arginines are located on the periphery of the protein at a turn between two transmembrane helices. Gating of the protein may be correlated with the repositioning of these residues within the membrane in response to changes in either the applied tension or membrane depolarization. The shift of these helices would then be coupled to changes in the interior of the protein that lead to gating of the channel.

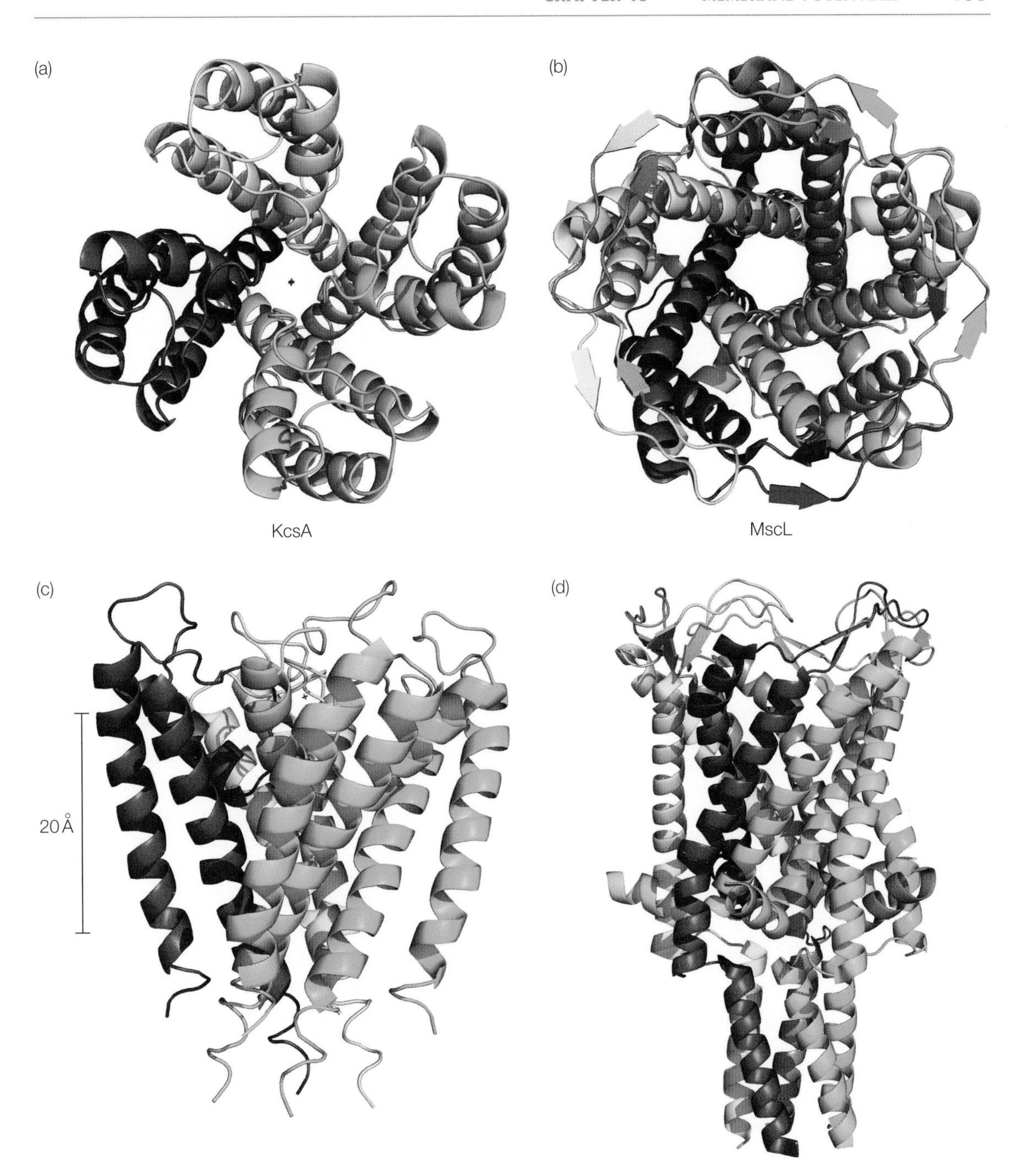

Figure 18.8 Structures of two ion channels. (a,c) The KcsA potassium channel, and (b,d) the MscL mechanosensitive channel. Roderick McKinnon won the Nobel Prize in Chemistry in 2003 for his structural studies on ion channels.

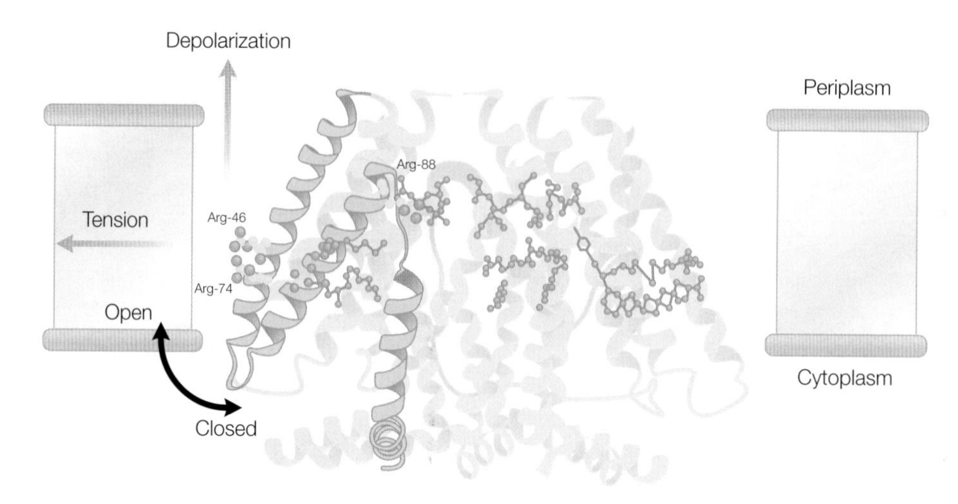

Figure 18.9 Model for the gating of mechanosensitive channels. Modified from Bass et al. 2002.

The potassium channel is a voltage-gated channel that can have up to six transmembrane helices. Only two are found for the bacterial channel KcsA, which is thought to resemble the pore region. The features of the remaining regions in eukaryotic channels are a voltage sensor, a tetramerization domain, and a signaling domain. The structures of the K^+ channel have shown common features that reveal motifs that facilitate the rapid and specific transfer of K^+ ions. The channels are composed predominantly of α helices surrounding a central channel. The channels have large entrances for ions, allowing ions to enter in a hydrated state and so not requiring the energetically unfavorable removal of water (Figure 18.10). The channel is narrow but the width cannot be used to explain the thousand-fold specificity for K^+, with an atomic radius of 1.33 Å, rather than Na^+, which has a smaller radius of 0.95 Å. Instead, the carbonyl atoms line the channel to precisely mimic the arrangement of water molecules surrounding a K^+ ion. To facilitate rapid transfer, ions are not moved individually through the channel. Rather, the channel has K^+ ions located throughout the channel. On entrance, a K^+ ion pushes electrostatically against the next and leads to the exit of another, in

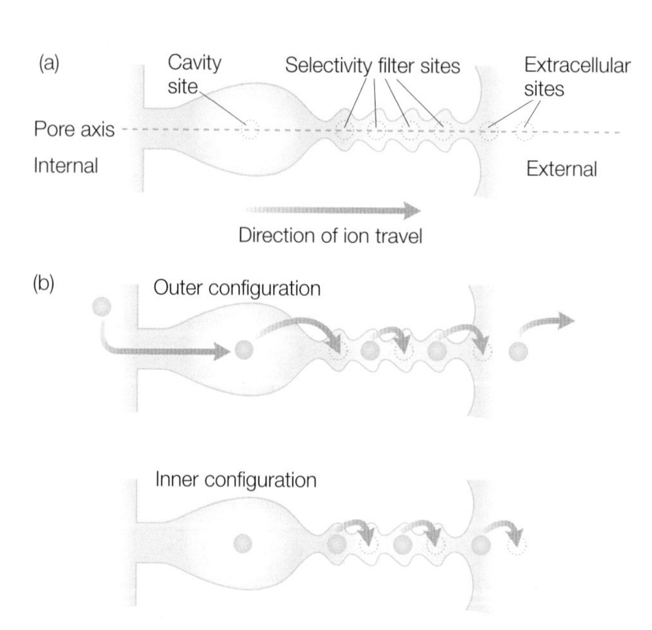

Figure 18.10 A schematic diagram of the central channel of the K^+ channel, with the domino process of ion transfer. (a) There are seven main sites for ions but (b) only half are occupied at any given time. Modified from Miller (2001).

a domino effect. In addition, the dipoles of small helices appear to contribute electrostatically to the selectivity, with the dipoles for the K^+ channel aligned but the homologous helices for the Cl^- channel being antiparallel.

An example of a ligand-gated channel is the acetylcholine receptor that mediates the transmission of nerve signals across synapses by the neurotransmitter acetylcholine. The arrival of a wave of membrane depolarization at the synapse leads to a release of acetylcholine into the synapse. The acetylcholine depolarizes the postsynaptic membrane by increasing the conductance of Na^+ and K^+ and correspondingly triggers an action potential. This change in ion permeability is defined by the acetylcholine receptor, which serves as a ligand-gated channel and the ion conductance, which is regulated by the binding of acetylcholine. The nicotinic acetylcholine receptor belongs to a superfamily that also contains receptors for the neurotransmitters serotonin, γ-aminobutyric acid (GABA), glycine, and glutamate. The N-terminal 230 or so amino acids of each receptor form the agonist-binding domain. This region, as well as binding neurotransmitters, also binds molecules of importance in human and veterinary medicine.

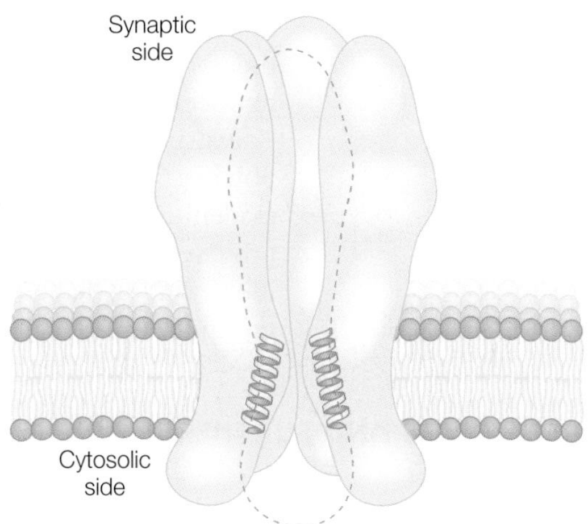

Figure 18.11 Overall structure of the nicotinic acetylcholine receptor showing the pentameric arrangement of the subunits and central channel. Modified from Unwin (1993).

The structure of the acetylcholine receptor has been determined to a limited resolution by electron microscopy (Figure 18.11). This receptor is a pentamer with two binding sites for acetylcholine. The receptor has a cylindrical appearance with an approximately 5-fold symmetry formed by the five protein subunits. At the center of the cylinder is a channel 10–20 Å in diameter. In response to the binding of acetylcholine, the channel opens. The transport of ions through the channel is determined by the presence of several negatively charged amino acids in the center of the channel that prevent the transport of anions and limit the transport of cations by size exclusion.

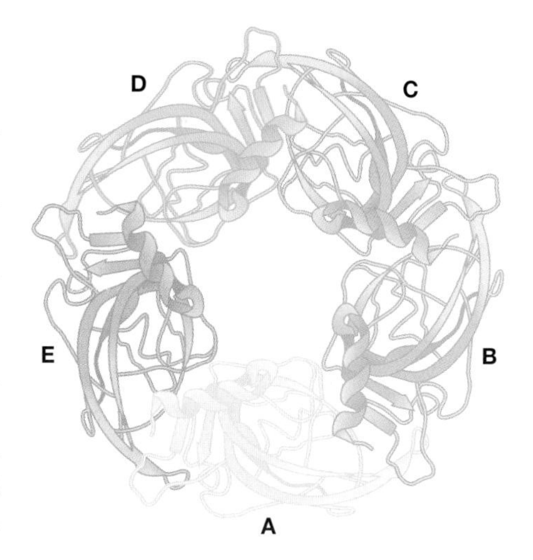

Figure 18.12 Structure of the water-soluble acetylcholine-binding homopentamer (individual subunits are labeled A through E). This protein is structurally homologous to the acetylcholine-binding extracellular domain of the acetylcholine receptor. Modified from Brejc et al. (2001).

Whereas the determination of the structure at an atomic level remains elusive, a structure has been determined for a water-soluble protein that binds acetylcholine (Figure 18.12). The acetylcholine protein is synthesized in glial cells located in the brains of snails and released in the synaptic cleft. The protein regulates interneurological transmission by binding acetylcholine but cannot serve as a channel since the protein does not possess any transmembrane domains. The sequence of this protein is homologous to the large extracellular N-terminal domain of the receptor that binds acetylcholine. The structure of the acetylcholine-binding protein shows that the single polypeptide chain is arranged in the same 5-fold symmetric arrangement, evident from the electron microscopy studies of the receptor. This structure provides the opportunity to investigate the mechanism of allosteric transitions that mediate activation and desensitization of the receptor as research groups continue their efforts to determine the structure of the receptor.

REFERENCES AND FURTHER READING

Abramson, J., Smirnova, I., Kasho, V. et al. (2003) Structure and mechanism of the lactose permease of *Escherichia coli*. *Science* **301**, 610–15.

Agee, P. and Kozono, D. (2003) Aquaporin water channels: molecular mechanisms for human disease. *FEBS Letters* **555**, 72–8.

Bass, R.B., Strop, P., Barclay, M., and Rees, D.C. (2002) Crystal structure of *Escherichia coli* MscS, a voltage-modulated and mechanosensitive channel. *Science* **298**, 1582–7.

Brejc, K., van Dijk, W.J., Klaassen, R.V. et al. (2001) Crystal structure of an ACh-binding protein reveals the ligand-binding domain of nicotinic receptors. *Nature* **411**, 269–76.

Gouaux, E. and MacKinnon, R. (2005) Principles of selective ion transport in channels and pumps. *Science* **310**, 1461–5.

Grutter, T. and Changeux, J.P. (2001) Nicotinic receptors in wonderland. *Trends in Biochemical Sciences* **26**, 459–63.

Huang, Y., Lemieux, M.J., Song, J., Auer, M., and Wang, D.N. (2003) Structure and mechanism of the glycerol-3-phosphate transporter from *Escherichia coli*. *Science* **301**, 616–20.

Hunte, C., Screpanti, E., Venturi, M. et al. (2005) Structure of a Na^+/H^+ antiporter and insights into mechanism of action and regulation by pH. *Nature* **435**, 1197–1201.

Karlin, A. (2002) Emerging structure of the nicotinic acetylcholine receptor. *Nature Reviews Neuroscience* **3**, 102–14.

Locher, K.P., Bass, R.B., and Rees, D.C. (2003) Breaching the barrier. *Science* **301**, 603–4.

Long, S.B., Campbell, E.B., and MacKinnon, R. (2005) Crystal structure of a mammalian voltage-dependent Shaker family K^+ channel. *Science* **309**, 897–903.

MacKinnon, R. (2003) Potassium channels. *FEBS Letters* **55**, 62–5.

Miller, C. (2001) See potassium run. *Nature* **414**, 23–4.

Miller, C. (2006) CLC chloride channels viewed through a transporter lens. *Nature* **440**, 484–9.

Miyazawa, A., Fujiyoshi, Y., Stowell, M., and Unwin, N. (1999) Nicotinic acetylcholine receptor at 4.6 Å resolution: transverse tunnels in the channel wall. *Journal of Molecular Biology* **288**, 765–86.

Roux, B., Allen, T., Berneche, S., and Im, W. (2004) Theoretical and computational models of biological ion channels. *Quarterly Reviews of Biophysics* **37**, 15–103.

Shi, N., Ye, S., Alam, A., Chen, L., and Jiang, Y. (2006) Atomic structure of a Na^+ and K^+ conducting channel. *Nature* **440**, 570–4.

Stroud, R.M., Savage, D., Mierke, L.J.W. et al. (2003) Selectivity and conductance among the glycerol and water conducting aquaporin family of channels. *FEBS Letters* **555**, 79–84.

Unwin, N. (1993) Nicotinic acetylcholine receptor at 9 Å resolution. *Journal of Molecular Biology* **229**, 1101–24.

Yamashita, A., Singh, S.K., Kawate, T., Jin, Y., and Gouaux, E. (2005) Crystal structure of a bacterial homologue of Na^+/Cl^- dependent neurotransmitter transporters. *Nature* **437**, 215–23.

Yernool, D., Boudker, O., Jin, Y., and Gouaux, E. (2004) Structure of a glutamate transporter homologue from *Pyrococcus horoikoshii*. *Nature* **431**, 811–18.

PROBLEMS

18.1 The lipid bilayer is permeable to what type of small molecules?

18.2 Consider a cell membrane at 298 K which has glucose at the concentration of 200 mM inside the cell and 2 mM outside the cell, with the voltage being 150 mV lower in the interior. Calculate the molar free energy difference for transporting the molecule from the interior to the exterior.

18.3 Consider a cell membrane at 295 K which has glucose at the concentration of 100 mM inside and 20 mM outside, with the voltage being 100 mV lower in the interior. Calculate the molar free energy difference for transporting the molecule from the exterior to the interior.

18.4 Consider a cell membrane at 298 K which has Na^+ at the concentration of 200 mM inside the cell and 20 mM outside the cell with the voltage being 100 mV higher in the interior. Calculate the molar free energy difference for transporting the molecule from the interior to the exterior.

18.5 A cell membrane at 298 K has a interior concentration of 1 mM Na^+ and an exterior concentration of 20 mM Na^+. Assume that equilibrium has been reached and the free energy difference is zero. Calculate the voltage of the interior relative to the exterior.

18.6 A cell membrane at 298 K has a interior concentration of 3 mM Na^+ and an exterior concentration of 60 mM Na^+. Assume that equilibrium has been reached and the free energy difference is zero. Calculate the voltage of the interior relative to the exterior.

18.7 Suppose that for every Na^+ transported from the interior to the exterior one K^+ is transported from the exterior to the interior. The voltage is 60 mV higher in the exterior. The interior concentrations are 100 mM for K^+ and 150 mM for Na^+ and the exterior concentrations are 2 mM for K^+ and 1 mM for Na^+. Calculate the molar free energy difference for each process and the net transport process.

18.8 For all of the following questions consider the Na^+/K^+-ATPase to be present in a membrane with the following conditions: Na^+ concentration inside, 20 mM; Na^+ concentration outside, 150 mM; K^+ concentration inside, 50 mM; K^+ concentration outside, 10 mM; V (inside)–V (outside), –50 mV; temperature, 310 K.

$$3Na^+ \text{ (in)} + 2K^+ \text{ (out)} + ATP + H_2O \leftrightarrow 3Na^+ \text{ (out)} + 2K^+ \text{ (in)} + ADP + P_i$$

(a) Calculate the free-energy difference for moving Na^+ from the inside to the outside.
(b) Calculate the free-energy difference for moving K^+ from outside to the inside.
(c) Calculate the free-energy difference for moving both Na^+ and K^+ in the observed stoichiometry.
(d) How should the free-energy difference for the hydrolysis of ATP compare to the value calculated in part (c) for the reaction to occur?

18.9 Suppose that for every Na^+ transported from the interior to the exterior one K^+ is transported from the exterior to the interior. The voltage is 65 mV higher in the exterior. The interior concentrations are 200 mM for K^+ and 200 mM for Na^+ and the exterior concentrations are 4 mM for K^+ and 2 mM for Na^+. Calculate the molar free energy difference for each process and the net transport process at a temperature of 298 K.

18.10 Suppose that for every Na^+ transported from the inside to the exterior one K^+ is transported from the exterior to the interior. The voltage is 60 mV higher in the interior. The interior concentrations are 100 mM for K^+ and 150 mM for Na^+ and the exterior concentrations are 2 mM for K^+ and 1 mM for Na^+. Calculate the molar free-energy difference for each process and the net transport process.

18.11 What is the biological role of transporters?

18.12 What is the alternating-access model of transporters?

18.13 What is a proposed role for short α helices in the Na^+/H^+ antiporter function?

18.14 What factors are proposed to control water flow in aquaporin?

18.15 Discuss briefly the similarities and differences between voltage-gated channels and nonvoltage-gated channels.

18.16 Compare and contrast voltage-gated potassium channels and the KcsA channel.

18.17 What is a ligand-gated channel?

18.18 What structural changes are proposed to induce channel gating in mechanosensitive channels?

18.19 How do electrostatics contribute to the ion flow?

18.20 What is the role of the large cavity at the entrance for ions in the potassium channel?

18.21 Explain how ion channels distinguish between Na^+ and K^+.

18.22 How can a FRET experiment be used to identify the voltage-induced structural changes of a voltage-gated ion channel?

18.23 Summarize the known structural properties of the acetylcholine receptor.

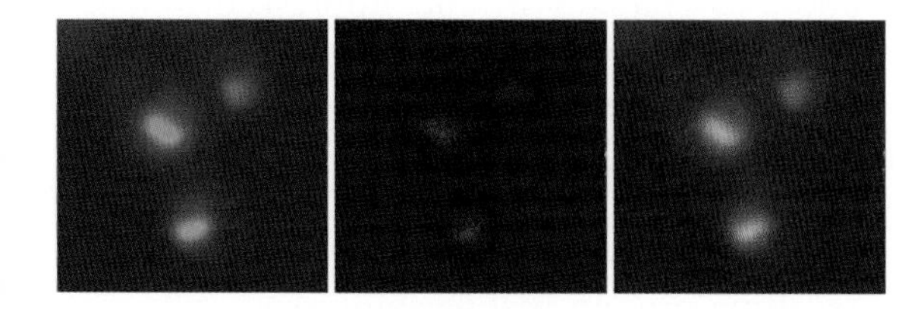

19 Molecular imaging

IMAGING IN CELLS AND BODIES

Fluorescence has been used for many years to visualize cellular components. Originally, small organic dyes were attached through the use of antibodies and by the use of fluorophores that recognized organelles. More recently, fluorescent proteins have provided the opportunity to probe gene expression, protein trafficking, and responses to signals. Spectroscopy using fluorescence has expanded rapidly as these tools have become more commonplace, allowing scientists to target specific sites even at the single-molecule level. Reporters for many key metabolites have been developed and the ability to perform genetic tagging provides scientists with many opportunities to probe cellular processes at a molecular level. One of the most versatile fluorescent probes is green fluorescent protein, which is described in the first section of this chapter. In addition to these developments at a cellular level, imaging techniques at the level of an organism are also becoming widely used. The second part of this chapter discusses how techniques such as positron emission tomography are giving healthcare providers the opportunity to probe cellular activities and identify markers for brain disorders that continue to be very difficult to treat.

GREEN FLUORESCENT PROTEIN

Green fluorescent protein (GFP) was discovered as a companion protein to aequorin, a chemiluminescent protein from the jellyfish *Aequorea,* and was found to be composed of a single polypeptide with 238 amino acid residues. GFP emits a green fluorescence when irradiated with UV light. The proper expression of the chromophore of GFP was found to require molecular oxygen but to be independent of any enzymes. Biosynthesis of the chromophore was found to be an exception compared to most chromophore-containing proteins as the protein was found to catalyze the

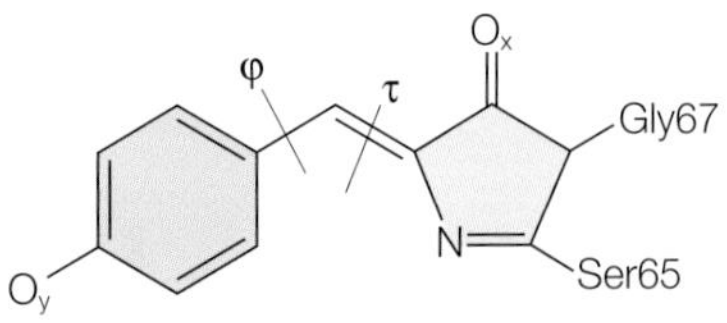

Figure 19.1 Structure of the chromophore of GFP formed after cyclization of the amino acid residues Ser-65, Tyr-66, and Gly-67.

synthesis of the pigment from the polypeptide chain. Digestion of GFP showed that the cofactor is formed by cyclization of residues Ser-65, Tyr-66, and Gly-67 (Figure 19.1), consistent with the observation of changes in the fluorescence when those amino acid residues were changed by mutagenesis. After formation the pigment is attached to the protein and, as a result, expression of the gene coding for GFP leads directly to the appearance of the color and fluorescence.

In most cases, the gene for GFP can be attached to a gene of interest and the resulting protein will be fluorescent, allowing tracking of genes using molecular biology. Such an approach was a significant improvement over the more traditional approach of labeling proteins with fluorescent compounds to determine their cellular localization and possibly conformational changes. Traditionally, such work was performed by first purifying the protein and then chemically modifying it to attach a reactive organic fluorophore. Such labeling studies required chemical attachment of the dye followed by reintroduction of the labeled protein into the cell. The use of GFP alleviates the necessity of purifying and manipulating the protein, offering a significant improvement for molecular imaging in cells at both a laboratory and industrial scale. For example, GFP has been used to optimize the expression of proteins by real-time monitoring of GFP fluorescence. GFP has also been used to study microbial growth and dispersion of organisms in natural environments.

The spectral range of fluorescent proteins expanded when new sources where considered. A number of mutants of GFP were identified in mutagenesis screens as having a color shift. The effect of specific amino acid residues was further probed using mutagenesis. Discovery of GFP-like proteins from the coral *Anthozoa* significantly expanded the wavelength region covered by these proteins as the new protein, DsRed, has a spectrum whose main absorption peak is shifted to the red compared to GFP. These discoveries produced a set of fluorescent proteins whose spectra ranged from green to red with emission maxima from 475 to 600 nm (Figure 19.2).

The structures of GFP and various mutants have been solved and shown to be highly homologous. GFP is an 11-stranded β barrel with an α helix that runs up the axis of the cylinder (Figure 19.3). The cylinder has a diameter of approximately 30 Å and a length of approximately 40 Å. The chromophore is attached to the α helix and buried in the center of the barrel. Folding of GFP into the barrel is crucial to the formation of the chromophore and its fluorescent properties. By enclosing the chromophore inside of the barrel, it is protected from the aqueous environment.

In wild-type GFP, the chromophore forms a well-defined structure that is stabilized by a hydrogen-bonding network involving the surrounding

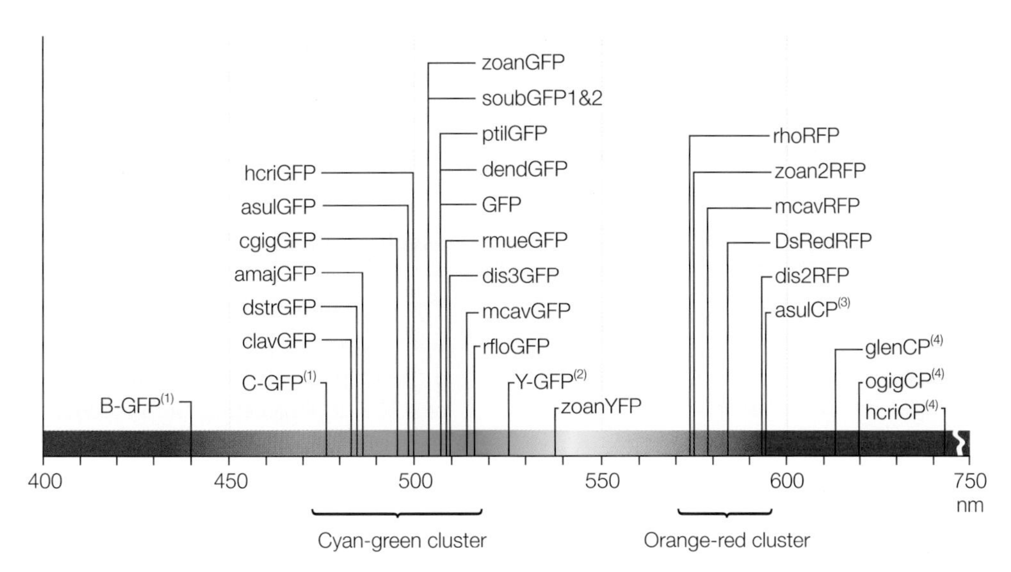

Figure 19.2 A wide range of GFP and GFP-related fluorescent proteins are now available. Modified from Matz et al. (2003).

protein as well as some bound water molecules (Figure 19.4). In some of the mutants, the amino acid residues that form the chromophore have been changed and the cofactor is still expressed. These variants give rise to different interactions between the pigment and the surrounding protein and hence different colors. For example, in the mutant with Tyr-66 substituted by a His residue, the cofactor is present in the same binding site and the protein is still fluorescent. When Ser-65 is changed to Thr, the chromophore is located in the pocket but is displaced compared to the wild type. The presence of the altered chromophore results in small displacements of the surrounding protein side chains. In addition to these structural changes, the hydrogen-bonding network surrounding the chromophore is significantly altered. The sum of these effects results in a chromophore with altered fluorescent properties. The sensitivity of the protein to point mutations and the desirability of engineering fluorescent proteins with altered optical properties has led to the availability of GFPs with many different optical properties. Indeed, one of the

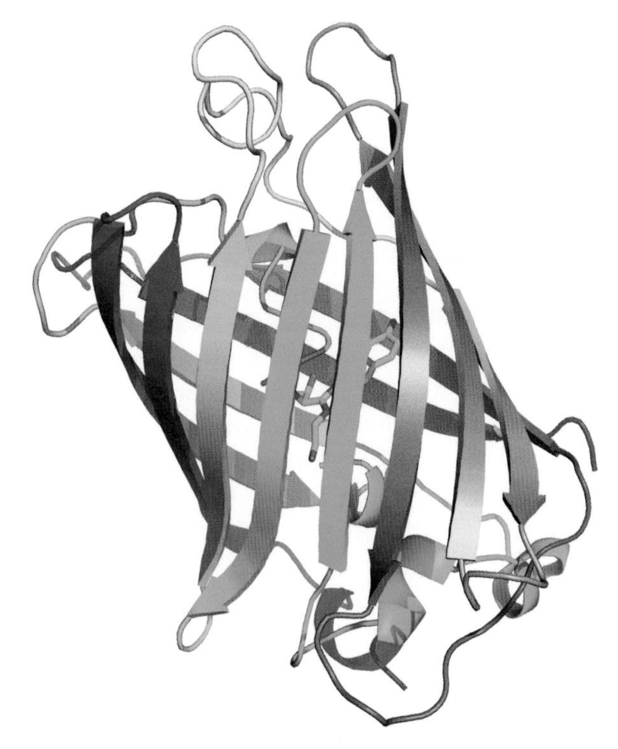

Figure 19.3 The structure of GFP showing the wrapping of the β-strands into a barrel with the chromophore located in the center of the protein.

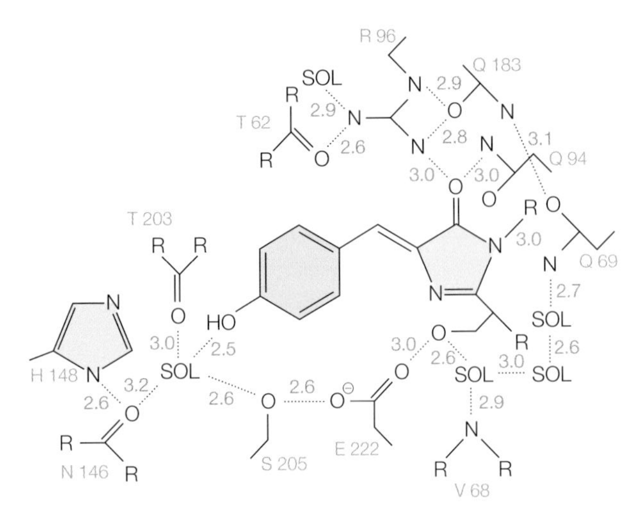

Figure 19.4 A schematic diagram showing the protein interactions between the chromophore and the protein environment for wild-type GFP. Modified from Wachter et al. (1997).

most commonly used GFPs is enhanced GFP, which produces a larger amount of fluorescence due to an improvement in its quantum efficiency, resulting from changes in the protein surrounding the pigment.

Mechanism of chromophore formation

The spectrum of the wild-type *Aequorea* protein shows the presence of two bands, with a major peak at 395 nm and a minor peak at 475 nm, suggesting the presence of more than one optical species. The optical spectrum is more complex at low temperature, with the absorption bands showing significantly more structure, indicating the presence of several trapped states that are close in energy. Using transient spectroscopy, the kinetics of the optical spectrum after excitation were shown to have a complex behavior. To investigate the reason for the two peaks, the optical properties of the chromophore have been probed. For example, significant spectral shifts are evident for a variant that is identified as YFP, or yellow fluorescent protein, which has one of the longest-wavelength emissions of all engineered variants. A pH titration of this protein shows a systematic behavior which can be modeled as arising from a pK_A associated with the chromophore (Figure 19.5).

The simplest interpretation is that the absorbance at 475 nm arises from GFPs that contain a deprotonated chromophore and that the 395-nm peak arises from those GFPs containing a protonated chromophore. These and other observations lead to a model that the chromophore of GFP has two states formed during the cofactor formation. The requirement of oxygen for fluorophore formation suggested that the involvement of oxygen represented the rate-limiting step. The folding of the peptide chain into the barrel establishes the protein environment surrounding residues Ser-65–Tyr-66–Gly-67, which initiates formation (Figure 19.6). Then cyclization occurs due to nucleophilic attack of the amide of Gly-67 on the carbonyl of Ser-65 followed by dehydration. The resulting conjugated molecule gives GFP its color and fluorescence for the first time. For DsRed, there is an additional oxidation step that leads to the extension of the conjugation. The resulting chromophore has an absorbance at a longer wavelength, as would be predicted using a particle-in-a-box model (Chapter 10).

The delineation of the cofactor-formation mechanism including the identification of the intermediates is being performed through a combination of functional and structural studies of mutants. Mutants that are defective because of changes that stop the assembly at intermediate points provide the opportunity to investigate intermediate states. For

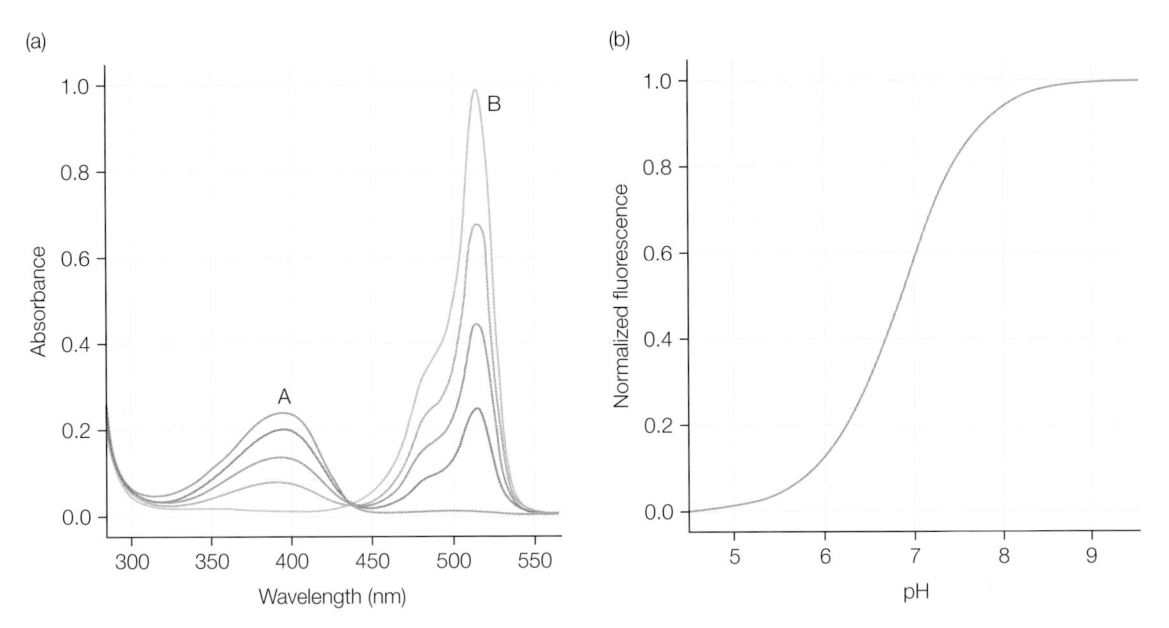

Figure 19.5 The pH dependence of the optical spectrum of YFP. (a) The amplitudes of the two absorption bands have a distinct pH dependence with an isosbestic point. (b) The relative amplitude can be fit to a pH dependence with a pK_A of 7. Modified from Elsliger et al. (1999).

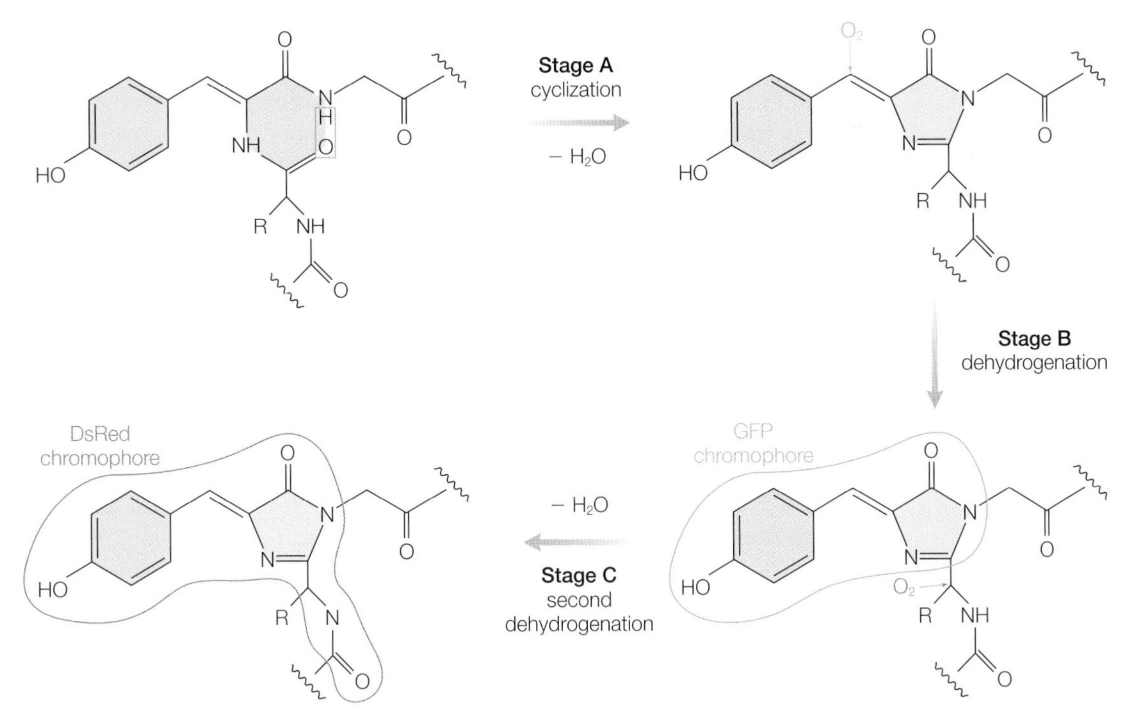

Figure 19.6 A general scheme for chromophore formation in GFPs consisting of a cyclization step followed by dehydration. In DsRed, the chromophore undergoes an additional step that results in a longer conjugated system. Modified from Matz et al. (2003).

example, replacement of Tyr-66 with Leu results in a colorless variant that represents a trapped intermediate of the pathway. The necessity of protein folding producing a tight conformation of Ser-65 and Gly-67 for the cyclization step was demonstrated by the properties of another colorless variant carrying the Ser-65-to-Gly and Tyr-66-to-Gly substitutions under anaerobic conditions.

Formation of the chromophore is a complex mechanism involving steps such as pre-organization of the protein, electrophilic and base catalysis, and the subsequent slow proton abstraction. The autocatalytic modifications that produce the cofactor in GFP are intrinsic properties as the formation requires only a single gene in contrast to the biosynthetic pathways required for assembly of cofactors in other proteins. Understanding the role of each amino acid residue may have a broad relevance for other proteins with similar chromophore-formation mechanisms that are identified not only through biochemical analysis but also by sequence-specific searches of the genomes of different organisms. The mechanism that produces the chromophore in GFP is not unique to this protein as the identification of proteins with prosthetic groups generated by post-translational modifications has expanded tremendously in recent years. The post-translational oxidation of tyrosine residues often plays an essential role, with examples being the formation of topaquinone in copper amine oxidases and the cross-linking of tyrosine to cysteine in galactose oxidase. In contrast to GFP, the spontaneous formation of such cofactors is usually facilitated by the presence of a metal cofactor. Whereas the general scheme for cofactor formation is known for GFP and these other proteins, the detailed mechanism is still under active investigation in a number of laboratories.

Fluorescence resonance energy transfer

Fluorescence resonance energy transfer, or FRET, is a non-radiative process by which energy from an *excited energy donor*, D*, is transferred to an *acceptor fluorphore*, A, resulting in an *excited acceptor*, A*, and a *ground-state donor*, D:

$$D^* + A \leftrightarrow D + A^* \tag{19.1}$$

The rate of the excitation transfer can be written in terms of a transition involving the *wavefunction of the acceptor* ψ_A *and excited donor,* ψ_D^*. The operator for the transition is given by the product of the transition dipole moments of the donor and acceptor divided by the *separation between the donor and acceptor*, r_{DA}, cubed:

$$k \propto \left[\int \psi_D^* \left(\frac{\mu_D \mu_A}{r_{DA}^3} \right) \psi_A \, d\tau \right]^2 \tag{19.2}$$

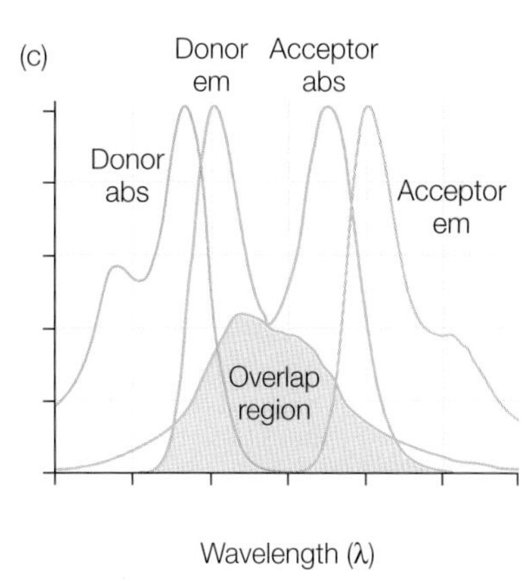

Figure 19.7 A FRET experiment can be used to measure the relative number of proteins forming complexes. (a) In both cases, the GFP is excited but subsequent emission arises from either the donor, when the proteins are far apart, or from the acceptor, when the proteins form a complex. The efficiency of the energy transfer has a sharp dependence upon (b) the distance and (c) requires a substantial spectral overlap. abs, absorption; em, emission. Modified from Bastiaens and Pepperkok (2000).

In terms of experimental information, the rate can be expressed using a number of parameters. Most critical is the extent of the spectral overlap. For the energy transfer to occur, the energy difference from D* to D must match that from A to A*. In spectral terms, the donor emission band must overlap with the acceptor absorption band (Figure 19.7). The second critical factor is the separation between the donor and acceptor, with the rate decreasing as r^{-6}. For many experiments, the measurement is performed under steady-state conditions and so the data provide not the rate of energy transfer by rather the efficiency of transfer, which can be expressed simply in terms of the separation:

$$\text{Efficiency of transfer} = \frac{r_0^6}{r_0^6 + r_{\mathrm{DA}}^6} \qquad (19.3)$$

The distance dependence for the efficiency is seen to result in a sharp dependence centered on the parameter r_0 that represents the distance at which the efficiency of energy transfer is 50%. This constant can be written in terms of the various nondistance factors such as the spectral overlap, quantum yield of the donor, and relative orientation of the dipole moments, κ^2:

$$r_0^6 \propto \kappa^2(\text{refractive index})^{-4} \times (\text{overlap}) \times (\text{yield}) \qquad (19.4)$$

In steady-state measurements of fluorescence intensity, the relative donor and acceptor emission ratios can be examined. Alternatively, the donor and acceptor emissions can be detected independently in the FRET measurement.

By combining the power of fluorescence microscopy with the optical properties of GFP, it is possible to probe intracellular processes and reactions in cells. FRET is very sensitive and can be used to determine whether two proteins come close together or whether two ends of a protein (or DNA) come close together. For example, consider the case of two proteins, identified as A and B, that under some circumstances can form a complex. Protein A is labeled with GFP that serves as the donor and protein B is labeled with an acceptor. When the two proteins are far apart then the excitation of GFP results in emission from GFP. However, when the two proteins form a complex, excitation of GFP is followed by energy transfer to the acceptor and emission from the acceptor. Thus, the relative amount of emission from GFP compared to the acceptor is directly related to the ratio of the proteins as monomers compared to those forming the complex.

Imaging of GFP in cells

Fluorescence imaging techniques can be performed not only under steady-state conditions but also as transient measurements. Among the possible approaches are photobleaching studies (Figure 19.8). In the presence of a strong light, pigments will undergo changes that result in loss of an absorption band and correspondingly loss of fluorescence. After the illumination beam is turned off, the molecules will recover, restoring the absorption and fluorescence if the bleaching is reversible. This photobleaching property provides an opportunity to specifically monitor a selected pool of GFP that is distinguished from other GFPs at different locations in the cell. In one case, an area of the cell is photobleached with a high-intensity laser pulse. The movement of the unbleached GFP from neighboring areas into the bleached area is monitored by recording the recovery of fluorescence in the photobleached area. The recovery of the fluorescence in the selected area will increase until all of the GFP has recovered. Alternatively, photobleaching can be performed repeatedly so that the GFP in the selected area does not have time to recover and is always bleached. With time, the bleached GFP will exchange with the GFP from the outside areas, diminishing the amount of fluorescence in the outside areas. Any GFP that enters the selected area is quickly photobleached so the fluorescence in the selected area remains minimal. As the

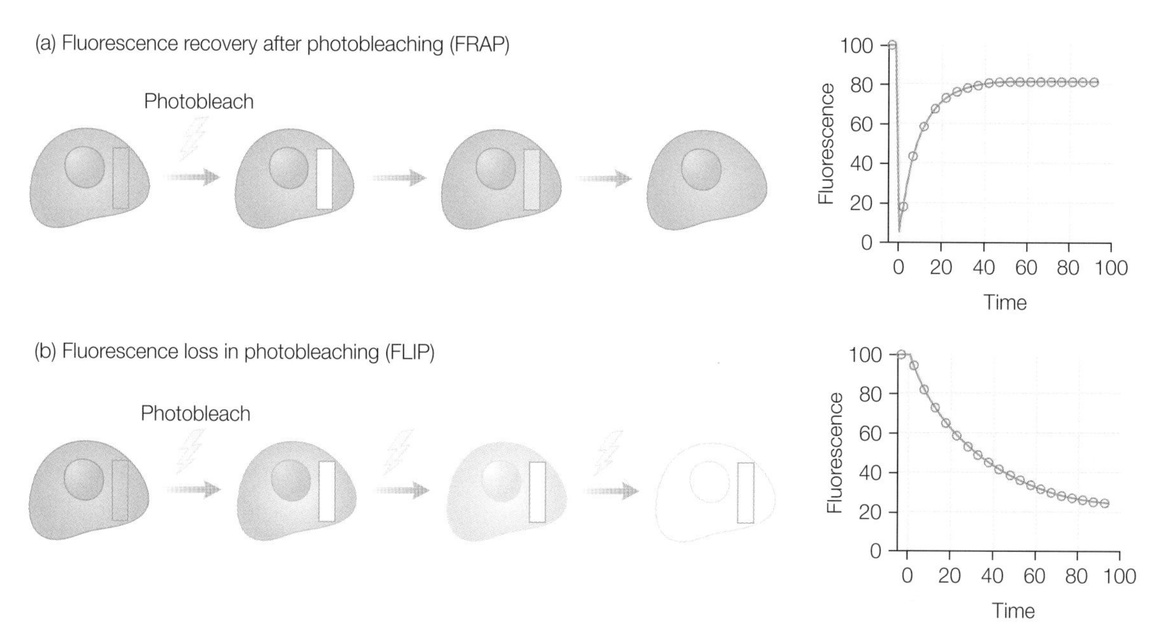

Figure 19.8 Kinetic microscopic technique for probing the movement of fluorescent molecules. (a) In the top view, a region of the cell is selectively bleached and the recovery of the fluorescent molecules into that region is assessed. (b) In the bottom view, a region of the cell is repeatedly photobleached and movement of the photobleached molecules out of the region can be monitored. Modified from Lippincott-Schwartz and Patterson (2003).

process continues, the entire cell will be photobleached. These techniques allow specific GFP to be identified and tracked in the cell without disrupting protein pathways or creating protein imbalances in the cell. The rates of fluorescence recovery or decay provide quantitative values for the movement of proteins within the cell, the dynamics of protein–protein complexes during different cellular states, and the interactions of proteins with various cellular components.

Eukaryotic cells are organized into a complex network of compartments with specialized function, separated by membranes. To understand the function of proteins in eukaryotic cells, the location of proteins needs to be identified and the role of each protein must be defined in each cellular compartment. Localization studies can be performed by attaching GFP as a fusion product to any given gene and monitoring the cellular positions that show fluorescence. Although this can be performed on an individual basis for any given protein, one of the goals of proteomics is to perform such studies on every protein of the cell. The complete genome has been sequenced for the yeast *Saccharomyces cerevisiae*, making this a well-characterized model system for proteomics. The gene for GFP was attached to the over 6000 open reading frames of *S. cerevisiae* using a library that represents most of the proteome (Figure 19.9). The identification was

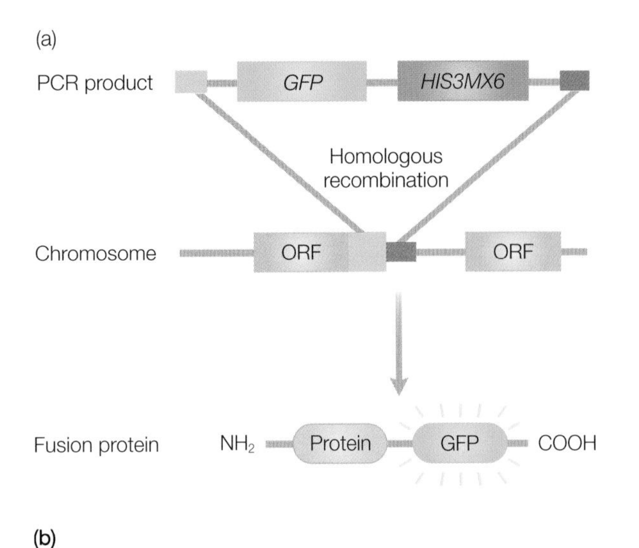

(b)

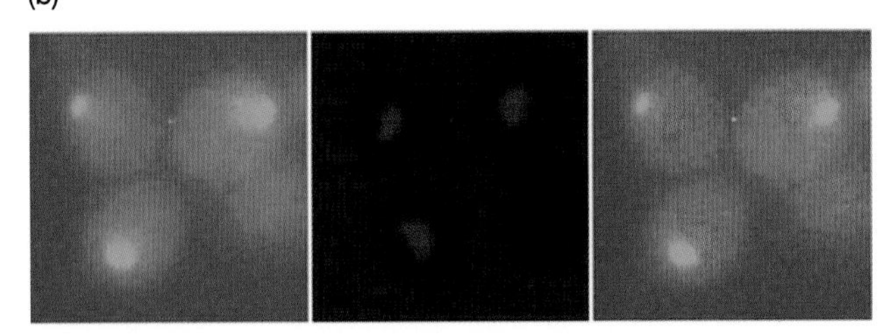

Figure 19.9 Localization studies using GFP. (a) The strategy for the library-based attachment of the GFP fusion products to over 6000 open reading frames of the *S. cerevisiae* genome. Inclusion of the His3MX gene permitted selection in histidine-free media. (b) Fluorescence microscopy of the expressed fusion products shows localization in different cellular areas for different proteins. Modified from Huh et al. (2003).

aided by fusing a GFP-like protein that fluoresces in the red to the cellular proteins that had been localized previously. The tagged open reading frames and proteins were examined at different cellular locations using fluorescence microscopy.

The fluorescence showed that the expression of the open reading frames could be organized into distinct subcellular categories. Possible protein–protein interactions of the open reading frames should be highly biased towards those proteins that share the cellular location. Cases where a protein is identified in more than one location reveal possible interactions that can be considered in the context of the entire cell. Multiple locations reflect functionally and physically related subcellular regions, including regions that undergo dynamic interchange of proteins. The availability of such proteomic libraries provides the opportunity to probe how proteins respond to external stimuli or growth conditions over a selected period of time. Complex protein interactions can be mapped out by examining the effect of the proteome localization due to alterations of specific proteins.

IMAGING IN ORGANISMS

Many of the spectroscopic techniques described in the previous chapters, such as MRI and fluorescence, are being used in clinical settings to visualize metabolic processes in the human body. Another imaging technique, positron emission tomography (PET), is being considered increasingly for patient treatment in the examination of molecular processes and their failure in disease. In this technique, a certain process is targeted and a suitable probe designed. The probes sensed in PET are drugs or analogs that have been labeled with radioisotopes. The measurements are sensitive so that very low doses can be applied to minimize unforeseen side effects. As the name implies, this technique is tomography-based, which means that the resulting information is measured in three dimensions so that it can be mapped onto the body.

Radioactive decay

There are several different radioactive processes that certain isotopes of elements can undergo. To understand these processes it is necessary to keep track of the specific isotopes involved. For the decay processes each isotope is denoted by the mass number, which equals the number of protons and neutrons, and the atomic number, which equals the number of protons. For example, the carbon isotope ${}^{12}_{6}\mathrm{C}$ has a mass number of 12 and an atomic number of 6, whereas the uranium isotope ${}^{238}_{92}\mathrm{U}$ has a mass number of 238 and an atomic number of 92. The different types of radioactive decay can be classified according to the particles that are emitted by the decay process. For example, α particles consisting of two protons and two neutrons, identified as ${}^{4}_{2}\alpha$, are emitted by radium and uranium:

$$\begin{aligned} &{}^{224}_{88}\mathrm{Ra} \rightarrow {}^{220}_{86}\mathrm{Rn} + {}^{4}_{2}\alpha \\ &{}^{238}_{92}\mathrm{U} \rightarrow {}^{234}_{90}\mathrm{Th} + {}^{4}_{2}\alpha \end{aligned} \tag{19.5}$$

As another example, a photon is emitted in the form of γ rays from the isotope ${}^{119}_{50}\mathrm{Sn}$. In this case, there can be no change in the isotopic composition, only in the energetics of the nucleus. In electron capture, an electron is captured by a nucleus and the atomic number can change along with a release of energy.

Of interest for its use in PET are isotopes than undergo β decay. Two examples of this process are the decay of the carbon isotope ${}^{14}_{6}\mathrm{C}$ into the nitrogen isotope ${}^{14}_{7}\mathrm{N}$ and the decay of ${}^{18}_{9}\mathrm{F}$ into ${}^{18}_{8}\mathrm{O}$:

$$\begin{aligned} &{}^{14}_{6}\mathrm{C} \rightarrow {}^{14}_{7}\mathrm{N} + {}^{0}_{-1}\beta \\ &{}^{18}_{9}\mathrm{F} \rightarrow {}^{18}_{8}\mathrm{O} + {}^{0}_{+1}\beta \end{aligned} \tag{19.6}$$

Since the β decay process results in a change of the atomic number by 1, there is an emission of a charged particle to conserve the overall charge. For ${}^{14}_{6}\mathrm{C}$, the conversion into ${}^{14}_{7}\mathrm{N}$ results in an increase in the atomic number; hence, the decay is accompanied by the emission of an electron to conserve charge. For ${}^{18}_{9}\mathrm{F}$, the decay into ${}^{18}_{8}\mathrm{O}$ results in a decrease in the atomic number; hence, charge conservation demands the emission of a positively charged particle with the same mass as an electron, namely a positron. As explained in Chapter 12, positrons are part of a whole family of antiparticles that match all particles but with some opposite properties. Positrons are the antimatter form of electrons, with a mass identical to, but charge opposite to, an electron. Positrons are not stable and will quickly combine with the corresponding particle to produce energy. Whereas there are several isotopes that will emit positrons, most have half-lives of only a few minutes and so are difficult to use in imaging techniques. Even the commonly used isotope ${}^{18}_{9}\mathrm{F}$ has a half-life

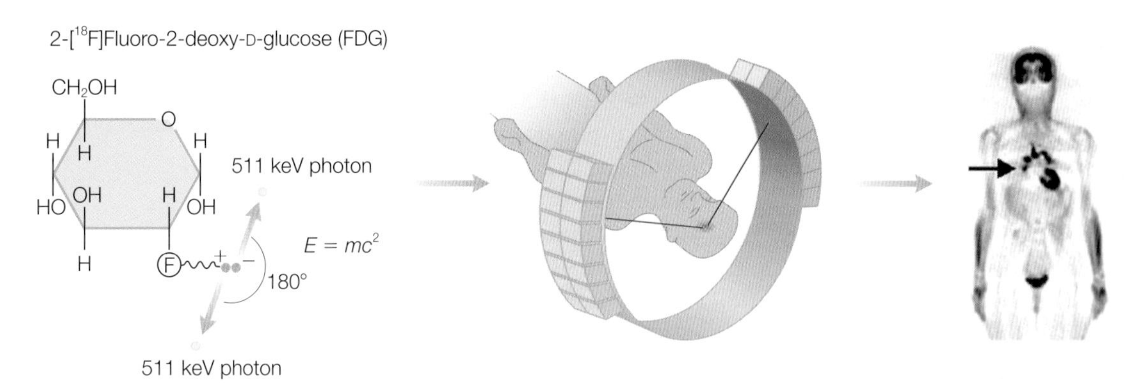

Figure 19.10 A schematic representation of PET. The positron-emitting radioisotope is injected intravenously into the body. When the ^{18}F decays a positron is emitted and rapidly annihilates with an electron releasing, two photons in opposite directions, which are detected. Modified from Phelps (2002).

of only 110 min and so the isotope must be generated at a cyclotron immediately prior to use.

PET

The technique of PET is an imaging technique that makes use of positron-emitting isotopes as probes of specific biochemical processes in vivo (Figure 19.10). The positron is not stable and will combine with an electron to produce energy that has a value equal to twice the electron mass, $2m_ec^2$, or twice 511 keV. The energy takes the form of two photons that, to conserve momentum, are directed along equal and opposite directions from the point where the positron and electron came together, with each photon having an energy of 511 keV. The detection of the two simultaneously emitted 511-keV photons occurs simultaneously and the common point of origin can be found by following the line of detection. For measurements of a patient, millions of pair combinations are recorded from many different angles around the subject. The combinations of all of these measurements provide reconstruction of the concentrations of the probe at different locations to generate the tomographic images.

In PET, the subject is injected with a molecule containing a label with an isotope that will emit positrons (Figure 19.11). One probe is 2-[^{18}F]fluoro-2-deoxy-D-glucose, which is used to locate glycolysis processes in the body. For these measurements, the end-product accumulation is proportional to the rate of glycolysis. Another probe is 3′-deoxy-3′-[^{18}F]fluorothymidine. This probe is a deoxy analog of thymidine. The probe can be transported and phosphorylated but the resulting modified probe is not dephosphorylated by phosphohydrolases and accumulates in the cell. Thus, both probes show where active transport and phosphorylation are

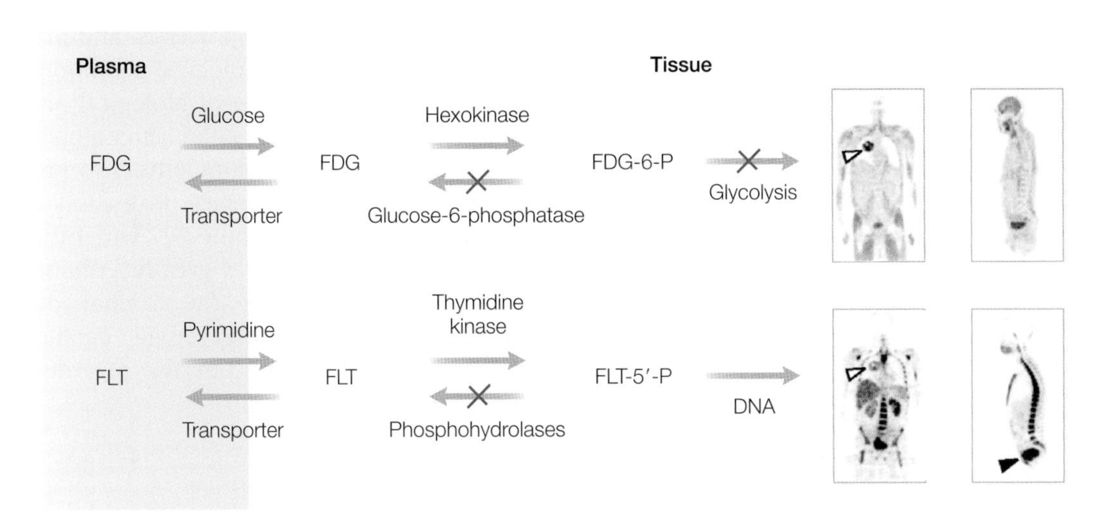

Figure 19.11 Models for the cellular transport and phosphorylation of the radioisotope probes, 2-[^{18}F]fluoro-2-deoxy-D-glucose and 3′-deoxy-3′-[^{18}F]fluorothymidine, used in PET measurements to locate areas actively undergoing glucose utilization and DNA replication. Modified from Phelps (2002).

occurring in the body. Once injected, the subject is then placed inside a detector and, after waiting for the probe to be transported and phosphorylated, the locations that have high concentrations of the probe are mapped. Areas that are undergoing high metabolism are shown by high use of glycolysis and correspondingly high concentrations of phosphorylated 2-[^{18}F]fluoro-2-deoxy-D-glucose. Such profiles of the brain have been found to be characteristic of certain diseases, such as Alzheimer's disease. The probe fluorothymidine is used to locate areas undergoing a high rate of DNA replication that is characteristic of cancer cells.

In addition to these clinical uses, PET can be used to characterize the stimulation responses of the brain. For example, specific areas of the brain are found to have high glucose metabolism in response to different sensory stimulations (Figure 19.12). PET provides a relatively benign approach for studying brain activity and to examine potential therapies. For example, by linking a PET reporter gene to a potential therapeutic gene, it is possible to track where the therapeutic gene is being translated and to correlate the gene activity with a physiological response.

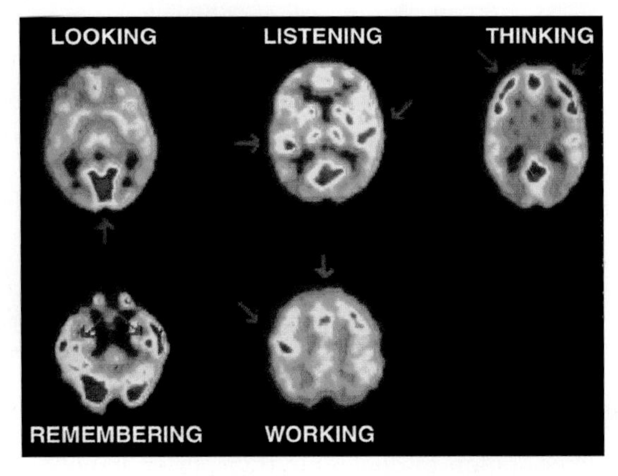

Figure 19.12 Comparison of PET scans. Modified from Phelps (2002).

By improving the technologies, it should be possible to track pharmaceuticals and provide improved monitors of the response of the body to therapeutics. Molecular imaging reveals biological processes and provides a means to quantify the processes under physiological conditions. New probes are being developed to expand the range of biological processes that can be studied as well as to improve specificity and better amplify the cellular signals. One of the goals of current developments is to provide diagnostics for the identification of the initial stages of diseases to allow treatment at early stages. Another goal is to combine PET with other imaging technologies as the resulting synergy should greatly enhance the capabilities of either technique alone. For example, the simultaneous measurement by PET and computer tomography should provide detailed information, not only about the metabolic processes, but also regarding vascular information.

Parkinson's disease

Parkinson's disease is the second most common neurodegenerative disease after Alzheimer's disease. The first warning sign is development of a tremor in one arm. As the disease progresses, voluntary movement becomes slower with some patients experiencing temporary loss of the ability to move. Currently there are no effective treatments to prevent the progression of the disease, including the deterioration and death of neurons that produce the neurotransmitter dopamine. The loss of the dopamine nerve terminals in the striatum section of the brain is the cause of most of the motor symptoms. The reason for the damage to the dopamine neurons is still under active investigation, with many factors proposed, such as exposure to abnormal proteins, protein aggregation, and oxidative stress. The disease also kills neurons that produce other neurological compounds such as norepinephrine, serotonin, and acetylcholine. Linked to the disease is the formation of Lewy bodies inside the neurons. Lewy bodies consist of filaments of the protein α-synuclein, which is misfolded and forms polymers. The role of the properly folded α-synuclein has not been determined unambiguously.

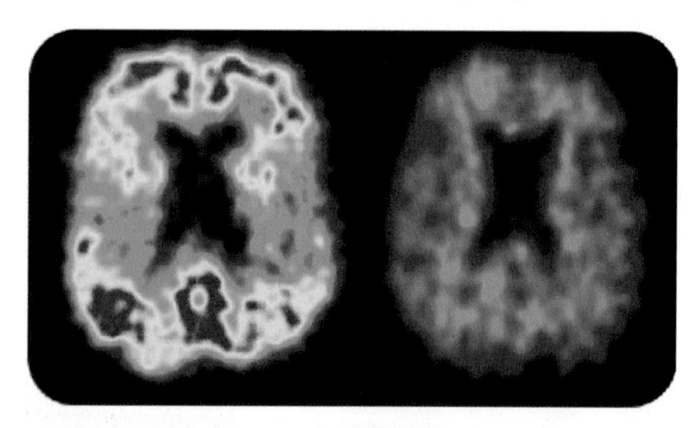

Figure 19.13 Compared to a healthy person (right), a patient with Parkinson's disease (left) will have a deterioration in dopamine-transporter activity, which is evident in PET scans. Printed with permission from the University of Pittsburgh Amyloid Imaging Group.

PET imaging of the brain can be used to identify the disease (Figure 19.13). Dopamine is produced in the body from the amino acid tyrosine. The aromatic ring L-tyrosine is modified with the addition of a hydroxyl group into L-dopa by tyrosine hydroxlase. This reaction is followed by the production of dopamine from L-dopa by dopacarboxylase. The activity of dopamine nerve terminals is measured by use of the probe [^{18}F]fluorodopa, which can be taken up by dopamine neurons and stored in their nerve terminals. Patients with Parkinson's disease show a sharp decrease in the amount of probe at the neuron terminals as detected by the PET images. Efforts are underway in a number of laboratories to utilize this technique to identify drugs that can slow the disease progression in Parkinson's patients.

REFERENCES AND FURTHER READING

Bastiaens, P.I.H. and Pepperkok, R. (2000) Observing proteins in their natural habitat: the living cell. *Trends in Biochemical Sciences* **25**, 631–7.

Chudakov, D.M., Lukyanov, S., and Lukyanov, K.A. (2005) Fluorescent proteins as a toolkit for in vivo imaging. *Trends in Biotechnology* **23**, 605–13.

Dove, A. (2005) The big picture. *Nature Medicine* **11**, 111–12.

Elsliger, M.A., Wachter, R.M., Hanson, G.T., Kallio, K., and Remington, S.J. (1999) Structural and spectral response of green fluorescent protein variants to changes in pH. *Biochemistry* **38**, 5296–301.

Giepmans, B.N.G., Adams, S.R., Ellsman, M.H., and Tsien, R.Y. (2006) The fluorescent toolbox for assessing protein location and function. *Science* **312**, 217–24.

Huh, W.K., Falvo, J.V., Gerke, L.C. et al. (2003) Global analysis of protein localization in budding yeast. *Nature* **425**, 686–90.

Lippincott-Schwartz, J. and Patterson, G.H. (2003) Development and use of fluorescent protein markers in living cells. *Science* **300**, 87–91.

March, J.C., Rao, G., and Bentley, W.E. (2003) Biotechnological applications of green fluorescent protein. *Applied Microbiology and Biotechnology* **62**, 303–15.

Matz, M.V., Lukyanov, K.A., and Lukyanov, S.A. (2003) Family of the green fluorescent protein: journey to the end of the rainbow. *BioEssays* **24**, 953–9.

Miller, G. (2006) A better view of brain disorders. *Science* **313**, 1376–9.

Phelps, M.E. (2002) Molecular imaging with positron emission tomography. *Annual Review of Nuclear Particle Science* **52**, 303–38.

Tsien, R. (1998) The green fluorescent proteins. *Annual Review of Biochemistry* **67**, 509–44.

Wachter, R.M. (2007) Chromogenic cross-link formation in green fluorescent protein. *Accounts of Chemical Research*, 40, 120–7.

Wachter, R.M., King, B.A., Hein, R. et al. (1997) Crystal structure and photodynamic behavior of the blue emission variant Y66H/Y145F of green fluorescent protein. *Biochemistry* **36**, 9759–65.

Zimmer, M. (2002) Green fluorescent protein (GFP): applications, structure, and related photophysical behavior. *Chemical Reviews* **102**, 759–81.

PROBLEMS

19.1 What is the cofactor in GFP?

19.2 How is the chromophore of GFP related to the gene sequence?

19.3 How was it shown that oxygen is required for cofactor formation?

19.4 Why do different forms of GFP have different spectra?

19.5 Why does wild-type GFP have absorption bands at 395 and 475 nm?

19.6 Explain why a mutation at Tyr-66 would result in altered optical properties.

19.7 Provide a molecular reason for why GFP is initially clear and becomes colored after a period of time.

19.8 Why does the spectrum of GFP have more than one absorption band that changes with pH?

19.9 What is the molecular interpretation of the pH dependence of the optical spectrum?
19.10 Explain how GFP is used in proteomics.
19.11 How can GFP be used in a FRET experiment to determine whether two proteins bind together?
19.12 What is the relationship between FRET efficiency and distance?
19.13 In addition to the separation distance, what other factors play a role in the FRET efficiency?
19.14 What is the difference between the chromophores of DsRed and GFP?
19.15 How can GFP be used to track cellular movement of proteins?
19.16 What are the different types of radioactive decay?
19.17 In PET, what is the source of the positrons?
19.18 In PET, why are two positrons always emitted simultaneously?
19.19 How can PET be used to monitor glucose metabolism?
19.20 How can PET be used with Alzheimer's and related diseases?

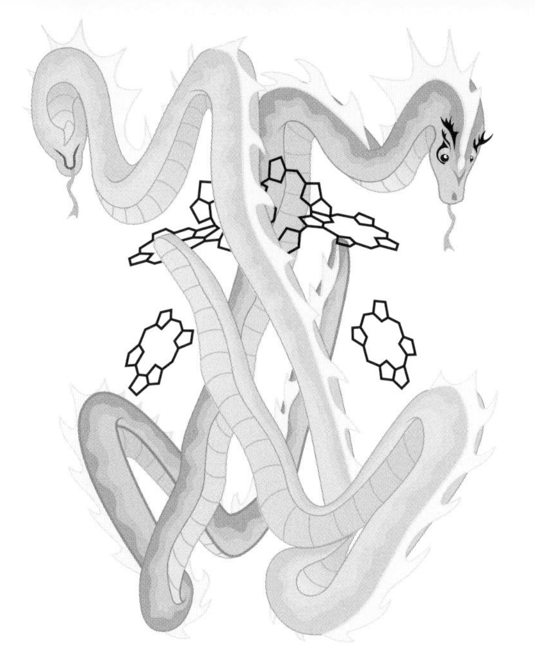

20

Photosynthesis

Photosynthesis is the biological process by which the energy of the sun is converted into energy-rich compounds that are used to drive cellular processes. The energy from this remarkable process has provided energy for essentially all life forms as well as having a dramatic effect on geological formations over time. The mechanism by which organisms convert light into chemical energy has been studied for hundreds of years. In the 1770s Joseph Priestly demonstrated that plants released a substance, we know now is oxygen, that could support life. In the 1930s Cornelis van Niel formulated the photosynthetic process in terms of oxidation/reduction reactions and Robert Hill established that chemical reactions led to oxygen evolution. Robert Emerson with William Arnold performed the first modern spectroscopic measurements on photosynthetic organisms that quantified the amount of oxygen evolution compared to the number of chlorophyll molecules. This was followed by many years of debate until the biochemical involvement of ATP and the chemiosomotic hypothesis were established (Chapter 9). In the late 1950s and early 1960s, Melvin Calvin identified the basic enzymatic reactions of photosynthetic organisms in what is now termed the Calvin cycle.

The biochemical isolation of the photosynthetic complexes was difficult as these complexes are located in cell membranes. By the early 1970s, the use of detergents to solubilize the proteins from the cell membrane (Chapter 7) led to the isolation of functional complexes. The availability of these purified complexes provided the chance for detailed biochemical and spectroscopic studies that have established the molecular mechanism of light conversion. The initial photochemical process, namely the capture of the light energy into a stable chemical state, is performed by large pigment–protein complexes that capture the light and then transfer the light energy to other complexes to perform electron-transfer reactions. These pigment–protein complexes make use of bacteriochlorophyll or chlorophyll cofactors, which are conjugated tetrapyrroles (Figure 20.1). The structures of these tetrapyrroles from different photosynthetic organisms are similar,

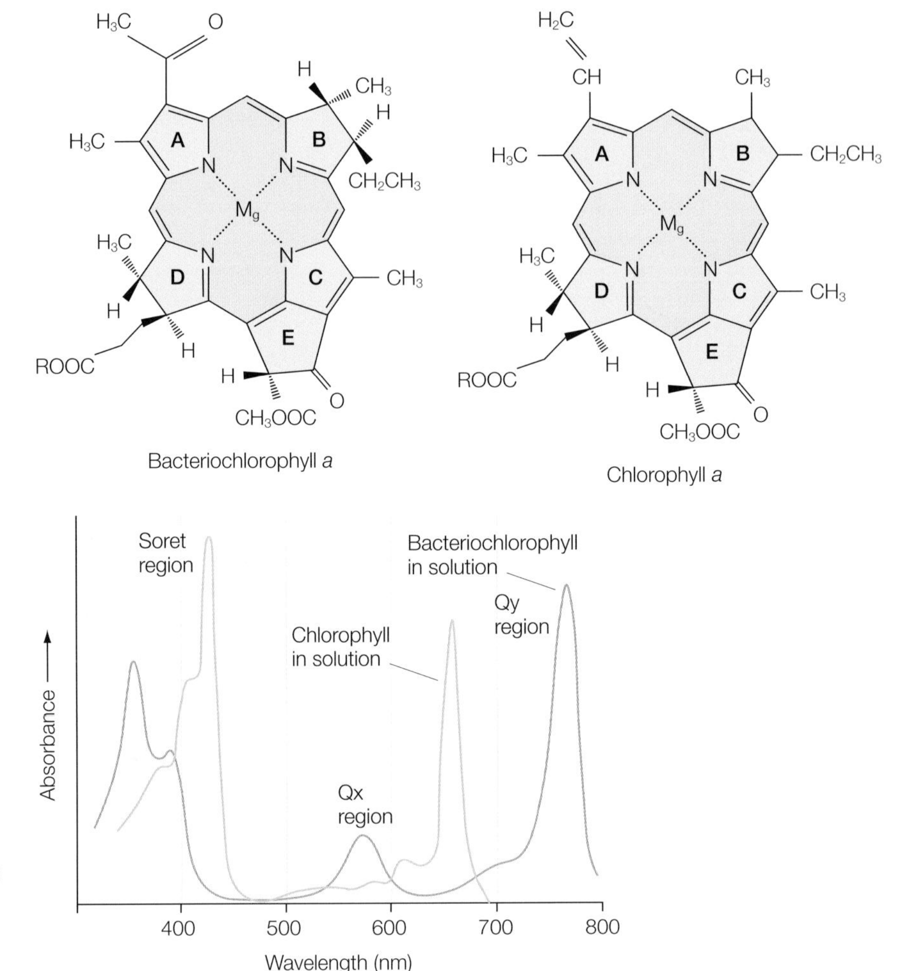

Figure 20.1
The structure and optical absorption spectra of bacteriochlorophyll *a* and chlorophyll *a*. Modified from Blankenship (2002).

with various substitutents. Chlorophyll *a* is found in all eukaryotic photosynthetic organisms. The structural difference between chlorophyll *a* and bacteriochlorophyll *a* lie in ring A on the C3 position and in ring B with a difference in the degree of saturation. These structural differences alter the conjugation and the electronic structure of the cofactor. The absorption spectrum of chlorophyll *a* shows a number of absorption bands with the lack of bands in the green region giving rise to their characteristic color. These absorption bands arise from π-to-π^* transitions of the conjugate macrocycle. The two lowest energy transitions are called the Q bands, with the extinction coefficient of chlorophyll *a* being 9.0×10^4 M^{-1} cm^{-1}. The absorption bands at higher energy are commonly termed the Soret bands. The spectrum of bacteriochlorophyll shows the same set of bands, with comparable extinction coefficients, but the peaks are shifted due to the difference in conjugation compared to chlorophyll.

Carotenoids are found in all known native photosynthetic organisms in hundreds of chemically distinct structures (Chapter 10). These molecules have extended delocalized π electrons that result in strong absorption bands in the green region. Carotenoids are accessory pigments in the absorption of sunlight that transfer the energy to the chlorophylls. Another function is a process termed photoprotection in which the carotenoids quench harmful excited states of chlorophylls. Carotenoids also are involved in the regulation of energy transfer through the xanthophyll cycle.

ENERGY TRANSFER AND LIGHT-HARVESTING COMPLEXES

One of the early outcomes of the experiments by Emerson and Arnold was that a saturating light flash produces only one oxygen molecule for about 2500 chlorophyll molecules. Although no mechanism for energy transfer was known at that time, this observation gave rise to the ideal of a photosynthetic unit, consisting of many pigments in which the light energy was trapped before conversion. This concept was disputed by several scientists, including James Franck and Edward Teller, but eventually the weight of experimental evidence led to its acceptance.

The reason for the development of a system of large pigments in the capture of sunlight can be understood from an energetic basis. The number of photons of light available for energy capture is relatively low. Full sunlight has an intensity, I, of about 1800 $\mu Em^{-2}\,s^{-1}$ in the visible region, corresponding to 10^{21} photons $m^{-2}\,s^{-1}$ or equivalently 10 photons $Å^{-2}\,s^{-1}$. For a molecule with a size of about 10 Å and an extinction coefficient of $10^5\,M^{-1}\,cm^{-1}$, only about 10 photons are estimated to be absorbed per second. Due to the low amount of light energy available, photosynthetic organisms optimize the capture of light by the presence of pigments that have been specifically designed for their ability to capture light energy and transfer that energy to specialized centers that perform the electron-transfer reactions. This role of the light-harvesting protein complexes is very similar to the function of large antenna dishes for radiowaves, which funnel the radiowaves to a central location where it is transformed into electrical energy.

Organisms use many different types of pigment and protein to perform the energy capture. In some cases the antennae are peripheral units and in other organisms the antennae are integral to the central protein that performs the photochemistry. The absorption spectra of the antenna varies in different photosynthetic organisms but in all cases the energy is funneled from the higher-energy pigments to the lowest-energy pigments. In the cell membranes of purple bacteria, the antennae form large ring structures with a diameter of approximately 65 Å (Figure 20.2). The ring for one type of antenna, called light-harvesting complex II, is composed of nine (sometimes eight) pairs of two-protein subunits arranged in a pseudo

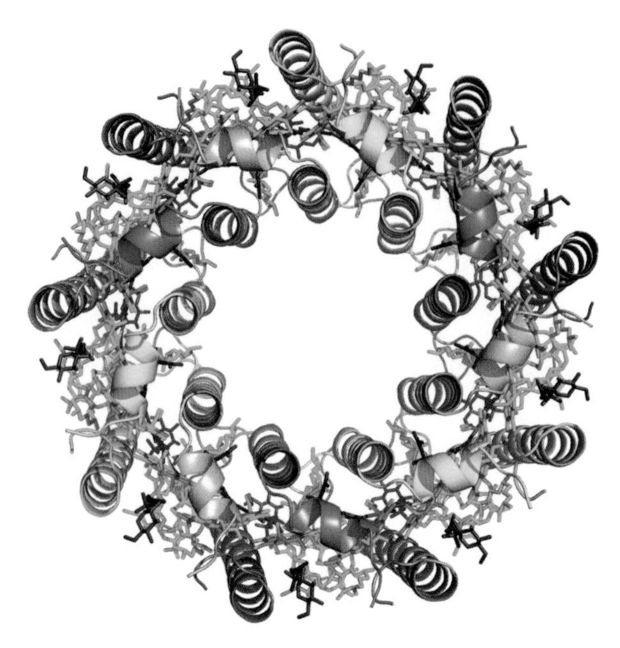

Figure 20.2 The three-dimensional structure of the light-harvesting complex II.

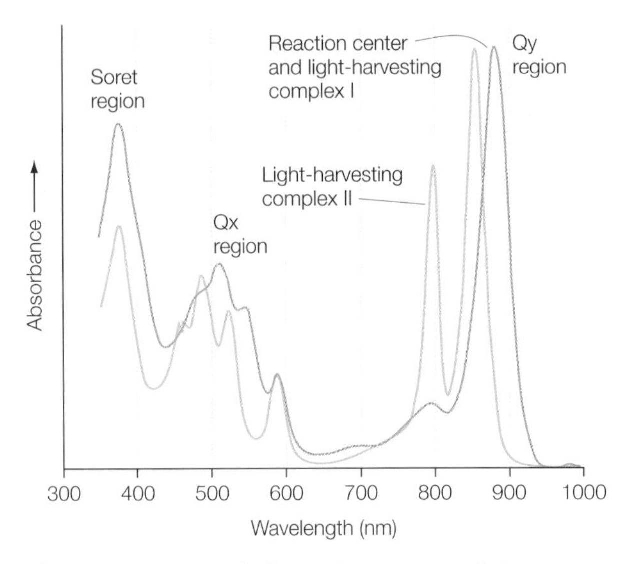

Figure 20.3 Optical absorption spectra of the light-harvesting complexes I and II.

9-fold symmetry around a central symmetry axis. Associated with each pair of subunits are three bacteriochlorophylls and one carotenoid that are located between the inner and outer rings of protein subunits.

Although all of the bacteriochlorophyll *a* molecules are chemically identical there are two spectrally distinct sets of bacteriochlorophyll *a* in the protein complex, with one set contributing to an absorption band at 800 nm and the other at 850 nm (Figure 20.3). The structure provides a clear explanation for this spectral division of the bacteriochlorophylls. The bacteriochlorophylls that absorb at 800 nm form a ring with the planes of the macrocycles perpendicular to the central symmetry axis. These pigments lie parallel to the plane of the membrane and are quite well separated from each other, with a distance of approximately 20 Å. Due to the large distance between neighboring pigments, the molecules interact very weakly and behave as independent molecules. The bacteriochlorophylls that absorb at 850 nm are arranged as closely interacting dimers with their planes approximately parallel to the symmetry axis. The dimers form a band of 18 (or 16) bacteriochlorophylls that are all highly interacting.

Energy transfer among the bacteriochlorophylls can occur by the Forster mechanism (Chapter 17). In this case, the bacteriochlorophylls serve as both energy donor and acceptor so the energies are well matched and the energy transfer has been measured to be very efficient. When the bacteriochlorophyll molecules are close then energy transfer can occur extremely rapidly with times of less than 100 fs. Energy transfer along the rings of bacteriochlorophylls has been measured using femtosecond spectroscopy, yielding the rates of energy transfer (Figure 20.4). Energy transfer from the 800-nm bacteriochlorophylls to the 850-nm bacteriochlorophylls occurs in 1 ps. The 850-nm bacteriochlorophylls are so strongly interacting that the energy is effectively delocalized among these molecules. In the photosynthetic membrane, the energy is transferred in 3 ps to a second protein called the light-harvesting complex I, which also has rings of proteins and bacteriochlorophylls. Again, the energy is effectively delocalized among the bacteriochlorophylls of that complex until transfer to the central site, the bacterial reaction center, occurs in 35 ps. The last step represents the slowest step of the entire process, presumably due to the relatively large distance of 35–40 Å to the bacteriochlorophylls of the reaction center.

ELECTRON TRANSFER, BACTERIAL REACTION CENTERS, AND PHOTOSYSTEM I

The reaction center can be considered to be an energy-transducing device that converts the energy of sunlight into chemical energy. Reaction centers are integral membrane proteins that are largely buried in the cell membrane. Each reaction center has two core protein subunits that surround a number of cofactors. For reaction centers from purple bacteria, reaction centers have four bacteriochlorophylls, two bacteriopheophytins

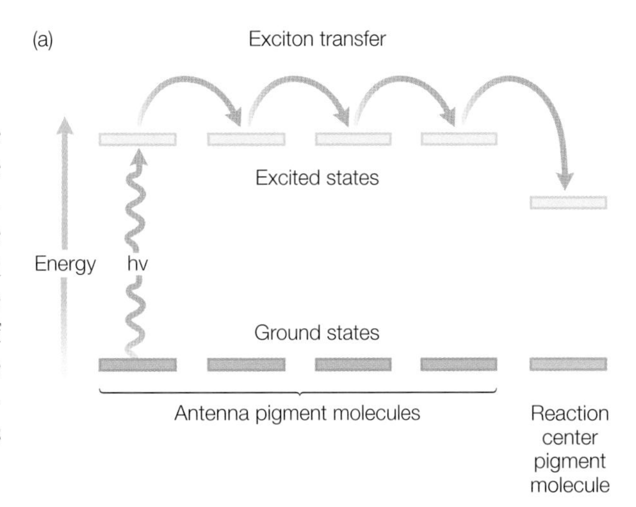

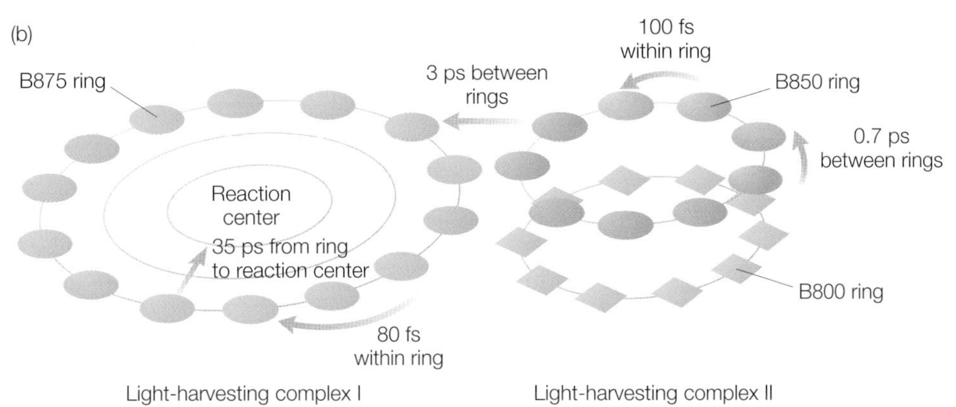

Figure 20.4 Energy transfer in photosynthetic bacteria. (a) Light absorption by the bacteriochlorophylls of the light-harvesting complexes results in the formation of an excited state followed by energy transfer, or excition transfer, among the different bacteriochlorophylls until the energy is transferred to the reaction center bacteriochlorophyll dimer. (b) A schematic view of the energy transfer process in the cell showing the transfer times, which range from 80 fs to 35 ps, between the different bacteriochlorophylls of the light-harvesting complexes I and II and the reaction center. Modified from Fleming and van Grondelle (1997).

(which have the structure of bacteriochlorophyll without the central Mg), two quinones, and one nonheme iron. Spectroscopic studies have revealed many aspects of the roles of the many different cofactors. In response to light the *electron donor,* D, becomes excited and transfers an electron to an initial *electron acceptor,* A:

$$DA \xrightarrow{h\nu} D^*A \rightarrow D^+A^- \tag{20.1}$$

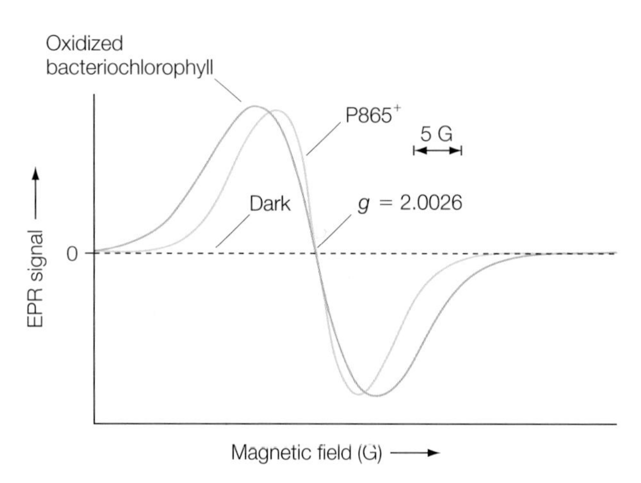

Figure 20.5 Light and dark EPR spectra and bacteriochlorophyll spectra. Light produces an oxidized donor, identified as $P865^+$, in reaction centers. Modified from Feher (1998).

The reaction center has no electron paramagnetic resonance (EPR) signal in the dark but upon illumination a signal is observed near $g = 2.0026$ (Figure 20.5). The signal was identified as arising from an oxidized bacteriochlorophyll based upon the g value that matches the g value observed for oxidized bacteriochlorophyll in solution but the linewidth of the EPR signal was much more narrow. It was soon realized that the width of the signal reflected the delocalization of the unpaired electron over two highly interacting bacteriochlorophylls. The linewidth ΔH is proportional to the number of nuclei, N, and the strength of the interaction between the nuclei and inpaired electron, A, due to heterogeneous broadening. Since the nuclei are equivalent, their contribution is a function of the square root of the number:

$$\Delta H \propto AN^{1/2} \tag{20.2}$$

A dimer has twice the number of nuclei but each electron spends on average half the time, which reduces the interaction strength by half. The ratio of the linewidths for the monomer and dimer is predicted to be:

$$\frac{\Delta H_{\text{monomer}}}{\Delta H_{\text{dimer}}} = \frac{A_{\text{monomer}}\sqrt{N}}{\frac{1}{2}A_{\text{dimer}}\sqrt{2N}} = \frac{2}{\sqrt{2}} = \sqrt{2} \tag{20.3}$$

Thus, EPR experiments showed that the primary electron donor was a bacteriochlorophyll dimer. In response to light, one of the absorption bands of the steady-state optical spectrum disappears and several band shift (Figure 20.6). The band that disappears, or bleaches, with the same kinetics observed for the EPR signals showing that the pigment associated with the 865-nm absorption band, was the primary electron donor, now termed P865. The two quinones could be biochemically removed from the reaction center and then bound back to the protein, providing the means for their identification as electron acceptors (Chapter 8). EPR studies of the nonheme iron were consistent with the iron always remaining in the Fe^{2+} state and thus not being an active participant in the electron-transfer process.

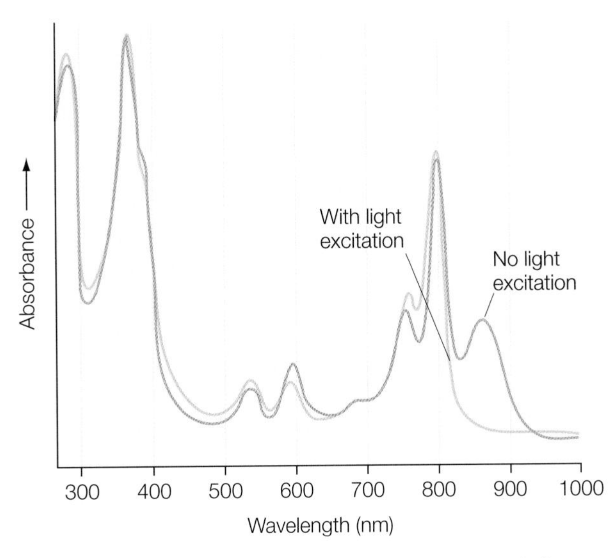

Figure 20.6 Light and dark optical spectra of the bacterial reaction center.

The optical spectrum has proved to be a rich means of characterizing the roles of the other cofactors of the reaction center and determining the electron-transfer rates. The absorption bands were seen to have different responses after illumination with a fast laser pulse (Figure 20.7). The band at 865 nm quickly bleached and then was unchanged due to rapid formation of the excited state, P865*, followed by the oxidized state, $P865^+$. The kinetics at other wavelengths, such as 755 nm, were more complex but could be interpreted using models of sequential reactions (Chapter 10) as arising due to one of the bacteriopheophytins serving as an intermediate electron acceptor.

These experimental studies established the kinetics and energetics of the photochemical and early secondary events (Figure 20.8). Excitation of the primary electron donor at 865 nm corresponds to an energy increase of 1.4 eV:

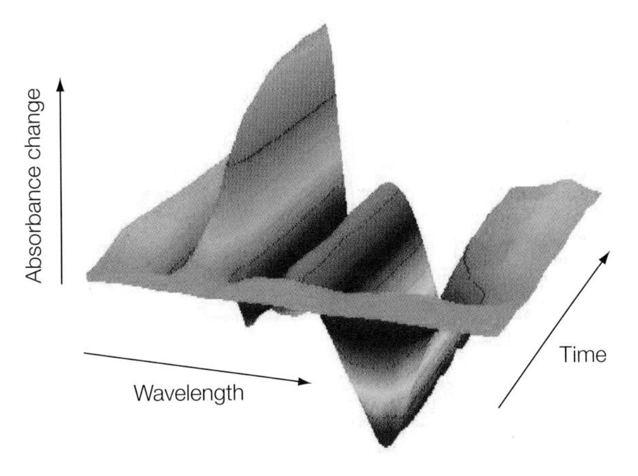

Figure 20.7 Transient optical spectra. Pronounced spectral changes are evident on the picosecond timescale. Figure courtesy of Su Luin.

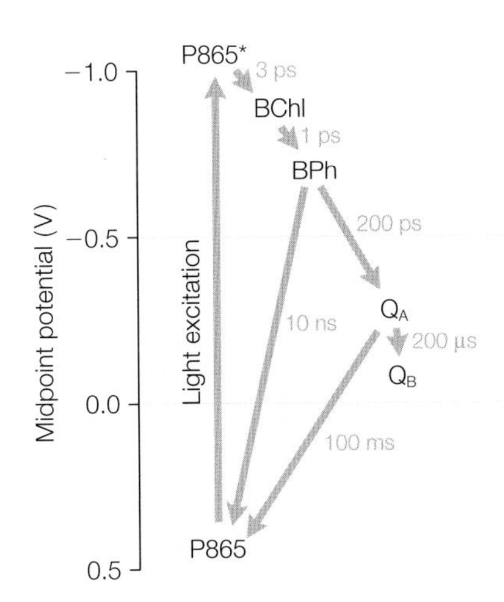

Figure 20.8 Energetics of the bacterial reaction center.

$$E = h\nu = \frac{hc}{\lambda} = \frac{(6.62 \times 10^{-24}\ \text{Js})(2.99 \times 10^{8}\ \text{m s}^{-1})}{8.65 \times 10^{-7}\ \text{m}} = 2.3 \times 10^{-9}\ \text{J} = 1.4\ \text{eV} \tag{20.4}$$

Electron transfer proceeds rapidly from the excited state to a bacteriochlorophyll monomer in 3 ps, and a bacteriopheophytin in 1 ps, the primary quinone with slower transfer to the two quinones (Figure 20.8). Since the rapid transfer from the bacteriochlorophyll monomer to bacteriopheophytin is faster than the transfer to the bacteriochlorophyll, the electrons are immediately depleted from the bacteriochlorophyll monomer, and a reduced state of the bacteriochlorophyll monomer is present in only very low proportions and is very difficult to observed experimentally. Since each step is energetically favorable with rapid forward electron-transfer rates, the yield of the transfer, termed the quantum yield, is nearly one, meaning that for every excitation of P865 an electron is transferred to the secondary quinone. The use of multiple electron acceptors does result in the loss of most of the initial excitation energy but the desired outcome is simply the transfer of an electron and subsequent proton transfer that can be used to make energy-rich compounds (Chapter 8).

A critical step in our understanding of the properties of reaction centers was the determination of their structures in the late 1980s. While protein crystallography was being used to solve the structures of thousands of water-soluble proteins (Chapter 15), the structures of integral membrane proteins proved to be elusive due to the difficulties in biochemically preparing and crystallizing these proteins in the presence of detergents. The first structure of an integral membrane protein was the reaction center from *Rhodopseudomonas viridis* (since renamed *Blastochloris viridis*) by Johann Deisenhofer, Hartmut Michel, and Robert Huber and coworkers (Deisenhofer et al. 1985). This structure was a landmark (awarded the Nobel Prize in Chemistry in 1988) as it demonstrated the feasibility of obtaining crystals of membrane proteins that were suitable for X-ray diffraction studies and has been followed, albeit somewhat slowly, by the elucidation of a number of other structures including the reaction center from *Rhodobacter sphaeroides* as well as unrelated proteins, including bacteriorhodopsin and ion channels (Chapters 17 and 18).

The reaction-center structure has a number of features that are conserved among all bacteria. Although the total number of protein subunits varies among the different bacterial species, there is always a central core pairing of two subunits with five transmembrane helices each (Figure 20.9). These two subunits are structurally related to each other by a pseudo 2-fold symmetry axis that runs down the center of the protein. The same symmetry axis also divides the cofactors into two branches that span most of the membrane from the periplasmic to the cytoplasmic side. The bacteriochlorophyll dimer, P870, is at one end, near the periplasmic side of the protein, and the primary and secondary quinones are at the other end with the nonheme iron located directly on the symmetry axis. The bacteriochlorophyll monomer and bacteriopheophytin that serve as intermediate electron acceptors are located on the A branch. The spectroscopic studies coupled with mutagenesis studies have established that electron transfer follows essentially exclusively along the A branch. The advantage of the two branches is that the quinones can serve different roles. The primary quinone serves an intermediate electron acceptor that shuttles the electrons to the secondary quinone. The electron stays on the secondary quinone until the second electron arrives and the quinone can leave with two electrons and two protons to be replaced with an unreduced quinone from the membrane pool.

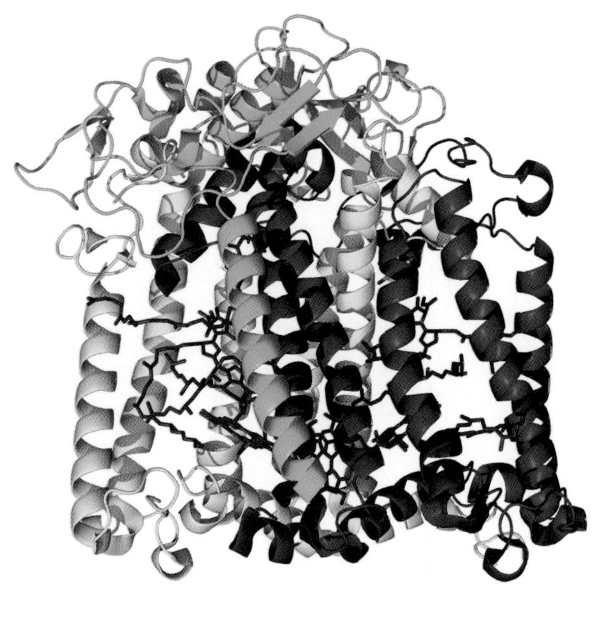

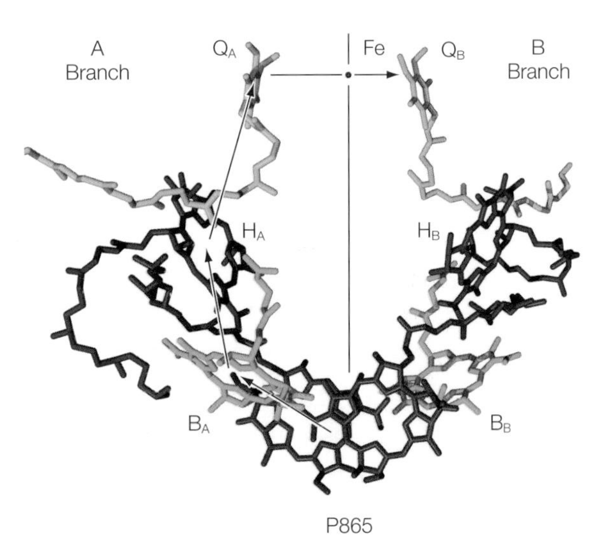

Figure 20.9 Three-dimensional structure of the bacterial reaction center.

The cofactors of the B branch of reaction centers have no direct role in the electron transfer in purple bacteria but may participate in the related complex photosystem I. Photosystem I is a much larger complex with at least 10 protein subunits in cyanobacteria and several more in eukaryotic organisms. These complexes have approximately 100 chlorophylls that mostly serve as antenna molecules to transfer the energy to the protein core. The general architecture of the protein core is similar to that of the reaction center, with two core protein subunits and two branches of cofactors (Figure 20.10). A chlorophyll dimer near the luminal side of the protein serves as the primary electron donor P700. Four additional chlorophylls and two quinones are also present. A unique feature is the presence of three iron–sulfur clusters that function as electron acceptors. One of the clusters, named F_X, is located along the central symmetry axis, as was found for the nonheme iron of the reaction center. The other two iron–sulfur clusters, called F_A and F_B, are bound to a small protein subunit attached to the core subunits.

The rate of electron transfer from the excited state, P700*, is very fast at 2 ps but the observed excited-state lifetime is significantly longer, at

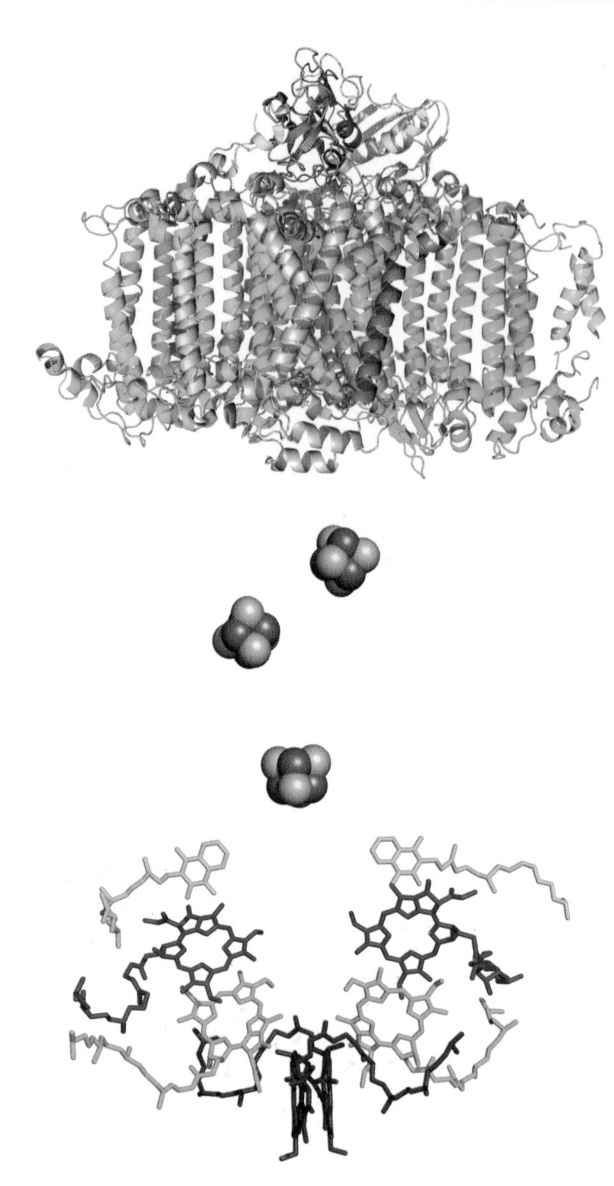

Figure 20.10 The structure of photosystem I showing the protein subunits (top) and the cofactors (bottom).

30 ps, because the excitation energy is not trapped on P700 but can be trapped on one of the many chlorophylls. Electron transfer proceeds to F_X in 200 ns with subsequent transfer most likely being sequential to F_A and then F_B despite the latter step being energetically unfavorable by 60 mV. This pathway of electron transfer from P700* to F_X results in a vectorial transfer of an electron along the symmetry axis. Photosystem I does not need to have a division of the branches to accommodate functionally distinctive quinones. This raises the question of whether the electron transfer is asymmetric along the two branches. The interpretation of many mutagenesis and spectroscopic studies in terms of the asymmetry of electron transfer remains an open question.

WATER OXIDATION

The ability of photosynthetic organisms to generate molecular oxygen from water, as originally demonstrated by Joseph Priestly in the 1770s, remains a fascinating area of research in photosynthesis. The ability of organisms to oxidize water represented a significant development in the Earth's history. When anoxygenic bacteria, the precursors to the modern purple bacteria, evolved into organisms that possessed this new capacity, the Earth's atmosphere changed from containing reduced gases to having a significant amount of oxygen. The presence of oxygen set the stage for the development of respiration in organisms as well as altering the oxidation states of minerals on the Earth. How plants and cyanobacteria generate oxygen remains an unsolved mystery at a molecular level, which is actively being investigated in many laboratories. Although water can be oxidized readily using electrodes, this reaction only occurs at a substantial energy expense, which makes the process unpractical for the economical delivery of hydrogen for use as fuel (Chapter 12). Therefore, an understanding of the biological process would have both intellectual and practical implications.

The site of water oxidation is photosystem II, or the water-oxidation complex. To oxidize water to molecular oxygen, the protein performs a four-electron process that is coupled with the transfer of four protons:

$$2H_2O \rightarrow O_2 + 4H^+ + 4e^- \qquad (20.5)$$

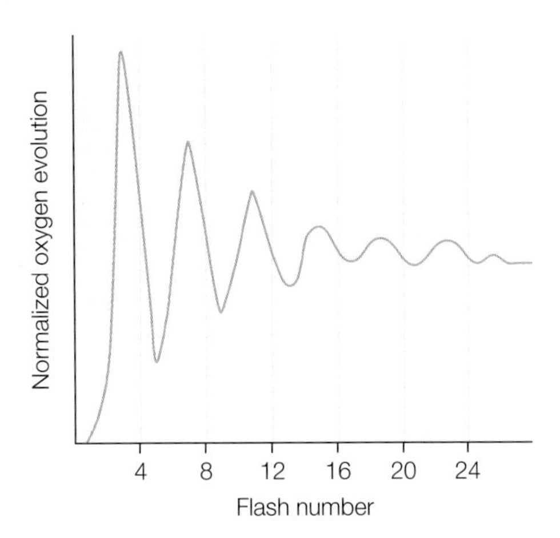

A series of experiments was performed in the 1960s by Pierre Joliot and later Bessel Kok using transient optical spectroscopy. These results demonstrated clearly that oxygen evolved only after the fourth light flash, giving rise to the proposed mechanism of the S-state model (Figure 20.11). For full oxygen evolution, photosystem II was shown to require the presence not only of manganese but also calcium.

Recently, the three dimensional structure of photosystem II has been reported (Figure 20.12). Whereas the structure reveals the overall organization of the

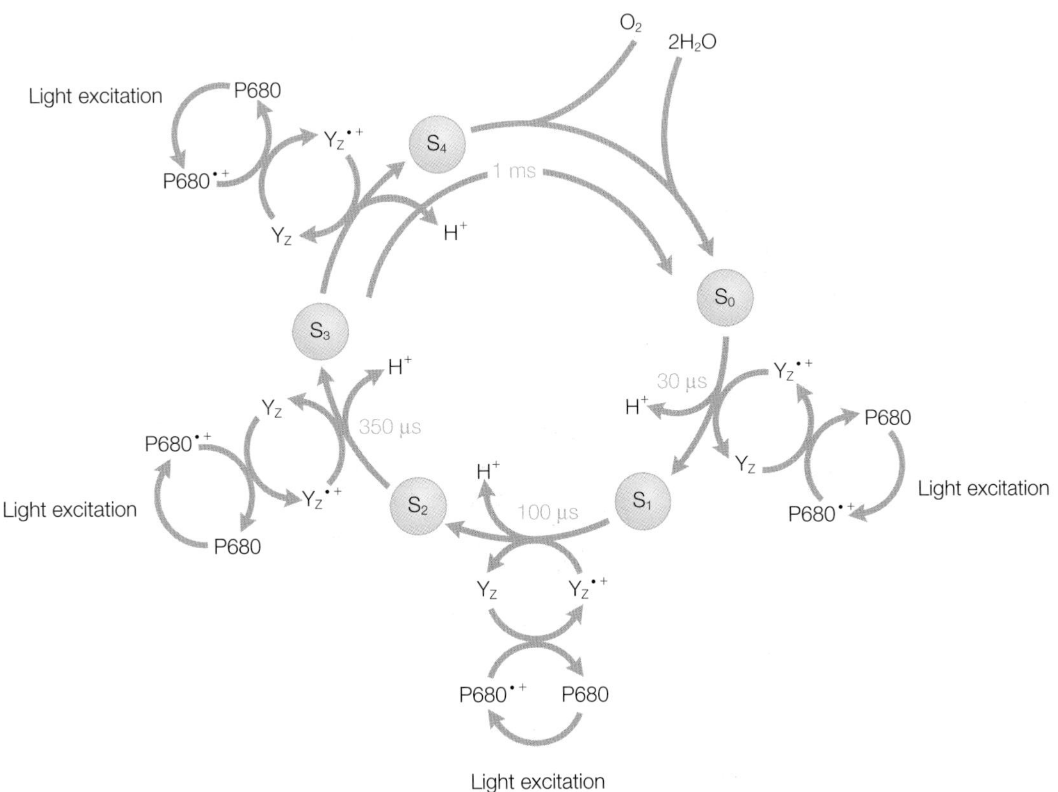

Figure 20.11 Flash photolysis revealed oscillations in oxygen production that gave rise to the concept of S states of photosystem II. Modified from Britt (1996).

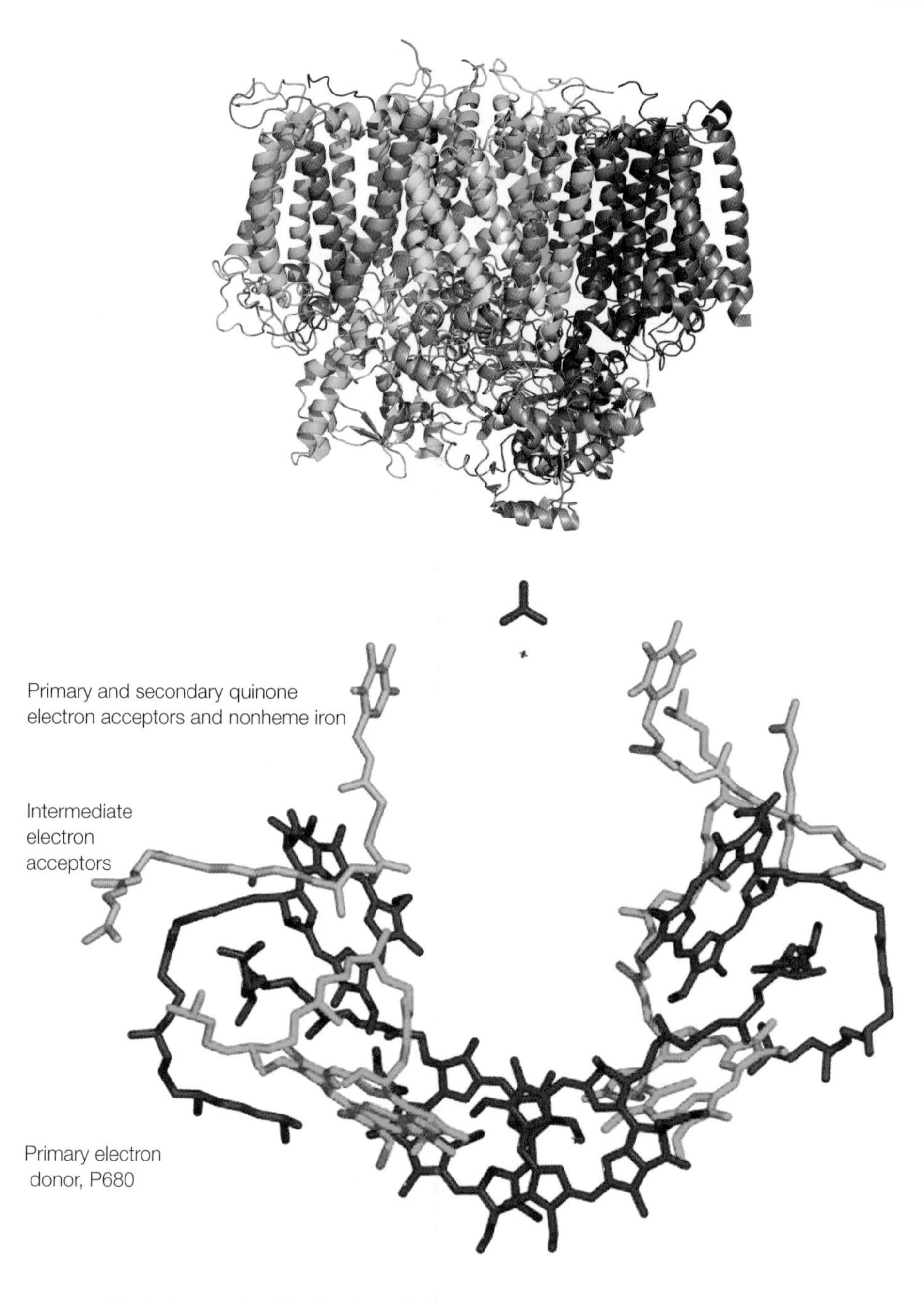

Figure 20.12 The three-dimensional structure of photosystem II.

complex, the resolution limit was limited and the three-dimensional structure of photosystem II could not be determined in detail. The core protein subunits and cofactors have been found to form the same symmetrical arrangement, as found in the other complexes. The position of the manganese cofactor has been identified but the molecular arrangement of this crucial cofactor has not been elucidated. Despite these limitations, however, the structure does provide a foundation for a discussion of the mechanism of water oxidation on a molecular level.

To oxidize water, photosystem II must have several specific chemical capabilities. First, water is a very stable molecule and, correspondingly, the oxidation/reduction midpoint potential at pH 7 is high, at +0.82 V, and is +0.93 V at pH 5, which is the pH of the thylakoid lumen where the reaction occurs (Chapter 6). To oxidize water, the primary electron donor, P680, must have a higher oxidation/reduction midpoint potential. Although the midpoint potential is too high to measure experimentally, P680 has been estimated to have a midpoint potential of about 1 V, making P680 the strongest oxidant of any natural biological system. To generate this potential, light is absorbed at 680 nm, which has an energy of 1.82 eV using eqn 20.4. Thus, both the excitation energy and the oxidation/reduction potential are much higher than those of the bacterial reaction center as $P680^+$ performs difficult oxidation reactions while the reaction center simply moves protons and electrons across the cell membrane. At a value of 1.1 V, the oxidation/reduction midpoint potential of P680 is much higher than that of chlorophyll *a* in organic solutions, which is less than 0.9 V. Presumably, the increase in potential of the chlorophyll *a* molecule in the protein is due to the presence of specific protein–chlorophyll interactions such as hydrogen bonds that can increase the potential of bacteriochlorophyll (Chapter 8). Compared to the midpoint potential of 0.93 V needed for water oxidation, the energy available is sufficient but with little room to spare as the difference in potentials of Y_Z (see below) and P680 is estimated to be 20–100 mV for the different S states.

Second, the electron-transfer process must be coupled to proton transfer. In photosystem II, the oxidized electron donor, $P680^+$, is reduced rapidly by a tyrosine residue, identified as Y_Z or Tyr_Z, that is located about 10 Å away from P680. Amino acid residues are normally thought of as passive components of proteins but in some cases they participate in oxidation/reduction reactions. The oxidation/reduction midpoint potential of Y_Z is high, at an estimated 0.9–1.0 V, as indeed it must be to retain the oxidation strength, but $P680^+$ is an even stronger oxidant and will take an electron from the tyrosine. By using an intermediate electron carrier, the electron transfer in photosystem II occurs with a high yield, as found in the bacterial reaction center. More importantly, protonated oxidized tyrosines are not energetically stable and the oxidation is coupled to the release of the phenolic proton, forming a neutral tyrosyl radical following mechanisms similar to those discussed for quinones (Chapter 5).

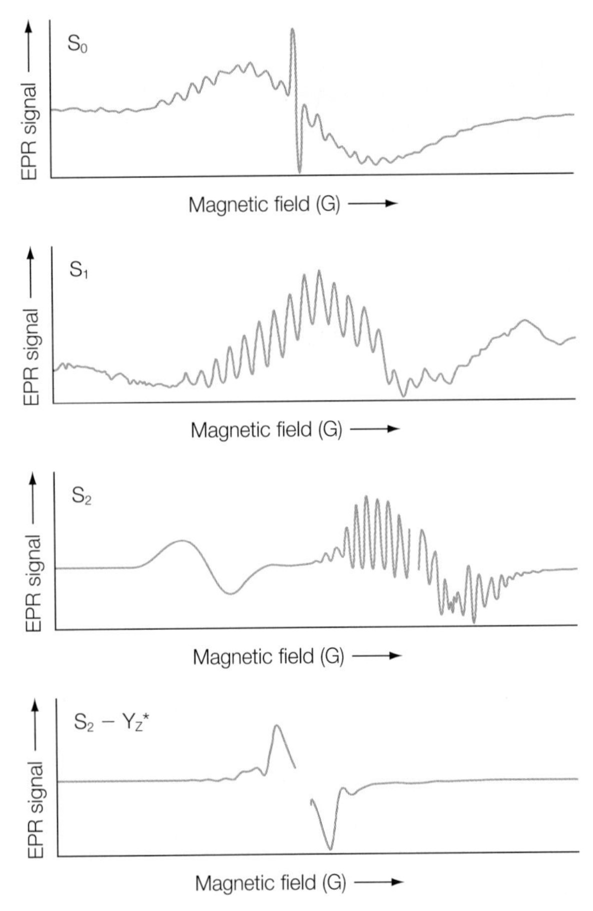

Figure 20.13 EPR of photosystem II showing different spectroscopic signatures for each S state. Modified from Britt et al. (2003).

Third, the generation of electrons is sequential as each absorption of a photon results in the transfer of a single electron. Therefore, the electron equivalents must be stored until the fourth electron is available to drive the reaction to completion. This function is performed by the manganese cluster, which consists of four manganese atoms. The manganese cluster is oxidized by Y_Z, with rates ranging from 20 to 300 ns for the different S states. As discussed above, the molecular arrangement of the four manganese has not been determined using the X-ray data due to the limited resolution. However, two spectroscopic techniques, extended X-ray absorption spectroscopy (EXAFS) and EPR, have been very revealing about the molecular arrangement.

As the manganese cluster undergoes the S-state transitions, the oxidation state of the cluster changes. The development of EPR spectroscopy has produced signals from all of the S states, which show different g values and degrees of resolved hyperfine interactions (Figure 20.13). The most characteristic EPR signal is for the S2 state that was found to be centered at $g = 2$ and contained multiple lines characteristic of hyperfine coupling between the unpaired electron and the manganese nuclei of the cluster. The ^{55}Mn isotope is 100% naturally abundant, with a nuclear spin of 1/2, so a single manganese will have an EPR spectrum with six lines ($= 2I + 1/2$). The abundance of lines can be modeled by assuming that the metal cluster must be more complex. Synthetic Mn(III)–Mn(IV) binuclear clusters have 16 hyperfine lines that resemble the spectrum of the S2 state. In addition to the hyperfine lines, the low-temperature EPR spectrum of the S2 state shows a signal at $g = 4.1$ that has no hyperfine structure. Comparison with model manganese systems led to the conclusion that the $g = 4.1$ signal was due to the presence of a tetranuclear cluster.

As discussed in Chapter 15, the advent of synchrotrons provided the opportunity to investigate metal clusters using EXAFS. The measurement of photosystem II using EXAFS showed the presence of three distinctive peaks corresponding the distances of 1.8, 2.7, and 3.3–3.4 Å (Figure 20.14). The distance of 2.7 Å is consistent with the presence of a di-μ-oxo-bridged manganese. The peak at 3.3 Å was assigned as representing a distance between two manganese atoms, other than the bridged manganese. In addition, a peak at 3.4 Å was assigned as representing a manganese-to-calcium distance. The shortest distances are from manganese ligands to oxygen and nitrogen, which cannot be resolved in these measurements. Based upon these distances a wide array of possible arrangements for the manganese cluster were possible. Although the possible configurations still remain open, the combination of the EXAFS, EPR, and X-ray data is most consistent with an asymmetric configuration with one manganese being somewhat distant from the other three.

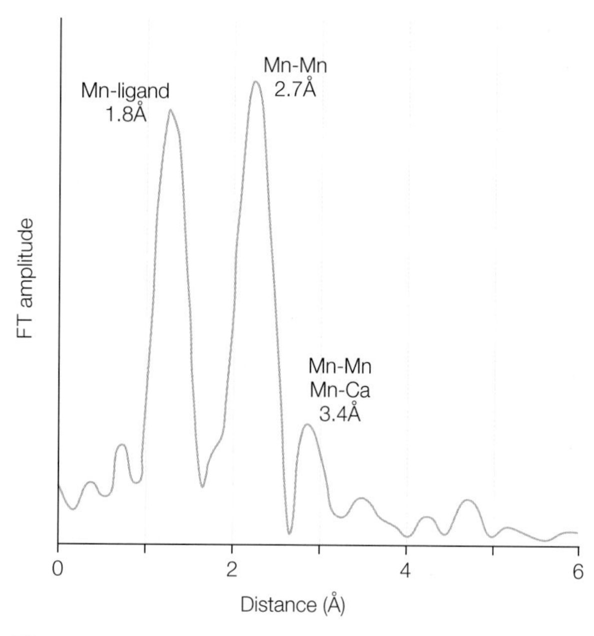

Figure 20.14 EXAFS spectrum of photosystem II revealing the distances in the Mn cluster. Modified from Yano et al. (2006).

As the manganese cluster becomes more oxidized, the positive charges could become very destabilizing. Since the cluster has four manganese, the charges can be distributed over the cluster to minimize any unfavorable oxidation state of any single manganese but the site would still be gaining charge. Mechanisms that increase the oxidation of the cluster would require bond rearrangements of the cluster that would trigger the water oxidation but such bond changes were not supported by the spectroscopic studies.

These difficulties were largely resolved with the development of the hydrogen-abstraction model, which invoked a direct involvement of the tyrosyl radical in both the electron and proton transfers. Consider a simple model which places Y_Z in close proximity to the manganese cluster with a bound water molecule (Figure 20.15). Light excitation causes formation of $P680^+$ (Figure 20.15a), which is rapidly reduced by Y_Z (Figure 20.15b) with a coupled release of the phenolic proton to a nearby base (Figure 20.15c). The proton does not stay at the base but leaves as the base is part of a proton pathway to the luminal surface (Figure 20.15d). The S-state advancement occurs as the electron is transferred from Y_Z to the manganese cluster, leaving a strong tyrosine base (Figure 20.15e). The tyrosine base abstracts a proton from the bound water molecule (Figure 20.15f), leaving a hydroxyl group bound to the manganese cluster. The tyrosine is then reset allowing the second proton to be abstracted after the next excitation of P680. By including a direct role of Y_Z in the process, electron and proton transfer are naturally coupled together in a manner that drives water oxidation. Since a proton is transferred away from the manganese cluster after each oxidation step, the overall charge of the cluster does not change during the S cycles.

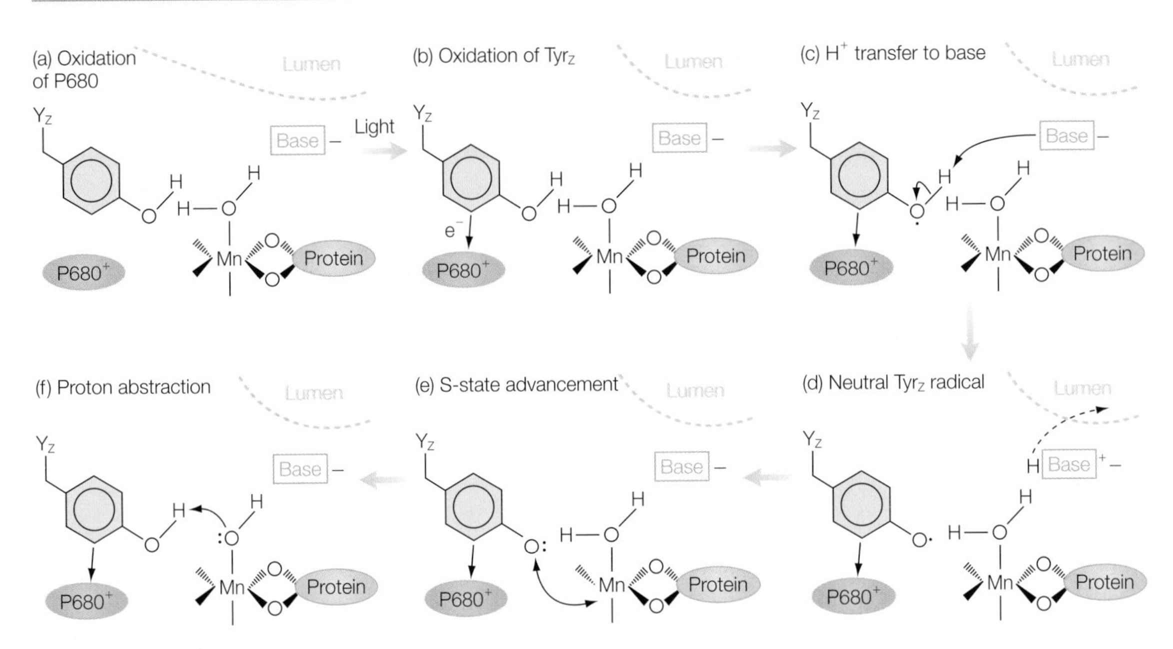

Figure 20.15 The hydrogen-abstraction model in which water oxidation is facilitated by the transfer of the hydrogen to the phenolic ring of the tyrosyl radical after oxidation. Modified from Britt (1996).

The hydrogen-abstraction model was highly innovative but appears to be somewhat over-simplified compared to the actual mechanism of water oxidation as derived from spectroscopic studies. For example, there is still discussion about the oxidation states of the manganese cluster and proton release at each of the S states. These results have led to modifications of the hydrogen-abstraction model, which are being subjected to ongoing experimental testing.

REFERENCES AND FURTHER READING

Allen, J.P., Feher, G., Yeates, T.O., Komiya, H., and Rees, D.C. (1987) Structure of the reaction center from *Rhodobacter sphaeroides* R-26: I. The cofactors. *Proceedings of the National Academy of Sciences USA* **84**, 5730–4.

Blankenship, R.E. (2002) *Molecular Mechanisms of Photosynthesis*. Blackwell Publishing, Oxford.

Blankenship, R.E. and Hartman, H. (1998) The origin and evolution of oxygenic photosynthesis. *Trends in Biochemcal Sciences* **23**, 94–7.

Blankenship, R.E., Madigan, M.T., and Bauer, C.E. (eds) (1995) *Anoxygenic Photosynthetic Bacteria*. Kluwer Academic Publishers, Dordrecht, The Netherlands.

Britt, R.D. (1996) Oxygen evolution. In *Oxygenic Photosynthesis: the Light Reactions* (Ort, D.R. and Yocum, C.F., eds). Kluwer Academic Publishers, Dordrecht, The Netherlands, pp. 137–64.

Britt, R.D., Peloquin, J.M., and Campbell, K.A. (2003) Pulsed and parallel-polarization EPR characterization of the photosystem II oxygen-evolving complex. *Annual Review of Biophysics and Biomolecular Structure* **29**, 463–95.

Debus, R.J. (2001) Amino acid residues that modulate the properties of tyrosine Y_Z and the manganese cluster in the water oxidizing complex of photosystem II. *Biochimica Biophysica Acta* **1503**, 164–86.

Deisenhofer, J., Epp, O., Miki, K., Huber, R., and Michel, H. (1985) Structure of the protein subunits in the photosynthetic reaction centre of *Rhodopseudomonas viridis* at 3 Å resolution. *Nature* **318**, 618–24.

Feher, G. (1998) Three decades of research in bacterial photosynthesis and the road leading to it. A personal account. *Photosynthesis Research* **55**, 1–40.

Ferreira, K.N., Iverson, T.M., Maghlaoui, K., Barber, J., and Iwata, S. (2004) Architecture of the photosynthetic oxygen-evolving center. *Science* **303**, 1831–8.

Fleming, G.R. and van Grondelle, R. (1997) Femtosecond spectroscopy of photosynthetic light-harvesting systems. *Current Opinion in Structural Biology* **7**, 738–48.

Jordan, P., Fromme, P., Witt, H.T. et al. (2001) Three-dimensional structure of cyanobacterial photosystem I at 2.5 Å resolution. *Nature* **411**, 909–17.

Kurisu, G., Zhang, H., Smith, J.L., and Cramer, W.A. (2003) Structure of the cytochrome b_6f complex of oxygenic photosynthesis: tuning the cavity. *Science* **302**, 1009–14.

McDermott, G., Prince, S.M., Freer, A.A. et al. (1995) Crystal structure of an integral membrane light-harvesting complex from photosynthetic bacteria. *Nature* **374**, 517–21.

Ort, D.R. and Yocum, C.F. (eds) (1996) *Oxygenic Photosynthesis: the Light Reactions*. Kluwer Academic Publishers, Dordrecht, The Netherlands.

Sauer, K. and Yachandra, V.K. (2002) A possible evolutionary origin for the Mn_4 cluster of the photosynthetic water oxidation complex from natural MnO_2 precipitates in the early ocean. *Proceedings of the National Academy of Sciences USA* **99**, 8631–6.

Stowell, M.H.B., McPhillips, T.M., Rees, D.C., Soltis, S.M., Abresch, E., and Feher, G. (1997) Light-induced structural changes in photosynthetic reaction center: implications for mechanism of electron–proton transfer. *Science* **276**, 812–16.

Tommos, C. and Babcock, G.T. (2000) Proton and hydrogen currents in photosynthetic water oxidation. *Biochimica Biophysica Acta* **1458**, 199–219.

Wydrzynski, T. and Satoh, K. (eds) (2005) *Photosystem II: the Light-driven Water: Plastoquinone Oxidoreductase*. Kluwer Academic Publishers, Dordrecht, The Netherlands.

Yano, J., Kern, J., Saer, K. et al. (2006) Where water is oxidized to dioxygen: structure of the photosynthetic Mn_4Ca cluster. *Science* **314**, 821–5.

Zouni, A., Witt, H.T., Kern, J. et al. (2001) Crystal structure of photosystem II from *Synechococcus elongatus* at 3.8 Å resolution. *Nature* **409**, 739–43.

PROBLEMS

20.1 Plants make chlorophyll whereas bacteria synthesize bacteriochlorophyll. What are the differences between these molecules that give rise to the difference in the optical spectra?

20.2 Based upon the optical spectra, how much light energy is absorbed by the primary electron donor of reaction centers from purple bacteria compared to P680 of photosystem II?

20.3 Contrast the organization of the proteins and cofactors in the light-harvesting 2 complex compared to the bacterial reaction center.

20.4 Why do the light-harvesting 1 and 2 complexes have different optical spectra?

20.5 Discuss the advantages and disadvantages of having two different light-harvesting complexes in the cell.

20.6 Why is the EPR linewidth more narrow for a dimer than for a monomer?

20.7 Provide a possible reason why electron transfer proceeds down only one branch of cofactors despite the symmetry of the structure.

20.8 How are the properties of the iron–sulfur clusters in photosystem I different from the quinones in reaction centers?

20.9 Why does oxygen evolution show a cyclic pattern with a series of laser flashes?

20.10 Based upon the S cycle, what is the maximum rate of oxygen production possible in plants?

20.11 Why does P680 have a high oxidation/reduction midpoint potential? What type of reactions can P680 perform that P865 cannot?

20.12 Suggest a possible reason as to how P680 can achieve a significantly higher midpoint potential than that observed for chlorophyll in solution.

20.13 What role does tyrosine play in the water-oxidation process?

20.14 What gives rise to the many hyperfine lines observed for photosystem II?

20.15 Why do the S states have different EPR spectra?

20.16 How many EXAFS lines are expected for a Mn_4 cluster? Given the observed spectrum, estimate the number of possible configurations of the Mn_4 cluster that are consistent with the EXAFS data.

20.17 What does the hydrogen-abstraction model predict about the number of protons transferred during each step of the S cycle?

20.18 Propose a chemical model of the water-oxidation process that predicts the changes in the oxidation states of each Mn in the Mn_4 cluster during the S cycle.

20.19 In 1957, Barry Commoner reported the presence of an EPR signal from a plant leaf. What signal do you think that he saw?

Answers to problems

CHAPTER 1

1.1 $P = \frac{\text{Force}}{\text{Area}} = \frac{F}{A}$

$$P = \frac{mg}{A} = \frac{\rho Vg}{A} = \frac{(1\ \text{kg m}^{-3})(0.1\ \text{m})^3(9.81\ \text{m s}^{-2})}{(0.1\ \text{m})^2}$$

$$= 9.81 \times 10^{-1}\frac{\text{kg}}{\text{m s}^{-2}}$$

$$= 0.981\ \text{Pa}$$

$$P = \frac{mg}{A} = \frac{\rho Vg}{A} = \frac{(1\ \text{kg m}^{-3})(0.01\ \text{m})^3(9.81\ \text{m s}^{-2})}{(0.01\ \text{m})^2}$$

$$= 9.81 \times 10^{-2}\frac{\text{kg}}{\text{m s}^{-2}}$$

$$= 0.0981\ \text{Pa}$$

1.2 $P_{initial} = P_{final}(V_{final}/V_{initial}) = (6\ \text{atm})(2\ \text{L}/4\ \text{L}) = 3\ \text{atm}$

1.3 (a) $202.65\ \text{Pa}\frac{1\ \text{atm}}{101{,}325\ \text{Pa}} = 0.002\ \text{atm}$

(b) $202.6\ \text{Pa}\frac{1.01325\ \text{bar}}{101{,}325\ \text{Pa}} = 0.0020265\ \text{bar}$

1.4 Temperature (K) = temperature (°C) + 273.15°

(a) 0°C = 273.15 K
(b) −270°C = 3.15 K
(c) +100°C = 373.15 K

1.5 The absolute temperature is a measure of the average kinetic energy of the molecules.

1.6 $P = \dfrac{nRT}{V - nb} - a\left(\dfrac{n}{V}\right)^2$

$$= \frac{(1\text{ mol})(8.3\text{ J}/(\text{K mol}))(298\text{ K})}{25\text{ L} - (0.043\text{ L mol}^{-1})(1\text{ mol})} - 3.60\text{ L}^2\text{ atm}^{-1}\text{ mol}^{-2}\left(\frac{1\text{ mol}}{25\text{ L}}\right)^2$$

$$P = \frac{2448.5\text{ J}}{24.96 \times 10^{-3}\text{ m}^3} \times \frac{1\text{ atm}}{1.01 \times 10^5\text{ Pa}} - 0.00576\text{ atm}$$

$$= 0.9711\text{ atm} - 0.00576\text{ atm} = 0.965\text{ atm}$$

1.7 $P = \dfrac{nRT}{V} = \dfrac{(0.1\text{ mol})(8.31\text{ J}/(\text{K mol}))(298\text{ K})}{10 \times 10^{-3}\text{ m}^3} = 24.8\text{ kPa}$

1.8 $n = \dfrac{PV}{RT} = \dfrac{(1\text{ atm})(2\text{ L})}{(8.21 \times 10^{-2}\text{ L atm}/(\text{K mol}))(283\text{ K})} = 0.086\text{ mol}$

1.9 $\dfrac{V_{initial}}{T_{initial}} = \dfrac{V_{final}}{T_{final}} \rightarrow T_{final} = T_{initial}\left(\dfrac{V_{final}}{V_{initial}}\right) = (298\text{ K})\left(\dfrac{0.9\text{ L}}{1.0\text{ L}}\right) = 283\text{ K}$

1.10 $n(\text{Nitrogen}) = 0.755\dfrac{100\text{ g}}{28.0\text{ g mol}^{-1}} = 2.69\text{ mol}$

$$n(\text{Oxygen}) = 0.232\frac{100\text{ g}}{32.0\text{ g mol}^{-1}} = 0.725\text{ mol}$$

$$n(\text{Argon}) = 0.013\frac{100\text{ g}}{39.9\text{ g mol}^{-1}} = 0.033\text{ mol}$$

Since the total is 3.45 mol, and the total pressure is 1 atm, the partial pressures are given by:

$$p(\text{Nitrogen}) = \frac{2.69\text{ mol}}{3.45\text{ mol}}1\text{ atm} = 0.78\text{ atm}$$

$$p(\text{Oxygen}) = \frac{0.725\text{ mol}}{3.45\text{ mol}}1\text{ atm} = 0.21\text{ atm}$$

$$p(\text{Argon}) = \frac{0.033\text{ mol}}{3.45\text{ mol}}1\text{ atm} = 0.01\text{ atm}$$

1.11 Cell membranes are largely lipids but they also have steroids, proteins, and sugars.

1.12 The lipids are arranged in a bilayer, with the hydrophobic chains in the interior and the polar groups on the exterior.

1.13 The fluid-mosaic model states that the components of a membrane are free to move within the membrane.

1.14 Amino acids have the fundamental unit of a carbonyl group and an amino group bonded to a carbon atom.

1.15 There are 20 different amino acid side chains found in proteins and these differ in structure, size, and charge.

1.16 Each atom of an amino acid is identified uniquely by the atoms of the side chain being designated by the Greek letters β, γ, δ, and ε, proceeding from the α carbon (the oxygen, nitrogen, and other carbon do not have letters assigned).

1.17 Only L-enantiomers of amino acids are found in proteins.

1.18 (a) Arginine, lysine, and histidine. (b) Aspartic acid and glutamic acid.

1.19 Proline.

1.20 Tyrosine, tryptophan, and phenylalanine.

1.21 This residue can make a disulfide bridge with another cysteine residue.

1.22 In DNA the sugar is deoxyribose and the bases are the purines adenine (A) and guanine (G), and the pyrimidines thymine (T) and cytosine (C).

1.23 DNA has two chains running in opposite directions down a central axis forming a double helix with a diameter of 20 Å.

1.24 Thymine is always paired with adenine as both of these bases can form two hydrogen bonds. Cytosine is always paired with guanine, forming three hydrogen bonds.

1.25 The backbone of RNA is formed by phosphodiester linkages as for DNA, but the connections are to riboses, not deoxyriboses.

CHAPTER 2

2.1 The internal energy will increase more if the system is at constant volume because, at constant pressure, some of the energy input as heat goes into work done on the surroundings instead of all going into the kinetic energy of the gas molecules.

2.2 Since the expansion is isothermal and of an ideal gas, the change in internal energy is zero. This means $q = -w$ and for a compression, w is positive. Therefore q must be negative.

2.3 The change in internal energy of any closed adiabatic system at constant volume is zero, assuming that only pressure/volume work is possible.

2.4 The internal energy is a constant.

2.5 The ITC profile reflects the binding of the drug to the target.

2.6 A favorable drug candidate would have an ITC profile that changes at low concentrations of the drug, reflecting a tight binding to the target of interest.

2.7 Since the expansion is adiabatic, $q = 0$. Since the external pressure is constant, $w = -P\Delta V = -(1\text{ atm})(6\text{ L}) = -6\text{ atmL} = -606\text{ J}$. The change in the internal energy is the sum of the heat and work, or -606 J.

2.8 $$w = -nRT \ln \frac{V_{final}}{V_{initial}} = -(0.5\text{ mol})(8.314\text{ J/(Kmol)})(303\text{ K}) \ln \frac{20\text{ L}}{10\text{ L}}$$
$$= -873.6\text{ J}$$

2.9 (a) The number of moles of carbon dioxide is $5\text{ g}/(44\text{ g mol}^{-1}) = 0.114$ mol. The maximum pressure that can develop is:

$$P = \frac{nRT}{V} = \frac{(0.114\text{ mol})\left(0.08206\frac{\text{Latm}}{\text{Kmol}}\right)(298\text{ K})}{0.03\text{ L}} = 92.63\text{ atm}$$

(b) The final volume of the expanded gas is:

$$V_{final} = V_{initial}\frac{P_{initial}}{P_{final}} = (0.03\text{ L})\frac{92.63\text{ atm}}{1\text{ atm}} = 2.78\text{ L}$$

$$w = -P\Delta V = -(2.75\text{ atm})(2.78\text{ L} - 0.03\text{ L}) = -278\text{ J}$$

2.10 Number of moles, $n = \frac{PV}{RT} = \frac{(1\text{ atm})(1\text{ L})}{\left(0.08206\frac{\text{Latm}}{\text{Kmol}}\right)(298\text{ K})} = 0.0409\text{ mol}$

$$T_{final} = T_{initial}\frac{P_{final}}{P_{initial}} = 298\text{ K}\frac{2\text{ atm}}{1\text{ atm}} = 596\text{ K}$$

$$q = nC_V\Delta T = (0.0409\text{ mol})(25\text{ J/(Kmol)})(596\text{ K} - 298\text{ K}) = 305\text{ kJ}$$

The change in the internal energy is equal to the heat transferred since no expansion work was performed.

2.11 The change in enthalpy will be 1.0 J.

2.12 The internal energy will not change.

2.13 $$w = -nRT \ln \frac{V_{final}}{V_{initial}} = -(1\text{ mol})(8.314\text{ J/(Kmol)})(298\text{ K}) \ln \frac{4\text{ L}}{2\text{ L}} = -5.76\text{ kJ}$$

For an ideal gas, the internal energy depends only upon the temperature so the change in internal energy is zero. The change in internal energy is the sum of the heat and work so the work is +5.76 kJ.

2.14 $q = nC_V\Delta T = (1.5\text{ mol})(25\text{ J/(Kmol)})(35\text{ K}) = 1.31\text{ J}$

Since there is no work performed, the heat equals the change in internal energy.

2.15 $w = -nRT\ln\frac{V_{final}}{V_{initial}} = -(1\text{ mol})(8.314\text{ J/(Kmol)})(293\text{ K})\ln\frac{30\text{ L}}{10\text{ L}} = -5.36\text{ kJ}$

For an isothermal ideal gas, the change in internal energy is zero when the temperature is fixed. Since the change in internal energy is the sum of the heat and work, the heat is equal and opposite to the work.

2.16 $C_{12}H_{12}O_{11}$ (solid) + $12O_2$ (gas) $\rightarrow$ $12CO_2$ (gas) + $11H_2O$ (liquid)

$\Delta H = 12(-393.5\text{ kJ mol}^{-1}) + 11(-285.8\text{ kJ mol}^{-1}) - 1(-2222\text{ kJ mol}^{-1})$
$= -5644\text{ kJ mol}^{-1}$

2.17 The temperature of the Earth is not uniform and so the overall temperature increase represents a sum between decreasing and increasing contributions.

2.18 The amount of carbon dioxide in the atmosphere represents a balance among the many different natural processes. With the introduction of carbon dioxide by human activity, the balance is no longer maintained and the amount in the atmosphere has increased significantly.

2.19 $T^4 = (\text{constant})\frac{SR^2(1-A)}{d_{es}^2} \rightarrow T_2^4 = \frac{d_1^2}{d_2^2}T_1^4 = \frac{d_1^2}{(0.95d_1)^2}T_1^4 = 1.11T_1^4$

or $T_2 = 1.03T$ or 3%

2.20 $T^4 = (\text{constant})\frac{SR^2(1-A)}{d_{es}^2} \rightarrow \frac{T_1^4}{R_1^2} = \frac{T_2^4}{R_2^2}$

Setting $T_2 = 1.1T_1$ yields a 20% increase in the amount of solar radiation.

$$R_2^2 = \frac{T_2^4}{T_1^4}R_1^2 = \frac{(1.1T_1)^4}{T_1^4}R_1^2 = 1.46R_1^2 \rightarrow R_2 = 1.21R_1$$

CHAPTER 3

3.1 The entropy change of the system will always increase.

3.2 (a) $\Delta S = \frac{q}{T} = \frac{2.5\times10^4\text{ J}}{(273+20)\text{K}} = 85.3\text{ J K}^{-1}$

(b) $\Delta S = \frac{q}{T} = \frac{2.5\times10^4\text{ J}}{(273+200)\text{K}} = 52.8\text{ J K}^{-1}$

3.3 For a spontaneous process, the change in the Gibbs energy will be negative.

3.4 The process is endothermic.

3.5 $\Delta S = nR \ln \frac{V_{final}}{V_{initial}} = nR \ln 0.5 = -0.69nR$

3.6 $\Delta S = nR \ln \frac{V_{final}}{V_{initial}} = 0$

3.7 (a) $\text{Efficiency} = 1 - \frac{T_{cold}}{T_{hot}} = 1 - \frac{293\text{ K}}{373\text{ K}} = 0.21$

(b) $\text{Efficiency} = 1 - \frac{T_{cold}}{T_{hot}} = 1 - \frac{333\text{ K}}{373\text{ K}} = 0.11$

3.8 The entropy change is 30 J K^{-1} and the enthalpy change is −85 J.

3.9 The entropy change of the cold block is opposite in sign and larger in magnitude than the entropy change of the hot block.

3.10 The change in the Gibbs energy is negative.

3.11 The change in entropy will always be greater than the ratio of the heat divided by the temperature.

3.12 The total entropy change is zero.

3.13 The change in internal energy is zero since the process is adiabatic and no work is performed. The process of burning is spontaneous so the change in entropy is positive. Since the system is adiabatic the change in entropy for the surroundings is zero.

3.14 $C_{12}H_{12}O_{11}$ (solid) + $12O_2$ (gas) → $12CO_2$ (gas) + $11H_2O$ (liquid)

$\Delta H = 12(-393.5\text{ kJ mol}^{-1}) + 11(-285.8\text{ kJ mol}^{-1}) - 1(-2222\text{ kJ mol}^{-1})$
$= -5644\text{ kJ mol}^{-1}$

3.15 $\Delta S = \frac{\Delta H}{T} = \frac{5 \times 10^5\text{ J mol}^{-1}}{(273.15 + 80)\text{K}} = 1.42\text{ J/(Kmol)}$

3.16 Hybrid cars make use of regenerative braking to charge batteries, use electric motors when the gasoline engine is inefficient, and turn off of the gasoline engine when idle.

3.17 Drugs that are selected based upon enthalpic optimization are usually more selective with a higher binding affinity than that obtained using entropic optimization.

3.18 Although the reaction is favorable, it is a complex multi-electron process that involves a large number of intermediate steps.

3.19 The metal surface favors a stressed conformation for the dinitrogen molecule.

3.20 Nitrogenase is the enzyme that catalyzes the formation of ammonia from molecular nitrogen.

CHAPTER 4

4.1 Yes.

4.2 No.

4.3 The presence of the salt stablilizes the water phase, and hence decreases the freezing temperature, as the enhanced molecular randomness opposes the tendency to freeze.

4.4 Phospholipids have two or three hydrocarbon chains attached to a polar head group.

4.5 The separation of the molecules into phases. When $B > B_{ideal}$ the chemical potential decreases more rapidly than the ideal case and the system behaves as a good solvent. When $B \ll 0$ the chemical potential decreases less rapidly and the system behaves as a poor solvent and phase separation occurs. For lipids and detergents, this phase separation can be in the form of the micelles, bilayers, or other aggregates.

4.6 In a micelle, the hydrophobic chains are in the interior and the polar head groups are exposed to the water. A liposome is an aggregate of lipids or detergent molecules in which the polar surfaces are exposed to the solvent and the hydrophobic chains are buried in a bilayer arrangement.

4.7 The detergent concentration at which micelles will form.

4.8 The tendency for forming micelles would be increased and so the critical micelle concentration would decrease.

4.9 The cubic lipid phase can be thought of as arising from a very high concentration of liposomes that aggregate into the cubic lipid phase.

4.10 Lipid rafts are regions of cell membranes that are rich in certain lipids and proteins. They have been proposed to play a role in cellular functions.

4.11 The rafts are small and the possibility that the preparation of the fractionation experiments led to the fractionation cannot be excluded.

4.12 The function of these proteins may be triggered by the formation of dimers in the raft.

4.13 $h = \frac{2\gamma}{\rho g r} = \frac{2(72.75\ \text{mN m}^{-1})}{(997.1\ \text{kg m}^{-3})(9.81\ \text{m s}^{-2})(2 \times 10^{-3}\ \text{m})} = 7.3\ \text{mm}$

4.14 The ratio of γ/ρ must be twice as large for the red liquid.

4.15 The detergent starts forming micelles and so the addition of detergent no longer disrupts the interaction of water molecules.

4.16 The enthalpy change is zero.

4.17 The Gibbs energy of mixing has a minimum when both the reactant and product are present.

4.18 The enthalpy of mixing is positive.

4.19 The entropy change is negative.

4.20 The change in buffer concentration is given by the ratio of solutions:

(a) Final buffer concentration

$$= \text{initial buffer concentration} \frac{\text{sample volume}}{\text{reservoir volume}}$$

$$= 0.1\ \text{M} \frac{10\ \text{ml}}{100\ \text{ml}} = 0.01\ \text{M}$$

(b) Final buffer concentration

$$= \text{initial buffer concentration} \frac{\text{sample volume}}{\text{reservoir volume}}$$

$$= 0.1\ \text{M} \frac{10\ \text{ml}}{1000\ \text{ml}} = 0.001\ \text{M}$$

4.21 Initially, the formation of a large aggregate is unfavorable, but once a certain size is created then the attachment of additional proteins is favorable.

4.22 A change in only the protein concentration corresponds to a vertical line.

4.23 The total entropy change is positive.

CHAPTER 5

5.1 The energy is at a minimum.

5.2 The change in the Gibbs energy will be negative for a spontaneous process.

5.3 The energy is at a minimum.

5.4 There is a relatively large entropic component compared to the enthalpic component.

5.5 There is a relatively small entropic component compared to the enthalpic component.

5.6 CH_3COO^- is acting as a base, H_2O is acting as an acid, CH_3COOH is acting as an acid, and OH^- is acting as a base.

5.7 (a) $pH = -\log(10^{-5}) = 5$
(b) $pH = -\log(10^{-9}) = 9$

5.8 The standard reaction Gibbs energy is greater than zero and equal to $-RT \ln K$.

5.9 At equilibrium the Gibbs energy of a reaction is always zero.

5.10 The total volume is 1 L. The 0.01 mol of NaOH will convert 0.01 mol of acetic acid to acetate, yielding final concentrations of 0.02 and 0.01 M for the acid and base forms, respectively. The pH is then equal to:

$$pH = pK_A + \log\frac{\text{base}}{\text{acid}} = 4.75 + \log\frac{0.1\ \text{M}}{0.2\ \text{M}} = 4.45$$

5.11 After adding the 0.03 mol of NaOH to 0.1 mol of acetic acid the solution has 0.07 mol of acetic acid and 0.03 mol of acetate ions. Since the total volume is 1 L, the molarities are 0.03 and 0.07 for the acetic acid and acetate ions, respectively. The pH is then given by the Henderson–Hasselback equation:

$$pH = pK_A + \log\frac{\text{base}}{\text{acid}} = 4.75 + \log\frac{0.03}{0.07} = 4.38$$

5.12 Adding 0.01 M of HCl to distilled water poises the H^+ concentration at 0.01 or 10^{-2} M (compared to the initial concentration of 10^{-7} M). The pH of this solution is $\log(10^{-2})$ or 2.

5.13 Glycine is an amino acid with the neutral structure: NH_2CII_2COOH. The amine group on one side of the molecule is a weak base and can be protonated to form a quaternary amine cation R-NH_3^+ and the carboxyl group on the other side can deprotonate to form the carboxylic anion R-COO^-. The pK_A of the amine group is about 9.8 (this is for dissociation of the conjugate base; the protonated amine). The pK_A for the carboxylic acid is about 4.5. The isoelectric pH is just the average of the two pK_A values.

5.14 The initial pH value is approximately 14.

5.15 Tris would have equal molar concentrations of Tris and its conjugate base at a pH equal to the pK_A, or 8.3, which is the optimal pH for buffering action.

5.16 The coupling allows the overall process to proceed in a neutral fashion.

5.17 The formation is determined by the proton-transfer step rather than the electron-transfer step.

5.18 Tyrosine can accept one proton on its nitrogen atom and can donate two protons, one from the carboxyl group and the other from the hydroxyl group. With increasing pH tyrosine changes from H_3Tyr^+, to H_2Tyr, to $HTyr^-$, and then to Tyr^{2-}.

5.19 The reduced tyrosine in very energetically unfavorable compared to the neutral tyrosyl.

5.20 Protons can travel long distances by following proton pathways involving hydrogen-bonding networks.

5.21 Recent data suggest that protons can travel through tunneling mechanisms that are normally thought to be restricted to electrons.

CHAPTER 6

6.1 Oxidation occurs at the positive electrode, the anode.

6.2 The overall reaction is the sum of the two half reactions:

$Zn\ (solid) \leftrightarrow Zn^{2+} + 2e^-$ $\qquad E_m = +0.76$ V
$Cu^{2+} + 2e^- \leftrightarrow Cu\ (solid)$ $\qquad E_m = +0.34$ V
$Zn\ (solid) + Cu^{2+} \leftrightarrow Cu\ (solid) + Zn^{2+}$ $\qquad E_m = +1.1$ V

6.3 The higher the midpoint potential, the better an oxidant the protein is.

6.4 The energy to make one ATP is 2.1×10^{-19} J. A single 700-nm photon has enough energy.

6.5 At the midpoint potential, the concentrations of the oxidized and reduced species are equal.

6.6
$$\log\frac{[\mathrm{ATP}]}{[\mathrm{ADP}][\mathrm{P_i}]} = -\frac{\Delta G^\circ}{2.3\mathrm{RT}} - n\,\Delta\mathrm{pH}$$
$$= -\frac{(8000\ \mathrm{cal\ mol^{-1}})}{2.3(2.0\ \mathrm{cal/(Kmol)})(298\ \mathrm{K})} - 4(2.5)$$
$$= -(5.9 - 10) = +4.1$$
$$\log\frac{[\mathrm{ATP}]}{[\mathrm{ADP}](0.01)} = \log\frac{[\mathrm{ATP}]}{[\mathrm{ADP}]} + 2.0 = +4.1$$
$$\rightarrow \log\frac{[\mathrm{ATP}]}{[\mathrm{ADP}]} = 2.1 \text{ or } \frac{[\mathrm{ATP}]}{[\mathrm{ADP}]} = 126$$

6.7 (a)
$$E = E^0 - \frac{RT}{nF}\ln\frac{[A_{oxidized}]}{[A_{reduced}]} = 125\ \mathrm{mV} + \frac{59.2\ \mathrm{mV}}{\ln 10}\ln(10) = 184.2\ \mathrm{mV}$$

(b)
$$E = E^0 - \frac{RT}{nF}\ln\frac{[A_{oxidized}]}{[A_{reduced}]} = 125\ \mathrm{mV} + \frac{59.6\ \mathrm{mV}}{\ln 10}\ln(1) = 125\ \mathrm{mV}$$

(c)
$$E = E^0 - \frac{RT}{nF}\ln\frac{[A_{oxidized}]}{[A_{reduced}]} = 125\ \mathrm{mV} + \frac{59.6\ \mathrm{mV}}{\ln 10}\ln(0.1) = 65.4\ \mathrm{mV}$$

6.8 A fit of the data yields a midpoint potential of 250 mV. The protein is cytochrome *c*.

6.9 $$\Delta G = \Delta G^\circ + RT \ln \frac{[\text{ADP}][\text{P}_i]}{[\text{ATP}]}$$

$$= -30{,}500\ \text{J}/(\text{Kmol}) + 8.315\ \text{J}/(\text{Kmol})(298\ \text{K}) \ln \frac{(0.00025)(0.00175)}{(0.00225)}$$

$$= -51.8\ \text{kJ mol}^{-1}.$$

6.10 The change in the Gibbs energy is the sum of the individual changes: $-61.0 + 30.5 = -31.4$ kJ mol^{-1}.

6.11 The change is the sum of the two values, or -64.8 kJ mol^{-1}.

6.12 The midpoint changes from 450 to 210 mV with an apparent pK_A of 5.

6.13 P865 of the bacterial reaction center has a higher midpoint potential than the heme and so an electron can be transferred from the cytochrome to the reaction center.

6.14 Photosystem II has a higher midpoint potential so electron transfer from the cytochrome to the photosystem II is energetically favorable.

6.15 $$I = \frac{1}{2} m_{\text{Na}^+} z_{\text{Na}^+}^2 + \frac{1}{2} m_{\text{Cl}^-} z_{\text{Cl}^-}^2 = \frac{1}{2}(0.5)(1)^2 + \frac{1}{2}(0.5)(-1)^2 = 0.5$$

6.16 $$I = \frac{1}{2} m_{\text{Mg}^{2+}} z_{\text{Mg}^{2+}}^2 + \frac{1}{2} m_{\text{Cl}^-} z_{\text{Cl}^-}^2 = \frac{1}{2}(0.5)(2)^2 + \frac{1}{2}(1.0)(-1)^2 = 1.5$$

6.17 Since the motion occurs during the series of electron-transfer steps that occur in the protein, the change in the distance of the Fe_2S_2 cluster to the other cofactors changes and subsequently the rate of electron-transfer changes (Chapter 10).

6.18 The observation of the rotation in three discrete steps would provide a very consistent model with the biochemical data.

CHAPTER 7

7.1 M s^{-1}

7.2 Nothing can be said about the individual rates, but the rate constant for the forward reaction is 10 times bigger than that of the reverse reaction.

7.3 There will be no change.

7.4 Four half-lives are required, 600 s.

7.5 $t_{1/2} = 0.69/k = 150$ s; four half-lives are required, 600 s.

7.6 They are both equal to 0.05 M.

7.7 [B]/[C] = 10 and [B] + [C] = 0.1 M

10[C] + [C] = 11[C] = 0.1 M → [C] = 0.009 M and [B] = 0.09

7.8 (a) $\frac{d[A]}{dt} = -k_{f1}[A][B] \quad \frac{d[B]}{dt} = -k_{f1}[A][B]$

$$\frac{d[C]}{dt} = k_{f1}[A][B] - k_{f2}[C] + k_{b2}[D] \quad \frac{d[D]}{dt} = k_{f2}[C] - k_{b2}[D]$$

(b) $\frac{d[A]}{dt} = -(2.0\ M^{-1}\ s^{-1})(0.1\ M)(0.1\ M) = -0.02\ Ms^{-1}$

$$\frac{d[B]}{dt} = -(2.0\ M^{-1}\ s^{-1})(0.1\ M)(0.1\ M) = -0.02\ Ms^{-1}$$

$$\frac{d[C]}{dt} = (2.0\ M^{-1}\ s^{-1})(0.1\ M)(0.1\ M) - (1.0\ s^{-1})(0.1\ M) + (5.0\ s^{-1})(0.1\ M) = 0.42\ Ms^{-1}$$

$$\frac{d[D]}{dt} = (1.0\ s^{-1})(0.1\ M) - (5.0\ s^{-1})(0.1\ M) = -0.40\ Ms^{-1}$$

(c) At equilibrium the concentrations of A and B are zero since the reactions are irreversible. The sum of the concentrations must be equal to 0.4 M and the ratio of the amounts of C and D is given by the ratio of the two rate constants:

$$K = \frac{[D]}{[C]} = \frac{k_{f2}}{k_{b2}} = 0.2 \rightarrow [D] = 0.2[C] \quad \text{and} \quad [C] + [D] = 0.4\ M$$

$$[C] + [D] = [C] + 0.2[C] = 0.4\ M \rightarrow [C] = 1/3\ M$$

$$[D] = 0.2[C] = 1/15\ M$$

7.9 $A = A_0 e^{-kt} \rightarrow 0.5\ M = e^{-k(10s)}$ or $k = -\frac{\ln 0.5}{10\ s} = 0.0693\ s^{-1}$

7.10 (a) $\frac{d[A]}{dt} = -k_f[A] \quad \frac{d[B]}{dt} = +k_f[A]$

(b) $\frac{d[A]}{dt} = -k_f[A] = -(1\ s^{-1})(0.1\ M) = -0.1\ Ms^{-1}$

$$\frac{d[B]}{dt} = +k_f[A] = (1\ s^{-1})(0.1\ M) = 0.1\ Ms^{-1}$$

(c) Since the reaction is irreversible, the final concentration of A is zero and that of B is 0.2 M.

7.11 (a) $\frac{d[A]}{dt} = -k_f[A] + k_b[B]$ $\frac{d[B]}{dt} = +k_f[A] - k_b[B]$

(b) $\frac{d[A]}{dt} = -(1\,s^{-1})(0.1\,M) + (1\,s^{-1})(0.1\,M) = 0$

$\frac{d[B]}{dt} = +(1\,s^{-1})(0.1\,M) - (1\,s^{-1})(0.1\,M) = 0$

(c) $\frac{[A]}{[B]} = \frac{1\,s^{-1}}{1\,s^{-1}} = 1$ so $[A] = [B] = 0.1\,M$

7.12 (a) $\frac{d[A]}{dt} = -k_{f1}[A]$ $\frac{d[B]}{dt} = +k_{f1}[A] - k_{f2}[B]$ $\frac{d[C]}{dt} = -k_{f2}[B]$

(b) $\frac{d[A]}{dt} = -k_{f1}[A] = -(0.1\,s^{-1})(0.1\,M) = -0.1\,Ms^{-1}$

$\frac{d[B]}{dt} = +k_{f1}[A] - k_{f2}[B] = +(0.1\,s^{-1})(0.1\,M) - (0.1\,s^{-1})(0.1\,M) = 0$

$\frac{d[C]}{dt} = -k_{f2}[B] = (0.1\,s^{-1})(0.1\,M) = 0.1\,Ms^{-1}$

(c) Since the reactions are irreversible, the concentrations of A and B are zero and that of C is 0.3 M.

7.13 $\ln k_2 - \ln k_1 = -\frac{E_A}{R}\left(\frac{1}{T_2} - \frac{1}{T_1}\right) \rightarrow E_A = -\frac{R(\ln k_2 - \ln k_1)}{1/T_2 - 1/T_1}$

$$E_A = -\frac{R(\ln k_2 - \ln k_1)}{1/T_2 - 1/T_1} = \frac{0.008314\text{ kJ/(Kmol)}(\ln 20 - \ln 10)}{(1/330\text{ K}) - (1/298\text{ K})}$$

$= 17.62$ kJ mol^{-1}

7.14 $\ln k_2 - \ln k_1 = -\frac{E_A}{R}\left(\frac{1}{T_2} - \frac{1}{T_1}\right) \rightarrow E_A = -\frac{R(\ln k_2 - \ln k_1)}{1/T_2 - 1/T_1}$

$$E_A = -\frac{R(\ln k_2 - \ln k_1)}{1/T_2 - 1/T_1} = \frac{0.008314\text{ kJ/(Kmol)}(-10.41 + 11.87)}{(1/293\text{ K}) - (1/303\text{ K})}$$

$= 107.8$ kJ mol^{-1}

7.15 $\ln k_2 - \ln k_1 = -\frac{E_A}{R}\left(\frac{1}{T_2} - \frac{1}{T_1}\right) \rightarrow \ln k_2 - 6.91$

$$= -\frac{5\text{ kJ mol}^{-1}}{0.008314\text{ kJ/(Kmol)}}\left(\frac{1}{277\text{ K}} - \frac{1}{295\text{ K}}\right) \rightarrow k_2 = 878\text{ s}^{-1}$$

7.16 $\ln k_2 - \ln k_1 = -\frac{1}{RT}(E_{A1} - E_{A2}) \rightarrow (E_{A1} - E_{A2}) = RT(\ln k_2 - \ln k_1)$

$$\begin{aligned}(E_{A1} - E_{A2}) &= RT(\ln k_2 - \ln k_1)\\ &= 0.008314 \text{ kJ/(Kmol)}(298 \text{ K})(\ln 1 - \ln 1000)\\ &= -17.11 \text{ kJ mol}^{-1}\end{aligned}$$

7.17 It is a competitive inhibitor.

7.18 It is an uncompetitive or mixed inhibitor.

7.19 The initial velocity is reaching a value of 280 μM min^{-1} at high substrate concentrations so this is assigned as the maximum velocity. The half velocity would be 140 μM min^{-1}, which occurs at 1×10^{-5} M and is assigned as the K_m.

7.20 The temperature in the exponent will decrease, making the free-energy dependence much sharper than at room temperature. There is also a weak temperature dependence in the coefficient.

7.21 As the free-energy difference approaches the reorganization energy, the activation energy decreases until the process becomes activationless when the rate reaches a maximum. At larger free energies the difference between the free-energy difference and reorganization energy increases again and the rate decreases.

CHAPTER 8

8.1 Probability = number of spades/number of cards = 13/52 = 1/4

8.2 Probability = number of cards with a four/number of cards
= 4/52 = 1/13

8.3 There are 9 possible outcomes. (a) The probability of three heads is 1 of 9. (b) The probability of two heads and one tail is 3 of 9.

8.4 Number = (52)(51)(50) = 132,600

8.5 (a) Probability = $1/4^3$ = 1 out of 64; (b) probability = $1/4^3$ = 1 out of 64; (c) probability = $1/4^7$ = 1 out of 16,384; (d) probability = $1/4^{12}$ = 1 out of 16,777,216.

8.6 (a) Number of outcomes = 2^3 = 8; (b) number of energetically distinct states = 4; (c) the three up spins have the highest energy, followed in order by two up and one down, one up and two down, and three down.

8.7 (a) $e^{-(E_2-E_1)/k_BT} = e^{-(9\times10^{-24}\text{J}/(1.38\times10^{-23}\text{J K}^{-1})(298\text{ K})} = e^{-0.021} \approx 0.979$
(b) $e^{-(E_2-E_1)/k_BT} = e^{-(9\times10^{-24}\text{J}/(1.38\times10^{-23}\text{J K}^{-1})(4\text{ K})} = e^{-0.16} = 0.849$

8.8 (a) $e^{-(E_2-E_1)/k_BT} = e^{-(5\times10^{-27}\text{J}/(1.38\times10^{-23}\text{J K}^{-1})(298\text{ K})} = e^{-1.2\times10^{-6}} \approx 0.9999$
(b) $e^{-(E_2-E_1)/k_BT} = e^{-(5\times10^{-27}\text{J}/(1.38\times10^{-23}\text{J K}^{-1})(4\text{ K})} = e^{-9\times10^{-5}} \approx 0.999$

8.9 $q = \dfrac{1}{1-e^{-h\nu/k_BT}} \quad \dfrac{h\nu}{k_BT} = \dfrac{(6.626\times10^{-34}\text{ Js})(500\text{ cm}^{-1})(3.0\times10^{10}\text{ cm s}^{-1})}{(1.38\times10^{-23}\text{ J K}^{-1})(298\text{ K})}$
$= 2.4$

$q = \dfrac{1}{1-e^{-2.4}} = 1.10$

8.10 This simple equation represents the foundation for his scientific career.

8.11 (a) With only one configuration, ln 1 = 0 so $S = 0$; (b) with two configurations, ln 2 = 0.69 and S = $0.7k_B$; (c) with 10 configurations ln 10 = 2.3 and $S = 2.3k_B$.

8.12 The proteins can fold into intermediate states that lead to the final folded state and additional proteins can assist in the folding process.

8.13 A profile of the different energies for a protein in different configurations. The energy is minimal at the true, folded state although some proteins may have closely lying states.

8.14 In many cases, proteins immediately form elements of secondary structures, and the presence of the secondary structural elements contributes to folding of the overall protein into the proper state.

8.15 $\dfrac{N_{conformation1}}{N_{conformation2}} = e^{-\Delta E/RT} = e^{-(2\times10^3\text{ J mol}^{-1})/(8.31\text{ J/(Kmol)}(298\text{K})} = e^{-0.81} = 0.45$

8.16 Prion-based disease represented a new mechanism that was independent of other infectious diseases.

8.17 The formation of β sheets extended over multiple copies of prions provides a mechanism for aggregate formation.

8.18 Most treatments are based upon attacking invading agents such as bacteria. Such treatments are not effective towards stopping the misfolding of a protein.

CHAPTER 9

9.1 (a) $v = \dfrac{h}{m_e\lambda} = \dfrac{6.6\times10^{-34}\text{ Js}}{(9.1\times10^{-31}\text{ kg})(2.0\times10^{-10}\text{ m})} = 3.6\times10^6\text{ m s}^{-1}$

(b) $v = \dfrac{h}{m_e\lambda} = \dfrac{6.6\times10^{-34}\text{ Js}}{(9.1\times10^{-31}\text{ kg})(1.0\times10^{-10}\text{ m})} = 7.3\times10^6\text{ m s}^{-1}$

(c) $v = \dfrac{h}{m_e\lambda} = \dfrac{6.6\times10^{-34}\text{ Js}}{(9.1\times10^{-31}\text{ kg})(10.0\times10^{-10}\text{ m})} = 7.3\times10^5\text{ m s}^{-1}$

9.2 (a) $$v = \frac{h}{m_p\lambda} = \frac{6.6 \times 10^{-34}\ \text{Js}}{(1.7 \times 10^{-27}\ \text{kg})(2.0 \times 10^{-10}\ \text{m})} = 2.0 \times 10^{3}\ \text{m s}^{-1}$$

(b) $$v = \frac{h}{m_p\lambda} = \frac{6.6 \times 10^{-34}\ \text{Js}}{(1.7 \times 10^{-27}\ \text{kg})(1.0 \times 10^{-10}\ \text{m})} = 4.0 \times 10^{3}\ \text{m s}^{-1}$$

(c) $$v = \frac{h}{m_p\lambda} = \frac{6.6 \times 10^{-34}\ \text{Js}}{(1.7 \times 10^{-27}\ \text{kg})(1.0 \times 10^{-9}\ \text{m})} = 4.0 \times 10^{2}\ \text{m s}^{-1}$$

9.3 (a) $$\lambda = \frac{h}{p} = \frac{h}{mv} = \frac{6.6 \times 10^{-34}\ \text{Js}}{(9.1 \times 10^{-31}\ \text{kg})(2 \times 10^{5}\ \text{m s}^{-1})} = 3.6 \times 10^{-9}\ \text{m}$$

(b) $$\lambda = \frac{h}{p} = \frac{h}{mv} = \frac{6.6 \times 10^{-34}\ \text{Js}}{(9.1 \times 10^{-31}\ \text{kg})(1 \times 10^{7}\ \text{m s}^{-1})} = 7.25 \times 10^{-11}\ \text{m}$$

(c) $$\lambda = \frac{h}{p} = \frac{h}{mv} = \frac{6.6 \times 10^{-34}\ \text{Js}}{(9.1 \times 10^{-31}\ \text{kg})(1.6 \times 10^{8}\ \text{m s}^{-1})} = 4.5 \times 10^{-12}\ \text{m}$$

9.4 (a) $$\lambda = \frac{h}{p} = \frac{h}{mv} = \frac{6.6 \times 10^{-34}\ \text{Js}}{(1.7 \times 10^{-27}\ \text{kg})(2 \times 10^{5}\ \text{m s}^{-1})} = 1.9 \times 10^{-12}\ \text{m}$$

(b) $$\lambda = \frac{h}{p} = \frac{h}{mv} = \frac{6.6 \times 10^{-34}\ \text{Js}}{(1.7 \times 10^{-27}\ \text{kg})(1 \times 10^{7}\ \text{m s}^{-1})} = 3.9 \times 10^{-14}\ \text{m}$$

(c) $$\lambda = \frac{h}{p} = \frac{h}{mv} = \frac{6.6 \times 10^{-34}\ \text{Js}}{(1.7 \times 10^{-27}\ \text{kg})(1.5 \times 10^{8}\ \text{m s}^{-1})} = 2.5 \times 10^{-15}\ \text{m}$$

9.5 (a) $v = 10^{-3}\ \text{m h}^{-1} \times (1\ \text{h}/3600\ \text{s}) = 2.8 \times 10^{-7}\ \text{m s}^{-1}$

$$\lambda = \frac{h}{mv} = \frac{6.6 \times 10^{-34}\ \text{Js}}{(5 \times 10^{-3}\ \text{kg})(2.8 \times 10^{-7}\ \text{m s}^{-1})} = 4.7 \times 10^{-25}\ \text{m}$$

(b) $v = 10^{-6}\ \text{m h}^{-1} \times (1\ \text{h}/3600\ \text{s}) = 2.8 \times 10^{-10}\ \text{m s}^{-1}$

$$\lambda = \frac{h}{mv} = \frac{6.6 \times 10^{-34}\ \text{Js}}{(5 \times 10^{-3}\ \text{kg})(2.8 \times 10^{-10}\ \text{m s}^{-1})} = 4.7 \times 10^{-22}\ \text{m}$$

(c) $v = 10^{-1}\ \text{m h}^{-1} \times (1\ \text{h}/3600\ \text{s}) = 2.8 \times 10^{-5}\ \text{m s}^{-1}$

$$\lambda = \frac{h}{mv} = \frac{6.6 \times 10^{-34}\ \text{Js}}{(5 \times 10^{-3}\ \text{kg})(2.8 \times 10^{-5}\ \text{m s}^{-1})} = 4.7 \times 10^{-27}\ \text{m}$$

9.6 (a) $v = 10^{5}\ \text{m h}^{-1} \times (1\ \text{h}/3600\ \text{s}) = 28\ \text{m s}^{-1}$

$$\lambda = \frac{h}{mv} = \frac{6.6 \times 10^{-34}\ \text{Js}}{(5 \times 10^{-6}\ \text{kg})(28\ \text{m s}^{-1})} = 4.7 \times 10^{-30}\ \text{m}$$

(b) $v = 10^{5}\ \text{m h}^{-1} \times (1\ \text{h}/3600\ \text{s}) = 28\ \text{m s}^{-1}$

$$\lambda = \frac{h}{mv} = \frac{6.6 \times 10^{-34}\ \text{Js}}{(1 \times 10^{-3}\ \text{kg})(28\ \text{m s}^{-1})} = 2.4 \times 10^{-32}\ \text{m}$$

(c) $v = 10^5$ m h^{-1} × (1 h/3600 s) = 28 m s^{-1}

$$\lambda = \frac{h}{mv} = \frac{6.6 \times 10^{-34}\ \text{Js}}{(5\ \text{kg})(28\ \text{m s}^{-1})} = 4.7 \times 10^{-36}\ \text{m}$$

9.7 The classical theory predicts the long-wavelength dependence correctly. As the wavelength decreases, the energy density is predicted to always increase, in conflict with the data, which reaches a peak and then decreases to zero.

9.8 Quantum theory predicts that the energy of the oscillators will increase with decreasing wavelength and that, below a certain wavelength, the oscillators will not have enough thermal energy available to allow for the vibration (within the statistical probability).

9.9 Classical theory allows for atoms to have any energy values and there is no mechanism that predicts the emission of light at discrete energy values. The quantum theory predicts that the electrons occupy states with fixed energy and emit light only with values corresponding to the difference in energy between the final and initial states.

9.10

Experimental result	Classical or quantum effect?
(a) Light can diffract	Classical
(b) Electrons can diffract	Quantum
(c) For black-body radiation, the energy density is small at small wavelengths	Quantum
(d) For blackbody radiation, the energy density is small at small wavelengths	Classical
(e) Light has wavelength	Classical
(f) Electrons are in atomic orbitals	Quantum
(g) Electrons have mass	Classical
(h) Electrons have a wavelength	Quantum

9.11 Classical theory predicts that the kinetic energy is independent of the frequency, in conflict with the observed linear dependence.

9.12 Quantum theory predicts that the kinetic energy is linearly dependent on the frequency, in excellent agreement with the observed linear dependence.

9.13 (a) $$KE = \frac{hc}{\lambda} - \Phi = \frac{(6.62 \times 10^{-34}\ \text{Js})(2.99 \times 10^{8}\ \text{m s}^{-1})}{2 \times 10^{-7}\ \text{m}} \times \frac{1\ \text{eV}}{1.60 \times 10^{-19}\ \text{J}}$$

$$- 2.0\ \text{eV} = 4.18\ \text{eV} = 6.7 \times 10^{-19}\ \text{J}$$

(b) $KE = \frac{hc}{\lambda} - \Phi = \frac{(6.62 \times 10^{-34}\ \text{Js})(2.99 \times 10^{8}\ \text{m s}^{-1})}{2.5 \times 10^{-7}\ \text{m}} \times \frac{1\ \text{eV}}{1.60 \times 10^{-19}\ \text{J}}$

$- 2.0\ \text{eV} = 2.97\ \text{eV} = 4.75 \times 10^{-19}\ \text{J}$

(c) $KE = \frac{hc}{\lambda} - \Phi = \frac{(6.62 \times 10^{-34}\ \text{Js})(2.99 \times 10^{8}\ \text{m s}^{-1})}{3.5 \times 10^{-7}\ \text{m}} \times \frac{1\ \text{eV}}{1.60 \times 10^{-19}\ \text{J}}$

$- 2.0\ \text{eV} = 1.55\ \text{eV} = 2.47 \times 10^{-19}\ \text{J}$

9.14 (a) $KE = \frac{hc}{\lambda} - \Phi = \frac{(6.62 \times 10^{-34}\ \text{Js})(2.99 \times 10^{8}\ \text{m s}^{-1})}{2 \times 10^{-7}\ \text{m}} \times \frac{1\ \text{eV}}{1.60 \times 10^{-19}\ \text{J}}$

$- 2.3\ \text{eV} = 3.88\ \text{eV} = 6.2 \times 10^{-19}\ \text{J}$

(b) $KE = \frac{hc}{\lambda} - \Phi = \frac{(6.62 \times 10^{-34}\ \text{Js})(2.99 \times 10^{8}\ \text{m s}^{-1})}{2.5 \times 10^{-7}\ \text{m}} \times \frac{1\ \text{eV}}{1.60 \times 10^{-19}\ \text{J}}$

$- 2.3\ \text{eV} = 2.67\ \text{eV} = 4.27 \times 10^{-19}\ \text{J}$

(c) $KE = \frac{hc}{\lambda} - \Phi = \frac{(6.62 \times 10^{-34}\ \text{Js})(2.99 \times 10^{8}\ \text{m s}^{-1})}{3.5 \times 10^{-7}\ \text{m}} \times \frac{1\ \text{eV}}{1.60 \times 10^{-19}\ \text{J}}$

$- 2.3\ \text{eV} = 1.25\ \text{eV} = 2.0 \times 10^{-19}\ \text{J}$

9.15 (a) $\lambda = \frac{hc}{\Phi} = \frac{(6.62 \times 10^{-34}\ \text{Js})(2.99 \times 10^{8}\ \text{m s}^{-1})}{2.25\ \text{eV} \times (1.6 \times 10^{-19}\ \text{J/1 eV})} = 5.52 \times 10^{-7}\ \text{m}$

(b) $\lambda = \frac{hc}{\Phi} = \frac{(6.62 \times 10^{-34}\ \text{Js})(2.99 \times 10^{8}\ \text{m s}^{-1})}{2.0\ \text{eV} \times (1.6 \times 10^{-19}\ \text{J/1 eV})} = 6.21 \times 10^{-7}\ \text{m}$

(c) $\lambda = \frac{hc}{\Phi} = \frac{(6.62 \times 10^{-34}\ \text{Js})(2.99 \times 10^{8}\ \text{m s}^{-1})}{2.5\ \text{eV} \times (1.6 \times 10^{-19}\ \text{J/1 eV})} = 4.95 \times 10^{-7}\ \text{m}$

9.16 The minimal frequency corresponds to a minimal energy ($h\nu$) needed to overcome the work function.

9.17 $\int_{V_o} \psi^* \psi \, d\tau$

9.18 According to classical theory, the energy is independent of frequency, but according to quantum theory, the energy is proportional to the frequency ($E = h\nu$).

9.19 $4m^3v = 4m^2p \rightarrow 4m^2(-i\hbar\nabla)$
$E\psi(r) = -4i\hbar m^2\nabla\psi(r) + V(r)\psi(r)$

9.20 (a) $\Delta x = \frac{\hbar}{2m\Delta v} = \frac{1.05 \times 10^{-34}\ \text{Js}}{2(1.67 \times 10^{-27}\ \text{kg})(10^{-9}\ \text{m s}^{-1})} = 31.4\ \text{m}$

(b) $\Delta x = \frac{\hbar}{2m\Delta v} = \frac{1.05 \times 10^{-34}\ \text{Js}}{2(1.67 \times 10^{-27}\ \text{kg})(10^{-10}\ \text{m s}^{-1})} = 314\ \text{m}$

(c) $\Delta x = \dfrac{\hbar}{2m\Delta v} = \dfrac{1.05 \times 10^{-34}\ \text{Js}}{2\,(1.67 \times 10^{-27}\ \text{kg})\,(10^{-7}\ \text{m s}^{-1})} = 0.314\ \text{m}$

9.21 (a) $\Delta x = \dfrac{\hbar}{2m\Delta v} = \dfrac{1.05 \times 10^{-34}\ \text{Js}}{2\,(9.11 \times 10^{-31}\ \text{kg})\,(10^{-9}\ \text{m s}^{-1})} = 57{,}630\ \text{m}$

(b) $\Delta x = \dfrac{\hbar}{2m\Delta v} = \dfrac{1.05 \times 10^{-34}\ \text{Js}}{2\,(9.11 \times 10^{-31}\ \text{kg})\,(10^{-10}\ \text{m s}^{-1})} = 576{,}300\ \text{m}$

(c) $\Delta x = \dfrac{\hbar}{2m\Delta v} = \dfrac{1.05 \times 10^{-34}\ \text{Js}}{2\,(9.11 \times 10^{-31}\ \text{kg})\,(10^{-7}\ \text{m s}^{-1})} = 576.3\ \text{m}$

9.22 (a) $\Delta x = \dfrac{\hbar}{2m\Delta v} = \dfrac{1.05 \times 10^{-34}\ \text{Js}}{2(5 \times 10^{-3}\ \text{kg})\,(10^{-1}\ \text{m s}^{-1})} = 1.1 \times 10^{-31}\ \text{m}$

(b) $\Delta x = \dfrac{\hbar}{2m\Delta v} = \dfrac{1.05 \times 10^{-34}\ \text{Js}}{2(5 \times 10^{-3}\ \text{kg})\,(10^{-2}\ \text{m s}^{-1})} = 1.1 \times 10^{-30}\ \text{m}$

(c) $\Delta x = \dfrac{\hbar}{2m\Delta v} = \dfrac{1.05 \times 10^{-34}\ \text{Js}}{2(5 \times 10^{-3}\ \text{kg})\,(10^{-6}\ \text{m s}^{-1})} = 1.1 \times 10^{-26}\ \text{m}$

9.23 (a) $-i\hbar\nabla$
(b) x
(c) $(-i\hbar\nabla)^2 = -\hbar^2\nabla^2$

9.24 (a) $-\dfrac{\hbar^2}{2m}\dfrac{d^2}{dx^2}\psi(x) + (Ax)\psi(x) = E\psi(x)$

(b) $-\dfrac{\hbar^2}{2m}\dfrac{d^2}{dx^2}\psi(x) + (Ax^2)\psi(x) = E\psi(x)$

(c) $-\dfrac{\hbar^2}{2m}\dfrac{d^2}{dx^2}\psi(x) + (Ax + Bx^2)\psi(x) = E\psi(x)$

9.25 (a) $-\dfrac{\hbar^2}{2m}\left(\dfrac{\partial^2}{\partial x^2} + \dfrac{\partial^2}{\partial y^2}\right)\psi(x,y) + (Ax^3 + By)\psi(x,y) = E\psi(x,y)$

(b) $-\dfrac{\hbar^2}{2m}\left(\dfrac{\partial^2}{\partial x^2} + \dfrac{\partial^2}{\partial y^2}\right)\psi(x,y) + (Ax + By^3)\psi(x,y) = E\psi(x,y)$

(c) $-\dfrac{\hbar^2}{2m}\left(\dfrac{\partial^2}{\partial x^2} + \dfrac{\partial^2}{\partial y^2}\right)\psi(x,y) + (Axy)\psi(x,y) = E\psi(x,y)$

9.26 (a) $-\dfrac{\hbar^2}{2m}\left(\dfrac{\partial^2}{\partial x^2} + \dfrac{\partial^2}{\partial y^2} + \dfrac{\partial^2}{\partial z^2}\right)\psi(x,y,z) + (Ax + By + Cz)\psi(x,y,z)$

$= E\psi(x,y,z)$

(b) $$-\frac{\hbar^2}{2m}\left(\frac{\partial^2}{\partial x^2}+\frac{\partial^2}{\partial y^2}+\frac{\partial^2}{\partial z^2}\right)\psi(x,y,z)+(Ax^3+By+Cz^2)\psi(x,y,z)$$

$$=E\psi(x,y,z)$$

(c) $$-\frac{\hbar^2}{2m}\left(\frac{\partial^2}{\partial x^2}+\frac{\partial^2}{\partial y^2}+\frac{\partial^2}{\partial z^2}\right)\psi(x,y,z)+(Axyz)\psi(x,y,z)=E\psi(x,y,z)$$

9.27 Classically, the particle would pass through either one slit or the other but in quantum mechanics the particle has a wave nature that represents the combination of passing through both slits.

9.28 According to the Heisenberg Uncertainity Principle, the uncertainties in the two parameters are coupled with the product being at least $\hbar/2$. The interpretation of this result is that the observer interferes with the object during the measurement. For example, when measuring the position of an object, the probe hits the object and causes a change in velocity.

9.29 $$1=\int_0^a \psi^*(x)\psi(x)\,\mathrm{d}x=\int_0^a A^2\,\mathrm{d}x=A^2a\rightarrow\psi(x)=A=\sqrt{1/a}$$

9.30 $$1=\int_0^a \psi^*(x)\psi(x)\,\mathrm{d}x=\int_0^a A^2x^2\,\mathrm{d}x=A^2\frac{a^3}{3}\rightarrow A=\sqrt{\frac{3}{a^3}}$$

9.31 First calculate the uncertainty of the ball's position:

$$\Delta x=\frac{\hbar}{2m\Delta v}=\frac{1.05\times10^{-34}\ \text{Js}}{2(1.05\times10^{-22}\ \text{kg})\times(0.5\times10^{-7}\ \text{m s}^{-1})}=1.0\times10^{-5}\ \text{m}$$

The probability is estimated by the ratio of Δx and the size of the glove:

$$\text{Probability}=\frac{l}{\Delta x}=\frac{1.0\times10^{-6}\ \text{m}}{1\times10^{-5}\ \text{m}}=0.1$$

9.32 To experience diffraction, the wavelength of the person must be comparable to the size of the opening:

$$\lambda=1\times10^{-7}\ \text{m}=\frac{h}{mv}=\frac{6.6\times10^{-34}\ \text{Js}}{(6.6\times10^{-18}\ \text{kg})\times(1\times10^{-9}\ \text{m s}^{-1})}$$

9.33 If an electron were confined to such a small volume then the uncertainty in the velocity would become very large and the motion of the electron would be subject to the Heisenberg Uncertainty Principle.

CHAPTER 10

10.1 For the first excited state, $n = 2$ so $\lambda = \dfrac{2L}{2} = L$.

10.2 $\lambda = \dfrac{2L}{n} = \dfrac{2(8\text{ Å})}{1} = 16\text{ Å}$

$$\lambda = \frac{2L}{n} = \frac{2(8\text{ Å})}{3} = \frac{16}{3}\text{ Å}$$

10.3 $\bar{p} = \dfrac{2}{L}\int_0^L \sin\left(\dfrac{\pi x}{L}\right)\left(\dfrac{\hbar}{i}\dfrac{\partial}{\partial x}\right)\sin\left(\dfrac{\pi x}{L}\right)\mathrm{d}x$

$$\frac{2}{L}\int_0^L \sin\frac{\pi x}{L}\left(\frac{\hbar}{i}\frac{\mathrm{d}}{\mathrm{d}x}\right)^2 \sin\frac{\pi x}{L}\,\mathrm{d}x$$

10.4 $\displaystyle\int_0^{L/4} \psi^*(x)\psi(x)\,\mathrm{d}x = \frac{2}{L}\int_0^{L/4} \sin^2\left(\frac{n\pi x}{L}\right)\mathrm{d}x$

10.5 $\displaystyle\int_0^{L/4} \psi^*(x)x^2\psi(x)\,\mathrm{d}x = \frac{2}{L}\int_0^{L/4} x^2\sin^2\left(\frac{n\pi x}{L}\right)\mathrm{d}x$

10.6 $-\dfrac{\hbar^2}{2m}\dfrac{\mathrm{d}^2}{\mathrm{d}x^2}\psi(x) = E\psi(x)$

$$\psi_1(x) = \sqrt{\frac{2}{L}}\sin\left(\frac{\pi x}{L}\right)$$

$$\frac{\mathrm{d}^2}{\mathrm{d}x^2}\left(\sqrt{\frac{2}{L}}\sin\frac{\pi x}{L}\right) = -\sqrt{\frac{2}{L}}\left(\frac{\pi}{L}\right)^2 \sin\frac{\pi x}{L}$$

$$-\frac{\hbar^2}{2m}\frac{\mathrm{d}^2}{\mathrm{d}x^2}\psi(x) = -\frac{\hbar^2}{2m}\left(-\sqrt{\frac{2}{L}}\left(\frac{\pi}{L}\right)^2 \sin\frac{\pi x}{L}\right) = \frac{\hbar^2}{2m}\left(\frac{\pi}{L}\right)^2 \psi(x) = E\psi(x)$$

$$E = \frac{\hbar^2}{2m}\left(\frac{\pi}{L}\right)^2 = \frac{\pi^2 h^2}{8mL^2}$$

10.7 $\displaystyle\int_{l/4}^{3l/4} \psi^*(x)\,\psi(x)\,\mathrm{d}x = \frac{2}{L}\int_{l/4}^{3l/4} \sin^2\left(\frac{\pi x}{L}\right)\mathrm{d}x$

$$= \frac{2}{L}\frac{1}{2}\left[x - \sin\left(\frac{2\pi x}{L}\right)\left(\frac{L}{2\pi}\right)\right]_{l/4}^{3l/4} = \frac{1}{2} + \frac{1}{\pi} = 0.82$$

10.8 $$\Delta E = \frac{(6.63 \times 10^{-34}\ \text{Js})^2}{8(9.11 \times 10^{-31}\ \text{kg})(10^{-10}\ \text{m})^2}(2^2 - 1^2) = 1.8 \times 10^{-17}\ \text{J}$$

$$\lambda = \frac{hc}{\Delta E} = \frac{(6.63 \times 10^{-34}\ \text{Js})(3 \times 10^8\ \text{m s}^{-1})}{1.8 \times 10^{-17}\ \text{J}} = 1.1 \times 10^{-8}\ \text{m} = 11\ \text{nm}$$

10.9 $$\Delta E = \frac{(6.63 \times 10^{-34}\ \text{Js})^2}{8(9.11 \times 10^{-31}\ \text{kg})(10^{-10}\ \text{m})^2}(5^2 - 1^2) = 1.4 \times 10^{-16}\ \text{J}$$

$$\lambda = \frac{hc}{\Delta E} = \frac{(6.63 \times 10^{-34}\ \text{Js})(3 \times 10^8\ \text{m s}^{-1})}{1.4 \times 10^{-16}\ \text{J}} = 1.4 \times 10^{-9}\ \text{m} = 1.4\ \text{nm}$$

10.10 (a) $n = 3$

(b) $\lambda = \frac{2L}{n} = \frac{2(10\ \text{nm})}{3} = \frac{20}{3}\ \text{nm}$

(c) $n = 4$

(d) $\lambda = \frac{2L}{n} = \frac{2(10\ \text{nm})}{4} = \frac{20}{4}\ \text{nm}$

(e) $$\Delta E = \frac{(6.63 \times 10^{-34}\ \text{Js})^2}{8(9.11 \times 10^{-31}\ \text{kg})(10^{-8}\ \text{m})^2}(4^2 - 3^2) = 4.2 \times 10^{-21}\ \text{J}$$

$$\lambda = \frac{hc}{\Delta E} = \frac{(6.63 \times 10^{-34}\ \text{Js})(3 \times 10^8\ \text{m s}^{-1})}{4.2 \times 10^{-21}\ \text{J}}$$

$$= 4.74 \times 10^{-5}\ \text{m} = 47{,}400\ \text{nm}$$

10.11 (a) The wavefunction is zero at $x \leq -a$ and $x \geq +3a$

(b) $-\frac{\hbar^2}{2m}\frac{d^2}{dx^2}\psi(x) = E\psi(x)$

(c) $\psi(x) = \sqrt{\frac{2}{4a}}\sin\left(\frac{n\pi(x + a)}{4a}\right) \quad n = 1, 2, 3 \ldots$

(d) $\lambda = \frac{2(4a)}{1} = 8a$

10.12 (a) $-\frac{\hbar}{2m}\frac{d^2}{dx^2}\psi(x) + V_0\psi(x) = E\psi(x)$

(b) $\psi(x) = 0$ at $x = b$ and $x = 3b$

(c) $\psi(x) = \sqrt{\frac{2}{2b}}\sin\left(\frac{n\pi(x - b)}{2b}\right)$

(d) $\lambda = 4b$

(e) $E = \frac{h^2\pi^2}{2m(2b)^2} + V_0$

10.13 (a) $-\frac{\hbar^2}{2m}\left(\frac{\partial^2}{\partial x^2}\psi(x,y)+\frac{\partial^2}{\partial y^2}\psi(x,y)\right)=E\psi(x,y)$

(b) $\psi(x,y)=X(x)Y(y)=\left(\sqrt{\frac{2}{10\text{ Å}}}\sin\left(\frac{\pi x}{10\text{ Å}}\right)\right)\left(\sqrt{\frac{2}{20\text{ Å}}}\sin\left(\frac{\pi y}{20\text{ Å}}\right)\right)$

(c) $E=\frac{h^2}{8m}\left[\left(\frac{n_1}{L_1}\right)^2+\left(\frac{n_2}{L_2}\right)^2\right]$

$$=\frac{(6.63\times10^{-34}\text{ Js})^2}{8(9.11\times10^{-31}\text{ kg})}\left[\left(\frac{1}{10\times10^{-10}\text{ m}}\right)^2+\left(\frac{1}{20\times10^{-10}\text{ m}}\right)^2\right]$$

(d) The first two electrons will be in the (1,1) level, which is equal to:

$$E_g=\frac{h^2}{8m}\left[\left(\frac{n_1}{L_1}\right)^2+\left(\frac{n_2}{L_2}\right)^2\right]=\frac{h^2}{8m}\left(\frac{1}{20\text{ Å}}\right)^2(4n_x^2+n_y^2)=5E_0$$

The two lowest-energy states are (n_1,n_2) equal to (1,1) and then (1,2). The longest wavelength transition is from the (1,1) state to the (1,2) and the change in energy is:

$$\Delta E=\frac{(6.6\times10^{-34}\text{ Js})^2}{8(9.1\times10^{-31}\text{ kg})}\left\{\frac{1}{(10^{-9}\text{ m})^2}\left[\left(\frac{2}{1.5}\right)^2+1^2\right]-\left[\left(\frac{1}{1.5}\right)^2+1^2\right]\right\}$$

$$=7.9\times10^{-20}\text{ J}$$

$$\lambda=\frac{hc}{\Delta E}=\frac{(6.6\times10^{-34}\text{ Js})(3\times10^{8}\text{ m s}^{-1})}{7.9\times10^{-20}\text{ J}}=2.5\times10^{-6}\text{ m}$$

10.14 (a) The wavefunction is zero for $x\le-2a$; $x\ge\pm3a$; $y\le-b$; $y\ge\pm2b$

(b) $-\frac{\hbar^2}{2m}\left(\frac{\partial^2}{\partial x^2}\psi(x,y)+\frac{\partial^2}{\partial y^2}\psi(x,y)\right)=E\psi(x,y)$

(c) $\psi(x,y)=\left(\sqrt{\frac{2}{5a}}\sin\left(\frac{n_x\pi(x+2a)}{5a}\right)\right)\left(\sqrt{\frac{2}{3b}}\sin\left(\frac{ny\pi(y+b)}{3b}\right)\right)$

10.15 $-\frac{\hbar^2}{2m}\left(\frac{\partial^2}{\partial x^2}\psi(x,y,z)+\frac{\partial^2}{\partial y^2}\psi(x,y,z)+\frac{\partial^2}{\partial z^2}\psi(x,y,z)\right)=E\psi(x,y,z)$

10.16 Optical spectroscopy has shown that the excited state of chlorophyll can be quenched by the presence of zeaxanthin, forming a transitory oxidized state of the carotenoid.

10.17 The length of the conjugation differs in the two carotenoids, giving them different energies for the electronic states.

10.18 Tunneling is the process by which a particle makes a transition between two states through an intermediate state that is classically forbidden. Tunneling depends upon distance, the size of the energy barrier, and the energy of the particle.

10.19 Scanning tunneling microscopy is a technique that makes use of the ability of an electron to tunnel through a vacuum to probe the surface of an object at an atomic level.

10.20 Atomic force microscopy is a technique that makes use of a cantilever to move a probe across a surface to probe the surface at an atomic level.

10.21 Many answers accepted.

10.22 Since the two acceptors are identical, the free-energy differences for the reactions should be about equal. However, the slower rate can be explained by assuming that one acceptor has a much higher reorganization energy.

10.23 The rate is seen to be nearing a maximal value at the upper range of the free-energy differences, so the reorganization energy is about 600 meV.

CHAPTER 11

11.1 $E = \frac{\hbar\omega}{2} = \frac{\hbar}{2}\sqrt{\frac{k}{m}}$

11.2 $\frac{1}{2}\hbar\sqrt{\frac{k}{m_e}}$

11.3 $\frac{5}{2}\hbar\sqrt{\frac{k}{m_e}}$

11.4 $\lambda = 2\pi c\sqrt{\frac{m_e}{k}}$

11.5 $\mu = \frac{m_C m_O}{m_C + m_O} = \frac{12 \times 16}{12 + 16} m_p = 1.14 \times 10^{-26}$ kg

11.6 $\mu = \dfrac{m_C m_O}{m_C + m_O} = \dfrac{13 \times 16}{13 + 16} m_p = 1.19 \times 10^{-26}$ kg

11.7 $\nu = \dfrac{1}{2\pi}\sqrt{\dfrac{600 \text{ N m}^{-1}}{1.67 \times 10^{-27} \text{ kg}}} = 9.5 \times 10^{13} \text{ s}^{-1}$

11.8 $\mu = \dfrac{m_C m_O}{m_C + m_O} = \dfrac{12 \times 16}{12 + 16} m_p = 1.14 \times 10^{-26}$ kg

$$\nu = \frac{1}{2\pi}\sqrt{\frac{k}{\mu}} = \frac{1}{2\pi}\sqrt{\frac{500 \text{ N m}^{-1}}{1.14 \times 10^{-26} \text{ kg}}} = 3.33 \times 10^{13} \text{ s}^{-1}$$

$$\bar{\nu} = \frac{\nu}{c} = 1111 \text{ cm}^{-1}$$

11.9 $\mu = \dfrac{m_C m_O}{m_C + m_O} = \dfrac{13 \times 16}{13 + 16} m_p = 1.19 \times 10^{-26}$ kg

$$\nu = \frac{1}{2\pi}\sqrt{\frac{k}{\mu}} = \frac{1}{2\pi}\sqrt{\frac{500 \text{ N m}^{-1}}{1.19 \times 10^{-26} \text{ kg}}} = 3.26 \times 10^{13} \text{ s}^{-1}$$

$$\bar{\nu} = \frac{\nu}{c} = 1087 \text{ cm}^{-1}$$

11.10 $\dfrac{\nu_{2m}}{\nu_m} = \dfrac{\sqrt{k/2m}}{\sqrt{k/m}} = \sqrt{\dfrac{1}{2}}$

11.11 $P = \displaystyle\int_1^{\infty} \frac{1}{\alpha\pi^{1/2}} e^{-y^2}\alpha \, dy = \frac{1}{\pi^{1/2}}\int_1^{\infty} e^{-y^2} \, dy$

11.12 H_2O $\quad \mu = \dfrac{16 \times 1}{17} m_H = 0.94 m_H \qquad \dfrac{\omega_{H_2O}}{\omega_{D_2O}} = \sqrt{\dfrac{k/\mu_{H_2O}}{k/\mu_{D_2O}}} \sqrt{\dfrac{1.78}{0.94}} = 1.38$

D_2O $\quad \mu = \dfrac{16 \times 2}{18} m_H = 1.78 m_H \qquad \lambda = \lambda_{H_2O} 1.38 = 4.1\,\mu\text{m}$

11.13 $\displaystyle\int_{-\infty}^{+\infty} (N_0 e^{-x^2/2\alpha^2})(x^2)(N_0 e^{-x^2/2\alpha^2}) \, dx$

11.14 $\displaystyle\int_{-\infty}^{+\infty} \left(2\frac{x}{\alpha}\right)(N_1 e^{-x^2/2\alpha^2})(x^2)\left(2\frac{x}{\alpha}\right)(N_1 e^{-x^2/2\alpha^2}) \, dx$

11.15 $$\frac{d}{dx}\psi_0(x) = N_0\frac{d}{dx}e^{-x^2/2\alpha^2} = N_0 e^{-x^2/2\alpha^2}\left(-\frac{x}{\alpha^2}\right)$$

$$\frac{d^2}{dx^2}\psi_0(x) = N_0\frac{d}{dx}\left(-\frac{x}{\alpha^2}e^{-x^2/2\alpha^2}\right) = N_0 e^{-x^2/2\alpha^2}\left(\frac{x^2}{\alpha^4}-\frac{1}{\alpha^2}\right)$$

$$= \psi_0(x)\left(\frac{x^2}{\alpha^4}-\frac{1}{\alpha^2}\right)$$

$$-\frac{\hbar^2}{2m}\left(\psi_0(x)\left(\frac{x^2}{\alpha^4}-\frac{1}{\alpha^2}\right)\right)+\frac{kx^2}{2}\psi_0(x) = E_0\psi_0(x)$$

$$\psi_0(x)\left[x^2\left(-\frac{\hbar^2}{2m\alpha^4}+\frac{k}{2}\right)+\left(\frac{\hbar^2}{2m\alpha^2}-E_0\right)\right] = 0$$

The term in the left parentheses is:

$$-\frac{\hbar^2}{2m\alpha^4}+\frac{k}{2} = -\frac{\hbar^2}{2m}\left(\frac{mk}{\hbar^2}\right)+\frac{k}{2} = 0$$

so the term on the right-hand side must be zero, yielding

$$E_0 = \frac{\hbar^2}{2m\alpha^2} = \frac{\hbar^2}{2m}\left(\frac{mk}{\hbar^2}\right)^{1/2} = \frac{\hbar}{2}\left(\frac{k}{m}\right)^{1/2} = \frac{\hbar\omega}{2}$$

11.16 (a) $$\mu = \frac{m_1m_2}{m_1+m_2} + \frac{(16)(16)}{16+16}m_p = 8m_p$$

(b) $$\mu = \frac{(15)(15)}{15+15}m_p = 7.5m_p$$

$$\frac{v_{16_0}}{v_{15_0}} = \frac{\sqrt{k/(8m_p)}}{\sqrt{k/(7m_p)}} = 0.97$$

11.17 $$-\frac{\hbar^2}{2m}\frac{d^2}{dx^2}\psi(x) + (Ax^2 + Be^{-x})\psi(x) = E\psi(x)$$

If B ≪ A, the wavefunctions will be the same as those for the harmonic oscillator.

11.18 $$-\frac{\hbar}{2m}\frac{d^2}{dx^2}\psi(x) + \left(\frac{k_1}{2}(+x)^2 + \frac{k_2}{2}(-x)^2\right)\psi(x) = E\psi(x)$$

11.19 $2\int_A^{\infty}(N_0 e^{-x^2/2\alpha^2})^2\,\mathrm{d}x$

11.20 $-\dfrac{\hbar^2}{2m}\dfrac{\mathrm{d}^2}{\mathrm{d}x^2}\psi(x)+\dfrac{k}{2}x^2\psi(x)=E\psi(x)$

$$\psi_1(y)=N_1(2y)e^{-y^2/2}$$

$$\frac{\mathrm{d}}{\mathrm{d}x}\psi_1(x)=\frac{2N_1}{\alpha}\frac{\mathrm{d}}{\mathrm{d}x}xe^{-x^2/2\alpha^2}=\frac{2N_1}{\alpha}\left(xe^{-x^2/2\alpha^2}\left(-\frac{x}{\alpha^2}\right)+e^{-x^2/2\alpha^2}\right)$$

$$\frac{\mathrm{d}^2}{\mathrm{d}x^2}\psi_1(x)=\frac{2N_1}{\alpha}\frac{\mathrm{d}}{\mathrm{d}x}\left(e^{-x^2/2\alpha^2}\left(-\frac{x^2}{\alpha^2}\right)+e^{-x^2/2\alpha^2}\right)$$

$$=\frac{2N_1}{\alpha}\left(e^{-x^2/2\alpha^2}\left(-\frac{2x}{\alpha^2}\right)+\left(-\frac{x^2}{\alpha^2}\right)\left(-\frac{2x}{\alpha^2}\right)e^{-x^2/2\alpha^2}+\left(-\frac{x}{\alpha^2}\right)e^{-x^2/2\alpha^2}\right)$$

$$=\frac{2N_1}{\alpha}e^{-x^2/2\alpha^2}\left(\frac{2x^3}{\alpha^4}-\frac{3x}{\alpha^2}\right)=\psi_1(x)\left(\frac{2x^2}{\alpha^4}-\frac{3}{\alpha^2}\right)$$

$$-\frac{\hbar^2}{2m}\left(\psi_1(x)\left(\frac{2x^2}{\alpha^4}-\frac{3}{\alpha^2}\right)\right)+\frac{kx^2}{2}x^2\psi_1(x)=E_1\psi_1(x)$$

$$\psi_1(x)\left[x^2\left(-\frac{\hbar^2}{m\alpha^4}+\frac{k}{2}\right)+\left(\frac{3\hbar^2}{2m\alpha^2}-E_1\right)\right]=0$$

$$-\frac{\hbar^2}{2m\alpha^4}+\frac{k}{2}=-\frac{\hbar^2}{2m}\left(\frac{mk}{\hbar^2}\right)+\frac{k}{2}=0$$

$$E_1=\frac{3\hbar^2}{2m\alpha^2}=\frac{3\hbar^2}{2m}\left(\frac{mk}{\hbar^2}\right)^{1/2}=\frac{3\hbar}{2}\left(\frac{k}{m}\right)^{1/2}=\frac{3\hbar\omega}{2}$$

11.21 (a) $-\dfrac{\hbar^2}{2m}\dfrac{\mathrm{d}^2}{\mathrm{d}x^2}\psi(x)+a(x^2+bx^3)\psi(x)=E\psi(x)\quad |x|\le L/2$

$$-\frac{\hbar^2}{2m}\frac{\mathrm{d}^2}{\mathrm{d}x^2}\psi(x)+cL^2\,\psi(x)=E\psi(x)\quad |x|\ge L/2$$

(b) $V(x)$ is continuous at $L/2$.

$$a\left(\frac{L^2}{4}+b\frac{L^2}{8}\right)=cL^2\quad\text{or}\quad\frac{a}{4}+\frac{b}{8}=c$$

(c) $\Delta v = \dfrac{\hbar}{2m\Delta x} = \dfrac{1.054 \times 10^{-34}\ \text{Js}}{2(1.054 \times 10^{-34}\ \text{kg})(0.5 \times 10^{-11}\ \text{m})} = 1.0 \times 10^{11}\ \text{m s}^{-1}$

(d) The region beyond $L/2$ is classically forbidden and so the probability of finding the particle in this region is very small, although non-zero.

(e) The wavefunctions will become those for the simple harmonic oscillator.

11.22 (a) $\mu = \dfrac{m_1 m_2}{m_1 + m_2} = \dfrac{(14)(14)}{14+14} m_p = 7m_p$

(b) $-\dfrac{\hbar^2}{2\mu}\dfrac{d^2}{dx^2}\psi(x) + \dfrac{k}{2}x^2\psi(x) = E\psi(x)$

(c) $\int_{2Å}^{\infty} \psi^*(x)\psi(x)\,dx = \int_{2Å}^{\infty} N_0^2 e^{-x^2/\alpha^2}dx$ for the ground state

(d) $F = -k(x_n - x_{n-1}) + k(x_{n+1} - x_n)$

(e) $-\dfrac{\hbar^2}{2m}\dfrac{d^2}{dx_n^2}\psi_n(x) + \dfrac{k}{2}[(x_n - x_{n-1})^2 + (x_{n+1} - x_n)^2]\psi_n(x) = E\psi_n(x)$

CHAPTER 12

12.1 $V(r) = \dfrac{-e^2}{4\pi\varepsilon_0 r}$

12.2 (a) The ground state is $n = 1$

$$E = -\frac{hcR_h}{1^2} = -13.6\ \text{eV}$$

(b) The first excited state is $n = 2$

$$E = -\frac{hcR_h}{2^2} = -3.4\ \text{eV}$$

12.3 $E = -\dfrac{hcR_h}{3^2} = -1.5\ \text{eV}$

12.4 $\psi_{200} = \dfrac{1}{4}\sqrt{\dfrac{1}{2\pi a_0^3}}\left(2 - \dfrac{r}{a_0}\right)e^{-r/2a_0}$

$2 - \dfrac{r}{a_0} = 0$ so $r = 2a_0 = 1.058 \times 10^{-10}$ m

12.5 $$\int_0^\infty \psi^* r \psi \, d\tau = \int_0^\infty r\left(\frac{1}{\pi a_0^3}\right) e^{-2r/a_0} 4\pi r^2 \, dr = \frac{4}{a_0^3}\int_0^\infty r^3 e^{-2r/a_0} \, dr$$

$$\frac{4}{a_0^3}\int_0^\infty r^3 e^{-2r/a_0} \, dr = 4a_0 \int_0^\infty x^3 e^{-2x} \, dx = 4a_0 e^{-2x}\left(-\frac{x^3}{2} - \frac{3x^2}{4} - \frac{3x}{4} - \frac{3}{8}\right)_0^\infty$$

$$= \frac{3a_0}{2} = 7.9 \times 10^{-11} \text{ m}$$

12.6 $$0 = \frac{d}{dr}(4\pi r^2 \psi^* \psi)$$

$$0 = \frac{8}{a_0^3} e^{-2r/a_0} r \left(1 - \frac{r}{a_0}\right) \quad \text{or} \quad r = a_0 = 5.29 \times 10^{-10} \text{ m}$$

12.7 Principal, $n = 1, 2, 3, \ldots$; angular momentum, $l = 0, 1, 2, \ldots, n - 1$; magnetic $m_l = l, l - 1, l - 2, \ldots, -l$.

12.8 $$\Delta E = -hcR_h\left(\frac{1}{3^2} - \frac{1}{1^2}\right) = \frac{hc}{\lambda}$$

$$\lambda = \frac{9}{8R_h} = \frac{9}{8(1.1 \times 10^5 \text{ cm}^{-1})} = 1.03 \times 10^{-5} \text{ cm}$$

12.9 n, Quantization of energy; l, quantization of total angular momentum; m_l: quantization of the z component of angular momentum.

12.10 $$\Delta E = -Z^2 hcR_h\left(\frac{1}{3^2} - \frac{1}{1^2}\right) = \frac{hc}{\lambda}$$

$$\lambda = \frac{9}{8Z^2R_h} = \frac{9}{8(29)^2(1.1 \times 10^5 \text{ cm}^{-1})} = 1.22 \times 10^{-8} \text{ cm} = 1.22 \text{ Å}$$

12.11 $$\int_0^\infty \psi^* r \psi \, d\tau = \int_0^\infty r \frac{1}{16} \frac{1}{2\pi a_0^3}\left(2 - \frac{r}{a_0}\right)^2 e^{-r/a_0} 4\pi r^2 \, dr$$

$$= \frac{1}{8a_0^3}\int_0^\infty r^3 \left(2 - \frac{r}{a_0}\right)^2 e^{-r/a_0} \, dr$$

12.12 $$\int \frac{1}{2\sqrt{2}}\sqrt{\frac{Z^3}{4\pi a_0^3}}\left(2 - \frac{2r}{a_0}\right) e^{-r/2a_0}\left(\frac{\hbar}{i}\nabla\right)\frac{1}{2\sqrt{2}}\sqrt{\frac{Z^3}{4\pi a_0^3}}\left(2 - \frac{2r}{a_0}\right) e^{-r/2a_0} r^2 \, dr \, d\theta \, d\phi$$

12.13 $$\int_{3a_0}^\infty \left[\frac{1}{2\sqrt{2}}\sqrt{\frac{1}{4\pi a_0^3}}\left(2 - \frac{r}{a_0}\right) e^{-r/2a_0}\right]^2 (4\pi r^2 \, dr)$$

12.14 $\psi_{p_x} = -\frac{1}{\sqrt{2}}(\psi_{p+} - \psi_{p-}) = xf(r)$

12.15 $\int\left(\sqrt{\frac{1}{\pi a_0^3}}e^{-r/a_0}\right)^2 4\pi r^2\,\mathrm{d}r = \frac{4}{a_0^3}\int e^{-2r/a_0}r^2\,\mathrm{d}r$

let $x = r/a_0$ then $\frac{4}{a_0^3}\int e^{-2r/a_0}r^2\,\mathrm{d}r = 4\int e^{-2x}x^2\,\mathrm{d}x = -e^{-2x}(2x^2 + 2x + 1)$

The limits are $r = 1.058$ or $x = 2$ and infinity so the probability is equal to:

$0 - (-e^{-4}(2 \times 4^2 + 2 \times 4 + 1)) = 0.75$

12.16 $\int\left(\sqrt{\frac{1}{\pi a_0^3}}e^{-r/a_0}\right)^2 4\pi r^2\,\mathrm{d}r = \frac{4}{a_0^3}\int e^{-2r/a_0}r^2\,\mathrm{d}r$

let $x = r/a_0$ then $\frac{4}{a_0^3}\int e^{-2r/a_0}r^2\,\mathrm{d}r = 4\int e^{-2x}x^2\,\mathrm{d}x = -e^{-2x}(2x^2 + 2x + 1)$

The limits are $r = 0.5$ Å or $x = 1$ and $r = 1.058$ Å or $x = 2$, so the probability is:

$-4e^{-4}(2 \times 4 + 4 + 1) + 4e^{-2}(2 + 2 + 1) = -0.44 + 0.68 = 0.24$

12.17 The most probable radial position of the electron is simply the peak position of this term. We can find the peak by setting the derivative equal to zero:

$$0 = \frac{\mathrm{d}}{\mathrm{d}r}(4\pi r^2\psi^*\psi)$$

$$0 = \frac{\mathrm{d}}{\mathrm{d}r}\left(4\pi r^2\,\frac{1}{\pi a_0^3}\,e^{-2r/a_0}\right) = \frac{4}{a_0^3}\frac{\mathrm{d}}{\mathrm{d}r}(r^2e^{-2r/a_0})$$

$$0 = \frac{8}{a_0^3}e^{-2r/a_0}r\left(1 - \frac{r}{a_0}\right) \quad \text{or} \quad r = a_0$$

12.18 $\int_0^\infty \psi^* r\psi\,\mathrm{d}\tau = \int_0^\infty r\left(\frac{1}{\pi a_0}\right)e^{-2r/a_0}4\pi r^2\,\mathrm{d}r = \frac{4}{a_0}\int_0^\infty r^3e^{-2r/a_0}\,\mathrm{d}r$

$$\frac{4}{a_0}\int_0^\infty r^3e^{-2r/a_0}\,\mathrm{d}r = 4a_0\int_0^\infty x^3e^{-2x}\,\mathrm{d}x = 4a_0e^{-2x}\left(-\frac{x^3}{2} - \frac{3x^2}{4} - \frac{3x}{4} - \frac{3}{8}\right)_0^\infty$$

$$= \frac{3a_0}{2}$$

12.19 A wavefunction can have a node where the value is exactly equal to zero. For the 2s orbital this occurs at:

$$\psi_{200} = \frac{1}{2\sqrt{2}}\sqrt{\frac{Z^3}{4\pi a_0^3}}\left(2 - \frac{\rho}{2}\right)e^{-\rho/4} = 0 \text{ when } \rho = 4$$

12.20 $\Phi(\phi + 2\pi) = \Phi(\phi)$

$Ae^{im_l\phi} = Ae^{im_l(\phi+2\pi)} = Ae^{im_l\phi}e^{im_l2\pi}$

$1 = e^{im_l2\pi} = \cos(2\pi m_l) + i\sin(2\pi m_l)$

with $m_l = 0, \pm 1, \pm 2, \pm 3, \ldots$.

12.21 $$\frac{\mathrm{d}}{\mathrm{d}\phi}\Phi(\phi) = \frac{\mathrm{d}}{\mathrm{d}\phi}(Ae^{im_l\phi}) = A\,(im_l)\,e^{im_l\phi}$$

$$\frac{\mathrm{d}^2}{\mathrm{d}\phi^2}\Phi(\phi) = \frac{\mathrm{d}}{\mathrm{d}\phi}(A(im_l)e^{im_l\phi}) = A(im_l)^2e^{im_l\phi} = -m_l^2\Phi(\phi)$$

12.22 $$\sin\theta\frac{\mathrm{d}}{\mathrm{d}\theta}\left[\sin\theta\frac{\mathrm{d}}{\mathrm{d}\theta}(B\cos\theta)\right] + [1(1+1)\sin^2\theta - 0](B\cos\theta) = 0$$

$$\sin\theta\frac{\mathrm{d}}{\mathrm{d}\theta}[\sin\theta(-B\sin\theta)] + 2B\cos\theta\sin^2\theta = 0$$

$$-B\sin\theta\frac{\mathrm{d}}{\mathrm{d}\theta}[\sin^2\theta] + 2B\cos\theta\sin^2\theta = 0$$

$$-B\sin\theta(2\sin\theta\cos\theta) + 2B\cos\theta\sin^2\theta = 0$$

$$0 = 0$$

12.23 $$\frac{\mathrm{d}}{\mathrm{d}r}\Pi(r) = r(-\alpha)e^{-\alpha r} + e^{-\alpha r}$$

$$\frac{\mathrm{d}^2}{\mathrm{d}r^2}\Pi(r) = -\alpha(-\alpha re^{-\alpha r} + e^{-\alpha r}) - \alpha e^{-\alpha r} = \alpha^2re^{-\alpha r} - 2\alpha e^{-\alpha r}$$

$$\alpha^2re^{-\alpha r} - 2\alpha e^{-\alpha r} + \frac{2m}{\hbar^2}\left(\frac{e^2}{4\pi\varepsilon_0 r} + E\right)re^{-\alpha r} = 0$$

$$re^{-\alpha r}\left(\alpha^2 + \frac{2mE}{\hbar^2}\right) + e^{-\alpha r}\left(-2\alpha + \frac{e^2}{4\pi\varepsilon_0}\frac{2m}{\hbar^2}\right) = 0$$

The second term is zero since:

$$\alpha = \frac{m}{\hbar^2}\frac{e^2}{4\pi\varepsilon_0}$$

This leaves the first term and:

$$E = -\frac{me^4}{32\pi^2\varepsilon_0\hbar^2}$$

12.24 The boundary surface represents the 90% probability value.

12.25 Angular momentum must be conserved and photons carry angular momentum. Thus the change in l must be 1.

12.26 (a) $V(r,\theta,\varphi) = \dfrac{-e^2}{4\pi\varepsilon_0}\dfrac{\cos\theta}{r^3}$

$$-\frac{\hbar^2}{2m}\nabla^2(r,\theta,\varphi)\,\psi(r,\theta,\varphi) + \frac{-e^2\cos\theta}{4\pi\varepsilon_0 r^3}\,\psi(r,\theta,\varphi) = E\psi(r,\theta,\varphi)$$

(b) The wavefunctions with different angular dependences will no longer be degenerate, and the dependence of the energy on distance is now weaker, so the wavefunctions will be further from the nucleus.

12.27 $$\int_0^\infty \psi^* r^2 \psi \,d\tau = \int_0^\infty r^2 \frac{1}{16}\frac{1}{2\pi a_0^3}\left(2 - \frac{r}{a_0}\right)^2 e^{-r/a_0}\, 4\pi r^2\, dr$$

$$= \frac{1}{8a_0^3}\int_0^\infty r^4\left(2 - \frac{r}{a_0}\right)^2 e^{-r/a_0}\, dr$$

$$\frac{1}{8a_0^3}\int_0^\infty (a_0x)^4\left(2 - \frac{a_0x}{a_0}\right)^2 e^{-a_0x/a_0} a_0\, dx = \frac{a_0^2}{8}\int_0^\infty x^4(2-x)^2 e^{-x}\, dx$$

$$= \frac{a_0^2}{8}\int_0^\infty (x^6 - 4x^5 + 4x^4)e^{-x}\, dx$$

$$\frac{a_0^2}{8}\int_0^\infty (x^6 - 4x^5 + 4x^4)e^{-x}\, dx = \frac{a_0^2}{8}(6! - 4(5!) + 4(4!)) = 34.5a_0^2$$

12.28 $n^2 = 5^2 = 25$

12.29 An electron experiences a shielded nuclear charge due to the presence of the other electrons. This can be expressed by assigning an effective nuclear charge that is lower than the atomic number.

12.30 $$Z_{eff}^{atom} = \sqrt{\frac{I^{atom}}{I^H}} = \sqrt{\frac{2370\text{ kJ mol}^{-1}}{1312\text{ kJ mol}^{-1}}} = 1.34$$

12.31 $E^{atom} = (Z_{eff}^{atom})^2 E^H = (2.4)^2 1312\text{ kJ mol}^{-1} = 7557\text{ kJ mol}^{-1}$

12.32 $$-\frac{\hbar^2}{2m}(\nabla_1^2 + \nabla_2^2)\,\psi(r_1r_2) - \frac{e^2}{4\pi\varepsilon_0}\left[\frac{2}{r_1} + \frac{2}{r_2} - \frac{1}{r_{12}}\right]\psi(r_1r_2) = E\psi(r_1r_2)$$

12.33 The outer electrons are both in 2s orbitals and beryllium has a larger nuclear charge.

12.34 The placement of the outer electron for boron in a 2p orbital compared to the 2s orbital for beryllium results in a lower ionization energy despite the increase in the nuclear charge.

12.35 Spin is a relativistic term that has no classical analog. Since Schrödinger's equation is based upon classical physics, spin does not appear.

12.36 Two components represent spin (up or down) and the other two components represent positive or negative energy, or equivalently matter and antimatter.

12.37 Nitrogen has seven electrons. The first four fill the 1s and 2s orbitals. The remaining three electrons fill the 2p orbitals, with the electrons occupying three different 2p orbitals to minimize the interactions between the electrons.

CHAPTER 13

13.1 $\frac{p_1^2}{2m_e} + \frac{p_2^2}{2m_e}$

13.2 (a) $-\frac{e^2}{4\pi\varepsilon_0}\left[\frac{1}{r_{A1}} + \frac{1}{r_{A2}} + \frac{1}{r_{B1}} + \frac{1}{r_{B2}} - \frac{1}{r_{12}} - \frac{1}{r_{AB}}\right]$

(b) $-\frac{e^2}{4\pi\varepsilon_0}\left[\frac{1}{r_{A1}} + \frac{1}{r_{A2}} + \frac{1}{r_{A3}} + \frac{1}{r_{B1}} + \frac{1}{r_{B2}} + \frac{1}{r_{B3}} - \frac{1}{r_{12}} - \frac{1}{r_{23}} - \frac{1}{r_{13}} - \frac{1}{r_{AB}}\right]$

13.3 $-\frac{\hbar^2}{2m}[\nabla_1^2 + \nabla_2^2]\psi(r_1,r_2) - \frac{e^2}{4\pi\varepsilon_0}\left[\frac{1}{r_{A1}} + \frac{1}{r_{A2}} + \frac{1}{r_{B1}} + \frac{1}{r_{B2}} - \frac{1}{r_{12}}\right]\psi(r_1,r_2)$

$= E\psi(r_1,r_2)$

13.4 $-\frac{\hbar^2}{2m}(\nabla_1^2 + \nabla_2^2 + \nabla_3^2)\psi(r_1r_2r_3)$

$-\frac{e^2}{4\pi\varepsilon_0}\left[\frac{2}{r_{1A}} + \frac{2}{r_{2A}} + \frac{2}{r_{3A}} + \frac{2}{r_{1B}} + \frac{2}{r_{2B}} + \frac{2}{r_{3B}} - \frac{1}{r_{12}} - \frac{1}{r_{23}} - \frac{1}{r_{13}} - \frac{4}{r_{AB}}\right]\psi(r_1r_2r_3)$

$= E\psi(r_1r_2r_3)$

13.5 The wavefunction must be written considering all possible combinations of the electrons in each state.

13.6 Two atoms will have an attractive interaction due to the London dispersion interaction in which random fluctuations give rise to transient dipole moments.

13.7 An increase in coupling will increase the energy difference between the molecular orbitals.

13.8 In valence bond theory, a bond is formed when electrons from two atoms pair up, with σ and π orbitals corresponding to s and p atomic orbitals, respectively.

13.9 A peptide bond is formed by the carboxyl group of one amino acid being joined to the amino group of another amino acid, resulting in the loss of a water molecule.

13.10 The higher dielectric constant decreases the effect of the interactions in water so the strength of the electrostatic interactions is much greater in benzene.

13.11 Proteins and DNA are stabilized by the presence of the cumulative effect of many interactions.

13.12 The (ϕ, ψ) values fall into relatively restricted regions of the plot as the sterically allowed conformations are limited, with most values found for the regions corresponding to α helices and β strands. The values for glycine are usually excluded because they frequently fall outside of the expected ranges.

13.13 Membrane proteins form well-defined secondary structures in the membrane, usually α helices, which have hydrogen bonds naturally formed along the backbone.

13.14 Hydrogen bonds stabilize secondary structure.

13.15 The surrounding amino acid side chains and water molecules will respond to the presence of a charge and will effectively minimize the electrostatic interactions.

13.16 Answers can include myoglobin and hemoglobin.

13.17 The binding of oxygen to the four hemes in hemoglobin is regulated allosterically by interactions among the four polypeptide chains.

13.18 For any given conformation, the interactions can be calculated and the energy of that state plotted.

13.19 These are high-energy states initially formed during protein folding in which the hydrophobic amino acids collapse together, forming what will become the interior of the protein.

13.20 Velocities are generated by assigning a certain temperature to the system and then generating a distribution of velocities, the average kinetic energy of which matches the kinetic energy.

CHAPTER 14

14.1 $A = \varepsilon cl$

where the proportionality constant ε is the molar absorption coefficient, or extinction coefficient, the concentration of the absorber is c, the pathlength of the sample is l, the intensity of the incident light is I, and the intensity of the exiting light is I_0.

14.2 (a) $\nu = \dfrac{c}{\lambda} = \dfrac{3 \times 10^8 \text{ m s}^{-1}}{1 \times 10^{-9} \text{ m}} = 3 \times 10^{17} \text{ m} \rightarrow$ X-ray

(b) $\nu = \dfrac{c}{\lambda} = \dfrac{3 \times 10^8 \text{ m s}^{-1}}{5 \times 10^{-7} \text{ m}} = 6.0 \times 10^{14} \text{ m} \rightarrow$ visible

(c) $\nu = \dfrac{c}{\lambda} = \dfrac{3 \times 10^8 \text{ m s}^{-1}}{3 \times 10^{-2} \text{ m}} = 1.0 \times 10^{10} \text{ m} \rightarrow$ microwave

14.3 (a) $\lambda = \dfrac{c}{\nu} = \dfrac{3 \times 10^8 \text{ m s}^{-1}}{1 \times 10^{18} \text{ s}^{-1}} = 3 \times 10^{-10} \text{ m} \rightarrow$ X-ray

(b) $\lambda = \dfrac{c}{\nu} = \dfrac{3 \times 10^8 \text{ m s}^{-1}}{1 \times 10^{6} \text{ s}^{-1}} = 3 \times 10^{2} \text{ m} \rightarrow$ radiowave

(c) $\lambda = \dfrac{c}{\nu} = \dfrac{3 \times 10^8 \text{ m s}^{-1}}{1 \times 10^{12} \text{ s}^{-1}} = 3 \times 10^{-4} \text{ m} \rightarrow$ infrared

14.4 (a) $A = 0.27$

(b) $c = \dfrac{1}{2.7} \times 0.1\% = 0.037\%$

14.5 (a) $A = 0.0154$

(b) $c = \dfrac{A}{\varepsilon l} = \dfrac{10^{-1}}{(1.54 \times 10^4 \text{ M}^{-1} \text{ cm}^{-1})(1 \text{ cm})} = 6.5 \times 10^{-6} \text{ M} = 6.5\,\mu\text{M}$

14.6 $c = \dfrac{A}{\varepsilon l} = \dfrac{10^{-1}}{10^3 \text{ M}^{-1} \text{ cm}^{-1} \times 1 \text{ cm}} = 0.1 \text{ mM}$

14.7 (a) $A = c\varepsilon l = (1 \times 10^{-6} \text{ M})(1 \times 10^4 \text{ M}^{-1} \text{ cm}^{-1})(1 \text{ cm}) = 0.01$
(b) $A = c\varepsilon l = (2 \times 10^{-4} \text{ M})(1 \times 10^4 \text{ M}^{-1} \text{ cm}^{-1})(1 \text{ cm}) = 2$

14.8 $[Tyr] = \dfrac{(5380 \text{ M}^{-1})(0.717) - (1960 \text{ M}^{-1})(0.239)}{(11{,}300 \text{ M}^{-1})(5380) - (1500 \text{ M}^{-1})(1960)} = 5.85 \times 10^{-5} \text{ M}$

$[Trp] = \dfrac{(11{,}300 \text{ M}^{-1})(0.239) - (1500 \text{ M}^{-1})(0.717)}{(11{,}300 \text{ M}^{-1})(5380) - (1500 \text{ M}^{-1})(1960)} = 2.81 \times 10^{-5} \text{ M}$

14.9 $w' = B'\rho$

14.10 Since the photon carries a spin of 1, the angular momentum of the electron must change by 1 to conserve momentum.

14.11 Only the stimulated process has an incident photon required for the transition.

14.12 The stimulated process has two identical photons output for every transition, leading to amplification, but for a spontaneous process there is no coordination of the light.

14.13 The fluorescent spectrum is always red, shifted relative to the absorption spectrum.

14.14 The factors that give rise to line width in biological samples are the interactions between the pigment and the protein surroundings, a distribution of pigment conformations, and vibrational states.

14.15 (a) $$\text{Efficiency} = \frac{r_0^6}{r_0^6 + r_{DA}^6} = \frac{(10\ \text{Å})^6}{(10\ \text{Å})^6 + (1\ \text{Å})^6} \approx 1$$

(b) $$\text{Efficiency} = \frac{r_0^6}{r_0^6 + r_{DA}^6} = \frac{(10\ \text{Å})^6}{(10\ \text{Å})^6 + (10\ \text{Å})^6} = 0.5$$

(c) $$\text{Efficiency} = \frac{r_0^6}{r_0^6 + r_{DA}^6} = \frac{(10\ \text{Å})^6}{(10\ \text{Å})^6 + (20\ \text{Å})^6} = \frac{10^6}{64 \times 10^6} = \frac{1}{64}$$

14.16 $$k_{obs} = k_{nonradiative} + k_{radiative} \rightarrow k_{nonradiative} = k_{obs} - k_{radiative}$$
$$= 1\ \text{ns}^{-1} - 0.1\ \text{ns}^{-1} = 0.9\ \text{ns}^{-1}$$

14.17 $$\text{Quantum yield} = \frac{k_{radiative}}{k_{observed}} = \frac{k_{radiative}}{k_{nonradiative} + k_{radiative}}$$

$$10^{-4} = \frac{10^8\ \text{s}^{-1}}{10^8\ \text{s}^{-1} + k_{nonradiative}} \rightarrow k_{nonradiative} = 10^{12}\ \text{s}^{-1}$$

14.18 The chlorosomes contain pigments that can funnel light energy into the reaction center. The efficiency of the chlorosomes allows the organisms to live under very low light intensity if necessary.

14.19 An off-diagonal peak shows that the pigments associated with the two absorption bands on the diagonal are coupled together.

14.20 The single-molecule measurements allow direct measurement of the individual spectral properties as well as of the individual reaction rates.

14.21 The bulk rates represent an average whereas the single-molecule measurements probe the individual rates.

14.22 Excitation of Cy3 results in energy transfer only in state c. With time, fluorescence is observed from Cy3 or Cy5, depending on which state is present at that time. The duration of any peak width corresponds to the time that the junction is in that state. Notice that the fluorescence from Cy5 jumps up when the fluorescence from Cy3 drops.

CHAPTER 15

15.1 To resolve the atomic structure the wavelength must be ≈1 Å.

15.2 X-ray microscopes cannot be constructed because lenses suitable for very small wavelengths cannot be made. The properties of materials are dramatically different for wavelengths of less than 100 Å than for visible light.

15.3 The scattering factor at zero angle is related to the number of electrons present and hence the atomic number. The scattering factor for all atoms drops off sharply with angle according to $(\sin\theta)/\lambda$.

15.4 (a) $\sin\theta = \frac{n\lambda}{2d} = \frac{1(0.1\,\text{nm})}{2(5\,\text{nm})} = 0.01 \rightarrow \theta = 0.57^\circ$

(b) $\sin\theta = \frac{n\lambda}{2d} = \frac{1(0.1\,\text{nm})}{2(0.5\,\text{nm})} = 0.1 \rightarrow \theta = 5.7^\circ$

15.5 A crystal is an array of repeating molecules that bind together in a regular manner. The crystal can be thought of as a molecule forming a structural motif built upon a lattice. The lattice can be classified according to the symmetry that it possesses and the type of Bravais lattice category. A protein crystal is a precise array of proteins that has translation symmetry about the unit cell constants. At least one protein is located at each of the lattice points of the space group describing the crystalline array. Located in the regions surrounding the proteins is solvent that comprises 30–80% of the total volume.

15.6 Salt crystals are composed only of the salts with the atoms at the cell edge. A protein crystal is composed of only 30–60% of protein – the rest is the crystallizing solution.

15.7 In an X-ray diffraction experiment the intensity of each diffraction point is measured. The square root of the intensity gives the amplitude of the structure factor. However, the structure factor is a complex number and all information about the phase is lost and must be determined through other measurements.

15.8 The diagram shows how the vector analysis can be applied to determine the phase associated with each structure factor.

15.9 In isomorphous replacement the protein is modified with the addition of a heavy metal whereas for anomalous dispersion the diffraction from an existing metal is measured at different wavelengths.

15.10 The Ramachandran plot shows whether the amino acid residues of a model fall into the allowed regions. Those that do not fall into the allowed region have improper geometry that has been altered, unless they are in a strained position in the protein.

15.11 Amino acids forming proteins are only of one isomer; the other is never present. In contrast, both isomers are often present when inorganic molecules are synthesized.

15.12 $$\Delta E = -Z^2hcR_h\left(\frac{1}{3^2} - \frac{1}{1^2}\right) = \frac{hc}{\lambda}$$

$$\lambda = \frac{9}{8Z^2R_h} = \frac{9}{8(29)^2(1.1 \times 10^5\ \text{cm}^{-1})} = 1.22 \times 10^{-8}\ \text{cm} = 1.22\ \text{Å}$$

15.13 $$\Delta E = -Z^2hcR_h\left(\frac{1}{3^2} - \frac{1}{1^2}\right) = \frac{hc}{\lambda}$$

$$\lambda = \frac{9}{8Z^2R_h} = \frac{9}{8(42)^2(1.1 \times 10^5\ \text{cm}^{-1})} = 5.8 \times 10^{-9}\ \text{cm} = 0.58\ \text{Å}$$

15.14 Synchrotrons can generate very brilliant radiation at a range of wavelengths.

15.15 In addition to making the density less pronounced compared to the background level, the lack of data beyond a certain resolution limit can artificially introduce negative density that can cancel true density, as was found for nitrogenase.

15.16 Different answers are possible, such as the presence of a nonamino acid residue, providing the opportunity for ligand coordinations that are not possible with the restriction of amino acid residues.

15.17 The different conformations are associated with the electron transfer and ATP hydrolysis steps, with the conformation leading to a shortened distance for electron transfer and presumably a much faster rate.

15.18 There is a single-bond distance associated with the cluster, presumably due to a single metal coordinated by a certain type of amino acid; for example, a copper held in place by histidines.

CHAPTER 16

16.1 (a) $$\nu = \frac{\gamma B}{2\pi} = \frac{1}{2\pi}(26.75 \times 10^7\ \text{T}^{-1}\ \text{s}^{-1}) \times (7\ \text{T})$$

$$= 2.13 \times 10^8\ \text{s}^{-1} = 298\ \text{MHz}$$

(b) $$\nu = \frac{\gamma B}{2\pi} = \frac{1}{2\pi}(26.75 \times 10^7\ \text{T}^{-1}\ \text{s}^{-1}) \times (20\ \text{T})$$

$$= 8.51 \times 10^8\ \text{s}^{-1} = 851\ \text{MHz}$$

16.2 $$\nu = \frac{\gamma B}{2\pi} = \frac{1}{2\pi}(1.93 \times 10^7\ \mathrm{T^{-1}\,s^{-1}}) \times (20\ \mathrm{T}) = 6.1 \times 10^6\ \mathrm{s^{-1}} = 61.3\ \mathrm{MHz}$$

$$\nu = \frac{\gamma B}{2\pi} = \frac{1}{2\pi}(1.93 \times 10^7\ \mathrm{T^{-1}\,s^{-1}}) \times (9\ \mathrm{T}) = 2.76 \times 10^7\ \mathrm{s^{-1}} = 27.6\ \mathrm{MHz}$$

16.3 $$\mu_{new} = \frac{e\hbar}{2m_{pnew}} = \frac{(1.6 \times 10^{-19}\ \mathrm{C})(1.05 \times 10^{-34}\ \mathrm{Js})}{2(4 \times 10^{-29}\ \mathrm{kg})} = 2.05 \times 10^{-25}\ \mathrm{JT^{-1}}$$

16.4 Large chemical shifts usually arise from small shielding and low electron density.

16.5 Large chemical shifts usually arise from small shielding and low electron density. The shielding constant is a measure of the extent of the shielding.

16.6 The separation of the spectra improves with increasing frequency.

16.7 For two spins A and B coupled by J, the lowest energy state is:

$$E = -\frac{1}{2}h\nu_A - \frac{1}{2}h\nu_B + \frac{1}{4}hJ$$

16.8 There are $n + 1 = 5$ lines.

16.9 An inhomogeneous field that increases in strength across the sample.

16.10 The spins will flip along the direction of the field of the spin pulse and then decay back to the direction along the external field with a time constant T_2.

16.11 $I = 1$ so there are three lines.

16.12 $I = 1/2$ so there are two lines.

16.13

	Chemical shift	**Split by**
C_α H	4.52	C_β H_2
C_β H_2	2.15 and 2.01	C_α H and C_γ H_2
C_γ H_2	2.64 and 2.64	C_β H_2
C_ε H_3	2.13	

16.14 The spectrum should have six diagonal peaks and five off-diagonal peaks.

16.15 By the proper combination of pulses, the spins will realign after being flipped, creating a large NMR signal termed an echo.

16.16 In a homogeneous field, a single peak is observed, reflecting the resonance frequency of the water. For an inhomogeneous field, three peaks are present at the relative locations of the beakers.

16.17 MRI is a non-invasive technique for showing the presence of fluids. For an accident, MRI would show the build-up of fluids in the brain at the trauma point.

16.18 (a) $$\nu = \frac{g_e \mu_B B}{h} = \frac{(2.0023)(9.27 \times 10^{-24}\ \text{JT}^{-1})(0.4\ \text{T})}{6.62 \times 10^{-34}\ \text{Js}}$$

$$= 11.5 \times 10^9\ \text{s}^{-1} = 11.5\ \text{GHz}$$

(b) $$\nu = \frac{g_e \mu_B B}{h} = \frac{(2.0023)(9.27 \times 10^{-24}\ \text{JT}^{-1})(0.3\ \text{T})}{6.62 \times 10^{-34}\ \text{Js}}$$

$$= 8.4 \times 10^9\ \text{s}^{-1} = 8.4\ \text{GHz}$$

16.19 $$g = \frac{h\nu}{\mu_B B} = \frac{(6.62 \times 10^{-34}\ \text{Js})(9 \times 10^9\ \text{Hz})}{(9.27 \times 10^{-24}\ \text{JT}^{-1})(0.3\ \text{T})} = 2.14$$

16.20 $$\nu_{EPR} = \nu_{NMR} \times \frac{\mu_B}{\mu_N} \times \frac{B_{EPR}}{B_{NMR}} = 200\ \text{MHz} \times \frac{1.67 \times 10^{-27}\ \text{kg}}{9.1 \times 10^{-31}\ \text{kg}} \times \frac{0.3\ \text{T}}{10\ \text{T}}$$

$$= 11\ \text{GHz}$$

16.21 The splitting arises from the unpaired electron interacting with a nucleus of $I - 1$.

16.22 The EPR spectrum will split into two lines that are separated by the hyperfine coupling constant.

16.23 The oxidation state determines whether there are five or six electrons in the d orbitals. The five electronic levels are divided in energy into a lower group of three and an upper group of two depending upon coordination. The relative energies of these five levels determines the distributions of the electrons.

16.24 There are six electrons in five electronic levels that are divided in energy into a lower group of three and upper group of two. For high spin, the spacing between the two groups is small, so one electron is in each level except for the lowest, which has two. This results in a net spin of 2. The low-spin value is zero.

16.25 The low-spin value is 1/2. There are five electrons in five electronic levels, which are divided in energy into a lower group of three and an upper group of two. For low spin, the spacing between the two groups is large so all electrons are in the lowest three levels with each level having two electrons except for the highest, which has 1. The presence of the single unpaired electron results in a net spin of 1/2.

16.26 The Fe^{2+} of the heme will give rise to a spectrum around $g = 2$ for low spin and $g = 6$ for high spin.

16.27 The oxidation state determines whether there are five or six electrons in the d orbitals. The five electronic levels are divided in energy into a lower group of three and an upper group of two depending upon coordination. The relative energies of these five levels determines the distributions of the electrons.

16.28 The signals should be centered around $g = 2$ with a width determined by the hyperfine coupling. For tyrosines the conjugated system leads to a broad and complex signal.

16.29 The EPR signal would probably be centered around $g = 2$ and would depend upon the redox state of each Fe^{2+}; the presence of two electronic spins would lead to the presence of multiple peaks in the spectrum.

16.30 The EPR signal has the characteristic g value and linewidth of a tyrosyl radical that is present in the protein near the iron cluster.

16.31 Deoxyribonucleotides are derived from the corresponding ribonucleotides by direct reduction at the 2′ carbon atom of the ribose. The reaction is driven by the action of two cysteines, one of which forms a transient radical before formation of a disulfide bond between the two cysteines (which is subsequently reduced by thioredoxin). The tyrosyl radical does not play a direct role in this mechanism, but may be involved in a regulatory or initiation role for the enzyme.

CHAPTER 17

17.1 They have the membrane protein rhodopsin, which contains a pigment that responds to light.

17.2 Transducin binds to rhodopsin, releases GDP, and takes up GTP; then transducin binds to phosphodiesterase. The phosphodiesterase–transducin complex is active, causing cGMP to convert into GMP. Both reactions deplete cGMP, closing channels regulated by cGMP.

17.3 The initial light-induced structural change in bacteriorhodopsin is isomerization of the retinal.

17.4 The absorption spectrum of the retinal changes as it undergoes the conformational changes of the photocycle.

17.5 The retinal in both cases undergoes an isomerization but the direction differs: *cis* to *trans* in rhodopsin and *trans* to *cis* in bacteriorhodopsin.

17.6 The isomerization and other conformational changes resulted in characteristic shifts of the infrared spectrum.

17.7 The initial three-dimensional crystals were not suitable for X-ray diffraction but the two-dimensional crystals yield useful electron-microscopy data.

17.8 The phase can be determined directly since a real image can be obtained with the diffraction image.

17.9 The cubic lipid phases were used to crystallize the protein.

17.10 Whereas the overall structure and the position of the retinal were identified, the detailed mechanism of how the light-induced changes lead to the transfer of a proton were still unknown.

17.11 Water molecules are important in establishing the hydrogen-bonding network in bacteriorhodopsin.

17.12 The initial isomerization causes a proton to be transferred.

17.13 The three-dimensional structures have the same overall fold with a retinal in the center.

17.14 Rhodopsin has a complex photocycle, which is coupled to a variety of membrane-associated proteins including phosphodiesterase and transducin. Bacteriorhodopsin is a closely related bacterial protein that also contains retinal, but performs a proton transfer rather than undergoing structural changes that lead to alterations of the membrane potential.

17.15 Transducin is one of the components of the G-coupled process; it binds to the light-induced structure of rhodopsin, resulting in the release of GDP and uptake of GTP.

17.16 The transfer process is coupled to a rearrangement of the protein, which allows the movement of the chloride ion.

17.17 The isomerization of retinal results in dipole movement of the retinal, and this triggers the proton-transfer process.

17.18 Whereas both cases represent light-induced switches, the isomerization-induced dipole motion of the retinal results in the positively charged proton and negatively charged chloride ion moving in opposite directions.

17.19 Melanopsin is a protein that is thought to have a fold similar to rhodopsin and to bind *trans*-retinal. Melanopsin has an absorption maximum at 480 nm. Light results in a *trans*-to-*cis* isomerization that triggers a signal cascade as found for the other rhodopsins. Despite this similarity of the photocycle, the cellular response is probably very different as the protein is not in the rod or cone cells but is rather located in special ganglion cells that are coupled to the optical nerve. In addition to questions concerning the response mechanism, the reason why why vertebrates respond to blue light remains an open question.

17.20 The proton pathway is critical to function and loss of Asp-85 breaks the proton pathway, which must have a series of ≈2-Å steps.

17.21 The retinal binds at a site very similar to that of bacteriorhodopsin. Nine of the surrounding residues are conserved and water is found, as was true for bacteriorhodopsin. A single chloride ion was found near the retinal at a location corresponding to that of Asp-85. Ion translocation is proposed to occur through a mechanism related to that of bacteriorhodopsin. The ion is proposed to be driven by ion–dipole interactions involving the NH group of the retinal before it is released toward the cytoplasm and a new chloride ion enters the transport site. Only the replacement of the negatively charged Asp-85 as a proton acceptor in bacteriorhodopsin by chloride in halorhodopsin changes the kinetic preference and therefore the ion specificity. Thus in both cases the retinal serves as a switch for the movement of the proton or ion.

CHAPTER 18

18.1 The lipid bilayer is basically impermeable to ions and polar molecules. Water, in contrast, readily traverses membranes. The permeability of small molecules ranges over several orders of magnitude and is correlated with the solubility in nonpolar solvents relative to water.

18.2 $\Delta G = 2.3\,RT \log \frac{2}{2 \times 10^{+2}} = -2.7 \text{ kcal mol}^{-1}$

18.3 $\Delta G = 2.3\,RT \log \frac{100}{20} = 2.3\,RT \times (+0.7) = +0.95 \text{ kcal mol}^{-1}$

18.4 $\Delta G = 2.3\,RT \log \frac{20}{200} + (1)(96.5 \text{ kJ mol}^{-1})(+0.10)$

$= (-2.7 + 14.5) \text{ kcal mol}^{-1} = +11.8 \text{ kcal mol}^{-1}$

18.5 $\Delta V = -58.6 \text{ meV} \log \frac{c_{out}}{c_{in}} = -58.6 \text{ meV} \log \frac{20}{1} = -76.2 \text{ meV}$

18.6 $\Delta V = -58.6 \text{ meV} \log \frac{c_{out}}{c_{in}} = -58.6 \text{ meV} \log \frac{60}{3} = -76.2 \text{ meV}$

18.7 $\Delta G_{\text{Na}} = RT \ln \frac{c_{out}}{c_{in}} + zF\,\nabla V$

$= (0.0083 \text{ kJ molK}^{-1})(295 \text{ K}) \ln \frac{1}{150} + (1)(96.5 \text{ kJ molV}^{-1})(+0.06 \text{ V})$

$= -12.3 + 5.8 \text{ kJ mol}^{-1} = -6.5 \text{ kJ mol}^{-1}$

$\Delta G_{\text{K}} = RT \ln \frac{c_{out}}{c_{in}} + zF\nabla V$

$= (0.0083 \text{ kJ molK}^{-1})(295 \text{ K}) \ln \frac{100}{2} + (1)(96.5 \text{ kJ molV}^{-1})(-0.06 \text{ V})$

$= 9.6 - 5.8 \text{ kJ mol}^{-1} = +3.8 \text{ kJ mol}^{-1}$

$\Delta G_{NET} = \Delta G_{\text{Na}} + \Delta G_{\text{K}} = -2.7 \text{ kJ mol}^{-1}$

18.8 (a) $\Delta G_{\text{Na}} = RT \ln \frac{c_{out}}{c_{in}} + zF\nabla V$

$= (0.0083 \text{ kJ molK}^{-1})(310 \text{ K}) \ln \frac{0.15}{0.02} + (1)(96.5 \text{ kJ molV}^{-1})(+0.05 \text{ V})$

$= 10.0 \text{ kJ mol}^{-1}$

(b) $\Delta G_{\text{K}} = RT \ln \frac{c_{out}}{c_{in}} + zF\nabla V$

$= (0.0083 \text{ kJ molK}^{-1})(310 \text{ K}) \ln \frac{0.05}{0.01} + (1)(96.5 \text{ kJ molV}^{-1})(-0.05 \text{ V})$

$= -0.7 \text{ kJ mol}^{-1}$

(c) $\Delta G_{NET} = 3\Delta G_{\text{Na}} + 3\Delta G_{\text{K}} = 28.6 \text{ kJ mol}^{-1}$

(d) Need $\Delta G_{ATP} - \Delta G_{NET} < 0$

18.9 $\Delta G_{\text{Na}} = RT \ln \frac{c_{out}}{c_{in}} + zF\nabla V$

$= (0.0083 \text{ kJ molK}^{-1})(298 \text{ K}) \ln \frac{2}{200} + (1)(96.5 \text{ kJ molV}^{-1})(+0.065 \text{ V})$

$= -11.3 + 6.2 \text{ kJ mol}^{-1} = -5.1 \text{ kJ mol}^{-1}$

$\Delta G_{\text{K}} = RT \ln \frac{c_{out}}{c_{in}} + zF\nabla V$

$= (0.0083 \text{ kJ molK}^{-1})(298 \text{ K}) \ln \frac{200}{4} + (1)(96.5 \text{ kJ molV}^{-1})(-0.065 \text{ V})$

$= +9.6 - 6.2 \text{ kJ mol}^{-1} = +3.4 \text{ kJ mol}^{-1}$

$\Delta G_{NET} = \Delta G_{\text{Na}} + \Delta G_{\text{K}} = -1.6 \text{ kJ mol}^{-1}$

18.10 $\Delta G_{\text{Na}} = RT \ln \frac{c_{out}}{c_{in}} + zF\nabla V$

$$= (0.0083 \text{ kJ molK}^{-1})(295 \text{ K}) \ln \frac{1}{150} + (1)(96.5 \text{ kJ molV}^{-1})(-0.06 \text{ V})$$

$$= -12.3 - 5.8 \text{ kJ mol}^{-1} = -18.1 \text{ kJ mol}^{-1}$$

$$\Delta G_{\text{K}} = RT \ln \frac{c_{out}}{c_{in}} + zF\nabla V$$

$$= (0.0083 \text{ kJ molK}^{-1})(295 \text{ K}) \ln \frac{100}{2} + (1)(96.5 \text{ kJ molV}^{-1})(-0.06 \text{ V})$$

$$= 9.6 + 5.8 \text{ kJ mol}^{-1} = 15.4 \text{ kJ mol}^{-1}$$

$$\Delta G_{NET} = \Delta G_{\text{Na}} + \Delta G_{\text{K}} = -2.7 \text{ kJ mol}^{-1}$$

18.11 Transporters move specific molecules across a membrane.

18.12 In the model the opening of a transporter shifts so that the substrate has access initially to one side and then the other, with simultaneous access to both sides forbidden.

18.13 The helices are proposed to flex as the pH increases, causing the channel to open.

18.14 The factors include channel size, electrostatic interactions, and dipole reorientation.

18.15 The channel regions of the proteins are very similar but the voltage-gated channels contain additional domains that change conformation in response to a voltage across the membrane; this conformational change is thought to open and close the channel region.

18.16 Potassium channel: a voltage-gated channel has six transmembrane helices. Only two are found for the bacterial KcsA, which is thought to resemble the pore region. The organization of the remaining regions are a S1–S4 voltage sensor, a T1 tetramerization domain, and a signaling domain.

18.17 A ligand-gated channel is a channel with an opening that is gated by the binding of a specific substrate.

18.18 The peripherial helices are proposed to shift in response to a change in tension or membrane polarization, leading to conformation changes that open or close the channel.

18.19 The ion flow is partially controlled by dipoles from the surrounding protein that favor the flow of a specific change in one direction.

18.20 The cavity allows the ions to enter without losing their soluated water molecules until they enter the filter region.

18.21 The filter region has several positions that each select for a particular ion by providing a coordination from the protein that is specific for a given ion through the outermost electrons of the ion.

18.22 Different parts of the ion channel are labeled with a donor and acceptor and the distance between these is calculated by measuring the efficiency of energy transfer from the donor to the acceptor, which has a dependence of $1/r^6$. These data are then measured for voltage differences to determine the voltage-induced structural changes.

18.23 This receptor is a pentamer with two binding sites for acetylcholine. The channel has approximately 5-fold symmetry and roughly forms a cylinder with a channel 10–20 Å in diameter. The transport of ions through the channel is determined by the presence of several negatively charged amino acids in the center of the channel, which prevent the transport of anions and limit the transport of cations by size exclusion. A structure has been determined for a water-soluble protein that binds acetylcholine and serves as a model for the receptor. This structure shows that acetylcholine binds at the interface between two subunits formed by several loops.

CHAPTER 19

19.1 GFP emits a green fluorescence when irradiated with ultraviolet light. Digestion of GFP shows that the cofactor is formed by cyclization of residues Ser-65, Tyr-66, and Gly-67.

19.2 The chromophore is formed by a modification of three residues, 65, 66, and 67.

19.3 Measurement of cells grown under either aerobic or anaerobic conditions showed that the GFP polypeptide is expressed under both conditions but that the cofactor is formed only when oxygen is present.

19.4 The cofactor has different interactions with the surrounding protein and the molecular structure of the cofactor changes when the amino acids (residues 65–67) that form the cofactor change.

19.5 The two absorption bands arise from two states of the pigment as it is formed.

19.6 With Ser-65 and Gly-67, this residue forms the chromophore. Any changes of this residue results in an altered pigment, if it is formed at all.

19.7 The modification involves cyclization, oxygenation, and proton transfer; the GFP becomes colored only after this process takes place.

19.8 The absorption band has contributions from different states of the cofactor, the relative populations of which change with pH.

19.9 There is a protonation change of the chromophore.

19.10 GFP can be tagged onto any gene in a genome and the locations of the protein product can be located in the different parts of the cell by fluorescence measurements.

19.11 One protein is labeled with GFP and the second is labeled with an energy acceptor. If the proteins bind, excitation of GFP is followed by energy transfer to the acceptor and fluorescence is measured from the acceptor. If the proteins do not bind, excitation of GFP is followed by fluorescence from GFP because the acceptor is too far away for energy transfer.

19.12 Efficiency of transfer $= \dfrac{r_0^6}{r_0^6 + r_{DA}^6}$

where the parameter r_0 represents the distance at which the efficiency of energy transfer is 50%.

19.13 There are several factors, such as the spectral overlap, that influence the efficiency.

19.14 In DsRed, the chromophore undergoes an additional dehydrogenation that results in a more red color.

19.15 If GFP is fused to a gene of interest, the encoded protein can be tracked by monitoring the GFP fluorescence in the cell.

19.16 The decay particles can be α, β, or γ radiation, depending upon the isotope.

19.17 For PET, the subject is injected with a molecule containing a label with an isotope that will emit positrons.

19.18 The positron is not stable and will combine with an electron to produce energy that has value equal to twice the electron mass, $2m_ec^2$, or twice 511 meV. The energy takes the form of two photons that, to conserve momentum, are directed along equal and opposite directions from the point where the positron and electron came together, with each photon having an energy of 511 meV.

19.19 The probe 2-[^{18}F]fluoro-2-deoxy-D-glucose is used to locate glycolysis processes in the body.

19.20 Areas that are undergoing high metabolism are shown by high use of glycolysis and correspondingly high concentrations of 2-[^{18}F]fluoro-2-deoxy-D-glucose. Such profiles of the brain have been found to be characteristic of certain diseases, such as Alzheimer's disease. Alternatively, use of fluorothymidine, is used to located areas that are undergoing high DNA replication, which is characteristic of cancer cells.

CHAPTER 20

20.1 The different substitutents result in differences in the nature of the conjugated system of the macrocycle.

20.2 Bacteriochlorophyll a absorbs at 865 nm, corresponding to an energy of 1.2 eV, while P680 absorbs at 680 nm, corresponding to 1.8 eV.

20.3 The cofactors of the bacterial reaction center are well buried within the protein whereas for the light-harvesting complex II the protein forms two concentric rings with the cofactors found in the middle, also forming rings, with bacteriochlorophylls in close contact despite being associated with different protein subunits.

20.4 The differences in the optical spectra reflect the different interactions of the bacteriochlorophylls with each other as well as with the surrounding protein.

20.5 Different answers are accepted. For example, the availability of two complexes provides the cell with the opportunity to regulate the amount of one complex, the light-harvesting complex II, without disturbing the energy transfer into the reaction center.

20.6 $$\frac{\Delta H_{monomer}}{\Delta H_{dimer}} = \frac{A_{monomer}\sqrt{N}}{\frac{1}{2}A_{monomer}\sqrt{2N}} = \sqrt{2} = 1.4$$

20.7 Although the structure is symmetrical, the energetics of the cofactors may be determined by the local protein environment and hence the rates of electron transfer may differ due to energetic considerations. Other answers are possible.

20.8 Both cofactors are electron acceptors but the properties of their electronic states differ considerably, including their oxidation/reduction potentials.

20.9 The conversion of light energy involves the sequential formation of the S states. Each absorption of light gives rise to the transfer of one electron. After the absorption of four photons the water-oxidation complex is capable of the four-electron process of water oxidation into molecular oxygen.

20.10 With the last step of the S cycle having a characteristic time of 1 ms, the rate of oxygen production could proceed at a rate of 1000 molecules per second.

20.11 The high potential allows P680 to perform more redox reactions, including the oxidation of water.

20.12 A higher oxidation/reduction potential is achieved by changing the protein environment to make the surroundings favor the ground state. This can be achieved by increasing the hydrophobicity or by introducing positively charged amino acid residues.

20.13 Tyrosine serves as an intermediate electron donor between $P680^+$ and the manganese cluster and as a participant in the associated proton-transfer processes.

20.14 The manganese ions are highly interacting and in a suitable oxidation state.

20.15 The oxidation states of the manganese cluster change in the S states.

20.16 The number of lines can vary depending on the configuration and potentially there could be many more than observed. Given the observed spectrum, there are many allowed configurations.

20.17 It predicts that each step is associated with the transfer of one proton.

20.18 Many answers are possible.

20.19 He saw the longed-lived state of photosystem II arising from an oxidized tyrosine.

Index

Fundamental constants

Constant	Symbol	Value	SI units	Non-SI units (cgs units)
Speed of light in a vacuum	c	2.99792	10^8 m s^{-1}	10^{10} cm s^{-1}
Avogadro's number	N_A	6.0221	10^{23} mol^{-1}	10^{23} mol^{-1}
Planck's constant	h	6.6262	10^{-34} Js	10^{-27} erg s
Boltzmann's constant	k_B	1.38066	10^{-23} J K^{-1}	10^{-16} erg K^{-1}
Gas constant	R	8.3144	J K^{-1} mol^{-1}	10^7 erg K^{-1} mol^{-1}
Faraday's constant	F	9.6487	10^4 C mol^{-1}	
Bohr radius	a_0	5.29177	10^{-11} m	10^{-9} cm
Rydberg constant	R_H	1.09737	10^7 m^{-1}	10^5 cm^{-1}
Elementary charge	e	1.60219	10^{-19} C	10^{-10} $cm^{3/2}$ $g^{1/2}$ s^{-1} mol^{-1} 4.80298 10^{-10} $cm^{3/2}$ $g^{1/2}$ s^{-1} mol^{-1}
Electron rest mass	m_e	9.1095	10^{-31} kg	10^{-28} g
Proton rest mass	m_p	1.67265	10^{-27} kg	10^{-24} g
Gravitational constant	G	6.6720	10^{-11} N m^2 kg^{-2}	10^{-8} dyn cm^2 g^{-2}
Permittivity of a vacuum	ε_0	8.85419	10^{-12} kg^{-1} m^{-3} s^4 A^2	
Bohr magneton	μ_B	9.27401	10^{-24} JT^{-1}	
Nuclear magneton	μ_N	5.05078	10^{-27} JT^{-1}	
Atomic mass unit		1.6605	10^{-27} kg	10^{-24} g
π		3.14159		
e		2.71828		
$\log_e 10 = \ln 10$		2.30258	($\ln x = 2.30258 \log_{10} x$)	

cgs, centimeter-gram-second.

Conversion factors for energy units

To convert from energy in units shown in the left-hand column to units shown in the top row, multiply by the factor at their intersection.

	eV	cm^{-1}	kcal mol^{-1}	kJ mol^{-1}	K	erg
1 ev	1	8065	23.06	96.48	1.160×10^4	1.602×10^{-12}
1 cm^{-1}	1.240×10^{-4}	1	2.859×10^{-3}	1.196×10^{-2}	1.439	1.986×10^{-16}
1 kcal mol^{-1}	4.336×10^{-2}	349.8	1	4.184	503.2	6.948×10^{-14}
1 kJ mol^{-1}	1.036×10^{-2}	83.60	0.239	1	120.3	1.661×10^{-14}
1 K	8.617×10^{-5}	0.6950	1.987×10^{-3}	8.314×10^{-3}	1	1.381×10^{-16}
1 erg	6.241×10^{11}	5.034×10^{15}	1.439×10^{13}	6.022×10^{13}	7.243×10^{15}	1

The periodic table

Legend	
Metal	
Semimetal	
Nonmetal	

6 — Atomic number
C — Symbol
12.01 — Atomic weight

Period	1	2	3	4	5	6	7	8	9	10	11	12	13	14	15	16	17	18
1	1 **H** 1.008																	2 **He** 4.003
2	3 **Li** 6.941	4 **Be** 9.012											5 **B** 10.81	6 **C** 12.01	7 **N** 14.01	8 **O** 16.00	9 **F** 19.00	10 **Ne** 20.18
3	11 **Na** 22.99	12 **Mg** 24.31											13 **Al** 26.98	14 **Si** 28.09	15 **P** 30.97	16 **S** 32.07	17 **Cl** 35.45	18 **Ar** 39.95
4	19 **K** 39.10	20 **Ca** 40.08	21 **Sc** 44.96	22 **Ti** 47.88	23 **V** 50.94	24 **Cr** 52.00	25 **Mn** 54.94	26 **Fe** 55.85	27 **Co** 58.93	28 **Ni** 58.69	29 **Cu** 63.55	30 **Zn** 65.39	31 **Ga** 69.72	32 **Ge** 72.61	33 **As** 74.92	34 **Se** 78.96	35 **Br** 79.90	36 **Kr** 83.80
5	37 **Rb** 85.47	38 **Sr** 87.62	39 **Y** 88.91	40 **Zr** 91.22	41 **Nb** 92.91	42 **Mo** 95.94	43 **Tc** 98.91	44 **Ru** 101.1	45 **Rh** 102.9	46 **Pd** 106.4	47 **Ag** 107.9	48 **Cd** 112.4	49 **In** 114.8	50 **Sn** 118.7	51 **Sb** 121.8	52 **Te** 127.6	53 **I** 126.9	54 **Xe** 131.3
6	55 **Cs** 132.9	56 **Ba** 137.3	71 **Lu** 175.0	72 **Hf** 178.5	73 **Ta** 180.9	74 **W** 183.8	75 **Re** 186.2	76 **Os** 190.2	77 **Ir** 192.2	78 **Pt** 195.1	79 **Au** 197.0	80 **Hg** 200.6	81 **Tl** 204.4	82 **Pb** 207.2	83 **Bi** 209.0	84 **Po** 209.0	85 **At** 210.0	86 **Rn** 222.0
7	87 **Fr** 223.0	88 **Ra** 226.0	103 **Lr** 262.1	104 **Rf** 261.1	105 **Db** 262.1	106 **Sg** 263.1	107 **Bh** 264.1	108 **Hs** 265.1	109 **Mt** 268	110 **Uun** 269	111 **Uuu** 272	112 **Uub** 277	113 Uut	114 **Uuq** 289	115 Uup	116 **Uuh** 289	117 Uus	118 **Uuo** 293

Period														
6	57 **La** 138.9	58 **Ce** 140.1	59 **Pr** 140.9	60 **Nd** 144.2	61 **Pm** 146.9	62 **Sm** 150.4	63 **Eu** 152.0	64 **Gd** 157.3	65 **Tb** 158.9	66 **Dy** 162.5	67 **Ho** 164.9	68 **Er** 167.3	69 **Tm** 168.9	70 **Yb** 173.0
7	89 **Ac** 227.0	90 **Th** 232.0	91 **Pa** 231.0	92 **U** 238.0	93 **Np** 237.0	94 **Pu** 244.1	95 **Am** 243.1	96 **Cm** 247.1	97 **Bk** 247.1	98 **Cf** 251.1	99 **Es** 252.0	100 **Fm** 257.1	101 **Md** 258.1	102 **No** 259.1